Lecture Notes in Computer Science 5165

Commenced Publication in 1973
Founding and Former Series Editors:
Gerhard Goos, Juris Hartmanis, and Jan van Leeuwen

Boting Yang   Ding-Zhu Du   Cao An Wang (Eds.)

Combinatorial Optimization and Applications

Second International Conference, COCOA 2008
St. John's, Canada, August 21-24, 2008
Proceedings

 Springer

Volume Editors

Boting Yang
University of Regina, Department of Computer Science
Regina, Saskatchewan S4S 0A2, Canada
E-mail: boting@cs.uregina.ca

Ding-Zhu Du
University of Texas at Dallas, Department of Computer Science and Engineering
Richardson, TX 75083, USA
E-mail: dzdu@utdallas.edu

Cao An Wang
Memorial University of Newfoundland, Department of Computer Science
St. John's, Newfoundland, Canada A1B 3X5, Canada
E-mail: wang@cs.mun.ca

Library of Congress Control Number: Applied for

CR Subject Classification (1998): F.2, C.2, G.2-3, I.3.5, G.1.6, E.5

LNCS Sublibrary: SL 1 – Theoretical Computer Science and General Issues

ISSN   0302-9743
ISBN-10   3-540-85096-1 Springer Berlin Heidelberg New York
ISBN-13   978-3-540-85096-0 Springer Berlin Heidelberg New York

Springer is a part of Springer Science+Business Media

springer.com

© Springer-Verlag Berlin Heidelberg 2008
Printed in Germany

Typesetting: Camera-ready by author, data conversion by Scientific Publishing Services, Chennai, India
Printed on acid-free paper   SPIN: 12445752   06/3180   5 4 3 2 1 0

Preface

The papers in this volume were presented at the Second International Conference on Combinatorial Optimization and Applications (COCOA 2008), held August 21–24, 2008, in St. John's, Newfoundland, Canada. The topics cover most areas in combinatorial optimization and applications.

A total of 84 papers were submitted, of which 44 were accepted for presentation at the conference. The selection was based on the papers' originality, quality and relevance to combinatorial optimization. The papers were evaluated by an international Program Committee consisting of Tetsuo Asano, Sergey Bereg, Binay Bhattacharya, Hans Bodlaender, Leizhen Cai, Bo Chen, Zhixiang Chen, Francis Chin, Kyung-Yong Chwa, Andreas Dress, Ding-Zhu Du, Michael Fellows, Fedor Fomin, Gena Hahn, Pavol Hell, Wen-Lian Hsu, Minghui Jiang, Ming-Yang Kao, Evangelos Kranakis, Michael Langston, Guohui Lin, Rolf H. Möhring, Xuehou Tan, Dimitrios M. Thilikos, Adrian Vetta, Cao An Wang, Lusheng Wang, Todd Wareham, Peter Widmayer, Jinhui Xu, Yinfeng Xu, Boting Yang, Yinyu Ye, Guochuan Zhang, Xiao Zhou, and Binhai Zhu. It is expected that most of the accepted papers will appear in a more complete form in scientific journals.

We received papers from Algeria, Austria, Bangladesh, Belgium, Brazil, Canada, China, Egypt, France, Germany, Hong Kong, India, Indonesia, Iran, Israel, Italy, Japan, Korea, Norway, Oman, Pakistan, Romania, Russia, Singapore, South Africa, Spain, Sweden, Switzerland, Taiwan, UK, and USA. Each paper was evaluated by at least three Program Committee members, assisted in some cases by external referees. In addition to the selected papers, the conference also included two invited presentations by Francis Y.L. Chin and Vijay V. Vazirani.

We thank all the people who have made this meeting possible: the authors for submitting papers, the Program Committee members and external referees (listed in the proceedings) for their excellent work, and the two invited speakers. Finally, we thank the Memorial University of Newfoundland for their support and the local organizers and their colleagues for their assistance.

August 2008

Ding-Zhu Du
Cao An Wang
Boting Yang

Organization

General Chair

Ding-Zhu Du (University of Texas at Dallas, USA)

Program Chairs

Cao An Wang (Memorial University of Newfoundland, Canada)
Boting Yang (University of Regina, Canada)

Program Committee

Tetsuo Asano (Japan Advanced Institute of Science and Technology)
Sergey Bereg (University of Texas at Dallas)
Binay Bhattacharya (Simon Fraser University)
Hans Bodlaender (University of Utrecht)
Leizhen Cai (Chinese University of Hong Kong)
Bo Chen (University of Warwick)
Zhixiang Chen (University of Texas-Pan American)
Francis Chin (Hong Kong University)
Kyung-Yong Chwa (Korea Advanced Institure of Science and Technology)
Andreas Dress (Bielefeld University)
Michael Fellows (The University of Newcastle)
Fedor Fomin (University of Bergen)
Gena Hahn (University of Montreal)
Pavol Hell (Simon Fraser University)
Wen-Lian Hsu (Academia Sinica, Taiwan)
Minghui Jiang (Utah State University)
Ming-Yang Kao (Northwestern University)
Evangelos Kranakis (Carleton University)
Michael Langston (University of Tennessee)
Guohui Lin (University of Alberta)
Rolf H. Möhring (Technische Universität Berlin)
Xuehou Tan (Tokai University)
Dimitrios M. Thilikos (National and Kapodistrian University of Athens)
Adrian Vetta (McGill University)
Lusheng Wang (City University of Hong Kong)
Todd Wareham (Memorial University of Newfoundland)
Peter Widmayer (ETH, Switzerland)
Jinhui Xu (State University of New York at Buffalo)
Yinfeng Xu (Xi'an Jiaotong University)
Yinyu Ye (Stanford University)

Guochuan Zhang (Zhejiang University)
Xiao Zhou (Tohoku University)
Binhai Zhu (Montana State University)

Organizing Committee

Danny Dyer (Memorial University of Newfoundland, Co-chair)
Cao An Wang (Memorial University of Newfoundland)
Chris Worman (University of Regina)
Boting Yang (University of Regina, Co-chair)

External Referees

Sang Won Bae	Jia-Ming Chang	Xi Cheng
Gerard Cornuejols	Vida Dujmovic	Danny Dyer
John Eblen	Matthew Follett	Jason Gedge
Petr Golovach	Alexander Grigoriev	Tobias Harks
Wiebke Hoehn	Takehiro Ito	Jeremy Jay
Molham Kamel	Marcin Kaminski	Stavros Kolliopoulos
Nicole Megow	Daniel Meister	Hannes Moser
Nicolas Nisse	Peter Noel	Neil Olver
Andy Perkins	Charles Phillips	Stanislaw Radziszowski
Md. Saidur Rahman	Gary Rogers	Johan van Rooij
Frank Ruskey	Saket Saurabh	Sebastian Stiller
Mohammed Uddin	Yngve Villanger	Yulai Xie
Lei Xu	Deshi Ye	Heping Zhang
Yun Zhang	Feifeng Zheng	Yongding Zhu

Table of Contents

Going Weighted:
Parameterized Algorithms for Cluster Editing

Sebastian Böcker[1], Sebastian Briesemeister[2], Quang B. A. Bui[1],
and Anke Truss[1]

[1] Lehrstuhl für Bioinformatik, Friedrich-Schiller-Universität Jena, Ernst-Abbe-Platz
2, 07743 Jena, Germany
{boecker,bui,truss}@minet.uni-jena.de
[2] Div. for Simulation of Biological Systems, ZBIT/WSI, Eberhard Karls Universität
Tübingen, Germany
briese@informatik.uni-tuebingen.de

Abstract. The goal of the CLUSTER EDITING problem is to make the
fewest changes to the edge set of an input graph such that the resulting
graph is a disjoint union of cliques. This problem is NP-complete but
recently, several parameterized algorithms have been proposed. In this
paper we present a surprisingly simple branching strategy for CLUSTER
EDITING. We generalize the problem assuming that edge insertion and
deletion costs are positive integers. We show that the resulting search
tree has size $O(1.82^k)$ for edit cost k, resulting in the currently fastest
parameterized algorithm for this problem. We have implemented and
evaluated our approach, and find that it outperforms other parametrized
algorithms for the problem.

1   Introduction

The WEIGHTED CLUSTER EDITING problem is defined as follows: Let $G_w = (V, E)$ be an undirected graph. For every vertex pair $\{u, v\} \in \binom{V}{2} = \{\{u, v\} : u, v \in V, u \neq v\}$ we know the cost of deleting $\{u, v\}$ from G_w in case $\{u, v\} \in E$, or inserting $\{u, v\}$ into G_w in case $\{u, v\} \notin E$. Our task is to transform G_w into a transitive graph (a disjoint union of cliques) by applying edge modifications with minimum total cost. For our theoretical analysis we assume that all pairs have *non-zero integer weight*. In the unweighted CLUSTER EDITING problem, insertion or deletion cost are one for each vertex pair.

In application, the above task corresponds to clustering objects, that is, partitioning a set of objects into homogeneous and well-separated subsets. Clustering data still represents a key step of numerous biological and medical problems, such as class discovery for tissue identification using gene expression data. Here, a clustering corresponds to a vertex disjoint union of cliques. The input graph is corrupted and we have to clean (edit) the graph to reconstruct the clustering [13] under the *parsimony criterion*.

Previous work. NP-hardness of the unweighted CLUSTER EDITING problem [13] was proven by Křivánek and Morávek [10]. Several heuristics were developed for

B. Yang, D.-Z. Du, and C.A. Wang (Eds.): COCOA 2008, LNCS 5165, pp. 1–12, 2008.
© Springer-Verlag Berlin Heidelberg 2008

the WEIGHTED VARIANT or rely on its graph-theoretic intuition, including CLICK [14] and FORCE [16]. The problem is APX-hard, and has a constant-factor approximation of 2.5 [15]. To find exact solutions, Grötschel and Wakabayashi [7] formulated the problem as an Integer Linear Program. The parameterized complexity of unweighted CLUSTER EDITING, using the minimum number of edge modifications as the parameter k, is well-studied: Until recently, the fastest implemented algorithm had running time $O(2.27^k + n^3)$ on an n-vertex graph [6,4], while in theory, the best known algorithm has running time $O(1.92^k + n^3)$ [5]. Guo [8] presented a linear problem kernel. In contrast, the fixed-parameter tractability of CLUSTER EDITING with "don't care edges", that is, edges whose modification cost is zero, is still an open problem [3].

For WEIGHTED CLUSTER EDITING, the authors presented a problem kernel in [1] and, in particular, introduced a new data reduction technique of merging vertices. Furthermore, we provided two branching strategies with search trees of size $O(3^k)$ and $O(2.42^k)$, respectively, where parameter k is the minimum total cost of edge modifications. We found that merging vertices significantly reduces running times and, in contrast to what theoretical bounds suggest, the $O(3^k)$ strategy consistently outperformed the $O(2.42^k)$ strategy. An experimental evaluation of exact methods for CLUSTER EDITING, including the branching strategy presented in this paper, can be found in [2].

Our contributions. We concentrate on the case that edge insertion and deletion costs are positive integers, and sketch how to adopt our results for real-valued graphs where necessary. We present a new branching strategy that is surprisingly simple, and show that the resulting search tree is of size $O(2^k)$. We then refine our analysis and show that by accurately choosing edges to branch on, we obtain running time $O(1.82^k + n^3)$. Our algorithm is the fastest known for unweighted CLUSTER EDITING and improves on the $O(1.92^k + n^3)$ algorithm in [5].

In Sec. 5 we compare running times of our algorithm to the previously best known results for a parameterized algorithm for WEIGHTED CLUSTER EDITING [1], and we observe improvements of several orders of magnitude. In our comparison, we use both simulated graphs and graphs that stem from protein similarity data and aim at clustering homologous proteins.

2   Preliminaries

Let V be the set of objects to be clustered, corresponding to the vertices of the graph. In this work, we consider only undirected graphs without self-loops and multiple edges. For brevity, we write uv as shorthand for an unordered pair $\{u, v\} \in \binom{V}{2}$. Let $s : \binom{V}{2} \to \mathbb{Z}$ be a *weight function* that encodes the input graph: For $s(uv) > 0$ a pair uv is an edge of the graph and has deletion cost $s(uv)$, while for $s(uv) < 0$, the pair uv is not an edge of the graph (we call it a *non-edge*) and has insertion cost $-s(uv)$. If $s(uv) = 0$, we call uv a *zero-edge*. Note that there are no zero-edges in the input graph, so that each pair of vertices is either an edge or a non-edge whose edit cost (deletion or insertion cost) is a positive integer. This is necessary solely to achieve provable running times. Nonetheless, zero-edges can appear in the course of computation and require additional attention when analyzing the algorithm.

When analyzing connected components we regard zero-edges as non-existing. Throughout this paper we assume that circles and paths do not contain zero-edges. A circle of length three is also called a *triangle*. We say that $C \subseteq V$ is a *clique* in an integer-weighted graph if all pairs $uv \in \binom{C}{2}$ are edges. If all vertex pairs of a connected component are either edges or zero-edges, we call it a *weak clique*. If all connected components of a graph are weak cliques, it is called *transitive*. Weak cliques in a transitive graph are also called *clusters*. An unweighted graph $G = (V, E)$ is transitive if and only if there exists no *conflict triple* in G, that is, three vertices vuw such that $vu, uw \in E$ but $vw \notin E$. Unfortunately, there exists no direct analogue of this statement for integer-weighted graphs. Vertices vuw form a *conflict triple* in an integer-weighted graph G_w if uv and uw are edges of G_w but vw is either a non-edge or a zero-edge. We distinguish two types of conflict triples vuw: if vw has weight zero then the conflict triple is called *weak*, whereas if vw is a non-edge then the conflict triple is called *strong*. In case the integer-weighted graph G_w contains no conflict triples then G_w is transitive. But the converse is obviously not true, as the example of a single weak conflict triple shows. A graph that does not contain any strong conflict triple is not necessarily transitive: For $V = \{u, v, w, x\}$ let uv, vw, wx be edges, let uw, vx be zero-edges, and let ux be a non-edge. The resulting graph is connected and contains no strong conflict triple, but is *not* a weak clique.

To solve WEIGHTED CLUSTER EDITING we first identify all connected components of the input graph and calculate the best solutions for all components separately, because an optimal solution never connects disconnected components. Furthermore, if the graph is decomposed during the course of the algorithm, then we recurse and treat each connected component individually. Our fixed-parameter algorithms often require a cost limit k: In case a solution with cost $\leq k$ exists, the algorithm finds this solution; otherwise, "no solution" is returned. To find an optimal solution we call the algorithm repeatedly, increasing k.

An unweighted CLUSTER EDITING instance can be encoded by assigning weights $s(uv) \in \{+1, -1\}$. In the resulting graph, all conflict triples are strong. During data reduction and branching, we may set pairs uv to "forbidden" or "permanent", meaning that the status of uv cannot be changed in the future. In fact, permanent edges can be merged immediately: As introduced in [1], *merging* uv means replacing the vertices u and v with a single vertex u', and, for all vertices $w \in V \setminus \{u, v\}$, replacing pairs uw, vw with a single pair $u'w$. In this context, we say that we *join* vertex pairs uw and vw. The weight of the joined pair is $s(u'w) = s(uw) + s(vw)$. In case one of the pairs is an edge while the other is not, the parameter k is reduced by $\min\{|s(uw)|, |s(vw)|\}$. Note that we may join any combination of two edges, non-edges, or zero-edges when merging two vertices. We stress that joined pairs can be zero-edges.

Throughout this paper, let $n := |V|$. To decrease input size, we introduced kernelization rules for WEIGHTED CLUSTER EDITING in [1]. For unweighted CLUSTER EDITING, Guo [8] uses the concept of *critical cliques* to construct a kernel of size $4k_{\mathrm{opt}}$ for unweighted CLUSTER EDITING. Critical cliques are cliques in the input graph which share the same neighborhood. In unweighted graphs, all vertices of a critical clique must end up in the same cluster, so we can always merge critical cliques. This idea does not apply directly to WEIGHTED CLUSTER EDITING but it is possible to adapt the concept by considering cliques with similar neighborhood [2]. When given an unweighted instance of cluster editing, we merge all critical cliques to transform the graph into an integer-weighted instance. The resulting graph has at most $4k_{\mathrm{opt}}$ vertices. The weighted graph can be constructed from the critical clique graph that, in turn, can be easily constructed in $O(m+n)$ time [9] for an n-vertex and m-edge graph. The weight of any tuple uv is simply the product of the corresponding critical clique sizes $|C_u| \cdot |C_v|$.

3   Edge Branching

We now describe a recursive algorithm for integer-weighted CLUSTER EDITING, following the bounded search tree paradigm. In this algorithm, we identify a conflict triple and then branch into two sub-cases to repair this conflict. By this, we invoke recursive calls on "simplified" instances of the problem where parameter k is decreased by some constants a, b. For branching vector (a, b) we can compute a branching number using the characteristic polynomial, and this branching number in turn governs the asymptotic size of the search tree, see e.g. [11] for details.

The **edge branching strategy** is as follows: Let uv be an edge of a (weak or strong) conflict triple vuw. Then, (a) set uv to forbidden, or (b) merge uv.

Let us first analyze this very simple strategy. One can easily check that this recursive procedure will at some point generate an optimal solution, because in every step we resolve a conflict triple. In the following we will analyze the size of the search tree. When deleting an edge uv we decrease the parameter by $s(uv)$. When merging vertices u, v, for each vertex $w \in V \setminus \{u, v\}$ we join the pairs uw and vw into a single pair with weight $s(uw) + s(vw)$. If $s(uw) \neq -s(vw)$ then parameter k can be lowered by $\min\{s(uw), -s(vw)\}$. In case $s(uw) = -s(vw)$ the new pair is a zero-edge, and this would prevent us from decreasing our parameter when joining the zero-edge in a later stage of the algorithm. To circumvent this problem, we assume that joining uw and vw with $s(uw) = -s(vw)$ only reduces the parameter by $\min\{s(uw), -s(vw)\} - \frac{1}{2} = |s(uw)| - \frac{1}{2} \geq \frac{1}{2}$. If at a latter stage we join this zero-edge with another pair, we decrease our parameter by the remaining $\frac{1}{2}$. Using this bookkeeping trick, our edge branching strategy has a branching vector of $(1, \frac{1}{2})$ that leads to a search tree of size $O(2.62^k)$.

We can easily improve this branching strategy by choosing a "good" edge uv, as follows: Choose the particular edge $uv \in E$ that *minimizes* the branching number of the corresponding branching step. The branching number is computed from branching vector (a, b) where a is the cost of deleting edge uv, while b is the cost of merging this edge. If one of these costs is zero, we say that the edge has infinite branching number. Using the bookkeeping trick introduced above, an edge uv with finite branching number is not necessarily part of *any* conflict triple: joining a zero-edge uw with a vertex pair vw generates cost $\frac{1}{2}$ irrespective of whether vw is an edge, non-edge, or zero-edge. So, even the edge with minimum branching number might not be part of any conflict triple.

The following is a simple observation regarding unweighted graphs:

Lemma 1. *Given a connected, unweighted graph G. If every edge of G is part of at most one conflict triple, then G is either a clique or a clique minus a single edge.*

Proof. If $G = (V, E)$ contains no conflict triple then G is a clique. Assume that there is at least one conflict triple vuw in G with $uv, uw \in E$ and $vw \notin E$. We constructively show that G is a clique minus the edge vw. If another vertex $x \in V \setminus \{u, v, w\}$ is adjacent to v then $ux \in E$ must hold, too: otherwise, uv is part of two conflict triples vuw and uvx contrary to our assumptions. Similarly, $ux \in E$ implies $vx \in E$. In conclusion, $ux \in E$ if and only if $vx \in E$. The same holds replacing v by w, and we infer that if some vertex x is adjacent to one of u, v, or w then it is adjacent to all of u, v, and w.

Next, consider two vertices x, y adjacent to all u, v, w. If $xy \notin E$ then vxw and xvy are two conflict triples containing the edge xv which conflicts with our assumptions, so $xy \in E$ must hold. Finally, consider vertices x, z where x is adjacent to u, v, w while z is not adjacent to u, v, w, and assume $xz \in E$. Now the edge vx is part of the two conflict triples vxw and vxz, again a contradiction to our assumptions. So, any vertex $x \in V \setminus \{v, w\}$ must be adjacent to all other vertices in G. $\square$

Lemma 2. *For an integer-weighted graph, the edge branching strategy that chooses an edge with minimum branching number has branching vector at least $(1, 1)$.*

Proof. Recall that if we create a zero-edge, this reduces k by at least $\frac{1}{2}$; and if we join a zero-edge, this reduces k by $\frac{1}{2}$. Let uv be the edge with minimum branching number. Note that removing uv induces cost $s(uv) \geq 1$, and let δ be the cost of merging uv. If $\delta \geq 1$ then we are done, so assume $\delta < 1$. This implies that at most one zero-edge was created or joined. In particular, uv is part of at most one conflict triple vuw, and there cannot be an edge that is part of two conflict triples. We transform the input graph into an unweighted graph G, where zero-edges and non-edges in the input graph are not present in G. By Lemma 1 above, the connected component containing vuw must be a clique minus vw in G. Regarding the weighted graph, all vertex pairs are edges except vw that may be a non-edge or a zero-edge. If vw is a zero-edge then our branching will stop when merging uv, so assume that vw is a non-edge. We now show that for this case, we can omit our bookkeeping trick of delayed parameter decrease.

We now either delete uv with cost $s(uv) \geq 1$, or merge uv. We distinguish the cases $s(uw) \geq -s(vw)$ and $s(uw) < -s(vw)$. If $s(uw) \geq -s(vw)$ holds then the joined pair has weight $s(uw) + s(vw) \geq 0$, the resulting connected component is a clique that can be removed from the graph, and we reduce the parameter

k by $\min\{s(uw), -s(vw)\} \geq 1$. For $s(uw) < -s(vw)$ the joined pair has weight $s(uw) + s(vw) < 0$, so we have not generated a zero-edge. We can assume in our analysis that parameter k is reduced by the full $\min\{s(uw), -s(vw)\} \geq 1$. So, the branching vector is at least $(1, 1)$ as claimed.   □

Hence, edge branching results in a search tree of size $O(2^k)$ for integer-weighted graphs.

4   Refined Edge Branching

We now refine our edge branching and present a sketch of proof showing that the search tree has size $O(1.82^k)$. This results in the fastest known algorithm for unweighted CLUSTER EDITING: the previously best-known branching strategy by Gramm et al. [5] results in a search tree of size $O(1.92^k)$. This algorithm uses complicated branching rules (more than 1300 cases) and has never been implemented. To the best of our knowledge, the fastest implementation for unweighted CLUSTER EDITING has running time $O(2.27^k + n^3)$ using 11 branching cases [6,4]. In contrast, our branching strategy is both fast and simple using only two branching cases.

Theorem 1. *For an integer weighted graph that contains no zero-edges, the* WEIGHTED CLUSTER EDITING *problem can be solved in* $O(1.82^k + n^3)$ *time.*

We modify the order in which edges are processed by the edge branching strategy, what allows for a simpler analysis of the running time behavior. We conjecture that Theorem 1 is also true for edge branching where edges are sorted with respect to branching number, but this requires many more case distinctions.

Let G_w be an integer-weighted and connected graph. We say that we *branch on* an edge uv by setting uv to forbidden and recursing, and merging uv and recursing. To deal with zero-edges, we use the above bookkeeping trick: Creating a zero-edge induces cost $\geq \frac{1}{2}$, and resolving a zero-edge induces the remaining cost $\frac{1}{2}$. We choose an edge to branch on according to the following order:

(A) If there is an edge with branching vector $(1, \frac{3}{2})$ or better then we branch on this edge.
(B) If there is an edge xy and a vertex z in G_w such that x,y,z form a triangle, and if there exist two additional vertices v_1, v_2 such that for both v_1, v_2 one of the following conditions holds (where x and y may be exchanged):
 (B1) xv_i is an edge and yv_i is a non-edge
 (B2) xv_i is a zero-edge and yv_i is a zero-edge
 (B3) xv_i is a zero-edge and yv_i is a non-edge, and zv_i is an edge or a zero-edge
 (B4) xv_i is an edge and yv_i is a zero-edge, and zv_i is a non-edge or a zero-edge
  Then branch on xy.

If no such edge exists, we stop the recursion. We will show below that the remaining graph must be a clique, a clique minus one edge (where the last edge is either a zero-edge or a non-edge), a path, a circle, or contains only 4 vertices. We will also show how to solve this remaining instance in polynomial time. See Fig. 1 for the four initial cases of condition (B). To be more precise, there are ten different subcases of condition (B) which are combinations of (B1),...,(B4), taking into account that we can exchange x and y. We denote them by (B11) to (B44).

If there exists an edge satisfying condition (A) then branching on this edge has branching number 1.76. The following lemma corresponds to condition (B) of edge sorting, and shows how we analyze two branching steps together: The first branching step can in fact result in a branching vector of $(1,1)$ but the next branching steps result in better branching vectors, leading to an overall branching number as desired.

Lemma 3. *Let G_w be an integer-weighted and connected graph, and assume that there is an edge xy that satisfies condition (B). Then, branching on xy and performing another branching step where edges to branch on are chosen according to the edge sorting, results in a branching vector of $(2, \frac{5}{2}, 2, 3)$ with branching number ≤ 1.82.*

Proof. Branching on edge xy leads to a branching vector of $(1,1)$: Deleting xy induces cost at least 1, and merging xy results in cost at least $2 \cdot \frac{1}{2} = 1$ since for each v_i a conflict triple or a zero-edge will be resolved. We will now show that after setting xy to "forbidden" there exists an edge with branching vector $(1, \frac{3}{2})$ and after merging xy there exists an edge with branching vector $(1,2)$. These are the worst-case branching vectors for the edge which is chosen in the next branching step.

First we analyze the case where xy is set to "forbidden", see Fig. 2: We show that now one of the edges xz or yz has branching vector $(1, \frac{3}{2})$. Setting xz or yz to forbidden results in cost 1. Merging xz or yz resolves the conflict triple xzy, resulting in cost 1 since xy is forbidden. If condition (B2), (B3), or (B4) holds then in addition, a zero-edge is resolved when merging xz or yz. If condition (B1) holds we distinguish two cases: If v_1z is a non-edge or a zero-edge, then we branch on xz which either resolves an additional zero-edge, or resolves the conflict triple v_1xz. If v_1z is an edge then we branch on yz which resolves the conflict triple v_1zy. Hence, either xz or yz have merging costs $\frac{3}{2}$.

Second we consider the case where x, y have been merged, see Fig. 3. Let w_{xy} be the vertex resulting from merging xy: We show that now, the edge $w_{xy}z$ has branching vector $(1,2)$. Deleting $w_{xy}z$ induces cost of 2 as $s(w_{xy}z) \geq 2$. Merging

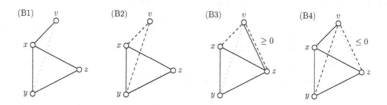

Fig. 1. Conditions (B1) to (B4) of edge sorting. Solid lines are edges, dashed lines are zero-edges, dotted lines are non-edges.

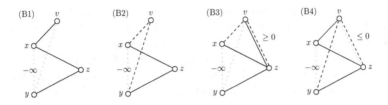

Fig. 2. Branching conditions (B1) to (B4) after xy is set to "forbidden"

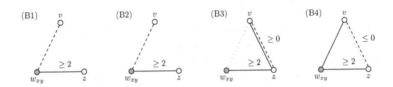

Fig. 3. Conditions (B1) to (B4) after merging xy

$w_{xy}z$ induces cost of $\frac{1}{2}$ for each v_i: If condition (B1) holds for v_i then $w_{xy}v_i$ is a zero-edge. Otherwise, we infer $s(xv_i) > 1$ or $s(yv_i) < -1$, so the initial branching on xy would have resulted in a branching vector of $(1, \frac{3}{2})$. Merging $w_{xy}z$ resolves this zero-edge. If condition (B2) holds then $w_{xy}v_i$ clearly is a zero-edge. For conditions (B3) and (B4) merging $w_{xy}z$ either resolves a zero-edge v_iz, or resolves a conflict triple $w_{xy}zv$ or $vw_{xy}z$. These observations hold both for v_1 and v_2, so merging $w_{xy}z$ results in total cost $\frac{2}{2}$.

We cannot guarantee that the edge branching strategy will actually branch on edges xz or yz (after xy has been set to forbidden) and $w_{xy}z$ (after merging xy) in the next step of the branching. But we have shown that edges with branching numbers 1.755 and 1.6191 exist after the first step of the branching. With regards to the first case, one can easily check that all possible branching vectors with branching number ≤ 1.755 are of the form $(a, b/2)$ for integers $a \geq 1$ and $b \geq 3$. Similarly, all branching vectors with branching number ≤ 1.6191 are of the form $(a, b/2)$ for integers $a \geq 1$ and $b \geq 4$, or $a \geq 2$ and $b \geq 2$. This shows that even if we pick other edges in the second step of our branching, we still can guarantee branching vector $(2, \frac{5}{2}, 2, 3)$ with branching number 1.82. $\qquad\square$

The following is again an observation regarding unweighted graphs:

Lemma 4. *Let G be a connected, unweighted graph. Assume that there is no edge in G that is part of three conflict triples, and there exists no triangle uvw in G such that uv is part of two conflict triples. Then G is a clique, a graph with at most one non-edge, a $K_{1,3}$, a path, or a circle.*

Proof. Assume that $G = (V, E)$ contains at least one edge xy that is part of two conflict triples: Otherwise, Lemma 1 guarantees that G is a clique or a clique minus a single edge. Let u and w be two vertices involved in conflict triples for

xy. This implies that either xu or yu is an edge, and that either xv or yv is an edge. Assume there exists another vertex $z \notin \{x, y, u, v\}$ with $xz \in E$: If $yz \in E$ then xyz is a triangle as excluded by our assumptions, and if $yz \notin E$ then xy is part of three conflict triples. So, no such z can exist and neither x nor y can be connected to any other vertex.

We distinguish two cases: the two conflict triples are either of the form uxy and uxv (asymmetric case), or uxy and xyv (symmetric case). For the asymmetric case, we can exchange u and v. Assume there exists another vertex $w \notin \{x, y, u, v\}$ with $uw \in E$. Then the edge xu is part of two conflict triples yxu and xuw and an additional edge xv exists. If $uv \notin E$ then xu is part of three conflict triples. If $uv \in E$ then the edge xu is part of a triangle xuv that is excluded by our assumptions. This implies that no such vertex w can exist, and the connected graph G is a $K_{1,3}$.

For the symmetric case, assume that there exists another vertex $w \notin \{x, y, u, v\}$ with $uw \in E$. Now, the edge xu is part of two conflict triples yxu and xuw, again in symmetric arrangement. If some $z \notin \{x, y, u, v, w\}$ exists with uz, we can show again that xu is part of three conflict triples or part of a triangle excluded by our assumptions. The same holds true for a vertex w with $vw \in E$. Repeating this argument we show that all vertices in the connected graph G have degree one or two, so G is a path or a circle.   □

Let us now assume that there is no edge that satisfies branching conditions (A) or (B). Again, we transform the integer-weighted graph into an unweighted graph G where zero-edges of the integer-weighted graph are transformed into non-edges in G. Clearly, G does not contain an edge that is part of three conflict triples. Using Lemma 4 we infer that G is either one of the graph structures described there, or there exists an edge xy that is part of a triangle xyz and that is part of two conflict triples. In the first case, we have reduced the weighted graph as claimed: The weighted graph is a clique, a clique minus one edge, a path, a circle, or contains only four vertices. In the second case there is an edge xy which is contained in a triangle and two conflict triples for which branching condition (B) does not apply. It can be shown by rather technical analysis that in all cases the weighted graph is a weak clique or a graph with exactly one non-edge. We defer the details to the full paper.

If the remaining graph is a (weak) clique, we are finished. If it is a graph with one non-edge uv, we can solve it in polynomial time by calculating a minimum u-v-cut. In case the cost of the cut is higher than $-s(uv)$, we insert uv and are finished, otherwise we cut the graph according to the minimum u-v-cut and obtain two (weak) cliques. If the remaining graph is a path or a circle, it can be solved in polynomial time with dynamic programming. Again, we defer the details to the full paper. If the graph has at most four vertices, we can easily try all possibilities of solving it.

Proof (Theorem 1). From the above we infer that our search tree has size $O(1.82^k)$. This results in a total running time of $O(1.82^k \cdot k^8 + n^3)$: Initially, we run the parameter-dependent data reduction from [1] in time $O(n^3)$. This data reduction results in a problem kernel with $O(k^2)$ vertices. For every edge

Table 1. Average running times for artificial data, edge branching and $O(3^k)$ branching strategy from [1]. Ten instances per bucket for sizes 10–50, five instances for sizes 60–100. For size 70 (80, 90, 100) one (four, all five, all five) instances did not stop after 20 days of computation using the $O(3^k)$ strategy. For size 90 (100) two (three) instances did not stop after 20 days of computation using the edge branching strategy. For average running times, we ignored these unfinished instances (*).

Size of instance	10	20	30	40	50	60	70	80	90	100
average # edit	8.3	28.1	66.7	115.5	183.2	263.0	351.6	459.0	594.0	728.6
3^k strategy [1]	10 ms	54 ms	1.0 s	29 s	7.6 min	27 h	58 h*	19 days*	n/a*	n/a*
edge branching	4 ms	16 ms	238 ms	2.5 s	18.2 s	5.5 h	17.7 h	13.8 h	34.8 h*	17.1 h*
with reduction [2]	3 ms	14 ms	163 ms	1.2 s	1.6 s	32 s	43 s	23 s	166 s	36 s

we compute the branching number that results from deleting and merging this edge in total time $O(k^6)$. Similarly, we can check for the substructures for branching condition (B) in time $O(k^8)$. In fact, we can get rid of the polynomial factor: We use interleaving [12] by performing data reduction repeatedly during the course of the search tree algorithm whenever possible. This reduces the total running time to $O(1.82^k + n^3)$. The remaining structures can be solved in polynomial time. □

Regarding WEIGHTED CLUSTER EDITING instances with real-valued weights, the edge branching strategy is also guaranteed to find the optimal solution. Let k be the cost parameter, we want to decide whether there is a solution of cost at most k. To estimate the worst-case running time we have to assume that all vertex pairs have weight at least one [1]. We redo our simple analysis from Sec. 3: Whenever joining two pairs of vertices results in a pair with absolute weight smaller than one, we put aside $\frac{1}{2}$ using our bookkeeping technique. This pair may later be part of a conflict triple, and when editing this pair we decrease k by $\frac{1}{2}$ we put aside earlier because the absolute weight of this pair can be arbitrarily small. A similar analysis to that given in this section, shows that the worst-case branching vector reduces to $(\frac{1}{2}, 2, 2)$ and the size of the search tree is $O(2.39^k)$. We defer the details to the full paper.

5   Computational Results

We have implemented the edge branching algorithm with search for the edge with maximum branching number in C++. We apply our data reduction from [1] to every instance in advance and when traversing the search tree. The program accepts nonnegative real values as edge modification costs. All running times were measured on an AMD Opteron-275 2.2 GHz with 6 GB of memory running Solaris 10.

We want to explore the performance of our algorithms and compare it to the previously fastest branching strategy for WEIGHTED CLUSTER EDITING from [1]. As reported there, branching strategies that do not merge vertices are clearly and consistently outperformed by those that do so, and unlike what theoretical running times suggest, the $O(2.42^k)$ was consistently outperformed by the $O(3^k)$ strategy. For our evaluation, we use artificial data. We generate artificial instances by first constructing a transitive graph with n vertices by uniformly drawing clique sizes in $\{1, \ldots, n\}$ until all vertices have been used up. Next, we perturb this graph: for each pair uv we delete or insert an edge uv with probability 0.15. Running times are reported in Table 1. We also run experiments on the protein similarity data used in [1] and observed similar results. As one can see, edge branching is much faster than the previously fastest branching algorithm, and performance is increased by several orders of magnitude. For comparison, we also report running times of the FPT algorithm from [2] that uses the same edge branching strategy but, in addition, employs new parameter-independent reduction rules to cut down instance sizes before branching, and further heuristic improvements.

6   Conclusion

We have presented a surprisingly simple branching strategy that lead to the fastest known parameterized algorithm for (integer-weighted) CLUSTER EDITING with respect to theoretical running time bounds. We believe that we can prove even better worst-case running times for this same strategy, using a refined, automated analysis similar to [5].

We implemented our algorithm and evaluated its performance. Together with further improvements reported in [2], our algorithm allows to solve weighted CLUSTER EDITING instances with several hundred edge modifications in a matter of seconds. This clearly proves the practical usefulness of our approach and constitutes a huge improvement over [4] where unweighted instances with 50 edge modifications required several hours of computation. Wittkop et al. [16] recently demonstrated the power of WEIGHTED CLUSTER EDITING for clustering homologous proteins, so algorithm both fast in theory and efficient in practice are highly desirable.

Acknowledgments

We thank Svenja Simon and Thilo Muth for helping with the implementation. S. Briesemeister gratefully acknowledges financial support from LGFG Promo-tionsverbund "Pflanzliche Sensorhistidinkinasen" at the University of Tübingen.

References

1. Böcker, S., Briesemeister, S., Bui, Q.B.A., Truß, A.: A fixed-parameter approach for weighted cluster editing. In: Proc. of Asia-Pacific Bioinformatics Conference (APBC 2008). Series on Advances in Bioinformatics and Computational Biology, vol. 5, pp. 211–220. Imperial College Press (2008)
2. Böcker, S., Briesemeister, S., Klau, G.W.: Exact algorithms for cluster editing: Evaluation and experiments. In: McGeoch, C.C. (ed.) WEA 2008. LNCS, vol. 5038, pp. 289–302. Springer, Heidelberg (2008)

3. Bodlaender, H.L., Cai, L., Chen, J., Fellows, M.R., Telle, J.A., Marx, D.: Open problems in parameterized and exact computation — IWPEC 2006. Technical Report UU-CS-2006-052, Department of Information and Computing Sciences, Utrecht University (2006)
4. Dehne, F., Langston, M.A., Luo, X., Pitre, S., Shaw, P., Zhang, Y.: The cluster editing problem: Implementations and experiments. In: Bodlaender, H.L., Langston, M.A. (eds.) IWPEC 2006. LNCS, vol. 4169, pp. 13–24. Springer, Heidelberg (2006)
5. Gramm, J., Guo, J., Hüffner, F., Niedermeier, R.: Automated generation of search tree algorithms for hard graph modification problems. Algorithmica 39(4), 321–347 (2004)
6. Gramm, J., Guo, J., Hüffner, F., Niedermeier, R.: Graph-modeled data clustering: Fixed-parameter algorithms for clique generation. Theor. Comput. Syst. 38(4), 373–392 (2005)
7. Grötschel, M., Wakabayashi, Y.: A cutting plane algorithm for a clustering problem. Math. Program. 45, 52–96 (1989)
8. Guo, J.: A more effective linear kernelization for Cluster Editing. In: Chen, B., Paterson, M., Zhang, G. (eds.) ESCAPE 2007. LNCS, vol. 4614, pp. 36–47. Springer, Heidelberg (2007)
9. Hsu, W.-L., Ma, T.-H.: Substitution decomposition on chordal graphs and applications. In: Hsu, W.-L., Lee, R.C.T. (eds.) ISA 1991. LNCS, vol. 557, pp. 52–60. Springer, Heidelberg (1991)
10. Křivánek, M., Morávek, J.: NP-hard problems in hierarchical-tree clustering. Acta Inform. 23(3), 311–323 (1986)
11. Niedermeier, R.: Invitation to Fixed-Parameter Algorithms. Oxford University Press, Oxford (2006)
12. Niedermeier, R., Rossmanith, P.: A general method to speed up fixed-parameter-tractable algorithms. Inform. Process. Lett. 73, 125–129 (2000)
13. Shamir, R., Sharan, R., Tsur, D.: Cluster graph modification problems. Discrete Appl. Math. 144(1–2), 173–182 (2004)
14. Sharan, R., Maron-Katz, A., Shamir, R.: CLICK and EXPANDER: a system for clustering and visualizing gene expression data. Bioinformatics 19(14), 1787–1799 (2003)
15. van Zuylen, A., Williamson, D.P.: Deterministic algorithms for rank aggregation and other ranking and clustering problems. In: Proc. of Workshop on Approximation and Online Algorithms (WAOA 2007). LNCS, vol. 4927, pp. 260–273. Springer, Heidelberg (2008)
16. Wittkop, T., Baumbach, J., Lobo, F., Rahmann, S.: Large scale clustering of protein sequences with FORCE – a layout based heuristic for weighted cluster editing. BMC Bioinformatics 8(1), 396 (2007)

Parameterized Graph Editing with Chosen Vertex Degrees

Luke Mathieson and Stefan Szeider

Department of Computer Science
University of Durham, UK
{luke.mathieson,stefan.szeider}@durham.ac.uk

Abstract. We study the parameterized complexity of the following problem: is it possible to make a given graph r-regular by applying at most k elementary editing operations; the operations are vertex deletion, edge deletion, and edge addition. We also consider more general annotated variants of this problem, where vertices and edges are assigned an integer cost and each vertex v has assigned its own desired degree $\delta(v) \in \{0, \dots, r\}$. We show that both problems are fixed-parameter tractable when parameterized by (k, r), but $W[1]$-hard when parameterized by k alone. These results extend our earlier results on problems that are defined similarly but where edge addition is not available. We also show that if edge addition and/or deletion are the only available operations, then the problems are solvable in polynomial time. This completes the classification for all combinations of the three considered editing operations.

1 Introduction

Deciding whether a given graph has a regular subgraph is a well studied problem. Chvátal et al. [5] give one of the earliest results, showing that the CUBIC SUBGRAPH problem is NP-complete. Plesník [17] proves that it remains NP-complete even when restricted to a planar bipartite graph with maximum degree 4. In the same paper he also shows that the r-REGULAR SUBGRAPH problem for $r \geq 3$ is NP-complete even on bipartite graphs of degree at most $r + 1$. Cheah and Corneil [4] show that a similar result holds for general graphs of degree at most $r + 1$. A series of results for further constraints is given by Stewart [18,19,20]. Bodlaender et al. [2] give a polynomial-time algorithm for producing a Δ-regular supergraph of a graph with maximum degree Δ, using at most $\Delta + 2$ additional vertices. Moser and Thilikos [15] give a series of results for certain parameterized versions, showing that when parameterized by the size of the regular subgraph, the problem is $W[1]$-hard, but when parameterized by both the number k of vertices to remove to make the graph regular and the regularity r, the problem is fixed-parameter tractable. In previous work [14] we show that when parameterized by k alone, the problem is $W[1]$-hard (a question left open by Moser and Thilikos). We also introduce a generalized version of the problem where vertices and edges are weighted, and each vertex has a degree function that specifies the number of edges to be incident on the vertex, rather than simply a fixed number for all vertices. When parameterized by the number k of edges and vertices to remove to obtain the regular graph, the problem is $W[1]$-hard, but when parameterized by k and the bound r on the

B. Yang, D.-Z. Du, and C.A. Wang (Eds.): COCOA 2008, LNCS 5165, pp. 13–22, 2008.
© Springer-Verlag Berlin Heidelberg 2008

degree function, the problem is fixed-parameter tractable. Interestingly the latter result improves that of Moser and Thilikos even though it is generalized, and additionally allows edge deletion.

In this paper we extend the editing operations available to include edge addition besides vertex and edge deletion (see Section 2.1 for precise definitions), thus giving the following problems.

EDIT TO REGULAR GRAPH
Instance: A graph $G = (V, E)$, two nonnegative integers k and r.
Question: Is there an r-regular graph H obtainable from G by at most k edit operations?

WEIGHTED EDIT TO CHOSEN DEGREE GRAPH
Instance: A graph $G = (V, E)$, nonnegative integers k and r, a weight function $\rho : V \cup E \rightarrow \{1, \ldots, k+1\}$ and a degree function $\delta : V \rightarrow \{0, \ldots, r\}$.
Question: Is there a graph H obtainable from G by edit operations of total cost at most k such that $\sum_{e \in E(v)} \rho(e) = \delta(v)$ holds for each vertex in H?

Variants of the above two problems with only one or two of the three editing operations available are defined similarly.

We previously demonstrated [14] that these two problems are $W[1]$-hard when parameterized by k. In this paper we complete the classification and show that they are both fixed-parameter tractable with parameter (k, r). We also give a simpler proof that the weighted edit problem is $W[1]$-hard when parameterized by k alone, via the $W[1]$-hardness of the related subproblem EDGE REPLACEMENT SET (see Section 4). We also prove that EDGE REPLACEMENT SET is NP-complete, giving an indication that a polynomial-time kernelization of the form previously used for the deletion version of the problems is unlikely to exist for the edit version. Additionally, we show that WEIGHTED EDGE EDIT TO CHOSEN DEGREE GRAPH, and the unweighted counterpart EDGE EDIT TO REGULAR GRAPH, where the edit operations are edge addition and deletion, both have polynomial-time algorithms. The results are summarized in Table below.

2 Preliminaries

2.1 Graph Modification

Graph modification or graph editing problems are widespread throughout the literature appearing in various forms in such areas as bioinformatics [6], electronic commerce [9] and graph theory [1]. Three fundamental operations for graph editing are edge deletion, vertex deletion and edge addition. For any combination of these three operations Cai [3] demonstrates fixed-parameter tractability for graph properties with finitely many obstructions in the induced order.

In this paper we consider simple, undirected graphs (whether weighted or unweighted). The edge between two vertices u and v is denoted uv (or equivalently vu). The degree of a vertex u is denoted $d(u)$.

Problem	Operations	Parameter	
		k	(k, r)
Uniform	v	$W[1]$-hard†	FPT‡†
	e+v	$W[1]$-hard†	FPT†
	v+a	$W[1]$-hard†	FPT*
	e+v+a	$W[1]$-hard*†	FPT*
	a	P¶*	P¶*
	e	P¶*	P¶*
	e+a	P*	P*
Annotated	v	$W[1]$-hard†	FPT†
	e+v	$W[1]$-hard†	FPT†
	v+a	$W[1]$-hard†	FPT*
	e+v+a	$W[1]$-hard*†	FPT*
	a	P¶*	P¶*
	e	P¶*	P¶*
	e+a	P*	P*
ERS		$W[1]$-hard*	FPT*

Results shown in: * this paper, ‡ [15], † [14], $\P$ follows from results on f-factors [13].

The editing operations are codified as 'e' - edge deletion, 'v' - vertex deletion and 'a' - edge addition. The 'uniform' version of the problem is where all vertices and edges have weight 1 and the desired graph is r-regular.

The final row indicates the complexity of EDGE REPLACEMENT SET.

The following operations alter a graph $G = (V, E)$ into a new graph $G' = (V', E')$. Deleting an edge uv simply removes that edge from the graph (i.e., $E' = E \setminus \{uv\}$, and $V' = V$). Deleting a vertex u removes that vertex and all incident edges (i.e., $V' = V \setminus \{u\}$, $E' = E \setminus \{uv \mid v \in V\}$). Adding an edge uv of course inserts an edge between u and v (i.e., $E' = E \cup \{uv\}$, $V' = V$).

In this paper we also consider weighted versions of these operations, which are defined similarly. Given a weighted edge or vertex, the cost of deletion is simply that weight. Note particularly that the cost of deleting a vertex is the weight of the vertex alone, not the weight of the vertex plus the weights of the incident edges, even though they are also removed (this is consistent with the normal definition for unweighted graphs, where deleting a vertex counts as one step, regardless of any incident edges). Weighted edge addition works as defined, except where an edge already exists, which in the unweighted case would prevent addition. In the weighted case however, we simply increase the weight of the existing edge. Thus, in the presence of edge addition one needs to consider the weighted and the unweighted variants of a problem separately, as the former is not just a special case of the latter.

2.2   Basic Parameterized Complexity

Here we introduce some basic, relevant parameterized complexity theory. For a more in-depth coverage we refer to the books of Downey and Fellows [7], Flum and Grohe [11] and Niedermeier [16]. When considering the complexity of a problem in a classical, P vs. NP setting, the only measure available is n, the instance size (or some function thereof). Parameterized complexity adds a second measure, that of a parameter k, which is given as a special part of the input. If a problem has an algorithm that runs in time $O(f(k)p(n))$, where p is a polynomial and f is a computable function of k, then the problem is *fixed-parameter tractable*, or in the class FPT. Conversely, the demonstration of hardness for the class $W[t]$ for some $t \geq 1$ gives the intuition that the problem is unlikely to be fixed-parameter tractable. This is analogous to a problem being NP-hard in the classical set-up. For the sake of clarity any problem is understood to be a decision problem unless explicitly stated otherwise (and the parameterized complexity classes that are referenced are defined for decision problems).

Demonstration of $W[t]$-hardness is normally done via an FPT reduction, which is the parameterized complexity equivalent of a polynomial-time many-one reduction in classical complexity theory. Given two parameterized problems Π_1 and Π_2, an FPT reduction $\Pi_1 \leq_{FPT} \Pi_2$ is a mapping from Π_1 to Π_2 that maps an instance (I, k) of Π_1 to an instance (I', k') of Π_2 such that (i) $k' = h(k)$ for some computable function h, (ii) (I, k) is a YES-instance of Π_1 if and only if (I', k') is a YES-instance of Π_2 and (iii) the mapping can be computed in time $O(f(k)p(|I|))$, where f is some computable function of the parameter k alone and p is a polynomial.

Then if Π_2 is in FPT, so is Π_1, and if Π_1 is $W[t]$-hard, so is Π_2.

The classes $W[t]$, $t = 1, 2, \ldots$, are defined as the classes of problems that can be FPT-reduced to certain weighted satisfiability problems. The classes form the chain FPT $\subseteq W[1] \subseteq W[2] \subseteq \ldots$, where all inclusions are believed to be strict (Flum and Grohe [11] in particular give detailed coverage of this hierarchy).

Reduction to a problem kernel, or *kernelization*, is one of the fundamental techniques for developing fixed-parameter tractable algorithms, and thus for demonstrating FPT membership. A problem is *kernelizable* if, given an instance (I, k) of the problem, where I is the input and k is the parameter, it is possible to produce in polynomial time an instance (I', k') where $|I'| \leq g(k')$ and $k' = h(k)$ for computable functions g and h, and (I, k) is a YES-instance if and only if (I', k') is a YES-instance. It can be shown that a problem is kernelizable in this sense if and only if it is fixed-parameter tractable. Kernelization is normally accomplished by the application of *reduction rules* to the instance. Further explanation of the theory can be found in Estivill-Castro et al.'s paper [10].

3   Easy Cases

Before moving to the general versions of the considered problems, let us examine restricted versions where *only* edge editing operations are allowed; we may not delete any vertices.

If only one of the operations is available, then the problems (weighted and uniform) can easily be seen to correspond the well-known polynomially solvable *f-factor problem* [13]. Although the f-factor problem does not explicitly include editing operations, edge deletions are dealt with implicitly as any f-factor of a graph has the same number of edges, we need merely then to compare the difference between this number and the total number of edges with the parameter. When the operation is edge addition, we simply use the complement of the input graph instead and modify the degree function appropriately. However it is not immediately apparent that these techniques may be directly applied to the case where we allow both edge addition and edge deletion. Hence we shall give a general construction for solving the WEIGHTED EDGE EDIT TO CHOSEN DEGREE GRAPH problem by application of Edmond's minimum weight perfect matching algorithm [13,8].

Let G, k be an instance of WEIGHTED EDGE EDIT TO CHOSEN DEGREE GRAPH. By allowing edge weights of 0, we may assume that G is a complete graph. Now solving the problem is clearly equivalent to finding an edge weight function $\rho' : E(G) \rightarrow \{0, 1, 2, \dots\}$ of G such that for each $v \in V(G)$ we have $\sum_{vv' \in E(G)} \rho'(vv') = \delta(v)$ and the cost of ρ', $\sum_{vv' \in E(G)} |\rho(vv') - \rho'(vv')|$, is at most k.

We construct a graph H with edge-weight function η as follows: For each vertex v of G we introduce in H a set $V(v)$ of $\delta(v)$ vertices. For each edge $vv' \in E(G)$ we add the following vertices and edges to H'.

1. We add two sets $V_{\text{del}}(v, v')$ and $V_{\text{del}}(v', v)$ of vertices, each of size $\rho(vv')$.
2. We add two sets $V_{\text{add}}(v, v')$ and $V_{\text{add}}(v', v)$ of vertices, each of size $\min(\delta(v), \delta(v'))$.
3. We add all edges uw for $u \in V(v)$ and $w \in V_{\text{del}}(v, v') \cup V_{\text{add}}(v, v')$, and all edges uw for $u \in V(v')$ and $w \in V_{\text{del}}(v', v) \cup V_{\text{add}}(v', v)$.
4. We add edges that form a matching $M_{vv'}$ between the sets $V_{\text{del}}(v, v')$ and $V_{\text{del}}(v', v)$. We will refer to these edges as *deletion edges*.
5. We add edges that form a matching $M'_{vv'}$ between the sets $V_{\text{add}}(v, v')$ and $V_{\text{add}}(v', v)$ and subdivide the edges of $M'_{vv'}$ twice, that is, we replace $xy \in M'_{vv'}$ by a path x, x_y, y_x, y where x_y and y_x are new vertices. We will refer to the edges of the form $x_y y_x$ as *addition edges*.

This completes the construction of H. It remains to assign deletion and addition edges e the weight $\eta(e) = 1$, and all other edges the weight 0. It can be verified that (G, k) a yes-instance of WEIGHTED EDGE EDIT TO CHOSEN DEGREE GRAPH if and only if H perfect matching of weight at most k, but owing to space restrictions, we omit the proof. If we remove the addition (deletion) edges from H, then we also have a construction that can be used to solve the edge deletion (addition) problem. Naturally this construction allows solutions for the uniform versions of the problems as well, as a subcase.

4   A Thorn in the Paw

Previously [14] we demonstrated the following result:

Theorem 1. WEIGHTED DELETION TO CHOSEN DEGREE GRAPH *is fixed parameter tractable for parameter* (k, r).

This was shown by reduction to problem kernel, with a kernel of size $O(kr(k+r))$. It is interesting to note that the result holds not only for DELETION TO REGULAR GRAPH, but also for WEIGHTED VERTEX DELETION TO CHOSEN DEGREE GRAPH, and that the generalization gives a smaller kernel than by using a similar method without the annotation.

Naturally we would like to achieve a similar result for the edit versions of the problems. However the kernelization for deletion problems heavily relies on the fact that if we delete a vertex in a clean region (defined below) or an edge incident with a vertex in a clean region, then we must delete the entire clean region. Consequently, we can shrink large clean regions as their specific structure is not relevant. This reasoning fails for editing problems where edge addition is allowed.

We prove in Section 5 that the edit version is indeed fixed-parameter tractable for (k, r), but obtaining a kernel is difficult, at least when approached in a similar manner to the previous investigation. Note that demonstration of fixed-parameter tractability guarantees that *some* kernelization exists, what we show here is that it is unlikely to take a certain (useful) form.

Firstly it is useful to define the notion of a *clean region* (first introduced by Moser and Thilikos [15]). Given a graph $G = (V, E)$, a function $\delta : V \rightarrow \{0, \ldots, r\}$, and a function $\rho : V \cup E \rightarrow \{1, 2, 3, \ldots\}$, we say a vertex v is *clean* if $\sum_{e \in E(v)} \rho(e) = \delta(v)$, where $E(v)$ denotes the set of edges incident on v. Then a clean region is a maximal connected subgraph of clean vertices. In the case where the graph is unweighted, we implicitly assume that $\delta(v) = r$ for all $v \in V$ and $\rho(x) = 1$ for all $x \in V \cup E$.

In both EDIT TO REGULAR GRAPH and WEIGHTED EDIT TO CHOSEN DEGREE GRAPH it may be necessary to delete a set of vertices from clean regions so that the edges that become available may be used to complete the degree of a vertex of insufficient degree (indeed there are easily constructable instances where this is the only way to solve the instance). This gives the following sub-problem:

EDGE REPLACEMENT SET
Instance: A graph $G = (V, E)$, two positive integers k and t.
Question: Does there exist a set $X \subseteq V$ such that $|X| \leq k$ and there are exactly t edges between vertices in X and vertices in $V \setminus X$?

Unfortunately, EDGE REPLACEMENT SET is NP-complete, thus making the possibility of obtaining a kernel in polynomial time by somehow identifying all relevant sets in the clean regions unlikely. The proof is by reduction from the following:

REGULAR CLIQUE
Instance: An r-regular graph $G = (V, E)$, a positive integer k.
Question: Does G contain a clique on k vertices?

REGULAR CLIQUE is NP-complete, and $W[1]$-complete for parameter k, but fixed-parameter tractable for parameter (k, r). We refer to previous work [14] for a detailed proof of these statements.

The proof of the following theorem requires that the regularity r of the input graph in the REGULAR CLIQUE instance be sufficiently large. It is possible to construct a "fixing gadget" that allows the degree of each vertex to be increased effectively arbitrarily, without introducing any non-trivial cliques. We refer again to previous work [14], and particularly to the proof of Lemma 3.1 contained therein for proof of this claim.

Theorem 2. EDGE REPLACEMENT SET *is* NP-*complete and* $W[1]$-*hard for parameter* k.

Proof. We shall concentrate on the $W[1]$-hardness proof; the NP-hardness follows from the same result.

Let $(G = (V, E), k)$ be an instance of REGULAR CLIQUE where G is r-regular. We may assume that r is not bounded in terms of k, since REGULAR CLIQUE is fixed-parameter tractable for parameter (k, r). For a set $X \subseteq V$ let $d(X)$ denote the number of edges $uv \in E$ with $u \in X$ and $v \in V \setminus X$. If X forms a clique in G then $d(X) = k(r - k + 1)$. Therefore we put $t = k(r - k + 1)$ and consider (G, k, t) as an instance of EDGE REPLACEMENT SET.

Let $X \subseteq V$ with $|X| \leq k$ and $d(X) = t$. We show that X has exactly k elements and forms a clique in G. Assume for the sake of contradiction that $|X| < k$. It follows that $d(X) \leq |X|r$ and consequently $r < r(k - |X|) \leq k^2 - k$. This contradicts the assumption that r is not bounded in terms of k. Hence we conclude $|X| = k$. Each vertex $x \in X$ has at most $k - 1$ neighbors in X and at least $r - k + 1$ neighbors in $V \setminus X$. Therefore, if at least one $x \in X$ had fewer than $k - 1$ neighbors in X, then $d(X) > k(r - k + 1) = t$. Since $d(X) = t$, it follows that X is a clique in G.   □

As the weighted edit problem contains EDGE REPLACEMENT SET as a subproblem we also have the following result:

Corollary 1. WEIGHTED EDIT TO CHOSEN DEGREE GRAPH *is* $W[1]$-*hard for parameter* k.

This can be observed by considering the following simple construction: an isolated vertex with $\rho(v) = k + 1$ and $\delta(v) = t \leq r$, along with a clique on y vertices inside a clean region, where each clique vertex has t/y 'outgoing' edges, low weight, and $k \geq y + t$. We cannot delete the isolated vertex, but the deleting the clique will give the requisite number of edges to 'fix' the isolated vertex.

Thus this proof demonstrates that a polynomial-time kernelization which relies upon identifying such candidate sets for deletion is unlikely to exist. Note also that the proof holds if we also demand in EDIT TO REGULAR GRAPH that the set X is connected.

5   Editing Is Fixed Parameter Tractable for Parameter (k, r)

To demonstrate that EDIT TO REGULAR GRAPH is fixed-parameter tractable we take a logical approach and apply the following meta-theorem which is due to Frick and Grohe [12].

Theorem 3 ([12]). *Let C be a polynomial-time decidable class of structures of effectively bounded local tree-width. Then the model checking problem for first-order logic on the class C is fixed-parameter tractable parameterized by the length of the first-order formula.*

More particularly we use their corollary that the parameterized model checking problem for first-order logic is fixed-parameter tractable for graphs of *bounded degree*. Stewart [21] pointed out that this result also holds if the degree bound is not global but depends on the parameter. Furthermore he indicated how this can be used to show that REGULAR SUBGRAPH with parameter (k, r) is fixed-parameter tractable. In the following we extend this approach to EDIT TO REGULAR GRAPH and further to WEIGHTED EDIT TO CHOSEN DEGREE GRAPH.

First we introduce the following reduction rule, used previously [14,15], that reduces an instance $(G, (k, r))$ of EDIT TO REGULAR GRAPH to another instance $(G', (k', r'))$ of EDIT TO REGULAR GRAPH with bounded degree:

Reduction Rule 1: If there exists a vertex v in G where $d(v) > k + r$, then $G' = G[V(G) \setminus \{v\}]$, $k' = k - 1$ and $r' = r$.

Therefore if we can formulate sentences ϕ_k, $k \geq 0$, of first-order logic such that ϕ_k is true for a graph G if and only if (G, k) is a YES-instance of EDIT TO REGULAR GRAPH, then we have established fixed-parameter tractability of EDIT TO REGULAR GRAPH, since by application of Reduction Rule 1 we have a graph of bounded degree. Note that the predicates Vx, Ey and Ixy mean that x is a vertex, y is an edge, and that y is incident on x, respectively. Furthermore we write $[n] = \{1, \ldots, n\}$.

The sentence is defined as

$$\phi_k = \bigvee_{k'+k''+k''' \leq k} \exists u_1, \ldots, u_{k'}, e_1, \ldots, e_{k''}, a_1, \ldots, a_{k'''}, b_1, \ldots, b_{k'''} (\phi'_k \wedge \forall v \, \phi''_k)$$

where ϕ'_k and ϕ''_k are defined below. ϕ'_k is the conjunction of the following clauses (1)...(4) that ensure that $u_1, \ldots, u_{k'}$ represent deleted vertices, $e_1, \ldots e_{k''}$ represent deleted edges, and a_i, b_i, $1 \leq i \leq k'''$ represent ends of added edges. Note that since added edges are not present in the given structure we need to express them in terms vertex pairs.

(1) $\bigwedge_{i \in [k']} V u_i \wedge \bigwedge_{i \in [k'']} E e_i$ "u_i is a vertex, e_i is an edge;"
(2) $\bigwedge_{i \in [k''']} V a_i \wedge V b_i \wedge a_i \neq b_i \wedge \bigwedge_j (u_j \neq a_i \wedge u_j \neq b_i)$ "a_i and b_i are distinct vertices and not deleted;"
(3) $\bigwedge_{i \in [k''']} \forall y (\neg I a_i y \vee \neg I b_i y)$ "a_i and b_i are not adjacent;"
(4) $\bigwedge_{1 \leq i < j \leq k''} (a_i \neq b_j \vee a_j \neq b_i) \wedge (a_i \neq a_j \vee b_i \neq b_j)$ "the pairs of vertices are mutually distinct."

The subformula ϕ''_k ensures that each vertex v has degree r after editing:

$$\phi''_k = (V v \wedge \bigwedge_{i \in [k']} v \neq u_i) \rightarrow \bigvee_{\substack{r', r'' \in [r] \\ r' + r'' = r}} \exists x_1, \ldots, x_{r'}, y_1, \ldots, y_{r''} \, \phi'''_k$$

where ϕ'''_k is the conjunction of the following clauses:

(5) $\bigwedge_{i \in [r']} I v x_i$ "v is incident with r' edges;"
(6) $\bigwedge_{1 \leq i < j \leq r'} x_i \neq x_j$ "the edges are all different;"
(7) $\bigwedge_{i \in [r'], j \in [k'']} x_i \neq e_j$ "the edges have not been deleted;"
(8) $\bigwedge_{i \in [r'], j \in [k']} \neg I u_j x_i$ "the ends of the edges have not been deleted;"
(9) $\forall x (I v x \rightarrow \bigvee_{i \in [r']} x = x_i \vee \bigvee_{i \in [k'']} x = e_i \vee \bigvee_i I x u_i)$ "v is not incident with any further edges except deleted edges;"
(10) $\bigwedge_{i \in [r'']} \bigvee_j (y_i = a_j \wedge v = b_j) \vee (y_i = b_j \wedge v = a_j)$ "v is incident with at least r'' added edges;"
(11) $\bigwedge_{j \in [r'']} (v = a_j \rightarrow \bigvee_i y_i = b_j) \wedge (v = b_j \rightarrow \bigvee_{j \in [r'']} y_i = a_j)$ "v is incident with at most r'' added edges."

By the above considerations, we have the following.

Theorem 4. EDIT TO REGULAR GRAPH *is fixed-parameter tractable for parameter* (k, r).

If we force k'' to be zero, then the same sentence suffices to prove that the variant with only edge addition and vertex deletion is also fixed-parameter tractable for parameter (k, r). This variant is $W[1]$-hard for parameter k by a previous result [14].

WEIGHTED EDIT TO CHOSEN DEGREE GRAPH can be classified by a similar approach, but first we must demonstrate that we can express the ρ and δ functions in first order logic. To this aim we introduce a series $W_i, 1 \le i \le k+1$, of weight predicates such that $W_i x$ is true for a vertex x if and only if $\rho(x) = i$, and a series $D_j, 0 \le j \le r$, of degree predicates such that $D_j x$ is true for a vertex x if and only if $\delta(x) = j$. We represent an edge of weight i by i parallel edges.

Hence we can formulate the following sentence to represent solutions of WEIGHTED EDIT TO CHOSEN DEGREE GRAPH.

$$
\psi_k = \bigvee_{\substack{k', k'', k''', l_1, \ldots, l_{k'} \in [k] \\ l_1 + \cdots + l_{k'} + k'' + k''' \le k}} \bigwedge_{i \in [k']} W_{l_i}(u_i) \wedge
$$

$$
\exists u_1, \ldots, u_{k'}, e_1, \ldots e_{k''}, a_1, \ldots, a_{k'''}, b_1, \ldots, b_{k'''} (\psi'_k \wedge \forall v \; \psi''_k)
$$

$$
\psi''_k = \bigvee_{j \in [r]} [(D_j v \wedge \bigwedge_{i \in [k']} v \ne u_i) \rightarrow \bigvee_{\substack{r', r'' \in [r] \\ r' + r'' = r}} \exists x_1, \ldots, x_{r'} \exists y_1, \ldots, y_{r''} \; \psi'''_k(j)]
$$

The subformula ψ'_k is the conjunction of the above clauses (1) and (2) (we omit (3) and (4) as we use multiple edges to encode edge weights) and $\psi'''_k(j)$ is obtained from the above subformula ϕ'''_k by setting r to j. Hence, as above, we conclude:

Theorem 5. WEIGHTED EDIT TO CHOSEN DEGREE GRAPH *is fixed-parameter tractable for parameter* (k, r).

By similar reasoning as before, this sentence demonstrates the fixed-parameter tractability of the variant without edge deletion for parameter (k, r). Again this variant is $W[1]$-hard for parameter k by a previous result [14].

6 Conclusion

We demonstrated that when parameterized by (k, r), the editing problems are fixed parameter tractable, but when parameterized by k, the problems are $W[1]$-hard. The only exceptions are when the editing operations are limited to edge addition and/or deletion, in which case the problems are solvable in polynomial time, thus completing the classification for all combinations of the three editing operations.

A change in complexity is also apparent when moving from the deletion only problems to the edit problems. It seems unlikely that a similar approach as used for the deletion problems can be used to develop a polynomial-time kernelization for the editing problems. For a feasible kernelization some new structural insight must be gained.

References

1. Bar-Yehuda, R., Rawitz, D.: Approximating element-weighted vertex deletion problems for the complete k-partite property. Journal of Algorithms 42(1), 20–40 (2002)
2. Bodlaender, H., Tan, R., van Leeuwen, J.: Finding a $\triangle$-regular supergraph of minimum order. Discrete Applied Mathematics 131(1), 3–9 (2003)
3. Cai, L.: Fixed-parameter tractability of graph modification problems for hereditary properties. Information Processing Letters 58(4), 171–176 (1996)
4. Cheah, F., Corneil, D.G.: The complexity of regular subgraph recognition. Discrete Applied Mathematics 27, 59–68 (1990)
5. Chvátal, V., Fleischner, H., Sheehan, J., Thomassen, C.: Three-regular subgraphs of four regular graphs. Journal of Graph Theory 3, 371–386 (1979)
6. Dehne, F., Langston, M., Luo, X., Pitre, S., Shaw, P., Zhang, Y.: The cluster editing problem: Implementations and experiments. In: Bodlaender, H.L., Langston, M.A. (eds.) IWPEC 2006. LNCS, vol. 4169, pp. 13–24. Springer, Heidelberg (2006)
7. Downey, R.G., Fellows, M.R.: Parameterized Complexity. Springer, Heidelberg (1999)
8. Edmonds, J.: Paths trees and flowers. Canadian Journal of Mathematics 17, 449–467 (1965)
9. Elkind, E.: True costs of cheap labor are hard to measure: Edge deletion and VCG payments in graphs. In: Riedl, J., Kearns, M.J., Reiter, M.K. (eds.) 6th ACM Conference on Electronic Commerce (EC-2005), pp. 108–116. ACM, New York (2005)
10. Estivill-Castro, V., Fellows, M., Langston, M., Rosamond, F.: FPT is P-TIME extremal structure I. In: Algorithms and Complexity in Durham 2005 (ACiD 2005), Texts in Algorithmics, pp. 1–41. College Publications (2005)
11. Flum, J., Grohe, M.: Parameterized Complexity Theory. Springer, Heidelberg (2006)
12. Frick, M., Grohe, M.: Deciding first-order properties of locally tree-decomposable structures. Journal of the ACM 48, 1184–1206 (2001)
13. Lovász, L., Plummer, M.D.: Matching Theory. Annals of Discrete Mathematics, vol. 29. North-Holland Publishing Co., Amsterdam (1986)
14. Mathieson, L., Szeider, S.: The parameterized complexity of regular subgraph problems and generalizations. In: Harland, J., Manyem, P. (eds.) Fourteenth Computing: The Australasian Theory Symposium (CATS 2008). CRPIT, vol. 77, pp. 79–86. ACS (2008)
15. Moser, H., Thilikos, D.: Parameterized complexity of finding regular induced subgraphs. In: Algorithms and Complexity in Durham 2006 (ACiD 2006), Texts in Algorithmics, pp. 107–118. College Publications (2006)
16. Niedermeier, R.: Invitation to Fixed-Parameter Algorithms. Oxford University Press, Oxford (2006)
17. Plesník, J.: A note on the complexity of finding regular subgraphs. Discrete Mathematics 49, 161–167 (1984)
18. Stewart, I.A.: Deciding whether a planar graph has a cubic subgraph is NP-complete. Discrete Mathematics 126(1–3), 349–357 (1994)
19. Stewart, I.A.: Finding regular subgraphs in both arbitrary and planar graph. Discrete Applied Mathematics 68(3), 223–235 (1996)
20. Stewart, I.A.: On locating cubic subgraphs in bounded-degree connected bipartite graphs. Discrete Mathematics 163(1–3), 319–324 (1997)
21. Stewart, I.A.: On the fixed-parameter tractability of parameterized model-checking problems. Information Processing Letters 106, 33–36 (2008)

Fixed-Parameter Tractability of Anonymizing Data by Suppressing Entries

Rhonda Chaytor[1], Patricia A. Evans[2], and Todd Wareham[3]

[1] School of Computing Science, Simon Fraser University, Vancouver BC, Canada
[2] Faculty of Computer Science, University of New Brunswick
Fredericton NB, Canada
[3] Department of Computer Science, Memorial University, St. John's NL, Canada
rhonda_chaytor@cs.sfu.ca, pevans@unb.ca, harold@cs.mun.ca

Abstract. A popular model for protecting privacy when person-specific data is released is k-*anonymity*. A dataset is k-anonymous if each record is identical to at least $(k-1)$ other records in the dataset. The basic k-anonymization problem, which minimizes the number of dataset entries that must be suppressed to achieve k-anonymity, is NP-hard and hence not solvable both quickly and optimally in general. We apply parameterized complexity analysis to explore algorithmic options for restricted versions of this problem that occur in practice. We present the first fixed-parameter algorithms for this problem and identify key techniques that can be applied to this and other k-anonymization problems.

1 Introduction

It is often desirable to make a dataset publicly available for research. One model of privacy protection which limits the risk of re-identification of individuals whose data is stored in released datasets is k-Anonymity. A dataset is k-anonymous if each record is identical to at least $(k$-$1)$ others in the dataset [8]. k-anonymization techniques minimize information loss while transforming a collection of records to be k-anonymous. Information loss must be minimized to make k-anonymized datasets useful in subsequent analyses; moreover, k-anonymization must be done efficiently to make it an attractive option for privacy protection software systems.

There are many types of k-anonymization (see [2] for a survey), the most basic of which suppresses the smallest possible number of dataset entries to achieve k-anonymity. Recently-derived complexity results [1,2,6] imply that efficient algorithms for optimally or even approximately solving the general version of this problem (and hence many other types of k-anonymization) probably do not exist. This has motivated privacy researchers to concentrate on fast heuristic algorithms that produce good sub-optimal results [3,9]. However, given the need to minimize information loss, efficient optimal algorithms that solve restricted versions of k-anonymization that occur in practice would be preferable. Questions about the existence and derivation of such algorithms are best addressed using the theory of parameterized complexity [4].

B. Yang, D.-Z. Du, and C.A. Wang (Eds.): COCOA 2008, LNCS 5165, pp. 23–31, 2008.
© Springer-Verlag Berlin Heidelberg 2008

In this paper, we give the first parameterized complexity analysis of a k-anonymization problem, namely the entry suppression problem. This analysis includes algorithms which demonstrate the fixed-parameter tractability of this problem under a number of practically-useful restrictions; underlying these algorithms are three general frameworks that may be applicable to other k-anonymization problems. This paper is organized as follows: Sections 2 and 3 give background on k-anonymization by entry suppression and parameterized complexity theory, Section 4 gives parameterized hardness and tractability results for this problem, and Section 5 discusses these results and directions for future research.

2   Problem Definition

Consider the following definitions adapted from [6, Section 2]. Represent a database D of n entities described by m attributes as a row-set $X = \{x_1, x_2, \ldots, x_n\}$, where $x_i \in \Sigma^m$ for some attribute value-set Σ. Let $x_i[j]$ be the jth element of x_i. A function $f : D \rightarrow (\Sigma \cup \{*\})^m$ is a **suppressor** for D if $\forall (x_i \in D) \forall (j \in \{1, 2, \ldots, m\})\{f(x_i[j]) \in \{x_i[j], *\}\}$. f thus transforms the database by replacing some entries by $*$, suppressing their original values. A transformed dataset $f(D)$ is k-**anonymous** if $\forall (x_i \in D)\{\exists (i_1, i_2, \ldots, i_{k-1} \in \{1, 2, \ldots, i, i + 1, \ldots, n\}) \mid f(x_{i_1}) = f(x_{i_2}) = \cdots = f(x_{i_{k-1}}) = f(x_i)\}$, so that each row is identical to at least $k - 1$ other rows. Our problem can now be stated as follows:

ENTRY SUPPRESSION (ESup)
Instance: An $n \times m$ dataset D over an alphabet Σ and integers $e, k \geq 0$.
Question: Can D be transformed into a k-anonymous dataset $f(D)$ by suppressing at most e entry values in D?

A number of complexity results have recently been derived for this problem. Meyerson and Williams [6] proved NP-hardness when $k \geq 3$ and gave a polynomial-time $O(m \log k)$-approximation algorithm. Improving upon these results, Aggarwal *et al.* proved NP-hardness when $k \geq 3$ and $|\Sigma| \geq 3$, and gave a polynomial-time $O(k)$-approximate algorithm using a graph-based representation [1]. Finally, Chaytor proved that polynomial-bounded absolute approximation or FPTAS algorithms do not exist for this problem unless $P = NP$ [2].

3   Parameterized Complexity Analysis

Given intractability results such as those at the end of the previous section that rule out optimal efficient algorithms for general versions of a problem, one may still be interested in algorithms whose non-polynomial running time is phrased purely in terms of an aspect x of that problem that is small in practice. *e.g.*, $O(2^x n^3 + m^2)$. Parameterized complexity theory [4] directly addresses such questions by defining parameterized problems, which break instances into a parameter k and a main part n, and fixed-parameter tractability such that an algorithm's running time can only be non-polynomial in the parameter, *i.e.*, the algorithm runs in $O(f(k)p(n))$ time, where f and p are arbitrary and polynomial functions, respectively.

Denote a parameterized problem X with parameter k by $\langle k \rangle$-X. Two basic techniques for deriving fixed-parameter algorithms [7] are **bounded search** and **kernelization** (in which a classical exhaustive search and the candidate solution-set, respectively, are bounded by a function of the parameter). Fixed-parameter intractability can be shown via hardness for some class of the W-hierarchy $= \{W[1], W[2], \ldots, W[P], \ldots, XP\}$, all of which seem (but have not been proven) to properly contain the class FPT of fixed-parameter tractable problems.

The analyses described above suffice in the case of problem restrictions encoded by single aspects; however, it is often of interest to look at restrictions operating over multiple aspects simultaneously. Such analyses can be simplified by the following relationships. For a problem of interest X, let S be a set of aspects for X.

- Given $S' \subseteq S'' \subseteq S$, if $\langle S' \rangle$-$X \in FPT$ then $\langle S'' \rangle$-$X \in FPT$.
- Given $S' \subseteq S'' \subseteq S$, if $\langle S'' \rangle$-$X \notin FPT$ then $\langle S' \rangle$-$X \notin FPT$.

4 The Parameterized Complexity of ENTRY SUPPRESSION

We investigate the parameterized complexity of ENTRY SUPPRESSION relative to various subsets of $S = \{n, m, |\Sigma|, k, e\}$. The following relationships between these aspects will be exploited below:

- As # values in any column cannot exceed # rows, $|\Sigma| \leq n$.
- As # suppressed entries cannot exceed # dataset entries, $e \leq mn$,
- As the size of any identical group cannot exceed # rows, $k \leq n$.
- If the input dataset is not already k-anonymous, then at least one group must need suppressed entries, so at least k entries must be suppressed. Thus $k \leq e$ unless the input dataset is either k-anonymous or cannot be k-anonymized with only e suppressed entries.

The last relationship has more general consequences. Checking a dataset to determine if it is k-anonymous can be done by sorting and then grouping the rows, in $O(mn \log n)$ time, so $\langle e \rangle$-ENTRY SUPPRESSION reduces to $\langle e, k \rangle$-ENTRY SUPPRESSION. Since the reverse reduction also holds, using $\langle e \rangle$ as a parameter is equivalent to using $\langle e, k \rangle$ as parameters.

4.1 Hardness Results

Theorem 4.1.1. $\langle k, |\Sigma| \rangle$-ESUP$\notin XP$ unless $P = NP$.

Proof: For a decision problem Π with aspect-set parameter S, if Π is NP-hard when all aspects in S are constants then $\langle S \rangle$-Π is not in XP unless $P = NP$ [10, Lemma 2.1.35]. Since ESUP is NP-hard when $k \geq 3$ and $|\Sigma| \geq 3$ [1], the result follows. □

4.2   Fixed-Parameter Tractability Results

While fixing k and $|\Sigma|$ is insufficient for fixed-parameter tractability, including either the number of rows or number of columns as a parameter does, in most cases, lead to fixed-parameter tractable algorithms. Note that limiting rows and columns simultaneously effectively fixes input size and hence is trivially in FPT (check all 2^{mn} entry-suppression combinations in $O(2^{mn}mn\log n)$ time). That being said, this basic exhaustive search algorithm forms part of the more complex algorithms given below.

Allowing the number of rows to be unlimited yields two cases, depending on whether or not all rows in the given dataset are distinct. The first case is fairly easy to deal with if m and $|\Sigma|$ form the parameter (as $n \leq |\Sigma|^m$, it can be solved in $O(2^{m|\Sigma|^m}nm\log n)$ time). Unfortunately, this will not typically hold, e.g., after all unique-value columns (such as name) have been removed as being identifying information. The second case, in which one has multiple copies of certain row, initially seems to be even easier, since the dataset is already somewhat anonymous. However, identical rows are not necessarily grouped together, and even k identical rows may need to be split up in order to form groups with other similar rows. For example, we can have a dataset that includes k copies of one row x, and k groups of $k-1$ copies each of other rows r_i $(1 \leq i \leq k)$, where each row r_i differs from row x in column i only. If the k copies of row x are kept together, then all other groups (of size k or more) must be formed from rows that differ in at least two columns, thus requiring the suppression of at least $2k^2 - 2k$ entries. If instead the x rows are broken up, with one x row grouped with each group of r_i rows, then only one entry must be deleted from each row, suppressing only k^2 entries. This example also illustrates some limitations on trying to kernelize datasets where the number of distinct rows is bounded.

To handle the situation where rows are not distinct, it suffices to add k to the previously-considered parameter. This is interesting in itself, because though k and $|\Sigma|$ do not by themselves yield fixed-parameter tractability, they do suffice when combined with m.

Theorem 4.2.1. $\langle m, |\Sigma|, k\rangle$-ESUP is solvable in $O(nm\log n + 2^{|\Sigma|^m km} \cdot |\Sigma|^m km \cdot (m\log|\Sigma| + \log k))$ time.

Proof: There will be at most $|\Sigma|^m$ distinct rows and, as discussed above, we cannot always group identical rows together. However, the number of groups needed will not be more than the number of distinct rows, since any row lost to another group must be needed in order to supplement a group of some other identical rows. Therefore the number of rows from any identical group that may be distributed independently to different groups is less than $|\Sigma|^m \cdot k$. If an identical group has more row copies than this, the remaining rows can be included in whichever group (that already includes some copies of them) requires them to have the least number of suppressions per additional row.

Algorithm 1 (see Figure 1) uses this principle to reduce and bound the number of rows that need to be partitioned through searching combinations of suppressed entries. There will be at most $2^{|\Sigma|^m km}$ entry combinations, and each combination

```
Algorithm 1:
    sort the rows and group identical rows
    for each identical group i
        if the number of copies c_i > |Σ|^m · k
            remove these excess copies
            x_i = c_i − |Σ|^m · k
    for all combinations of entries in the remaining rows
        suppress the selected combination of entries
        sort the rows and group identical rows
        if all groups are of size ≤ k
            t = total number of suppressions of grouped rows
            for all initial groups i that have excess copies
                find the grouped copy that has the fewest suppressions s_i
                if x_i < k
                    t = t + x_i · s_i
            if t ≤ e
                return true
    return false
```

Fig. 1. Algorithm 1. See Theorem 4.2.1 for details

requires $|\Sigma|^m k$ rows of length m to be sorted and grouped. Arriving at the total number of suppressions, including the rows that were not partitioned, is done for each original identical group by multiplying in the suppressions for those rows that were left out of the partition. Rows that are in groups of k or more should not be added in, since they are already part of a sufficiently large group. Thus Algorithm 1 runs in $O(nm \log n + 2^{|\Sigma|^m km} \cdot |\Sigma|^m km \cdot (m \log |\Sigma| + \log k))$ time.
□

Given the relationships between e and k mentioned earlier, the above also implies that $\langle m, |\Sigma|, e \rangle$-ESUP and $\langle m, |\Sigma|, e, k \rangle$-ESUP are in *FPT*.

Let us now consider limiting the number of rows.

Theorem 4.2.2. $\langle n \rangle$-ESUP is solvable in $O(nm \log m + 2^{n^{n+1}} n^{n+1} \log n)$ time.

Proof: If two columns produce an identical partitioning of the rows, then they need to have the same entries suppressed. If two identical columns have different entries suppressed, then they will partition the rows in different ways. If instead both columns have the same entries suppressed (choosing whichever had the fewest suppressed entries), then the partitioning produced by the other column will be removed from the overall partition (formed by overlaying the different column partitions). Removing a partitioning cannot decrease the size of the sets of rows in the overall partition, so making the columns identically suppressed will not increase the number of suppressions or break the k-anonymity condition. Thus the columns in the dataset can be restricted to the columns that produce different partitions of the rows, which is less than $|\Sigma|^n$. Each set of columns that produce the same partition can be represented by a single column, weighted by the size of the set, and there will be no more than $|\Sigma|^n$ such weighted columns.

Algorithm 2:
x for all columns j from 1 to m
        relabel the entries of column j, starting from 1, in order
    sort the columns, and group them into sets of identical columns
    for each set j of identical columns
        remove all but one representative
        weight it by s_j, the number of columns in the identical group
    for all combinations of entries in the at most $n \times |\Sigma|^n$ table
        Sum = sum, over all selected entries, of their columns' weights
        if $Sum \leq e$
            suppress all of these entries
            sort the table and group identical rows
            if all groups are of size $\geq k$
                return *true*
    return *false*

Fig. 2. Algorithm 2. See Theorem 4.2.2 for details

Algorithm 2 (see Figure 2) implements this approach, and runs in $O(nm \log m + 2^{n \cdot |\Sigma|^n} \cdot |\Sigma|^n n \log n) = O(nm \log m + 2^{n^{n+1}} n^{n+1} \log n)$ time (as $|\Sigma| \leq n$). $\square$ As the different partitions of the rows can be searched by identifying columns that partition the rows in the same way as that exploited above by kernelization, an analogous bounded search algorithm solves $\langle n \rangle$-ESUP in $O(2^n n^{n+1}(m + \log n))$ time (details omitted due to lack of space).

All of the algorithms above limit the number of symbols $|\Sigma|$, either directly or by limiting n. However, if we allow Σ to be arbitrarily large, we can still achieve fixed-parameter tractability by limiting m and e.

Theorem 4.2.3. $\langle m, e \rangle$-ESUP is in *FPT* and is solvable in $O(2^{2em}(e^2(e + k)2^e)^{e^2+e} mn \log n)$ time.

Proof (Sketch): This result holds courtesy of Algorithm 3 (see Figure 3). In order to be k-anonymizable, the number of rows in the original dataset that are in groups of less than k identical copies must be less than e, since each of these groups must have at least one entry suppressed per row. This produces a sub-dataset D'' of at most e rows, so there are at most 2^{em} different combinations of entries in that subset to consider suppressing and testing to determine if this suppression combination will work.

Once a suppression combination is to be tested, the other rows (already part of larger groups, D' in Algorithm 3) need to be considered to be added to the rows with suppressed entries. Since the suppression combination has determined which entries of D'' are suppressed, we consider rows from D' that are compatible with the groups of D''. We may need to consider more than just their excess rows; however, we can only suppress e entries, so we only need to consider at most e rows from each group.

```
Algorithm 3:
   sort and group identical rows
   select all rows in groups of size < k
   if more than e rows are selected
         return false
   D' = dataset − selected rows
   D'' = selected rows
   for all entry combinations in the selected rows D''
               where each row has at least one entry included
      suppress the combination of entries in D''
      e' = e − number of entries suppressed
      if e' ≥ 0
         sort and group identical rows from D''
         construct D''' from D' by:
            for all pairs i₁, i₂ of groups of identical rows in D'
               let d(i₁, i₂) = Hamming distance between i₁ and i₂ rows
            sort the (i₁, i₂) pairs by nondecreasing d(i₁, i₂)
            D''' = empty
            for each pair i₁, i₂ of groups of identical rows in D'. as sorted
               size(i) = number of rows in i
               m(i) = set of groups in D'' whose unsuppressed entries
                        match group i for all non-suppressed entries
               c(i) = number of groups j already in D''' with
                        m(j) = m(i) and (size(j) = size(i) or size(j) ≥ e' )
               if c(i₁) < e'
                  move min(e', size(i₁)) rows of group i₁ from D' to D'''
               if c(i₂) < e'
                  move min(e', size(i₂)) rows of group i₂ from D' to D'''
         for each set X of up to e' groups from D'''
            mark up to e' rows from D' that agree with all identical
                  columns in X
         add all marked rows to D'''
         for each selection of e' rows from D'''
            for all combinations of up to e' entries in the selected rows
               suppress that combination of entries
               re-build D by uniting all rows in D', D'' and D'''
               if at most e entries in D are suppressed
                  sort and group identical rows of D
                  if all groups of D are of size ≥ k
                     return true
   return false
```

Fig. 3. Algorithm 3. See Theorem 4.2.3 for details

Since $|\Sigma|$ is not bounded, there may be many groups in D' that are compatible with the groups of D''. However, groups in D' that are compatible with exactly the same groups in D'' are equivalent with respect to filling out the D'' groups. If they are also the same size then they are also equivalent with respect to how

Table 1. Summary of Parameterized Results for ENTRY SUPPRESSION. For each FPT result, the numbers of the algorithms implying this result are given in parentheses.

	$-$	k	e	k,e
$-$	NP-hard	$\notin XP$	???	???
$|\Sigma|$	$\notin XP$	$\notin XP$	???	???
m	???	???	$FPT(3)$	$FPT(3)$
n	$FPT(2)$	$FPT(2)$	$FPT(2)$	$FPT(2)$
$|\Sigma|,m$	???	$FPT(1)$	$FPT(1,3)$	$FPT(1,3)$
$|\Sigma|,n$	$FPT(2)$	$FPT(2)$	$FPT(2)$	$FPT(2)$

many rows will be left, and if a group has at least $e + k$ members then it can be used instead of a smaller group. We may need to include rows from more than one group to spread out the rows removed, but the number of groups needed to be considered is no more than $e(e+k)2^e$, since we will not be able to extend by more than e rows, and of the equivalent groups (2^e possible $m(i)$ sets, $(e + k)$ sizes) our only concern is how well they fit with each other. This key consideration is managed by adding first those groups that are the closest together, and thus will need the fewest entries suppressed to merge. Finally, since these pairs may need to be merged with other groups that agree with them, sufficient groups that agree with their common columns are also added.

This boundary dataset (D''' in Algorithm 3) will thus initially include at most $e^2(e+k)2^e$ rows. It will have no more than $(e(e+k)2^e)^{e+1}$ rows marked, leading to no more than $e^2(e + k)2^e + (e(e + k)2^e)^{e+1}$ rows in D'''. Up to e rows from D''' need to be selected to extend D''. The remaining rows of D''' need to be added to the groups of D'', back into their original groups in D', or put together to form new groups; trying a suppression pattern of entries from e rows selected from D''' will induce a complete grouping over all the rows. All of the rows then need to be reassembled and tested to determine if the suppression limit of e has been respected and the k-anonymity condition is met.

Given the above, Algorithm 3 runs in $O(mn \log n + 2^{em}mn + 2^{em}(e^2(e + k)2^e)^{e^2+e}e2^{em}mn \log n) = O(2^{2em}(e^2(e + k)2^e)^{e^2+e}mn \log n)$ time. $\square$

5   Conclusion

Table 1 summarizes the hardness and tractability results presented in this paper, including all results implied by pairwise parameter and parameter subset relationships. Note that as $k \leq e$, unless the dataset is already k-anonymous, the two rightmost columns are equivalent.

The algorithms underlying these FPT results illustrate techniques that can be applied to this problem, individually and together, and provide a general framework for algorithm development. Searching, either using a bounded search through the entire set or searching through a problem kernel, can consider all possible entry suppression combinations or all partitions. The results of such searches are equivalent, since partitioning the dataset into sets of rows will require any nonidentical columns in a group to have all their entries suppressed; similarly, an entry suppression combination will induce a partition of the rows into groups. Also, kernels can be found for problem variants where the number of rows or columns that differ or need to be considered can be bounded by a function of the parameters.

The most pressing direction for future research is to improve the running times of the given FPT algorithms. Though most of these algorithms are not themselves practical, they do indicate which aspect-combinations allow fixed-parameter tractability. Past experience has shown that once a problem is shown to be in FPT, more sophisticated techniques can often be applied to derive practical algorithms [7], and it is our hope that this will be the case for k-anonymization. Also, the complexity of some aspect-combinations is open. Of particular interest here is $\langle m, k \rangle$-ESUP, since datasets to be anonymized frequently have $m \leq 25$ and $k \leq 5$ [5].

Acknowledgments. The research described in this paper was supported by an NSERC PGS-D scholarship (RC) and NSERC research grants 204923 (PE) and 228104 (TW).

References

1. Aggarwal, G., Feder, T., Kenthapadi, K., Motwani, R., Panigrahy, R., Thomas, D., Zhu, A.: Anonymizing tables. In: Eiter, T., Libkin, L. (eds.) ICDT 2005. LNCS, vol. 3363, pp. 246–258. Springer, Heidelberg (2004)
2. Chaytor, R.: Utility Preserving k-Anonymity. Technical Report MUN-CS 2006-01, Dept.Computer Science, Memorial University of Newfoundland (2006)
3. Chaytor, R.: Allowing Privacy Protection Algorithms to Jump out of Local Optimums: An Ordered Greed Framework. In: Bonchi, F., et al. (eds.) PinKDD 2007. LNCS, vol. 4890, pp. 33–55. Springer, Heidelberg (2008)
4. Downey, R., Fellows, M.: Parameterized Complexity. Springer, Heidelberg (1999)
5. MacDonald, D.: Personal Communication (2005)
6. Meyerson, A., Williams, R.: On the complexity of optimal k-anonymity. In: Proc. of 23rd ACM Sym. on Principles of Database Systems (PODS 2004), pp. 223–228 (2004)
7. Niedermeier, R.: Invitation to Fixed-Parameter Algorithms. Oxford University Press, Oxford (2006)
8. Sweeney, L.: Achieving k-anonymity privacy protection using generalization and suppression. Int'l J. on Uncertainty, Fuzziness and Knowledge-Based Systems 10(5), 571–588 (2002)
9. Wang, K., Yu, P., Chakraborty, S.: Bottom-up generalization: a data mining solution to privacy protection. In: ICDM 2004, pp. 249–256 (2004)
10. Wareham, T.: Systematic Parameterized Complexity Analysis in Computational Phonology. Ph.D.thesis, Dept.Computer Science, University of Victoria (1999)

Multiple Hypernode Hitting Sets and Smallest Two-Cores with Targets[*]

Peter Damaschke

Department of Computer Science and Engineering
Chalmers University, 41296 Göteborg, Sweden
ptr@cs.chalmers.se

Abstract. The multiple weighted hitting set problem is to find a subset of nodes in a hypergraph that hits every hyperedge in at least m nodes. We extend the problem to a notion of hypergraphs with so-called hypernodes and show that it remains fixed-parameter tractable (FPT) for $m = 2$, with the number of hyperedges as the parameter. This is accomplished by a nontrivial extension of the known dynamic programming algorithm for usual hypergraphs. The result might be of independent interest for assignment problems, but here we need it as an auxiliary result to solve a different problem motivated by network analysis: We give an FPT algorithm that computes a smallest 2-core including a given set of target vertices in a graph, with the number of targets as the parameter. (A d-core is a subgraph where every vertex has degree at least d within the subgraph.) This FPT result is best possible, in the sense that an FPT algorithm for 3-cores cannot exist, for simple reasons.

1   Introduction and Contributions

Hitting set problems are fundamental in various branches of computer science and in combinatorial optimization. Since the problems are NP-hard, the complexity of several parameterizations became interesting. We refer to recent parameterized algorithms for hitting sets in hypergraphs of fixed rank [6] and for enumerations of all hitting sets [5].

In the MULTIPLE WEIGHTED HITTING SET problem, we are given a family of hyperedges (a hypergraph) on a set V of nodes with positive real weights, and an integer m, and we seek a subset of nodes with minimum total weight that intersects every hyperedge in at least m distinct nodes. (We are using the words "intersect" and "hit" interchangeably.) The unweighted case with $m = 1$ is the HITTING SET problem. We generalize the former problem as follows. Besides the hyperedges we are given another family of subsets of V that we call *hypernodes*. Every hypernode (rather than every node) has a positive weight. We seek a subset S of hypernodes with minimum total weight so that the union of hypernodes in S intersects every hyperedge in at least m nodes. We refer to this problem as

---

[*] Work supported by the Swedish Research Council (Vetenskapsrådet), grant no. 2007-6437, "Combinatorial inference algorithms – parameterization and clustering".

B. Yang, D.-Z. Du, and C.A. Wang (Eds.): COCOA 2008, LNCS 5165, pp. 32–42, 2008.

MULTIPLE WEIGHTED HYPERNODE HITTING SET. The MULTIPLE WEIGHTED HITTING SET problem is the special case where the the hypernodes are singleton sets, one for each node in V. In MULTIPLE WEIGHTED HYPERNODE HITTING SET we may assume without loss of generality that all hyperedges are pairwise disjoint: If they are not, we "disjunctify" them as follows. For each pair v, e of a node v and hyperedge $e \ni v$, replace v with a copy v_e of v that appears exclusively in e. While making the hypernodes larger, this transformation does not change the problem, and the blow-up is obviously polynomial in the size of the hypergraph. Disjoint hyperedges will simplify the notations in algorithms for the problem. Note that, if we apply the above transformation to a usual weighted hypergraph (an instance of MULTIPLE WEIGHTED HITTING SET), we also obtain disjoint hypernodes, as they come from single nodes. But we stress that hypernodes may still overlap in the general case of MULTIPLE WEIGHTED HYPERNODE HITTING SET. In the following, k always denotes the number of hyperedges which we denote $V(1), \ldots, V(k)$. For easier orientation we repeat the definition more formally:

MULTIPLE WEIGHTED HITTING SET
Input: a set V of nodes, k hyperedges, i.e., sets $V(1), \ldots, V(k) \subset V$ which can be assumed to be pairwise disjoint, another family of subsets of V called hypernodes, each with a positive cost, and an integer m.
Output: a set S of hypernodes with minimum total cost, so that at least m distinct nodes of every hyperedge are contained in some hypernode in S.

We assume that the reader is familiar with the theory of fixed-parameter tractable (FPT) problems; introductions can be found in [3,9]. HITTING SET is known to be $W[2]$-complete (thus probably not in FPT), but HITTING SET is in FPT in parameter k. (This was observed, e.g., in [7] in dual form, i.e., for the SET COVER problem). This result can be immediately extended to MULTIPLE WEIGHTED HYPERNODE HITTING SET with $m = 1$: The idea is to use dynamic programming on the subsets of index set $\{1, \ldots, k\}$. We process the hypernodes successively. Every new hypernode is used in the solution or not. We only have to keep track of the minimum weights of solutions that hit each subfamily of $V(1), \ldots, V(k)$. Thus the time bound is $O^*(2^k)$. We withhold the straightforward details, because we will treat the more general case $m = 2$ in Section 2.

The difficulty increases dramatically in case $m = 2$. Already for $m = 2$ it is not easy to see an FPT algorithm, even a bad one, let alone combined parameters (k, m). As opposed to HITTING SET, it does not help much to classify the hypernodes in 2^k types according to which hyperdeges they intersect. Since we are supposed to hit every hyperedge in at least two *distinct* nodes, we have to memorize the nodes already used up, but we cannot afford to do that for every possible combination of nodes in the already selected hypernodes, as the $|V(i)|$ are not bounded by any function of parameter k. In Section 2 we develop an FPT algorithm for MULTIPLE WEIGTHED HYPERNODE HITTING SET with $m = 2$, by a nontrivial extension of the known dynamic programming scheme for HITTING SET.

Hypernode hitting sets may be of independent interest in combinatorial optimization, e.g., for assignment problems with multiple demands. However, our motivation is another graph problem that we treat in the rest of the paper. To avoid confusion, we use the term *vertex* in graphs and *node* in hypergraphs.

For any fixed integer d, a d-core in a graph $G = (V, E)$ is a subset $C \subseteq V$ (or the induced subgraph) where every vertex $v \in C$ has at least d neighbors in C. Cores are interesting in network design (as robust subnetworks), but mainly in graph-theoretic approaches to clustering. A recent bioinformatics application of cores is the prediction of protein complexes in protein interaction networks [2].

A maximal d-core in any graph is uniquely determined and can be trivially computed by an elimination algorithm: Starting with $C = V$, any vertex in C that lacks enough neighbors in C is removed from C. Since the degree of a vertex in C can only get smaller, we never have to reinsert such vertices. However, the problem becomes difficult if we are interested in *minimum* d-cores that contain a given set of target vertices (similarly as in the STEINER TREE problem that seeks a minimal connected set containing a set of target vertices). We define:

MINIMUM d-CORE
Input: a graph $G = (V, E)$ and a set $T \in V$ of target vertices (or: targets).
Output: a d-core C such that $C \supseteq T$ and $|C|$ is minimized.

We call a d-core C *minimal* if C does not contain a smaller nonempty d-core. Likewise, $C \supseteq T$ is a *minimal d-core including T* if no d-core $C' \subset C$ includes T. Carefully distinghuish between *minimal* and *minimum* which refers to set inclusion and cardinality, respectively.

MINIMUM 0-CORE is a trivial problem. Minimal 1-cores in a graph are exactly the pairs of adjacent vertices. In an instance of MINIMUM 1-CORE, only vertices being isolated in T need another neighbor in C. Thus we may assume without loss of generality that T is an independent set, therefore MINIMUM 1-CORE becomes equivalent to the HITTING SET problem for thefamily of hyperedges $N(v)$, $v \in T$, where $N(v)$ denotes the neighborhood of vertex v. Since HITTING SET is in FPT, with the number of hyperedges as the parameter, it follows that MINIMUM 1-CORE is in FPT with parameter $t = |T|$, the number of target vertices. In Section 4 we will show that MINIMUM 2-CORE is still in FPT with parameter t, using the preceding FPT result for MULTIPLE WEIGTHED HYPERNODE HITTING SET with $m = 2$. Prior to this final contribution, Section 3 provides some preliminary results on the hardness of MINIMUM d-CORE. In particular, our FPT result for $d = 2$ is best possible in the sense that case $d = 3$ is not in FPT (unconditional).

Here it is worth mentioning another related hardness result: Finding a chordless cycle through two target vertices is W[1]-complete [8]. Note that a chordless cycle is a 2-core, but MINIMUM 2-CORE permits arbitrary 2-cores rather than only chordless cycles. (Cores are in general not even connected.) The STEINER TREE problem is known to be in FPT, with the number t target vertices as the parameter. It can be solved in $O^*(3^t)$ time using an old algorithm from [4]. As recently shown in [1], finding minimum d-cores without specified target vertices is W[1]-hard for any $d \geq 3$, with the size of the solution as parameter.

Our study is exploratory research in problem complexity, not derived from an immediate application. The use of cores and target ("seed") vertices in [2] is very different from the MINIMUM d-CORE problem, however one may use cores in similar inference tasks. Suppose we have a model of a protein interaction network with only a sparse set of confirmed interactions, and for some set T of proteins we know that they belong to the same functional group, moreover, in a functional group we expect each protein to interact with at least d others. Then it is sensible to ask what is a smallest possible vertex set including T, with this property.

2   Weighted Hypernode Hitting Sets with Two Hits

In this section we prove that MULTIPLE WEIGHTED HYPERNODE HITTING SET with $m = 2$ is in FPT, with parameter k. First we outline the algorithm.

Hypernodes are listed in any order. In each step of the algorithm we consider the next hypernode from list. We also maintain a family F of possible *partial solutions*. A partial solution is the union of the hypernodes we have selected so far. A family F of partial solutions is called *promising* if at least one member of F can be extended to an optimal final solution, by adding a suitable subset of hypernodes from the remainder of the list. Hence, as soon as all hypernodes have been considered, a promising F contains an optimal solution. The algorithm starts with F containing only one partial solution, namely the empty set. This initial F is (trivially) promising.

We use the following conventions to avoid many repetitions of similar definitions. Remember that $V(1), \ldots, V(k)$ are the (disjoint!) hyperedges. For any symbol X (an upper-case letter, perhaps with further symbols attached) that denotes a subset of nodes, we denote by $X(i)$ the set $X \cap V(i)$. If such $X(i)$ contains only one node, we denote it by $x(i)$ (using the corresponding lower-case letter). Conversely, by mentioning a node $x(i)$ we implicitly mean that this is the only node of the corresponding set $X(i)$.

The *signature* of a partial solution P is the vector s of numbers $s(i) = \min\{|P(i)|, 2\}$, for $i = 1, \ldots, k$. Any final solution can be written as $P \cup Q$, where P is the union of hypernodes chosen until the current step of the algorithm, and Q is the set of nodes added later. Note that $P \cap Q = \emptyset$ by definition. In the following we will always split solutions in this way. Given a signature s, a *1-index* is an index i with $s(i) = 1$, similarly we speak of *0- and 2-indices*. Let seq be any sequence of pairwise distinct indices, possibly the empty sequence. If index j appears in seq, we denote by seq_j the prefix of seq before j.

Let T denote the hypernode to be considered in the current step of the algorithm. We update F by adding all sets $P \cup \{T\}$, $P \in F$, to the old F. Since F was promising before this step, the new F is promising, too. The tricky part is to remove some partial solutions from F, in such a way that F remains promising, but the number of sets in F is kept below some threshold that may only depend on parameter k:

Lemma 1. *Any promising family F of partial solutions has a promising subfamily with at most $e^2 2^k k!$ different partial solutions. (Here e denotes Euler's number.) Such a subfamily can be computed from F in a time polynomial in the size of F.*

Proof. First we consider any fixed signature s, therefore we suppress s in our notation, for convenience. Let $G[]$ denote the family of members of F with signature s. Suppose $G \neq \emptyset$. Let $P[]$ be some partial solution in $G[]$ with minimum weight. Consider any optimal solution $P \cup Q$, $P \in G[]$, with the property that $Q(i) \setminus P[](i) \neq \emptyset$ for all 1-indices i. Then $P[] \cup Q$ is a valid solution, as we can easily see: For 0-indices i we have $|Q(i)| \geq 2$, and for 2-indices i we have $|P[](i)| \geq 2$. For 1-indices i, the above property guarantees $|P[](i) \cup Q(i)| \geq 2$. Furthermore, since $P[]$ has at most the weight of P, solution $P[] \cup Q$ is optimal, too. Hence, partial solutions $P \in G[]$ other than $P[]$ are only needed as part of possible optimal solutions $P \cup Q$ where $Q(i) \subseteq P[](i)$ holds for some 1-index i. Note that the latter condition implies $q(i) = p[](i)$ and $p(i) \neq p[](i)$.

This gives rise to the following inductive hypothesis, for any sequence seq of pairwise distinct 1-indices: Partial solutions $P \in G[seq]$ other than some minimum-weight partial solution $P[seq] \in G[seq]$ are only needed as part of possible optimal solutions $P \cup Q$ where $q(j) = p[seq_j](j)$ holds for all j in seq, and $p(i) \neq p[seq](i) = q(i)$ holds for some 1-index i not occuring in seq. (As we saw above, the hypothesis is true for the empty sequence seq. Note that some conditions are vacuously true in the base case.)

Accordingly, for each 1-index i not occuring in seq we define $G[seq, i]$ to be the family of all partial solutions $P \in G[seq]$ with $p(i) \neq p[seq](i)$. Let $P[seq, i]$ be a minimu-weight partial solution in $G[seq, i]$. Now we prove the induction hypothesis for the sequence seq, i, similarly as in the base case.

Consider any optimal solution $P \cup Q$, $P \in G[seq, i]$, with the properties that $q(j) = p[seq_j](j)$ for all j in seq, and $q(i) = p[seq](i)$, and $Q(j) \setminus P[seq, i](j) \neq \emptyset$ for all 1-indices j not occuring in seq, i. Then $P[seq, i] \cup Q$ is a valid solution: For 0-indices j we have $|Q(j)| \geq 2$, and for 2-indices j we have $|P[seq, i](j)| \geq 2$. For 1-indices j not in seq, i, the above properties obviously guarantee $|P[seq, i](j) \cup Q(j)| \geq 2$. As for the particular index i, observe that $p[seq, i](i) \neq p[seq](i) = q(i)$, hence $|P[seq, i](i) \cup Q(i)| \geq 2$ as well. For 1-indices j in seq we also have $p[seq, i](j) \neq q(j)$, this because $q(j) = p[seq_j](j)$ holds, and $P[seq, i] \in G[seq, i] \subset G[seq_j, j]$ implies $p[seq, i](j) \neq p[seq_j](j)$ by definition of $G[seq_j, j]$.

Furthermore, since $P[seq, i]$ has at most the weight of P, solution $P[seq, i] \cup Q$ is optimal among all solutions with the aforementioned properties. Consequently, partial solutions $P \in G[seq, i]$ other than $P[seq, i]$ are only needed as part of possible optimal solutions $P \cup Q$ where $q(j) = p[seq_j](j)$ for all j in seq, and $q(i) = p[seq](i)$, and $Q(j) \subseteq P[seq, i](j)$ for some 1-index j not occuring in seq, i. The latter condition also implies $|Q(j)| = 1$, $q(j) = p[seq, i](j)$, and $p(j) \neq p[seq, i](j)$. This concludes the induction step.

Our inductive definition yields families $G[seq]$ and partial solutions $P[seq] \in G[seq]$ (unless $G[seq] = \emptyset$, in which case seq is not extended further). As soon as a seq contains all 1-indices i, it suffices to keep only some optimal $P[seq]$ in

$G[seq]$. Other $P \in G[seq]$ cannot be parts of optimal $P \cap Q$ anymore, since no 1-indices j are left where $Q(j)$ could undesirably be contained in $P[seq](j)$.

Until now, s was any fixed signature. By construction, the family of all $P[seq]$, now for all sequences seq and all signatures s, is still a promising subfamily of F. Clearly, each $G[seq, i]$ and $P[seq, i]$ is obtained from $G[seq]$ and $P[seq]$ in polynomial time, in the size of F. It remains to bound the size of this promising subfamily. There exist $2^{k-t}\binom{k}{t}$ signatures with exactly t 1-indices, and for every such signature we can form $\sum_{i=0}^{t} \binom{t}{i}(t - i)! < et!$ sequences of distinct 1-indices. Finally observe $\sum_{t=0}^{k} 2^{k-t}\binom{k}{t}et! < e2^k \sum_{t=0}^{k} \binom{k}{k-t}t! < e^2 2^k k!$.   □

Perhaps the $k!$ term can be improved by using similarities of partial solutions for different index sequences. An intriguing question is whether an exponential bound with constant base is accomplishable. We also conjecture that the scheme can be extended to any constant m. However, at the moment we get:

Theorem 1. MULTIPLE WEIGHTED HYPERNODE HITTING SET *with $m = 2$ can be solved in $O^*(2^k k!)$ time.*

Proof. To summarize the given algorithm: We successively add the given hypernodes to the problem instance, in an arbitrary order, and maintain a promising family F of partial solutions, i.e., unions of selected hypernodes. Initially, F is the family with just the empty set. For every new hypernode T we insert all sets $P \cup \{T\}$, $P \in F$, in F. Then we use the procedure in Lemma 1 to extract a promising subfamily of F with $O(2^k k!)$ sets. When all hypernodes are included, an optimal solution in F is an optimal solution to the problem. All auxiliary computations are obviously polynomial.   □

3 Hardness of Minimum Cores Including a Target Set

The following result is quite easy to obtain, however it does not directly follow from the $d = 1$ case.

Theorem 2. MINIMUM d-CORE *is NP-complete for any constant d.*

Proof. (Sketch.) Case $d = 1$ is equivalent to the NP-complete HITTING SET problem, see Section 1. Now consider $d \geq 2$. Given any instance of HITTING SET where, without loss of generality, the hyperedges cover the whole set of nodes, we construct a graph as follows. Every node of the hypergraph is represented by a vertex of the graph. Furthermore, every hyperedge e is represented by a target vertex, adjacent to those vertices representing the nodes of e. Finally, we attach to every vertex v (of both types) a $(d+1)$-clique of new target vertices, and insert an edge between v and $d - 1$ vertices of this $(d + 1)$-clique. Since every target vertex not being in an attached clique needs yet another neighbor in the d-core, we obtain a one-to-one correspondence between d-cores and hitting sets. Note that every non-target vertex in the d-core is in fact adjacent to $d - 1$ vertices in its attached clique, and to another target vertex.   □

Due to NP-completeness, it is natural to study the parameterized complexity. As mentioned earlier, case $d = 1$ is fixed-parameter tractable in the number t of target vertices, and for $d = 2$ we show that in the next section. For $d \geq 3$ we cannot get FPT results, as the complexity is not even bounded by any function of t. Already for $d = 3$ and $t = 1$, the following construction gives arbitrarily large equivalent instances of HITTING SET. We construct a tree with the only target vertex at the root. The root has three children, and all other inner vertices have two children. Only the leaves are adjacent to further vertices. Obviously, any 3-core C that includes the target vertex must also include the whole tree. Furthermore, every leaf needs two more neighbors in C. Now it is not hard to construct an instance of *Hitting Set* that is equivalent to minimizing $|C|$.

4   Minimum Two-Cores Including a Target Set

In this section we develop an FPT algorithm for MINIMUM 2-CORE, with the number t of target vertices as the parameter. We use some standard notation: $N(v)$ is the neighborhood of vertex v. A subgraph *spanned* by an edge set consists of this edge set and all involved vertices (ignoring any additional edges between them!). The length of a path is the number of edges.

Consider a graph $G = (V, E)$ with target set $T \subset V$. Clearly, we may assume that all vertices in G have degree at least 2. We reduce our graph as follows. Every vertex v gets *valency* $\max\{2 - |N(v) \cap T|, 0\}$, that is, the valency of v is the number of non-target neighbors that v demands in a 2-core C, provided that $v \in C$. Target vertices of valency 0 are removed. Edges between target vertices are removed as well. For convenience we still use G and T to denote the reduced graph (with vertices labeled by their valencies) still by G, and the remaining target set, respectively.

For the moment we fix, simultaneously for every target vertex $v \in T$ with valency i, at least i neighbors of v that shall belong to C, and call these neighbors *ports*. Clearly, each port has valency 0 (if it is adjacent to several targets) or 1. Later we will discuss how we actually choose these ports, and this is the point where we will need the MULTIPLE WEIGHTED HYPERNODE HITTING SET algorithm for $m = 2$.

A path in G is said to be *regular* if all its inner vertices are non-targets and have valency 2. (In particular, trivial paths of length 0 or 1 are regular.) A subset F of edges is called *saturating* if, for every vertex v in the subgraph spanned by F, the number of edges of F incident to v is at least the valency of v. Note that $C \supset T$ is a 2-core iff some saturating edge set spans all vertices in $C \setminus T$ of valency greater than 0. Hence we can work in the following with saturating edge sets rather than 2-cores, and take advantage of their special structure:

Lemma 2. *Let F be a saturating edge set that spans at least the set of ports of valency 1 and is minimal with this property. Then the subgraph spanned by F consists of connected components of the following types:*

Star: *a vertex called the* center *is connected to distinct ports by internally vertex-disjoint regular paths (in particular, a star might consist of a single path to only one port),*

Loop: *a regular path (possibly of length 0) with a port v at one end, and a cycle of vertices of valency 2 attached to the other end. We call this structure a* loop *for vertex v.*

Proof. In the following, all vertex degrees are meant with respect to F, and removing a vertex means also to remove all incident edges from F.

Let u_0 be a vertex of degree larger than 2, and u_1 any of its neighbors ($u_0 u_1 \in F$). Since u_0 has valency at most 2, and F is minimal, we have $u_0 u_1 \in F$ only because the degree of u_1 equals its valency. Hence u_1 cannot have valency 0. If u_1 has valency and degree 1, it must be a port, otherwise we could remove u_1 from the component. If u_1 has valency and degree 2, let u_2 be its other neighbor. The degree of u_2 equals its valency, otherwise we could remove u_1. If u_2 has valency and degree 1, it must be a port, otherwise we could remove u_1 and u_2. If u_2 has valency and degree 2, let u_3 be its other neighbor, etc. This inductive argument gives a path $u_0, \ldots, u_k$ where all u_i, $0 < i < k$, have valency 2, and either u_k is a port or $u_k = u_0$ (cycle). If every path starting in u_0 ends in a port, the component is a star. Otherwise, there is at most one cycle in the component, since vertices of all further cycles (except u_0) could be removed, due to minimality. In this case we get a loop.

If u_0 as specified above does not exist, all vertex degrees are 1 or 2, hence the component is just a path or cycle. No two distinct vertices u, v of degree 2 but with smaller valency can exist, since otherwise we could remove the edge or subpath between u and v, leaving both u and v with at least one neighbor. Thus, a cycle component has exactly one port and no other vertices of valency smaller than 2, which yields a loop (consisting of a cycle and a path of length 0). A path component cannot end with a subpath of vertices with valency 2, since we could remove such a (maximal) subpath. We conclude that a path component has to connect two vertices of valency smaller than 2. Moreover, by similar reasoning as above, at most one inner vertex has valency smaller than 2. If such an inner vertex exists, both end vertices must be ports, otherwise we could again remove a subpath. Thus, we get a star with two paths. If all inner vertices have valency 2, at least one end vertex is a port, and we get a star with one or two paths and an appropriately chosen center. (If a star is merely a regular path connecting two ports, we can declare an arbitrary vertex the center. Note that a port may be the center of a star.)   □

For every port we can efficiently compute several items that we have specified in Lemma 2:

Lemma 3. *For every port u we can compute in polynomial time a shortest regular path from u to every vertex v, and a minimum-size loop for u (if one exists).*

Proof. We have to prove this for the loop only. For every v, where either $v = u$ or v has valency 2, we compute a shortest regular u, v-path, and a shortest cycle through v where all vertices (except v if $v = u$) have valency 2. Let us call it a *regular cycle*. For computing a shortest regular cycle we may check every neighbor w of v and determine a shortest regular w, v-path avoiding the edge vw. We claim that, for some v, combining a shortest regular u, v-path and a shortest regular cycle through v yields a minimum-size loop for u. To prove the claim, consider a smallest loop for u, and let v be the vertex where path and cycle intersect. If there are several smallest loops for u, we choose one where the u, v-path is as short as possible. Assume that some shortest regular u, v-path and some shortest regular cycle through v intersect in further vertices other than v. Then, obviously, there exists either a smaller loop for u, or a loop with the same vertices and edges, where path and cycle intersect in exactly one vertex $v' \neq v$ at a smaller distance to u. Both cases contradict the choice of the loop. This proves the claim for the specified v. It follows that we only need to pick a smallest loop among the shortest path-cycle combinations, for all v. $\qquad\square$

Now we construct in polynomial time an instance of MULTIPLE WEIGHTED HYPERNODE HITTING SET, based on Lemma 3.

Nodes and disjoint hyperedges: A hyperedge is assigned to each target vertex u, and the nodes in the hyperedge of u are the vertices in $N(u)$, however we "disjunctify" the neighborhoods of target vertices. That is, if a vertex v of G is adjacent to several targets u, we put one node for v in the hyperedge of every such u. Furthermore, if target u has valency 1, we represent each vertex in $N(u)$ by *two* nodes in the hyperedge of u.

Hypernodes from single vertices: For every vertex v of G being adjacent to more than one target, all nodes coming from v build a hypernode of weight 1.

Hypernodes from stars: Next we consider every vertex c of G as the center of several possible stars. In the following, the *regular distance* between two vertices of G is the length of a shortest regular path connecting them. In the neighborhood of every target vertex of valency 2, we determine two vertices with the smallest regular distances to c (ties are broken arbitrarily). Similarly, in the neighborhood of every target of valency 1, we determine one vertex with the smallest regular distance to c. These distinguished vertices will become ports in several stars with center c. We select some shortest regular path from c to each of the (at most $2t$) ports. Then we consider the (less than 4^t) unions of subsets of these selected regular paths starting in c. Among these unions we keep only the stars, i.e., unions of regular paths that are pairwise internally vertex-disjoint. Finally, for any such star we create a hypernode, consisting of the nodes corresponding to the ports. (Note that for any port adjacent to a target u of valency 1, we include both corresponding nodes in the hypernode, so that it hits the hyperedge of u twice.) The weight of a hypernode is the number of vertices in the corresponding star in G.

Hypernodes from loops: For each vertex v of G such that a loop for v exists, we create a hypernode corresponding to some minimum-size loop for v. It consists of all nodes created from v, and its weight is the number of vertices in the selected minimum-size loop.

This finishes the reduction. We fix $m = 2$. Now we establish the relationship between the problems:

Lemma 4. *From any optimal solution to the* MULTIPLE WEIGHTED HYPER-NODE HITTING SET *instance constructed above, we can compute in polynomial time a minimum 2-core including T in G.*

Proof. By construction, every hypernode corresponds to some star in G (with ports adjacent to targets), or some loop for one port (adjacent to some target in G), or a single vertex adjacent to more than one target. For convenience we refer to these single vertices as ports of valency 0. Recall that a port has valency 1 (0) if the port is neighbor of one (at least two) target(s).

Let S be an optimal solution to our instance of MULTIPLE WEIGHTED HYPERNODE HITTING SET with $m = 2$. Then the vertices in the corresponding stars and loops, and the ports of valency 0, extend the target set T to a 2-core. This follows from the shape of stars and loops, and from the fact that S hits every hyperedge at least twice. By definition of hypernode weights, the weight of S is at least the total number of vertices in the stars and loops, plus the number of ports of valency 0 (in other words, the number of non-target vertices in the 2-core).

We claim that the 2-core defined by S has already minimum cardinality.

Assume for contradiction that a smaller 2-core C including T exists, where C has minimum cardinality. Ports are in the following the vertices of C adjacent to any targets. Since C is a 2-core, some saturating edge set F spans the vertices of $C \setminus T$ of valency greater than 0. In particular, F spans at least the ports of valency 1 in C. We may assume that F is minimal with this property, since otherwise a proper subset of F is saturating and spans the same ports of valency 1, and together with the ports of valency 0 this gives a 2-core no larger than C. Due to minimality, F has the structure reported in Lemma 2, that is, ports of valency 1 are connected by stars and loops which are pairwise disjoint.

Consider any loop component of F, say, a loop for port v. We may assume that this loop has minimum size among all possible loops for v, since otherwise we may replace the loop with a smaller one for v and get a smaller 2-core, a contradiction. Hence there exists a hypernode for any loop component of F. Consider any star component of F, with center c. Let v be some port in this star, adjacent to target u. Note that v has valency 1 and is adjacent to no other target. If v is not one of the two closest ports v_1, v_2 in $N(u)$ selected in the reduction (those with smallest regular distances to c), we replace the path from c to v in the star, with a shortest regular path from c to v_1 or v_2, provided that one of them is not yet in another star or loop. If both v_1 and v_2 are already occupied, u has these two neighbors in C, and we can just remove the path from c to v. Thus, we either obtain a smaller 2-core (contradiction), or we need only those ports selected in the reduction. After these replacements, there exists a hypernode also for any star component in F.

Since C had already minimum size, F is still a minimal saturating edge set that spans the (possibly changed) ports of valency 1, thus it complies with the structure in Lemma 2. But now the connected components of F and the ports of valency 0 in C correspond to hypernodes, and form another solution S' to our MULTIPLE WEIGHTED HYPERNODE HITTING SET instance. Since the components are pairwise vertex-disjoint, the weight of S' is the number of vertices spanned by F, plus the number of ports of valency 0. This finally contradicts the minimality of the weight of S.   □

Now we can state our final result:

Theorem 3. MINIMUM 2-CORE *with t targets can be solved in $O^*(2^t t!)$ time.*

Proof. This follows from Theorem 1 and Lemma 4, since in the reduction to MULTIPLE WEIGHTED HYPERNODE HITTING SET we needed only $O^*(4^t)$ time to construct the weighted hypernodes.   □

References

1. Amini, O., Sau, I., Saurabh, S.: Parameterized complexity of the smallest degree-constrained subgraph problem. In: 3rd IWPEC 2008. LNCS, vol. 5018, pp. 13–29 (2008)
2. Bader, G.D., Hogue, C.W.V.: An automated method for finding molecular complexes in large protein interaction networks. BMC Bioinformatics 4(2) (2003)
3. Downey, R.G., Fellows, M.R.: Parameterized Complexity. Springer, Heidelberg (1999)
4. Dreyfus, S., Wagner, R.: The Steiner problem in graphs. Networks 1, 195–207 (1971)
5. Elbassioni, K., Hagen, M., Rauf, I.: Some fixed-parameter tractable classes of hypergraph duality and related problems. In: 3rd IWPEC 2008. LNCS, vol. 5018, pp. 91–102 (2008)
6. Fernau, H.: Parameterized algorithms for hitting set: The weighted case. In: Calamoneri, T., Finocchi, I., Italiano, G.F. (eds.) CIAC 2006. LNCS, vol. 3998, pp. 332–343. Springer, Heidelberg (2006)
7. Fomin, F.V., Kratsch, D., Woeginger, G.J.: Exact (exponential) algorithms for the dominating set problem. In: Hromkovič, J., Nagl, M., Westfechtel, B. (eds.) WG 2004. LNCS, vol. 3353, pp. 245–256. Springer, Heidelberg (2004)
8. Haas, R., Hoffmann, M.: Chordless paths through three vertices. Theoretical Computer Science 351, 360–371 (2006)
9. Niedermeier, R.: Invitation to Fixed-Parameter Algorithms. Oxford Lecture Series in Mathematics and Its Applications. Oxford University Press, Oxford (2006)

Parameterized Complexity of Candidate Control in Elections and Related Digraph Problems

Nadja Betzler* and Johannes Uhlmann*

Institut für Informatik, Friedrich-Schiller-Universität Jena,
Ernst-Abbe-Platz 2, D-07743 Jena, Germany
{betzler,uhlmann}@minet.uni-jena.de

Abstract. There are different ways for an external agent to influence the outcome of an election. We concentrate on "control" by adding or deleting candidates of an election. Our main focus is to investigate the parameterized complexity of various control problems for different voting systems. To this end, we introduce natural digraph problems that may be of independent interest. They help in determining the parameterized complexity of control for different voting systems including Llull, Copeland, and plurality votings. Devising several parameterized reductions, we provide a parameterized complexity overview of the digraph and control problems with respect to natural parameters.

1 Introduction and Preliminaries

The investigation of voting systems is an important field of interdisciplinary research. Besides obvious classical applications in political or other elections, voting systems also play an important role in multi-agent systems or rank aggregation. In addition to work that focuses on the problem to determine the winner of an election for different voting systems, there is a considerable amount of work investigating how an external agent or a group of voters can influence the election in favor or disfavor of a distinguished candidate. The studied scenarios are manipulation [3], electoral control [1,6,7,8], lobbying [2], and bribery [6]. In this work, we investigate the parameterized complexity of some variants of electoral control and closely related digraph problems. Before describing our results, we introduce the considered problems.

Problem statements. An *election* (V, C) consists of a set V of n votes and a set C of m candidates. A *vote* is an ordered preference list containing all candidates. To *control* an election, an external agent, traditionally called *chair*, can change the voting procedure to reach certain goals. The considered types of control are adding, deleting, or partitioning candidates or voters [1,8]. Further, one distinguishes between *constructive control* (CC), that is, the chair aims at making a distinguished candidate the winner, and *destructive control* (DC), that is, the chair wants to prevent a distinguished candidate from winning [8]. In this work,

---

* Supported by the DFG, research projects DARE, GU 1023/1 and PABI, NI 369/7.

B. Yang, D.-Z. Du, and C.A. Wang (Eds.): COCOA 2008, LNCS 5165, pp. 43–53, 2008.
© Springer-Verlag Berlin Heidelberg 2008

we focus on *candidate control*, that is, either deleting or adding candidates, for plurality and Copeland$^\alpha$ votings. In *plurality voting*, for every vote the candidate that is ranked first in the preference list gets one point. The *score* of a candidate is the total number of its points. A candidate with the highest score wins. Note that we still need the whole preference lists of the voters to see the effects of deleting or adding candidates. *Copeland$^\alpha$ voting* is based on pairwise comparisons between candidates: A candidate wins the pairwise head-to-head contest against another candidate if he is better positioned in more than half of the votes. The winner of a head-to-head contest is awarded one point and the loser receives no point. If the candidates are tied, both candidates get α points for $0 \leq \alpha \leq 1$. A *Copeland$^\alpha$* winner is a candidate with the highest score. Faliszewski et al. [6] devote their paper to the two important special cases $\alpha = 0$, denoted as *Copeland*, and $\alpha = 1$, denoted as *Llull*. Next, we introduce two digraph decision problems which are closely related to constructively controlling Copeland and Llull by deleting candidates.[1]

MAX-OUTDEGREE DELETION (MOD)
Given: A digraph $D = (W, A)$, a distinguished vertex $w_c \in W$, and an integer $k \geq 1$.
Question: Is there a subset $W' \subseteq W \setminus \{w_c\}$ of size at most k such that w_c is the only vertex that has maximum outdegree in $D[W \setminus W']$?

Analogously, we define MIN-INDEGREE DELETION (MID), where one wants to make a distinguished vertex to be the only vertex with minimum indegree. The correspondence to elections is based on the fact that the relations between the candidates can be depicted by a digraph where the candidates are represented by the vertices and there is an arc from vertex c to vertex d iff the corresponding candidate c defeats the corresponding candidate d in the head-to-head contest. Then, the deletion of a vertex one-to-one corresponds to the deletion of a candidate in the election. Further, the Copeland score of a candidate c is exactly the number of the out-neighbors of the corresponding vertex v_c and the Llull score is the total number of vertices minus the number of in-neighbors of v_c.

Known results. A series of publications [1,6,7,8] provides a complete picture of the classical computational complexity for four standard voting systems (approval, plurality, Condorcet, and Copeland$^\alpha$) for ten basic types of control.[2] Concerning candidate control, plurality and Copeland votings lead to NP-hardness results whereas all other voting systems are either immune or allow for polynomial-time solvability [6,7,8]. Regarding parameterized complexity, Faliszewski et al. [7] obtained some first results. They considered control of Copeland$^\alpha$ voting with respect to the parameters "number of candidates" and "number of votes". For candidate control they obtained fixed-parameter

---

[1] The digraph problems that are equivalent to adding candidates are omitted due to space restrictions.
[2] Besides the classification into P and NP-hard, a voting system can be classified as "immune" against a type of control if a non-winner can never be made a winner.

Table 1. Parameterized complexity of MAX-OUTDEGREE DELETION and MIN-INDEGREE DELETION

| | # deleted vertices k | | maximum degree d | | (k, d) | |
	MOD	MID	MOD	MID	MOD	MID
general digraphs	W[2]-c	W[2]-c	NP-c for $d \geq 3$	FPT	FPT	FPT
acyclic digraphs	W[2]-c	P	NP-c for $d \geq 3$	P	FPT	P
tournaments	W[2]-c	W[2]-c	-	-	-	-

tractability with respect to the parameter "number of candidates". The parameterized complexity with respect to the parameter "number of votes" was left open. To the best of our knowledge, there is no previous work dealing with the newly introduced digraph problems MOD and MID.

Motivation. First, from the "control person's" point of view, it is interesting to find efficient strategies to reach his goal. There are legal scenarios as for example persuading additional players to participate in a sport competition in order to make the favorite player the winner. Parameterized complexity analysis is meaningful in this context. Second, the fact that a voting system is susceptible to control or manipulation can be considered as an undesirable property. Thus, the goal of most publications is to show that, if control is not impossible, it is at least computationally hard (often showing NP-hardness). Although NP-hardness is not a sufficient criterion, as it does not imply hardness on the practically relevant average case, it is plausible to investigate whether there are any hard instances at all. However, as also noted by Conitzer et al. [3], such hardness results lose relevance if there are efficient fixed-parameter algorithms for realistic settings.

Our contributions. We provide a first study of the two natural digraph problems MOD and MID and show that they are closely related to the considered control problems. In Section 2, we investigate the computational complexity of MOD and MID for several special graph classes and parameters providing a differentiated picture of their parameterized complexity including algorithms and intractability (Table 1). The main technical achievement of this part is to show that MOD and MID are W[2]-complete in tournaments. Some of the considered special cases and parameterizations of the digraph problems map to realistic voting scenarios with presumably small parameters. Based on these connections and by giving further parameterized reductions, in Section 3 we provide an overview of parameterized hardness results for control problems (Table 2). Regarding the structural parameter "number of votes", we answer an open question of Faliszewski et al. [7] for Llull and Copeland votings by showing that even for a constant number of voters candidate control remains NP-hard. Due to the lack of space, we defer many details and proofs to a full version.

Preliminaries. In an election, we can either seek for a *winner*, that is, if there are several candidates who are best in the election, then all of them win, or for a *unique winner*. Note that a unique winner does not always exist. We only consider the unique winner case, but all our results can be easily modified to work for the winner case as well. We focus on control by adding candidates (AC) or deleting candidates (DC). Then, for example, we can define the decision problems of constructively controlling a Copeland$^\alpha$ election as follows:

CC-DC-COPELAND$^\alpha$

Given: A set C of candidates, a set V of votes with preferences over C, a distinguished candidate $c \in C$, and an integer $k \geq 1$.
Question: Is there a subset $C' \subseteq C$ of size at most k such that c is (unique) Copeland$^\alpha$ winner in the election $(V, C \backslash C')$?

CC-AC-COPELAND$^\alpha$

Given: Two disjoint sets C, D of candidates, a set V of votes with preferences over $C \cup D$, a distinguished candidate $c \in C$, and an integer $k \geq 1$.
Question: Is there a subset $D' \subseteq D$ of size at most k such that c is (unique) Copeland$^\alpha$ winner in the election $(V, C \cup D')$?

The other problems are defined analogously (see for example [6,8]). The *position* of a candidate a in a vote v is the number of candidates that are better than a in v plus one. That is, the leftmost (and best) candidate in v has position 1 and the rightmost has position m. Further, within every election we fix some arbitrary order over the candidates. Specifying a subset C' of candidates in a vote means that the candidates of C' are ordered with respect to that fixed order. An occurrence of $\overleftarrow{C'}$ in a vote means that the candidates of C' are ordered in reverse to that fixed order.

For a directed graph (digraph) $D = (W, A)$ and for a vertex $w \in W$, the set of *in-neighbors* of w is defined as $N_{in}(w) := \{u \in W \,|\, (u, w) \in A\}$ and the set of *out-neighbors* of w is given by $N_{out}(w) := \{u \in W \,|\, (w, u) \in A\}$. Moreover, the *indegree* (*outdegree*) of w is defined as $\mathrm{indeg}(w) := |N_{in}(w)|$ ($\mathrm{outdeg}(w) := |N_{out}(w)|$). Further, the *degree* is defined as $\deg(w) := \mathrm{indeg}(w) + \mathrm{outdeg}(w)$. In digraphs, we do not allow bidirected arcs and loops. An l-arc coloring $\mathcal{C} : A \to \{1, 2, \ldots, l\}$ is called *proper* if any two distinct arcs of the same color do not share a common vertex. A *tournament* is a digraph where, for every pair of vertices u and v, there is either (u, v) or (v, u) in the arc set.

A problem is called *fixed-parameter tractable (FPT)* with respect to a parameter k if it can be solved in $f(k) \cdot n^{O(1)}$ time, where n denotes the input size, and f is an arbitrary computable function. The first two levels of (presumable) parameterized intractability are captured by the complexity classes W[1] and W[2]. A *parameterized reduction* reduces a problem instance (I, k) in $f(k) \cdot |I|^{O(1)}$ time to an instance (I', k') such that (I, k) is a yes-instance if and only if (I', k') is a yes-instance and k' only depends on k but not on $|I|$.

As discussed in the introduction, there are parameterized reductions from MOD (MID) to CC-DC-COPELAND (CC-DC-LLULL) with respect to the parameters number of deleted vertices and candidates, respectively. The reverse

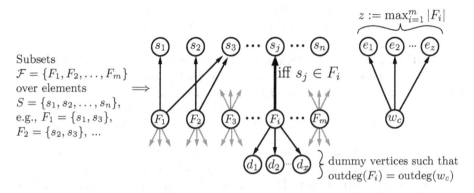

Fig. 1. Parameterized reduction from a HITTING SET-instance (left) to an MOD-instance (right). Deleting an "element-vertex" s_j in the digraph has the effect that for all "subset-vertices" corresponding to the subsets containing s_j the outdegree is decreased below the outdegree of the distinguished vertex w_c, that is, the corresponding subsets are "hit" in the HITTING SET-instance. Further, assume that there is a solution for the MOD-instance that contains a subset vertex F_i or one of its dummy neighbors. Then, instead of this vertex we can delete any subset-neighbor s_j of F_i. Based on these observations one can show that there is a hitting set of size k iff w_c can become vertex with maximum outdegree by deleting k vertices.

parameterized reductions can be obtained by a simple construction of Faliszewski et al. [6]. Thus, we say that the problems are *FPT-equivalent*.

2   Parameterized Complexity of MOD and MID

This section is concerned with the parameterized complexity of MOD and MID with respect to the parameters "number of deleted vertices" k and "maximum degree" d for different classes of graphs. Our results are summarized in Table 1. In the following, we only prove W[2]-hardness. Using the machinery of Downey and Fellows [4], it is not hard to also show containment in W[2].

Theorem 1. MAX-OUTDEGREE DELETION *is W[2]-complete with respect to the parameter "number of deleted vertices" in acyclic digraphs and NP-complete in acyclic digraphs with maximum degree three.*

The W[2]-hardness can be shown by a parameterized reduction from the W[2]-complete HITTING SET (HS) problem [5]. Given a subset family $\mathcal{F} = \{F_1, F_2, \ldots, F_m\} \subseteq 2^S$ of a base set $S = \{s_1, s_2, \ldots, s_n\}$ and an integer $k \geq 1$, the HITTING SET problem asks to decide whether there exists a subset $S' \subseteq S$ of size at most k such that for every $1 \leq i \leq m$ we have $S' \cap F_i \neq \emptyset$. We defer the formal proof of Theorem 1 to the full version of this paper. Here, we only illustrate the basic construction (see Fig. 1). The resulting digraph of the MOD-instance is acyclic, which gives the first part of the theorem. The second part directly follows from the fact that HITTING SET is NP-complete even if every subset has size two and every element occurs in exactly three subsets (3X-2-HITTING SET).

Proposition 1. *a)* MIN-INDEGREE DELETION *can be solved in polynomial time in acyclic digraphs. In general digraphs, it is fixed-parameter tractable with respect to the parameter "indegree of the distinguished vertex w_c".*

b) MAX-OUTDEGREE DELETION *is fixed-parameter tractable with respect to the combined parameters "outdeg(w_c)" and "number of deleted vertices k".*

Proof. (Sketch) a) First part (acyclic graphs): Since in acyclic graphs there always exists a vertex with indegree zero, w_c must have indegree zero to be the only minimum indegree vertex. Thus, one can iteratively delete all other vertices with indegree zero.

Second part (parameter indeg(w_c)): If one knows for an MID-instance which in-neighbors of the distinguished vertex w_c are part of a minimum solution, then the problem becomes trivial: One can delete these vertices and extend the resulting partial solution to a minimum-cardinality solution. For this, one iteratively adds all vertices of indegree smaller than the (new) indegree of w_c to the solution since all vertices of indegree smaller than the distinguished vertex must be deleted. Hence, exhaustively trying all subsets of in-neighbors of w_c yields an algorithm with running time $O(2^{\text{indeg}(w_c)} \cdot |W|^2)$.

b) Here, we give a simple branching strategy: Consider a vertex $u \in W \setminus \{w_c\}$ with outdegree at least outdeg(w_c). Furthermore, let $N \subseteq N_{out}(u)$ with $|N| = $ outdeg(w_c). Then, we have to delete one of the vertices in $(N \cup \{u\}) \setminus \{w_c\}$, that is, we can branch into at most outdeg$(w_c)+1$ cases. In each case, we can decrease the parameter k by one, leading to a search tree of size $O((\text{outdeg}(w_c)+1)^k)$. □

The following theorem is based on a parameterized reduction from the W[2]-complete DOMINATING SET problem [5]. The basic idea is similar to the HITTING SET reduction (Fig. 1), but the details are quite involved.

Theorem 2. MAX-OUTDEGREE DELETION *and* MIN-INDEGREE DELETION *are W[2]-complete with respect to the parameter "number of deleted vertices" even in the case that the input graph is a tournament.*

3   Parameterized Complexity of Candidate Control

In this section, we turn our attention to elections. For candidate control in Llull and Copeland votings we show NP-hardness for a constant number of votes. Further, we provide parameterized intractability results with respect to the number of deleted/added candidates for plurality and Copeland$^\alpha$ votings.

Number of votes as parameter. In many election scenarios there is only a small number of votes. For example, consider a human resources department where few people are deciding which job applicant gets the employment. An open question of Faliszewski et al. [7] regards the parameterized complexity of Copeland$^\alpha$ elections with respect to the parameter "number of votes". We answer this question for Llull and Copeland.

Theorem 3. CC-DC-COPELAND *is NP-complete for six votes,* CC-AC-COPELAND *is NP-complete for eight votes,* CC-DC-LLULL *is NP-complete for ten votes, and* CC-AC-LLULL *is NP-complete for six votes.*

Proof. (Sketch) For all problems NP-membership is obvious. We only give the NP-hardness proof for CC-DC-COPELAND to demonstrate the basic idea.

The proof consists of two phases. The first phase is a reduction from 3X-2-HITTING SET to MOD as depicted in Fig. 1. The digraph D of a resulting MOD-instance (D, w_c, k) has maximum degree three and the underlying undirected graph of D is bipartite. More precisely, one partition consists of the subset-vertices and w_c, and the other partition consists of the element-vertices and the neighbors of w_c. As we reduce from 3X-2-HITTING SET, we do not have any further dummy vertices. In the second phase we show that D can be encoded by an election with only six votes by exploiting this special structure of D.

Now, we describe the second phase. Due to König [9] we know that a bipartite graph is Δ-edge-colorable, where Δ denotes the maximum degree of the graph. Moreover, a corresponding proper Δ-edge coloring can be computed in polynomial time. Thus, for D there exists a proper 3-arc-coloring $\mathcal{C} : A \to \{\mathcal{R}, \mathcal{G}, \mathcal{B}\}$. Note that in the underlying undirected graph of D the edges of the same color class form a matching, that is, two arcs of the same color do not share a common vertex. Hence, the coloring $\mathcal{C}$ partitions the arc set into three classes of independent arcs. We next describe how the arcs of graph D can be encoded in an election with six votes. Let $A_\mathcal{R} = \{(r_1, r_1'), \ldots, (r_l, r_l')\}$ denote the arcs colored by $\mathcal{R}$. Furthermore, $W_{\overline{\mathcal{R}}}$ denotes the set of vertices that are not incident to any arc of $A_\mathcal{R}$. To encode $A_\mathcal{R}$, we add the two votes $r_1 > r_1' > r_2 > r_2' > \cdots > r_l > r_l' > W_{\overline{\mathcal{R}}}$ and $\overleftarrow{W_{\overline{\mathcal{R}}}} > r_l > r_l' > \cdots > r_2 > r_2' > r_1 > r_1'$ to the election. In the same way we add two votes for the arcs colored by $\mathcal{B}$ and $\mathcal{G}$, respectively. The correctness of the construction follows from two observations. First, since the arcs of the same color do not share common endpoints, in every vote all vertices occur exactly once and we have a valid election. Second, consider an arc $(w', w'') \in A$ with $\mathcal{C}((w', w'')) = X$ for any color $X \in \{\mathcal{R}, \mathcal{B}, \mathcal{G}\}$. Then, w' defeats w'' in the votes $v_{X,1}$ and $v_{X,2}$ and ties with w'' in the remaining four votes. Moreover, since every arc occurs in exactly one color class, all arcs are encoded, and, since all other candidates are tied in every pair of the votes, we have ties between all other pairs of candidates.

In summary, in the constructed Copeland election a candidate c can become the unique winner by deleting k candidates iff in D the corresponding vertex w_c can become the maximum outdegree vertex by deleting k vertices. $\qquad\square$

Number of deleted/added candidates as parameter. To control an election without raising suspicion one may add or delete only a limited number of candidates. Here, we investigate whether it is possible to obtain fixed-parameter algorithms

Table 2. Results in boldface are new. The results for Copeland$^\alpha$ hold for all $0 \leq \alpha \leq 1$. The W[2]-hardness results for CC-AC-Plurality and DC-AC-Plurality follow from the NP-completeness proofs [1,8]. The polynomial-time (P) results are from [6,7].

| | Copeland$^\alpha$ | | Plurality | |
	CC	DC	CC	DC
Adding Candidates (AC)	**W[2]-c**	P	W[2]-h	W[2]-h
Deleting Candidates (DC)	**W[2]-c**	P	**W[2]-h**	**W[1]-h**

under this assumption. More specifically, we consider the parameterized complexity of destructive and constructive control by adding or deleting a fixed number of candidates. Our results are summarized in Table 2. It turns out that all NP-complete problems are intractable from this parameterized point of view as well. This even holds true for plurality voting, which can be considered as the "easiest" voting system in terms of winner evaluation and for which the MANIPULATION problem can be solved optimally by a simple greedy strategy [3].

Copeland. For elections without ties in all pairwise head-to-head contests, CC-DC-Copeland$^\alpha$ coincides for all $0 \leq \alpha \leq 1$, since these problems only differ in the way ties are evaluated. As discussed in the introduction MOD and CC-DC-Copeland$^\alpha$ are FPT-equivalent. Using the same reductions one can show that MOD in tournaments is FPT-equivalent to CC-DC-Copeland$^\alpha$ without ties. Thus, the W[2]-hardness of CC-DC-Copeland$^\alpha$ without ties follows directly from Theorem 2.[3] For adding candidates we obtain W[2]-hardness using similar ideas.

Plurality. For plurality voting, the W[2]-hardness results for control by adding candidates follow from the reductions for the NP-hardness [1,8]. In contrast, the reductions used to show NP-hardness for control by deleting candidates [1,8] do not imply their parameterized hardness. Thus, we develop new parameterized reductions to show W[1]/W[2]-hardness.[4] For the constructive case we can show W[2]-hardness by a reduction from MOD. Note that the encoding of a MOD instance into a plurality election is more demanding than for Copeland voting and the other direction (encoding a plurality election into MOD) is not obvious.

Theorem 4. *Constructive control of plurality voting by deleting candidates is W[2]-hard with respect to the parameter "number of deleted candidates".*

Proof. (Sketch) We present a parameterized reduction from MOD. Given an MOD instance $(D = (W, A), w_c, k)$ with $W = \{w_1, w_2, \ldots, w_n\}$ and $w_c = w_1$, we construct an election (V, C) as follows: We have one candidate corresponding to every vertex, that is, $C' := \{c_i \mid w_i \in W\}$. The set of candidates C then consists of C' and an additional set F of "dummy" candidates (only used to "fill"

---

[3] Having no ties in the pairwise head-to-head contests is a realistic scenario. It is always the case for an odd number of votes and likely for a large number of votes. In contrast, the NP-hardness proofs of the considered problems rely on ties [6,7].

[4] The class containment for all kinds of candidate control in plurality voting is open.

positions that cannot be taken by other candidates in our construction). The set of votes V consists of two subsets V_1 and V_2. In V_1, for every $c_i \in C'$ we have outdeg(w_i) votes in which c_i is at the first position and with dummy candidates in the positions from 2 to $k+1$. Then, for every such vote, the remaining candidates follow in arbitrary order. In V_2, for every $c_i \in C'$ we have $|W|$ votes in which c_i is at the first position. For all candidates $c_j \neq c_i$ with $w_j \notin N_{in}(w_i)$, we observe that in exactly one of these $|W|$ votes c_j is at the second position. In all other of these votes, the second position is filled with a dummy candidate. Moreover, we add dummies to all positions from 3 to $k+1$. Concerning the dummies, in V_1 and V_2 we ensure that every dummy candidate $f \in F$ has a position better than $k+2$ only in one of the votes. This can be done such that the size of F is less than $(k+1) \cdot |V|$. The dummies exclude the possibility of "accidently" getting candidates in the first position. Note that by deleting k candidates only a candidate that is at one of the first $k+1$ positions in a vote has the possibility to increase his plurality score. Further, by construction, the dummy candidates fulfill the following two conditions. First, the score of a dummy candidate can become at most one. Second, it does never make sense to delete a dummy as by this only other dummies can get into the first position of a vote. Next, we prove the correctness of the reduction.

Claim: Candidate c_1 can become the plurality winner of (V, C) by deleting k candidates iff w_1 can become the only maximum-degree vertex in D by deleting k vertices.

"$\Rightarrow$": Denote the set of deleted candidates by R. We show that after deleting the set of vertices $W_R := \{w_i \mid c_i \in R\}$ the vertex w_1 is the only vertex with maximum degree. Before deleting any candidates, for every candidate c_i we have score(c_i) = score(c_1) + s_i with $s_i :=$ outdeg(w_i) − outdeg(w_1). After deleting the candidates in R, candidate c_1 is the winner. Hence, for $i = 2, \ldots, |W|$ we must have either that score(c_i) < score(c_1) or that c_i is deleted. For a non-deleted candidate c_i with $i > 1$ the difference between score(c_i) and score(c_1) must be decreased by at least $s_i + 1$. By construction, the only way to decrease the difference by one is to delete a candidate such that c_1 becomes first in one more vote and c_i does not increase its number of first positions. All candidates that can be deleted to achieve this correspond to vertices in $N_{in}(w_i) \backslash N_{in}(w_1)$. To improve upon c_i we must delete at least $s_i + 1$ candidates that fulfill this requirement. Hence, in D the outdegree of w_i is reduced to be less than the outdegree of w_1.

"$\Leftarrow$": Let $T \subseteq D$ denote the solution for MOD. We can show ("reverse" to the other direction) that by deleting the set of candidates $C_T := \{c_i \mid w_i \in T\}$ candidate c_1 becomes a plurality winner.   □

In contrast to Copeland$^\alpha$ elections, for plurality elections destructive control by deleting candidates is NP-hard [8]. We show that it is even W[1]-hard by presenting a parameterized reduction from the W[1]-complete CLIQUE problem [5]. Given an undirected graph $G = (W, E)$ and a positive integer k, the CLIQUE problem asks to decide whether G contains a complete subgraph of size at least k.

Theorem 5. *Destructive control of plurality voting by deleting candidates is W[1]-hard with respect to the parameter "number of deleted candidates".*

Proof. Given a CLIQUE instance $(G = (W, E), k)$, we construct an election as follows: The set of candidates is $C := C_W \uplus C_E \uplus \{c, w\} \uplus D$ with $C_W := \{c_u \mid u \in W\}$, $C_E := \{c_{uv} \mid \{u, v\} \in E\}$, and a set of dummy candidates D. In the following, the candidates in C_W and C_E are called *vertex candidates* and *edge candidates*, respectively. Further, we construct the votes in a way such that w is the candidate that we like to prevent from winning, c is the only candidate that can beat w, and D contains dummy candidates that can gain a score of at most one. In the set of votes V we have for every vertex $u \in W$ and for each incident edge $\{u, v\} \in E$ one vote of the type $c_u > c_{uv} > c > \dots$, that is, there are $2 \cdot |E|$ votes of this type, two for every edge. Additionally, V contains $|W| + k \cdot (k - 1)$ votes in which w is at the first position and $|W| + 1$ votes in which c is at the first position. In all votes, the remaining free positions between 2 and $k + \binom{k}{2} + 1$ are filled with dummies such that every dummy occurs in at most one vote at a position better than $k + \binom{k}{2} + 2$. This can be done using less than $|V| \cdot (k + \binom{k}{2} + 1)$ dummy candidates. In every vote the candidates that do not occur in this vote at a position less than $(k + \binom{k}{2} + 1)$ follow in arbitrary order.

Claim: Graph G contains a clique K of size k iff candidate c can become plurality winner by deleting $k' := k + \binom{k}{2}$ candidates.

"$\Rightarrow$": Delete the $k + \binom{k}{2}$ candidates that correspond to the vertices and edges of K. Then, for every of the $\binom{k}{2}$ deleted edge candidates we also deleted the two vertex candidates that correspond to the endpoints of this edge. Therefore, for every of the $\binom{k}{2}$ edges candidate c gets in the first position in two more votes. Hence, the score of candidate c is increased by $2 \cdot \binom{k}{2} = k \cdot (k - 1)$ and the score of candidate w is not affected. Thus, the total score of w is $|W| + k \cdot (k - 1)$ and the total score of c is $|W| + k \cdot (k - 1) + 1$; therefore, w is defeated by c.

"$\Leftarrow$": Note that, by construction, we cannot decrease the score of w and we cannot increase the score of a vertex candidate (which is at most $|W| - 1$). Further, by the deletion of at most k' candidates the score of a dummy candidate can become at most one, and the score of an edge candidate can become at most two. Hence, c is the only candidate that can prevent w from winning. Furthermore, as the deletion of at most k' dummies never moves c into a first position, we can assume that the solution deletes only edge and vertex candidates.

We omit the proof that the only way to increase the score of c by at least $k \cdot (k - 1)$ is to choose edge and vertex candidates that correspond to the vertices and edges of a clique of size k. $\qquad\square$

References

1. Bartholdi III, J.J., Tovey, C.A., Trick, M.A.: How hard is it to control an election. Mathematical and Computer Modelling 16(8-9), 27–40 (1992)
2. Christian, R., Fellows, M.R., Rosamond, F.A., Slinko, A.M.: On complexity of lobbying in multiple referenda. Review of Economic Design 11(3), 217–224 (2007)

3. Conitzer, V., Sandholm, T., Lang, J.: When are elections with few candidates hard to manipulate? Journal of the ACM 54(3), 1–33 (2007)
4. Downey, R.G., Fellows, M.R.: Threshold dominating sets and an improved version of W[2]. Theoretical Computer Science 209, 123–140 (1998)
5. Downey, R.G., Fellows, M.R.: Parameterized Complexity. Springer, Heidelberg (1999)
6. Faliszewski, P., Hemaspaandra, E., Hemaspaandra, L.A., Rothe, J.: Llull and Copeland voting broadly resist bribery and control. In: Proc. of 22nd AAAI 2007, pp. 724–730 (2007)
7. Faliszewski, P., Hemaspaandra, E., Hemaspaandra, L.A., Rothe, J.: Copeland voting fully resists constructive control. In: Proc. of 4th AAIM 2008 (2008)
8. Hemaspaandra, E., Hemaspaandra, L.A., Rothe, J.: Anyone but him: The complexity of precluding an alternative. Artificial Intelligence 171, 255–285 (2007)
9. König, D.: Über Graphen und ihre Anwendungen auf Determinantentheorie und Mengenlehre. Mathematische Annalen 77, 453–465 (1916)

A Parameterized Perspective
on Packing Paths of Length Two

Henning Fernau and Daniel Raible

Universität Trier, FB IV—Abteilung Informatik, 54286 Trier, Germany
{fernau,raible}@informatik.uni-trier.de

Abstract. We study (vertex-disjoint) packings of paths of length two (i.e., of P_2's) in graphs under a parameterized perspective. Starting from a maximal P_2-packing $\mathcal{P}$ of size j we use extremal combinatorial arguments for determining how many vertices of $\mathcal{P}$ appear in some P_2-packing of size $(j+1)$ (if it exists). We prove that one can 'reuse' $2.5j$ vertices. Based on a WIN-WIN approach, we build an algorithm which decides if a P_2-packing of size at least k exists in a given graph in time $\mathcal{O}^*(2.482^{3k})$.

1 Introduction and Definitions

Mathematical Motivation. We consider a natural generalization of the well-known matching problem in graphs. Recall that a maximum matching is a maximum cardinality set of vertex disjoint edges, i.e., a packing with paths of length one. We are going to study packings by paths of length two (abbreviated as P_2). More formally, we consider the following problem, called P_2-PACKING:

Given: A graph $G = (V, E)$, and the parameter k.
We ask: Is there a set of k vertex-disjoint P_2's in G?

P. Hell and D. Kirkpatrick [11,9] proved $\mathcal{NP}$-completeness for this problem. In fact, they showed that general MAXIMUM H-PACKING is $\mathcal{NP}$-complete. Here, H is a graph with at least three vertices in some connected component. Notice that P_2-PACKING attracts attention as it is $\mathcal{NP}$-hard, whereas the classical matching problem, which is P_1-PACKING, is solvable in polynomial time.

Parameterized interests. H. Fernau and D. F. Manlove [6] discovered a primal-dual relation to TOTAL EDGE COVER. Recall that an *edge cover* is a set of edges $EC \subseteq E$ that cover all vertices of a given graph $G = (V, E)$. An edge cover is called *total* if every component in $G[EC]$ has at least two edges. This type of constraint for covering problems is motivated by modelling clustering properties within cover sets, see [6]. By matching techniques, the problem of finding an edge cover of size at most k is solvable in polynomial time. However, the following Gallai-type identity [6] proves that finding total edge covers of size at most k is $\mathcal{NP}$-hard: The sum of the number of P_2's in a maximum P_2-packing and the size of a minimum total edge cover equals $n = |V|$. H. Fernau and D. F. Manlove [6] also showed that TOTAL EDGE COVER is fixed-parameter tractable

B. Yang, D.-Z. Du, and C.A. Wang (Eds.): COCOA 2008, LNCS 5165, pp. 54–63, 2008.

(or: lies in $\mathcal{FPT}$, for short). This is quite interesting since there are few natural, unrestricted problems where both the primal and the dual variant are known to lie in $\mathcal{FPT}$.

Applications. There is a strong link to the TEST COVER (TC) problem [1] with applications ranging from fault testing and diagnosis, pattern recognition to biological identification. The input to TC is a hypergraph $H = (G, E)$ and one wishes to identify a subset $E' \subseteq E$ (the *test cover*) such that, for any distinct $i, j \in V$, there is an $e' \in E'$ with $|e' \cap \{i.j\}| = 1$. TC models identification problems: Given a set of individuals and a set of binary attributes we search for a minimum subset of attributes that identifies each individual distinctly. For the special yet important case TCP2, where for all $e \in E$ we have $|e| \leq 2$, K. M. J. Bontridder *et al.* [1] could show the following two assertions. (1) If H has a test cover of size τ, then there is a P_2-packing of size $n - \tau - 1$ that leaves at least one vertex isolated. (2) If H has a maximal P_2-packing of size π that leaves at least one vertex isolated, then there is a test cover of size $n - \pi - 1$. This also establishes a close relation between TEST COVER and TOTAL EDGE COVER. So we can employ our algorithms to solve the TCP2 case of TEST COVER by using an initial catalytic branch that determines one vertex that should be isolated.

Discussion of Related Work. R. Hassin and S. Rubinstein [8] found a randomized $\frac{35}{67}$-approximation for finding a maximum P_2-packing. K. M. J. Bontridder *et al.* [1] studied deterministic approximation algorithms, considering a series of heuristics H_ℓ. H_ℓ starts from a maximal P_2-packing $\mathcal{P}$ and tries to improve it by replacing ℓ P_2's by $\ell + 1$ P_2's. The corresponding approximation ratios ρ_ℓ are as follows: $\rho_0 = \frac{1}{3}$, $\rho_1 = \frac{1}{2}$, $\rho_2 = \frac{5}{9}$, $\rho_3 = \frac{7}{11}$ and $\rho_\ell = \frac{2}{3}$ for $\ell \geq 4$.

As any P_2-PACKING instance can be transformed into a 3-SET PACKING instance one can use Y. Liu *et al.* [12] algorithm which needs $\mathcal{O}^*(4.61^{3k})$ steps, or the very recent algorithm of J. Wang and Q. Feng [14] running in time $\mathcal{O}(3.52^{3k})$. This is the culmination point of a sequence of papers subsequently improving on the running time of this problem. Alternatively, we can use randomized parameterized algorithms; the best published algorithms yields a running time of $\mathcal{O}^*(2.52^{3k})$, see [3]. Recently, we were informed by I. Koutis[1] that he has developed a randomized parameterized algorithm for this problem that runs in time $\mathcal{O}^*(2^{3k})$. The first paper to individually study P_2-PACKING under a parameterized view was E. Prieto and C. Sloper [13]. The authors were able to prove a $15k$-kernel. Via a clever midpoint search on the kernel they could achieve a deterministic run time of $\mathcal{O}^*(3.403^{3k})$. Another special case of 3-SET PACKING studied from a parameterized perspective is 3-DIMENSIONAL MATCHING, see [12] for a deterministic algorithm of run time $\mathcal{O}^*(2.77^{3k})$. Recently, J. Wang *et al.* [15] found a kernel of size $7k$ for P_2-PACKING, resulting in a deterministic $\mathcal{O}^*(2.61^{3k})$-algorithm for this problem.

Our Contributions. The main achievements of this paper are: (1) We present an algorithm which solves this problem in time $\mathcal{O}^*(2.482^{3k})$. (2) We exhibit an

---

[1] Personal communication; the corresponding paper will be presented at ICALP 2008.

extremal combinatorial argument to show that, given a P_2-packing of size j and provided that a larger packing exists, we can reuse $2.5j$ vertices of the known packing. This improves a similar result for general 3-SET PACKING [12] where only $2j$ elements are reusable. (3) Another novelty is that in this algorithm, the inductive augmentation step is interleaved with kernelization. This pays off not only heuristically but also asymptotically by a specific form of combinatorial analysis. Thereby we can completely skip the time consuming color-coding which was needed in Liu *et al.* [12] for 3-SET PACKING. (4) We show that WIN-WIN games can be played with two different brute-force algorithms to finally achieve the claimed running time. We believe that especially the idea of saving colors by extremal combinatorial arguments could be applied in other situations, as well.

Some Notations and Definitions. We only consider undirected graphs $G = (V, E)$. For a subgraph H of G, denote by $N(H)$ the set of vertices that are not in H but adjacent to at least one vertex on H, i.e., $N(H) = (\bigcup_{v \in H} N(\{v\})) \setminus H$. The subgraph H is *adjacent* to a vertex v if $v \in N(H)$. A P_2 in G is a path which consists of three vertices and two edges. For any path p of this kind we consider the vertices as numbered such that $p = p_1 p_2 p_3$ (where the roles of p_1 and p_3 might be interchanged). For a path p, $V(p)$ ($E(p)$, resp.) denotes the set of vertices (edges, resp.) on p. Likewise, for a set of paths $\mathcal{P}$, $V(\mathcal{P}) := \bigcup_{p \in \mathcal{P}} V(p)$ ($E(\mathcal{P}) := \bigcup_{p \in \mathcal{P}} E(p)$, resp.).

Kernelization. Based on the work [13] of E. Prieto and C. Sloper, Wang *et al.* exhibited the following result [15]:

Theorem 1. P_2-PACKING *admits a kernel with at most $7k$ vertices.*

That result was obtained by optimizing the use of fat and double crowns through local improvements, called **Rule 1** and **Rule 2**.

We mention here that the (more general results) of H. Fernau and D. Manlove [6] can be improved for the parametric dual (in the sense of the mentioned Gallai-type identity) TOTAL EDGE COVER, parameterized by k_d upperbounding the edge cover size:

Theorem 2. TOTAL EDGE COVER *admits a kernel with at most $1.5k_d$ vertices.*

Proof. Since we aim at a total edge cover, the largest number of vertices that can be covered by k edges is $1.5k$ (namely, if the edge cover is a P_2-packing). Hence, if the graph contains more than $1.5k$ vertices, we can reject. This leaves us with a kernel with at most $1.5k$ vertices.   □

This also allows us to state lower bounds for the kernel sizes, based on works of J. Chen *et al.* [2]:

Corollary 1. *Trivially, P_2-PACKING does not admit a kernel with less than $3k$ vertices.* TOTAL EDGE COVER *does not admit a kernel with less than $\alpha_a k_d$ vertices for any $\alpha_d < (7/6)$, unless $\mathcal{P} = \mathcal{NP}$.*

Proof. A P_2-packing of size k is only possible in a graph with at least $3k$ vertices. Due to Theorem 1 and [2, Theorem 3.1], there does not exist a kernel of size $\alpha_d k_d$ for TOTAL EDGE COVER under the assumption that $\mathcal{P} = \mathcal{NP}$ if $(7-1)(\alpha_d-1) < 1$. $\square$

2   Combinatorial Properties of P_2-Packings

This section is devoted to proving the following combinatorial result by extremal combinatorial arguments. Notice that $\mathfrak{Q}_{(2)}$ denotes a set of P_2-packings of size $(j + 1)$. The exact definition of $\mathfrak{Q}_{(2)}$ will be given later.

Theorem 3. *Let $\mathcal{P}$ be a maximal P_2-packing of size j. If there is a P_2-packing of size $(j+1)$, then there is also a packing $\mathcal{Q} \in \mathfrak{Q}_{(2)}$ with $|V(\mathcal{P}) \cap V(\mathcal{Q})| \geq 2.5j$.*

The combinatorial properties of $\mathcal{Q}$ will be used in the next section by the inductive step of our algorithm for P_2-PACKING. Among all maximal P_2-packings of size $(j + 1)$, we will consider those packings $\mathcal{Q}$ that maximize

$$\sum_{p \in \mathcal{P}} \sum_{q \in \mathcal{Q}} 1_{[E(p)=E(q)]}, \tag{1}$$

where $1_{[\,]}$ is the indicator function. We call the set of these packings $\mathfrak{Q}_{(1)}$. In $\mathfrak{Q}_{(1)}$, we find those packings $\mathcal{Q}$ that 'reuse' the maximum number of P_2's from the packing $\mathcal{P}$. From Liu et al. [12], we know:

Lemma 1. $|V(p) \cap V(\mathcal{Q})| \geq 2$ *for any $p \in \mathcal{P}$ and $\mathcal{Q} \in \mathfrak{Q}_{(1)}$.*

Proof. If there is $p \in \mathcal{P}$ with $|V(p) \cap V(\mathcal{Q})| = 1$, then replace the intersecting path of $\mathcal{Q}$ by p. In the case where $|V(p) \cap V(\mathcal{Q})| = 0$, simply replace an arbitrary $q \in \mathcal{Q} \setminus \mathcal{P}$, that must exist by pigeon-hole, by p. In both cases, we obtain a packing $\mathcal{Q}'$ of the same size as $\mathcal{Q}$, but $\sum_{p \in \mathcal{P}} \sum_{q \in \mathcal{Q}'} 1_{[E(p)=E(q)]} = \sum_{p \in \mathcal{P}} \sum_{q \in \mathcal{Q}} 1_{[E(p)=E(q)]} + 1$, contradicting $\mathcal{Q} \in \mathfrak{Q}_{(1)}$. $\square$

A slightly sharper version is the next assertion:

Corollary 2. *If $\mathcal{Q} \in \mathfrak{Q}_{(1)}$, then for any $p \in \mathcal{P}$ with $p \notin \mathcal{Q}$, there are $q_1, q_2 \in \mathcal{Q}$, $q_1 \neq q_2$, with $|V(p) \cap V(q_i)| \geq 1$ $(i = 1, 2)$.*

Proof. Suppose it exists $p \in \mathcal{P}$ and only one $q \in \mathcal{Q}$ with $|V(p) \cap V(q)| \geq 2$. Then $\mathcal{Q} \setminus \{q\} \cup \{p\}$ improves on priority (1), contradicting $\mathcal{Q} \in \mathfrak{Q}_{(1)}$. $\square$

We sharpen this combinatorial bound by considering from the set $\mathfrak{Q}_{(1)}$ only those P_2-packings $\mathcal{Q}'$ which maximize the following second property:

$$\sum_{p \in \mathcal{P}} \sum_{q \in \mathcal{Q}'} |E(p) \cap E(q)|. \tag{2}$$

The set of the remaining P_2-packings will be called $\mathfrak{Q}_{(2)}$. So, in $\mathfrak{Q}_{(2)}$ are those packings from $\mathfrak{Q}_{(1)}$ which cover the maximum number of edges in $E(\mathcal{P})$.

In contrast to the general situation with 3-SET PACKING, paths are more concrete objects that can be shifted or folded along the given graph. These geometric ideas will be used to finally prove our claimed combinatorial theorem.

We define $\mathcal{P}_i(\mathcal{Q}) := \{p \in \mathcal{P} \mid i = |p \cap V(\mathcal{Q})|\}$. A vertex $v \in V$ is a $\mathcal{Q}$-endpoint if there is a unique $q = q_1 \ldots q_3 \in Q$ such that $v = q_1$ or $v = q_3$. A vertex v is called $\mathcal{Q}$-midpoint if there is a $q = q_1 q_2 q_3 \in \mathcal{Q}$ with $q_2 = v$.

1. We call $q = q_1 q_2 q_3 \in \mathcal{Q}$ *foldable* on $p = p_1 p_2 p_3 \in \mathcal{P}$ if, for $q_2 \in V(p) \cap V(q)$, we have $p_s = q_2$, $s \in \{1, 2, 3\}$, and either $p_{s+1} \notin V(\mathcal{Q})$ or $p_{s-1} \notin V(\mathcal{Q})$, see Figure 1(a).
2. If q is foldable on p, then substituting q by $q \setminus \{q_i\} \cup \{p_{s\pm1}\}$ with $i \in \{1, 3\}$, will be called $(q_i, p_{s\pm1})$-*folding*, see Figure 1(b).
3. We call $q = q_1 q_2 q_3 \in \mathcal{Q}$ *shiftable* with respect to q_1 (q_3, resp.) on $p = p_1 p_2 p_3 \in \mathcal{P}$ if the following holds: $q_1 \in V(p) \cap V(q)$ ($q_3 \in V(p) \cap V(q)$, resp.) and either $p_{s+1} \notin V(\mathcal{Q})$ or $p_{s-1} \notin V(\mathcal{Q})$ where $p_s = q_1$ ($p_s = q_3$, resp.) and $s \in \{1, 2, 3\}$, see Figure 1(c).
4. If q is shiftable on p with respect to $t \in \{q_1, q_3\}$, then substituting q by $q \setminus \{g\} \cup \{p_{s+1}\}$ (or by $q \setminus \{g\} \cup \{p_{s-1}\}$, resp.), $g \in \{q_1, q_3\} \setminus \{t\}$, will be called (g, p_{s+1})-*shifting* $((g, p_{s-1})$-*shifting*, resp.), see Figure 1(d).

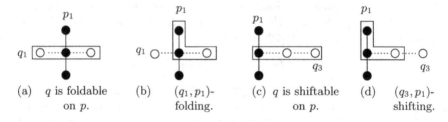

(a) q is foldable on p. (b) (q_1, p_1)-folding. (c) q is shiftable on p. (d) (q_3, p_1)-shifting.

Fig. 1. The black vertices and solid edges indicate the P_2-packing $\mathcal{P}$. The polygons contain the P_2's of the packing $\mathcal{Q}$.

Lemma 2. *If $q = q_1 q_2 q_3 \in \mathcal{Q}$ with $\mathcal{Q} \in \mathfrak{Q}_{(2)}$ is shiftable on $p \in \mathcal{P}$ with respect to q_1 (or q_3, resp.), then there is some $p' \in \mathcal{P}$ with $p' \neq p$ such that $\{q_3, q_2\} \in E(p')$ (or $\{q_2, q_1\} \in E(p')$, resp.).*

Proof. We examine the case where $V(p) \cap V(q) = \{q_1\}$ and, w.l.o.g., $p_{s+1} \notin V(\mathcal{Q})$. Now assume the contrary. Then by (q_3, p_{s+1})-shifting, we obtain a P_2-packing $\mathcal{Q}'$. Comparing $\mathcal{Q}$ and $\mathcal{Q}'$ with respect to priority 1, $\mathcal{Q}'$ is no worse than $\mathcal{Q}$. But $\mathcal{Q}'$ improves on priority 2, as we gain $\{p_s, p_{s+1}\}$. But this contradicts $\mathcal{Q} \in \mathfrak{Q}_{(2)}$.   □

Lemma 3. *If $\mathcal{Q} \in \mathfrak{Q}_{(2)}$, then no $q \in \mathcal{Q}$ is foldable.*

Proof. Suppose some $q \in \mathcal{Q}$ is foldable on p and, w.l.o.g., $p_{s+1} \notin V(\mathcal{Q})$. Then by (q_1, p_{s+1})-folding q we could improve on priority 2 (without weakening priority 1), contradicting $\mathcal{Q} \in \mathfrak{Q}_{(2)}$.   □

Suppose there is a path p with $|V(p) \cap V(\mathcal{Q})| = 2$. Then p shares exactly one vertex $p_{q'}, p_{q''}$ with paths $q', q'' \in \mathcal{Q}$ due to Corollary 2. In the following $p_{q'}$ and $p_{q''}$ will always refer to the two cut vertices of the paths $q', q'' \in \mathcal{Q}$ which cut a path p with $|V(p) \cap V(\mathcal{Q})| = 2$.

Lemma 4. *Let $\mathcal{Q} \in \mathfrak{Q}_{(2)}$. Consider $p \in \mathcal{P}$ with $|V(p) \cap V(\mathcal{Q})| = 2$ and neither $p_{q'}$ nor $p_{q''}$ are $\mathcal{Q}$-endpoints. Then one of q', q'' is foldable.*

Proof. Let $i, j \in \{1, 2, 3\}$ such that $p_{q'} = p_i$ and $p_{q''} = p_j$. Then for $f \in \{1, 2, 3\} \setminus \{i, j\}$, we have $p_f \notin V(\mathcal{Q})$. W.l.o.g., $\{p_i, p_f\} \in E(p)$. Then q' is (q_1', p_f)-foldable. □

Corollary 3. *Let $\mathcal{Q} \in \mathfrak{Q}_{(2)}$ and $p \in \mathcal{P}$ with $|V(p) \cap V(\mathcal{Q})| = 2$. Then one of $p_{q'}, p_{q''}$ must be a $\mathcal{Q}$-endpoint.*

Proof. Assume the contrary. Then using Lemmas 3 and 4 lead to a contradiction. □

Proof. (of Theorem 3) Suppose there is a path $p \in \mathcal{P}$ with $|V(p) \cap V(\mathcal{Q})| = 2$. By Corollary 3, w.l.o.g.. $p_{q'}$ is a $\mathcal{Q}$-endpoint. For $p_{q''}$ there are two possibilities: **a)** $p_{q''}$ is also a $\mathcal{Q}$-endpoint. Let $\{p_f\} = V(p) \setminus \{p_{q'}, p_{q''}\}$. Then, w.l.o.g., $\{p_{q'}, p_f\} \in E(p)$. Therefore $p_{q'}$ is shiftable. **b)** $p_{q''}$ is a $\mathcal{Q}$-midpoint.

Claim. $p_{q''} \neq p_2$: Suppose the contrary. Then w.l.o.g., $p_{q'} = p_1$ and thus q'' is foldable on p by a (q_1'', p_3)-folding. This contradicts Lemma 3. The claim follows. W.l.o.g., we assume $p_{q''} = p_1$. Then it follows that $p_{q'} = p_2$, as otherwise a (q_1'', p_2)-folding would contradict Lemma 3 again. From $p_{q'} = p_2$ and $p_3 \notin V(\mathcal{Q})$ we can derive that also in this case $p_{q'}$ is shiftable.

We now examine for both cases the implications of the shiftability of $p_{q'}$. W.l.o.g., we suppose that $p_{q'} = q_1'$. Due to Lemma 2 there is a $p' \in \mathcal{P}$ with $\{q_3', q_2'\} \in E(p')$. From Corollary 2, it follows that there must be a $\bar{q} \in \mathcal{Q} \setminus \{q'\}$ with $|V(p') \cap V(\bar{q})| = 1$. Hence, $|V(p') \cap V(\mathcal{Q})| = 3$. Note that q' is the only path in $\mathcal{Q}$ with $|V(q') \cap V(p')| = 2$. Summarizing, we can say that for any $p \in \mathcal{P}$ with $|V(p) \cap V(\mathcal{Q})| = 2$ we find a distinct $p' \in \mathcal{P}$ (via q') such that $|V(p') \cap V(\mathcal{Q})| = 3$. So, there is a total injection γ from $\mathcal{P}_2(\mathcal{Q})$ to $\mathcal{P}_3(\mathcal{Q})$. From $|\mathcal{P}_2(\mathcal{Q}) \cup \mathcal{P}_3(\mathcal{Q})| = j$ and the existence of γ we derive $|\mathcal{P}_2(\mathcal{Q})| \leq 0.5j$. This implies $|V(\mathcal{P}) \cap V(\mathcal{Q})| = 2|\mathcal{P}_2(\mathcal{Q})| + 3|\mathcal{P}_3(\mathcal{Q})| \geq 2.5j$. □

3   The Algorithm

We are going to discuss three main aspects of Algorithm 1: (1) how matching techniques can be used in the WIN-WIN-approach, (2) why the algorithm is yielding a correct solution, and (3) how the run time is estimated.

3.1   Used Matching Techniques

We would like to point out the following two facts about P_2-packings. First, if a graph has a P_2-packing $\mathcal{P} = \{p^1, \ldots, p^k\}$, then it suffices to know the set of midpoints $\mathcal{M}_\mathcal{P} = \{p_2^1, \ldots, p_2^k\}$ to construct a P_2-packing of size k (which is possibly

---

Algorithm 1. An Algorithm for finding a P_2-packing $\mathcal{P}$ with $|\mathcal{P}| \geq k$ if possible.

1: **repeat**
2:    {Apply the crown-based kernelization algorithm exhibited in [15].}
3:    Greedily extend $\mathcal{P}$ to a maximal packing using **Rule 1** and **Rule 2**.
4:    Try to find a double or fat crown.
5: **until** $\mathcal{P}$ is not changed
6: $j \leftarrow |\mathcal{P}|$.
7: **if** $j \geq k$ **then**
8:    **return** YES
9: $\mathcal{P}' \leftarrow \emptyset$
10: **for** $\ell{=}0$ to $0.3251j$ **do**
11:    **for all** $S_i \subseteq V(\mathcal{P})$, $S_o \subseteq V \setminus V(\mathcal{P})$ with $|S_i| = (j+1) - \ell$ and $|S_o| = \ell$ **do**
12:       Try to construct a P_2-packing $\mathcal{P}'$ with $S_i \cup S_o$ as midpoints.
13: **for** $\bar{\ell} = 0$ to $0.1749j + 3$ **do**
14:    **for all** $B_i \subseteq V(\mathcal{P})$, $B_o \subseteq V \setminus V(\mathcal{P})$ with $|B_i| = 2(j+1) - \bar{\ell}$ and $|B_o| = \bar{\ell}$ **do**
15:       **for all** possible endpoint pairs $(e_1^1, e_2^1), \ldots, (e_1^{j+1}, e_2^{j+1})$ from $B_i \cup B_o$ **do**
16:          Try to construct a P_2-packing $\mathcal{P}'$ with $(e_1^1, e_2^1), \ldots, (e_1^{j+1}, e_2^{j+1})$ as endpoint pairs.
17: **if** $\mathcal{P}' \neq \emptyset$ **then**
18:    $\mathcal{P} \leftarrow \mathcal{P}'$; goto 1.
19: **else**
20:    **return** NO

$\mathcal{P}$) in polynomial time. This fact was discovered by E. Prieto and C. Sloper [13] and basically can be achieved by bipartite matching techniques. Secondly, it also suffices to know the set of endpoint pairs $E_{\mathcal{P}} = \{(p_1^1, p_3^1), \ldots, (p_1^k, p_3^k)\}$ to construct a P_2-packing of size k in polynomial time. This is due to Lemma 3.3 of Jia et al. [10] as any P_2-packing instance also can be viewed as a 3-SET PACKING instance. This is the basic ingredient for the WIN-WIN strategy used in Alg. 1 to finally tune the running time.

Details on the mentioned matching techniques can be found in the following two propositions.

Proposition 1. Let the vertex set $\mathcal{M} = \{m_1, \ldots, m_j\}$ contain all the midpoints of some P_2-packing $\mathcal{P}$ in a graph $G(V, E)$. Then we can construct a P_2-packing $\mathcal{P}'$ of size j in polynomial time.

Proof. Use the following algorithm:

- Find a maximum matching M in the auxiliary bipartite graph $G' = (V', E')$, where $V' = A \cup B$ is the bipartition with $A = \mathcal{M} \times \{1,2\}$ and $B = V \setminus \mathcal{M}$, $E' = \{((u,i), w) \mid 1 \leq i \leq 2, u \in A, w \in B, \{u,w\} \in E\}$.
- If all elements of A are matched in M, then we have found a packing $\mathcal{P}'$ of G as follows: $\mathcal{P}' = \{(x, y, z) \mid \{((y,1), x\}, \{(y,2), z\}\} \subseteq M\}$.

Note that $M_{\mathcal{P}} = \{\{\{(p_2, 1), p_3\}, \{(p_2, 2), p_1\}\} \mid p_1 p_2 p_3 \in \mathcal{P}\}$ matches A into B in G'. Thus, $\mathcal{P}'$ must exist and is of size j. $\square$

Proposition 2. *Let the tuple set $\mathcal{E} = \{(e_1^1, e_2^1), \ldots, (e_1^j, e_2^j)\}$ contain all endpoint pairs of some P_2-packing $\mathcal{P}$ in $G(V, E)$. Then we can construct a P_2-packing $\mathcal{P}'$ of size j in polynomial time.*

Proof. Use the following algorithm:

- Find a maximum matching M in the auxiliary bipartite graph $G' = (V', E')$, where $V' = A \cup B$ is the bipartition with $A = \mathcal{E}$ and $B = V \setminus \{v \in V \mid \exists (e_1^h, e_2^h) \in \mathcal{E}$ with $v = e_1^h$ or $v = e_2^h\}$, $E' = \{\{(e_1^h, e_2^h), u\} \mid (e_1^h, e_2^h) \in A, u \in B, \{e_1^h, u\} \in E$ and $\{e_2^h, u\} \in E\}$
- If all elements of A are matched in M, then we have found a packing $\mathcal{P}'$ of G as follows: $\mathcal{P}' = \{(e_1^h, e_2^h, u) \mid \{(e_1^h, e_2^h), u\} \in M\}$

Note that $M_{\mathcal{P}} = \{\{(p_1, p_3), p_2\} \mid p_1 p_2 p_3 \in \mathcal{P}\}$ matches A into B in G'. Thus, $\mathcal{P}'$ must exist and is of size j. $\square$

3.2   Correctness

The correctness of the kernelization part is shown in [15].

If a P_2-packing $\mathcal{P}'$ with $|\mathcal{P}'| = j + 1$ exists, we can partition the midpoints $M_{\mathcal{P}'}$ in a part which lies within $V(\mathcal{P})$ and one which lies outside. We call them $M_{\mathcal{P}'}^i := M_{\mathcal{P}'} \cap V(\mathcal{P})$ and $M_{\mathcal{P}'}^o := M_{\mathcal{P}'} \cap O$, respectively with $O := V(\mathcal{P}') \setminus V(\mathcal{P})$. Theorem 3 yields $|O| \leq 0.5j + 3$ and thus $|M_{\mathcal{P}'}^o| \leq 0.5j + 3$. Basically, we can find an integer ℓ with $0 \leq \ell \leq 0.5j + 3$ such that $|M_{\mathcal{P}'}^i| = (j + 1) - \ell$ and $|M_{\mathcal{P}'}^o| = \ell$. In step 10 we run through every such ℓ until we reach $0.3251j$. For any choice of ℓ, in step 11 we cycle through all possibilities of choosing sets $S_i \subseteq V(\mathcal{P})$ and $S_o \subseteq V \setminus V(\mathcal{P})$ such that $|S_i| = (j + 1) - \ell$ and $|S_o| = \ell$. Here S_i and S_o are candidates for $M_{\mathcal{P}'}^i$ and $M_{\mathcal{P}'}^o$, respectively. For any choice of S_i and S_o we try to construct a P_2-packing. If we succeed once we can return the desired larger P_2-packing. Otherwise we reach the point where $\ell = 0.3251j$. At this point we change our strategy. Instead of looking for the midpoints of $\mathcal{P}'$ we focus on the endpoints. We do so because this will improve the run time as we will see later. O is the disjoint union of $M_{\mathcal{P}'}^o$ and the endpoints of $\mathcal{P}'$ which do not lie in $V(\mathcal{P})$ which we call $E_{\mathcal{P}'}^o$. At this point we must have $|M_{\mathcal{P}'}^o| > 0.3251j$ and therefore $|E_{\mathcal{P}'}^o| < 0.1749j + 3$ as $O \leq 0.5j + 3$. Now there must be an integer $\bar{\ell}$ with $0 \leq \bar{\ell} \leq 0.1785j + 3$ such that $|E_{\mathcal{P}'}^o| = \bar{\ell}$ and the number of endpoints within $V(\mathcal{P})$ (called $E_{\mathcal{P}'}^i$) must be $2(j + 1) - \bar{\ell}$. In step 13 we iterate through $\bar{\ell}$. In the next step we cycle through all candidate sets for $E_{\mathcal{P}}^o$ and $E_{\mathcal{P}}^i$ which are called B_i and B_o in the algorithm.

In step 15 we consider all possibilities $(e_1^1, e_2^1), \ldots, (e_1^{j+1}, e_2^{j+1})$ of how to pair the vertices in $B_i \cup B_o$. A pair of endpoints (e_r^s, e_{r+1}^s) means that both vertices should appear in the same P_2 of $\mathcal{P}'$. Finally, we try to construct $\mathcal{P}'$ from $(e_1^1, e_2^1), \ldots, (e_1^{j+1}, e_2^{j+1})$ by computing a matching according to [10].

3.3   Running Time

The only exponential run time contribution comes from the **for**-loops in Alg. 1. For any ℓ we execute step 10 at most $\binom{3j}{(j+1)-\ell}\binom{4j}{\ell} \in \mathcal{O}\left(\binom{3j}{j-\ell}\binom{4j}{\ell}\right)$ times, since

$|V(\mathcal{P})| = 3j$ and $|V \setminus V(\mathcal{P})| \leq 4j$ due to Theorem 1. Likewise, $\mathcal{O}\left(\binom{3j}{2j-\ell}\binom{4j}{\ell}\right)$ upperbounds step 13.

Lemma 5. *For any integer z with $0 \leq z \leq 0.5j - 1$ the following holds:*

1. $\binom{3j}{j-z}\binom{4j}{z} < \binom{3j}{j-(z+1)}\binom{4j}{z+1}$; and 2. $\binom{3j}{2j-z}\binom{4j}{z} < \binom{3j}{2j-(z+1)}\binom{4j}{z+1}$.

Proof. 1. We have $\binom{3j}{j-(z+1)}\binom{4j}{z+1} - \binom{3j}{j-z}\binom{4j}{z} = \frac{(3j)!(4j)!((j-z)(4j-z)-(2j+z+1)(z+1))}{(j-z)!(2j+z+1)!(z+1)!(4j-z)!}$.

Now it is enough to show $(j - z)(4j - z) - (2j + z + 1)(z + 1) > 0$ which evaluates to $4j^2 - 7jz - 2j - 2z - 1 > 0$. For the given z this always is true.

2. We have $\binom{3j}{2j-(z+1)}\binom{4j}{z+1} - \binom{3j}{2j-z}\binom{4j}{z} = \frac{(3j)!(4j)!((2j-z)(4j-z)-(j+z+1)(z+1))}{(2j-z)!(j+z+1)!(z+1)!(4j-z)!}$.

Then $((2j - z)(4j - z) - (j + z + 1)(z + 1)) = 8j^2 - 7jz - j - 2z - 1$ which for the given z is greater than zero. $\qquad\square$

With Lemma 5 step 10 is upperbounded by $\mathcal{O}\left(\binom{3j}{(0.6749)j}\binom{4j}{0.3251j}\right)$ and step 13 by $\mathcal{O}\left(\binom{3j}{1.8251j}\binom{4j}{0.1749j}\right)$. Both are dominated by $\mathcal{O}(15.285^j)$. Notice the asymptotic speed-up we achieve by changing the strategy (WIN-WIN).

Theorem 4. P_2-PACKING *can be solved in time* $\mathcal{O}^*(2.482^{3k})$.

4   Future Work

It would be nice to derive smaller kernels than $7k$ or $1.5k$ for P_2-PACKING or TOTAL EDGE COVER, resp., in view of the mentioned lower bound results [2].

A closely related problem is MAXIMUM P_3-PACKING for which R. Hassin and S. Rubinstein [7] found a $\frac{3}{4}$-approximation. We try to apply extremal combinatorial methods to save colors for P_d-packings for $d \geq 3$. First results seem to be promising. So, a detailed combinatorial (extremal structure) study of (say graph) structure under the perspective of a specific combinatorial problem seems to pay off not only for kernelization (see [5]), but also for iterative approaches.

Developing exact algorithms for MAXIMUM P_2-PACKING would be interesting. Dynamic programming yields an $\mathcal{O}^*(2^n)$-algorithm. By Theorem 2, TOTAL EDGE COVER can be solved in time $\mathcal{O}^*(2^{1.5k}) \subseteq \mathcal{O}^*(2.829^k)$. Improving on exact algorithmics would also improve on the parameterized algorithm for TOTAL EDGE COVER. Alternatively, find a search-tree algorithm for TOTAL EDGE COVER.

We finally mention that H. Fernau, J. Kneis and P. Rossmanith could show that also the general TEST COVER problem is in $\mathcal{FPT}$, a bit surprising in view of the fact that the quite similar FEATURE SET problem is W[2]-complete [4]. However, the general algorithm is far from practical and needs to be improved.

References

1. De Bontridder, K.M.J., Halldórsson, B.V., Halldórsson, M.M., Lenstra, J.K., Ravi, R., Stougie, L.: Approximation algorithms for the test cover problem. Math. Progr. Ser. B 98, 477–491 (2003)

2. Chen, J., Fernau, H., Kanj, Y.A., Xia, G.: Parametric duality and kernelization: lower bounds and upper bounds on kernel size. SIAM Journal on Computing 37, 1077–1108 (2007)
3. Chen, J., Lu, S., Sze, S.-H., Zhang, F.: Improved algorithms for path, matching, and packing problems. In: Bansal, N., Pruhs, K., Stein, C. (eds.) Symposium on Discrete Algorithms SODA, pp. 298–307. SIAM, Philadelphia (2007)
4. Cotta, C., Moscato, P.: On the parameterized complextiy of problems related with feature identification for gene expression data mining techniques. Bioinformatics 1, 1–8 (2002)
5. Estivill-Castro, V., Fellows, M.R., Langston, M.A., Rosamond, F.A.: FPT is P-time extremal structure I. In: Broersma, H., Johnson, M., Szeider, S. (eds.) Algorithms and Complexity in Durham ACiD 2005. Texts in Algorithmics, vol. 4, pp. 1–41. King's College Publications (2005)
6. Fernau, H., Manlove, D.F.: Vertex and edge covers with clustering properties: Complexity and algorithms. In: Algorithms and Complexity in Durham ACiD 2006, pp. 69–84. King's College, London (2006)
7. Hassin, R., Rubinstein, S.: An approximation algorithm for maximum of 3-edge paths. Information Processing Letters 63, 63–67 (1997)
8. Hassin, R., Rubinstein, S.: An approximation algorithm for maximum triangle packing. Discrete Applied Mathematics 154, 971–979; 2620 [Erratum] (2006)
9. Hell, P., Kirkpatrick, D.G.: Star factors and star packings. Technical Report 82-6, Computing Science, Simon Fraser University, Burnaby, B.C. V5A1S6, Canada (1982)
10. Jia, W., Zhang, C., Chen, J.: An efficient parameterized algorithm for m-set packing. Journal of Algorithms 50, 106–117 (2004)
11. Kirkpatrick, D.G., Hell, P.: On the completeness of a generalized matching problem. In: ACM Symposium on Theory of Computing STOC, pp. 240–245 (1978)
12. Liu, Y., Lu, S., Chen, J., Sze, S.-H.: Greedy localization and color-coding: improved matching and packing algorithms. In: Bodlaender, H.L., Langston, M.A. (eds.) IWPEC 2006. LNCS, vol. 4169, pp. 84–95. Springer, Heidelberg (2006)
13. Prieto, E., Sloper, C.: Looking at the stars. In: Downey, R., Fellows, M., Dehne, F. (eds.) IWPEC 2004. LNCS, vol. 3162, pp. 138–148. Springer, Heidelberg (2004)
14. Wang, J., Feng, Q.: An $O^*(3.52^{3k})$ parameterized algorithm for 3-set packing. In: Agrawal, M., et al. (eds.) Theory and Applications of Models of Computation TAMC. LNCS, vol. 4978, pp. 82–93. Springer, Heidelberg (2008)
15. Wang, J., Ning, D., Feng, Q., Chen, J.: An improved parameterized algorithm for a generalized matching problem. In: Agrawal, M., Du., D.-Z., Duan, Z., Li, A. (eds.) TAMC 2008. LNCS, vol. 4978, pp. 212–222. Springer, Heidelberg (2008)

New Algorithms for k-Center and Extensions

René Brandenberg and Lucia Roth*

Zentrum Mathematik, Technische Universität München,
Boltzmannstr. 3, D–85747 Garching b. München, Germany
{brandenb,roth}@ma.tum.de

Abstract. The problem of interest is covering a given point set with homothetic copies of several convex containers $C_1, \ldots, C_k$, while the objective is to minimize the maximum over the dilatation factors. Such k-containment problems arise in various applications, e.g. in facility location, shape fitting, data classification or clustering. So far most attention has been paid to the special case of the Euclidean k-center problem, where all containers C_i are Euclidean unit balls. New developments based on so-called core-sets enable not only better theoretical bounds in the running time of approximation algorithms but also improvements in practically solvable input sizes. Here, we present some new geometric inequalities and a Mixed-Integer-Convex-Programming formulation. Both are used in a very effective branch-and-bound routine which not only improves on best known running times in the Euclidean case but also handles general and even different containers among the C_i.

Keywords: approximation algorithms, branch-and-bound, computational geometry, geometric inequalities, containment, core-sets, k-center, diameter partition, SOCP, 2-SAT.

1 Introduction

The issue of the following is the k-containment problem, that is covering a given point set with homothetic copies of several convex containers $C_1, \ldots, C_k$, while the objective is to minimize the maximum over the dilatation factors used in the covering. k-Containment problems arise in various applications, for instance in facility location, shape fitting, data classification or clustering (see [2], [18], [31], and [35] for several examples).

The k-center problem (the k-containment problem with identical containers) is known to be NP-complete in general dimensions even when $k \geq 2$ and all containers are Euclidean unit balls (the *Euclidean k-center problem*) or $k \geq 3$ and all containers are l_∞ unit cubes [29]. Many approximation algorithms have been suggested for solving k-center problems (see [1] and the surveys [2], [31]). In many papers, the aim is improving complexity bounds and the presented algorithms are mostly of theoretical value. For practical purposes many purely

---

* Supported by the "Deutsche Forschungsgemeinschaft" through the graduate program "Angewandte Algorithmische Mathematik", Technische Universität München.

B. Yang, D.-Z. Du, and C.A. Wang (Eds.): COCOA 2008, LNCS 5165, pp. 64–78, 2008.
© Springer-Verlag Berlin Heidelberg 2008

heuristic approaches exist (see e.g. [3], [19], [23], or [35]). Although they behave well for many inputs, they fail to provide provable guarantees.

So far most attention has been paid to the Euclidean k-center problem. Until recently it was believed that even in this case there is little hope to solve bigger instances, i.e. $n \geq 3$ or $k \geq 3$ (see e.g. [31]). Therefore, the planar Euclidean 2-center problem has been studied separately, for instance in [13], [14], [21], [24], and [32]. Recent progress is due to so-called core-sets [11], which gain a polynomial time approximation scheme (PTAS) for Euclidean k-center. However, the proposed full enumeration of all partitions of possible core-sets quickly causes non-computability in practice, even for moderate approximation errors. A first simple branch-and-bound (B&B) algorithm was suggested in [27].

Nevertheless, non-Euclidean containers are of practical interest, too. For instance in data analysis, the goal is in finding "similar" data points. Usually, there is no inherent reason why the 2-norm should be the better choice to express relations between data points than e.g. the 1- or ∞-norm. This is noteworthy as the polytopal norms often simplify calculations, e.g. in pattern recognition [28]. Special cases of rectilinear k-center problems have been addressed in [6] and [22]. In facility location (see e.g. Sect. 6) and shape fitting (see e.g. [9]) even non-symmetric and/or different container shapes may occur. Our algorithms allow both, general shapes and different C_i's within one instance (see Fig. 2 and 3 for examples).

Sections 2 and 3 address the basic definitions and a fundamental B&B procedure with good practical performance. In Sect. 4, a Mixed Integer Convex Programming formulation is given and its relaxation is used for further performance improvements. Especially if $k = 2$, further progress is achieved by diameter partitioning algorithms. These are described in Sect. 5, also including a couple of new geometric inequalities guaranteeing good bounds and a 2-SAT formulation used for the 2-containment problem with different containers. Both Sects. 2 and 3 are enhanced by some examples and experiments.[1] We stress that the new methods apply to a wider class of problems, therefore state them in full generality and provide an example in Sect. 6 indicating the use of the extension.

2   Problem Formulation

A container C is a full dimensional, convex, and compact subset of $\mathbb{R}^n$ with $0 \in \text{int}(C)$. For any container C and any $x \in \mathbb{R}^n$ let $\|x\|_C := \min_{\rho \geq 0}\{x \in \rho C\}$. Furthermore, for any point set $P \subset \mathbb{R}^n$ let $R(P, C) := \min_{c \in \mathbb{R}^n} \max_{p \in P} \|p - c\|_C$ and $d(P, C) := \max_{p,q \in P} R(\{p, q\}, C)$.

For any container $C \subset \mathbb{R}^n$, let s_C denote the *Minkowski symmetry* of C, that is the maximal dilatation factor ρ such that some translate of $-\rho C$ is contained in C, or for short $s_C = 1/R(-C, C)$. Obviously, $s_C \leq 1$, and we say that C is symmetric if and only if $s_C = 1$. In the latter case C can be translated such

---

[1] The experiments are restricted to exemplary tests with balls and cubes as containers, in order to allow valuation of the running times. Note that our results may be more important for other container shapes, where no specialized methods such as the core-set results apply.

that $C = -C$, i.e. C is 0-symmetric. Furthermore, $s_C \geq 1/n$ follows from John's theorem [25] (and can easily be shown directly).[2]

If C is 0-symmetric, $\| \cdot \|_C : \mathbb{R}^n \to \mathbb{R}$ denotes the Minkowski norm with unit ball C. In this case, $R(P,C)$ and $d(P,C)$ denote the outer radius and half-diameter of P with respect to $\| \cdot \|_C$. Furthermore, if P *and* C are symmetric, $R(P,C) = d(P,C)$ [16]. However, be aware that $\|x\|_C \neq \| - x\|_C$ for some $x \in \mathbb{R}^n$ if $C \neq -C$. Now, the problem of interest can be stated. Let $k \in \mathbb{N}$, and for $1 \leq i \leq k$, let C_i^n be families of n-dimensional containers and $\mathcal{P}^n$ the family of finite point sets in $\mathbb{R}^n$.

MINIMAL k-CONTAINMENT PROBLEM UNDER HOMOTHETICS(k-**MCP**$_{Hom}^{\mathcal{P}}$)

Input: $n \in \mathbb{N}$, $m \in \mathbb{N}$, $P = \{p_1, \ldots, p_m\} \subset \mathcal{P}^n$, $C_1 \in \mathcal{C}_1^n, \ldots, C_k \in \mathcal{C}_k^n$.

Task: $\min \rho$, s. th. $P = \{p_1, \ldots, p_m\} \subset \bigcup_{1 \leq i \leq k}(c_i + \rho C_i)$, $c_1, \ldots, c_k \in \mathbb{R}^n$.

The optimal value ρ is denoted by $R(P, C_1, \ldots, C_k)$. If $k = 1$, we get the minimal 1-containment problem under homothetics, which indeed computes the outer radius $R(P,C)$ of P with respect to the (non-symmetric) norm $\|\cdot\|_C$. When solving k-containment problems for general containers C, many and therefore fast computations of $R(P,C)$ and especially $R(\{p,q\},C)$ with $\{p,q\} \subset P$ are needed. An overview on good solution or approximation techniques for different representations of the container C is given in [10]. k-MCP$_{Hom}^{\mathcal{P}}$ becomes the well known k-center problem when $C = C_1 = \ldots = C_k$:

k-CENTER PROBLEM

Input: $n \in \mathbb{N}$, $m \in \mathbb{N}$, $P = \{p_1, \ldots, p_m\} \in \mathcal{P}^n$, $C \in \mathcal{C}^n$.

Task: $\min \rho$, s. th. $\forall j \in \{1, \ldots, m\}\ \exists i \in \{1, \ldots, k\} : \|p_j - c_i\|_C \leq \rho$, $c_1, \ldots, c_k \in \mathbb{R}^n$.

In this case the optimal radius ρ is denoted by $R^k(P,C)$.

3   A Core-Set Based Branch-and-Bound Scheme

In this section, we describe a basic core-set based B&B algorithm for k-MCP$_{Hom}^{\mathcal{P}}$.

3.1   Core-Sets

Let $S \subset P$ such that all points of S are assigned consistently with an optimal solution of the full k-MCP$_{Hom}^{\mathcal{P}}$ instance. For each of the k parts $S_i \subset S$, let c_i denote a center in an optimal solution of the corresponding 1-containment problem. Let $\rho = \max_i R(S_i, C_i)$. If for all $p \in P$ an index i exists such that $p \in c_i + (1 + \varepsilon)\rho C_i$, we have

$$\rho \leq R(P, C_1, \ldots, C_k) \leq (1 + \varepsilon)\rho$$

---

[2] Note that s_C of vertex- or facet-presented polytopes C can be computed via linear programming [17].

implying an ε-approximate solution of k-MCP$^{\mathcal{P}}_{Hom}$. Any such S is called an ε-core-set of P (with respect to $C_1, \ldots, C_k$).

In [11] it was shown that if all C_i are Euclidean, the sizes of the core-sets depend only on ε and neither on n nor m. Helly's theorem [20] implies the existence of core-sets whose size is independent of the number of points in P for all container shapes. However, dimension independence does not hold true for general (non-symmetric) containers [10].[3] Furthermore, one should note that in l_∞-spaces every diametrical pair of points is a 0-core-set (see Sect. 5.1), but that the algorithm as proposed in [11] may construct a core-set of size depending on n [10].

3.2   Branch-and-Bound Scheme

At each node in the B&B tree, we regard a core-set $S \subset P$ already partitioned into clusters S_i which have to be covered by homothetic copies of the containers C_i. For the branching, a point $p^* \in P \setminus S$ not (yet) covered is chosen and added to each of the sets S_i consecutively. We choose the point p maximizing $\min_i \|p - c_i\|_{C_i}$[4], or, in case the maximum is too expensive to compute, any point p with $\|p - c_i\|_{C_i}$ bigger than the current $(1 + \varepsilon) \max_i \rho_i$. The remaining points play no further role in this step of the basic B&B procedure. (This will be improved in Sect. 4.)

For the branching, the clusters are sorted according to the distances $\|p - c_i\|_{C_i}$ and then p^* is assigned to the nearest cluster first. With this greedy-like strategy, good upper bounds are computed at an early stage of the algorithm, resulting in fast truncation of many branches and shorter overall running time. Solving the 1-center instances for each C_i and its assigned core-set points generates first lower bounds on the optimal value for the subtree below the current node.[5]

The algorithm returns an ε-core-set $S \subset P$ consisting of the points chosen at the nodes of an optimal branch, partitioned into k subsets $S_1, \ldots, S_k$, corresponding to the assignment of the points to the containers $C_1, \ldots, C_k$.

Algorithm 1[6]

> *initialize: set $S_i = \emptyset$, $\rho_i = 0$, c_i arbitrarily for all i,*
> *and $\bar{\rho}$ to an upper bound for $R(P, C_1, \ldots, C_k)$*
> *k-containment(S_i, ρ_i, c_i):*
> *update the global upper bound $\bar{\rho}$*
> *compute $\delta = \max_{p \in P \setminus \bigcup S_i} \min_i(\|p - c_i\|_{C_i})$*
> *let p^* the point where the maximum is attained*

---

[3] In case of general symmetric containers the existence of dimension independent ε-core-sets is open.

[4] In [27] p^* maximizes $\min_i(\|p - c_i\|_{C_i} - \rho_i)$, but our choice yields better results.

[5] It is recommendable to compute the $\|.\|_{C_i}$-distances between the new point and the points already assigned to S_i first to prevent unnecessary radius computations.

[6] The algorithm is written down recursively for better readability. However, to gain good running times, recursion in implementations should be avoided.

> *if $(1 + \varepsilon) \max_i \rho_i \geq \delta$: return*
> *else: sort cluster indices descending according to $\|p^* - c_i\|_{C_i}$*
>   *for $j = i_1, \ldots, i_k$:*
>   *recompute c_j and ρ_j for $S_j = S_j \cup p^*$*
>   *if $\max_i \rho_i \leq \bar{\rho}(1 + \varepsilon)$:*
>   *k-containment(S_i, ρ_i, c_i)*
> *return the best S_i, ρ_i, and c_i found*

Testing the $(1 + \varepsilon)$-containment condition at each node of the tree yields an approximation algorithm with a running time of $O(k^h nm)$, where h is the size of a maximal core-set constructed during the algorithm. It follows from [11] that for Euclidean k-center this B&B algorithm is a PTAS as $h = O(k/\varepsilon^2)$ in this case.

If an upper bound for the optimal radius is known, $\bar{\rho}$ can be initialized accordingly. Since the first k steps of Algorithm 1 (i.e. when each cluster contains exactly one point), match the first k steps of the greedy algorithm in [15] (assuming that the distance to an empty cluster is set to zero), in the *symmetric case* an approximation factor of at least 2 can be guaranteed at that stage.

According to [27], the implementation reported there is the first to practically solve huge k-center instances. The experiments in that paper show that the B&B algorithm used performs much better on practical data sets than the predicted worst case running times suggest. It is concluded that in dimensions 2 and 3, Euclidean k-center is practical for $\varepsilon \geq 0.01$ and $k \leq 4$, whereas computations in 3-space are significantly more expensive than in 2-space. The latter is caused in the fact, that though the upper bounds on core-set sizes are dimension independent, in practical computations the core-set sizes in lower dimensions are far from the upper bounds and grow noticeably with the dimension (and so do the running times of the B&B procedure). However, it is also reported that "some of the data sets [...] solved in 3D [...], ran for almost a week on an Intel Itanium system". Our implementation allows solving Euclidean k-center instances with bigger input sizes even in higher dimensions and for greater k values within some hours (at most) on an Intel Core 2 system[7]. Our realization of Algorithm 1 already substancially improves the running times as reported in [27] and further improvements are obtained by the methods presented in the following. In addition to that, our methods apply to *general k-containment problems*.

4   Convex Relaxation

In this section, a version of k-MCP$^{\mathcal{P}}_{Hom}$ with additional information is considered. It is assumed that the correct clusters are known for *some* of the points in P. This is a natural hypothesis in the context of a B&B scheme and enhances the chances to compute good upper and lower bounds for the optimal solution.

Especially good lower bounds are crucial for the performance of a B&B procedure. Whereas Algorithm 1 computes local lower bounds by determining the

---

[7] Both implementations use Matlab and comparable SOCP solvers.

radii of the current clusters, we now propose lower bounds taking both, assigned and unassigned points, into account. The new bounds are at least as good as the old ones, but usually much better.

4.1 A Mixed-Integer-Convex Program

Recall that the core-set $S = S_1 \cup \ldots \cup S_k$ denotes the assigned subset of P, i.e. $S_i \subset c_i + \rho C_i$ for some $c_i \in \mathbb{R}^n$ and $\rho > 0$, $i = 1, \ldots, k$. Now, let $S_0 \subset P \setminus S$ denote some of the unassigned points. Then the k-MCP$_{Hom}^{\mathcal{P}}$ with assigned points in $S_1, \ldots, S_k \neq \emptyset$ can be formulated as a mixed integer convex program with variables ρ, c_i and $\lambda_{ij} \in \{0, 1\}$. For this purpose for each $p_j \in S_0$ and each possible cluster S_i, a reference point $q_{ij} \in \mathrm{conv}(S_i)$ is fixed (see Sect. 4.2 for strategies for choosing these points).

$$
\begin{aligned}
\min \; & \rho \\
\|p_j - c_i\|_{C_i} &\le \rho \qquad && \forall p_j \in S_i, \; i = 1, \ldots, k \\
\|\lambda_{ij} p_j - c_i + (1 - \lambda_{ij}) q_{ij}\|_{C_i} &\le \rho \qquad && \forall p_j \in S_0, \; \forall i = 1, \ldots, k \\
\sum_{i=1}^{k} \lambda_{ij} &= 1 \qquad && \forall p_j \in S_0 \\
\lambda_{ij} &\in \{0, 1\} && \forall p_j \in S_0, \; \forall i = 1, \ldots, k
\end{aligned}
\tag{1}
$$

So, whenever $\lambda_{ij} = 0$ only the reference point q_{ij} has to be covered, a redundant condition. In contrast, p_j actually has to be contained in the homothetic copy of C_i if $\lambda_{ij} = 1$.

4.2 Relaxation

Relaxing the $\{0, 1\}$-condition on the multipliers λ_{ij} yields a convex program, providing a lower bound for $R(P, C_1, \ldots, C_k)$. A possible interpretation of the relaxation is including not the point p_j itself but a point on the line section between p_j and q_{ij} for all i, whereas the constraint $\sum_{i=1}^{k} \lambda_{ij} = 1$ enforces that not all of these points can be close to the reference points (see Fig. 1).

Picking q_{ij} such that the distance between q_{ij} and p_j is small gives the best bounds. However, the projection of p_j onto the convex hull of S_i causes elongation of overall computing time. Balancing between fast computations and a good choice of q_{ij}, the most successful strategy seems choosing q_{ij} as the point in S_i closest to p_j.

For polytopal C_i, the relaxation of (1) is a Linear Program; for Euclidean containers, we get a Second-Order-Cone Program (SOCP). Many other cases can be cast as SOCPs, too, for instance when the containers are intersections or Minkowski sums of Euclidean balls and polytopes (compare [10]).

Obviously, the more points from $P \setminus S$ belong to S_0, the better the lower bound on $R(P, C_1, \ldots, C_k)$ will be. However, as each $p \in S_0$ results in at least $k - 1$ additional variables and constraints, the relaxation of (1) is practical only when S_0 is not too big. Experiments show that even very small sets S_0 usually provide enough potential to reduce the number of nodes in the B&B tree significantly

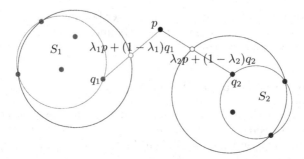

Fig. 1. The geometric meaning of the relaxed program. Optimal cluster radii with (black) and without (red resp. blue) considering the unassigned point p.

(compare Table 1, where 5 points have been chosen). There are different possible strategies to select points for S_0, e.g. randomly, maximizing the minimal distance to a current cluster, maximizing the distance to the latest core-set point, or maximizing the minimal distance to the unassigned points already chosen (which is what we do in Table 1). The solution of the convex program provides not only lower bounds. Upper bounds can easily be obtained by assigning the points $p_j \in P \setminus S$ to the clusters, e.g by $\min_i \|p_j - c_i\|_{C_i}$ or $\max_i \lambda_{ij}$ if $p \in S_0$.

The test results in Table 1 show that the MISOCP-relaxation significantly reduces the size of the B&B scheme for Euclidean k-center. Since solving the convex program at each node of the B&B tree is expensive, the improvement in the running time is still considerable but not as big as in the number of nodes. Further speedup should be possible by advanced strategies for the MISOCP-relaxation. In particular, we expect that improvements can be achieved through more elaborate techniques for determining the nodes at which to solve the convex program, the accuracy to which it should be solved, and the points in $|S_0|$. Moreover, practical solutions may be accelerated significantly by replacing the pure B&B algorithm by some kind of branch-and-cut routine.

5   Diameter Partitioning

Another possibility to improve the performance of Algorithm 1 is to consider $R(\{p,q\}, C_i)$ for all pairs of points $\{p,q\} \subset P$, and all $i = 1, \ldots, k$. The distances between point pairs provide information about optimal partitionings which can be used to compute bounds for $R(P, C_1, \ldots, C_k)$. The approach is useful especially when $k = 2$ and $R(\{p,q\}, C_i)$ can easily be computed. Surely, computing all pairwise distances is quadratic in the number of input points, so the approach is practical mainly for moderate point sets P.

5.1   Identical Containers

The information about the pairwise distances is captured in the ρ-distance graph:

Definition 1. *For every $\rho > 0$ we call the graph $G(\rho) = (P, E)$ with edges for every pair $\{p, q\}$ with $R(\{p, q\}, C) > \rho$ the ρ-distance graph of (P, C).*

Table 1. The B&B algorithm with and without SOCP bounds for Euclidean k-center and an approximation error of 0.01. The 3D geometric model data sets are comparable to the ones used in [27]. The 5D "rand. box" data sets refer to equally distributed points within boxes with randomly scaled axes. (We assume that this is more appropriate for k-center problems than, e.g., equally distributed points within the unit cube.) Sizes of the B&B tree and running times (in seconds) are listed – in case of the random data sets, the mean over samples of 20. We use a 2.0 GHz Intel Core 2 system running Matlab R2006B and SeDuMi [30], [33]. The code for Euclidean distance computations is provided by [12].

data set	m	n	k	pure B&B			B&B with relaxed MISOCP		
				nodes	leaves	time	nodes	leaves	time
cat	352	3	4	10353	2138	505.6	2380	144	207.7
shark	1744	3	4	649	126	26.1	225	27	13.6
seashell	18033	3	4	12718	2365	925.6	3266	479	371.0
dragon	437645	3	3	341	96	154.9	161	43	89.2
rand. box	1000	5	3	889.3	57.8	44.6	623.9	35.7	45.6
rand. box	1000	5	4	20919.9	3249.8	1272.6	6544.0	238.4	611.2
rand. box	10000	5	3	2595.1	167.6	166.6	1577.7	84.3	139.0
rand. box	10000	5	4	32611.9	3021.6	2273.7	13768.3	808.1	1459.4

The next algorithm computes the maximal ρ such that $G(\rho)$ is k-colorable. Finding a k-coloring of $G(\rho)$ corresponds to partitioning the point set P into k subsets, where no pair of points with $R(\{p, q\}, C) > \rho$ lies within one set.

Algorithm 2

for all l pairs $\{p, q\}$ of points in P:
  compute $\rho_j = R(\{p, q\}, C)$, $1 \leq j \leq l$
label such that $\rho_1 \geq \ldots \geq \rho_l$
for $j = 1, \ldots, l$:
  if $G(\rho_j)$ is not k-colorable
    break
  set $\rho = \rho_j$
return ρ

Deciding whether a graph is k-colorable is itself a hard problem if $k \geq 3$ and Algorithm 2 may not be polynomial. Still, good bounds may be obtained from heuristic coloring algorithms.

If $k = 2$, Algorithm 2 can be implemented by maintaining a 2-coloring of $G(\rho)$ while successively inserting new edges. One should note that since $G(\rho)$ may be not connected, more than two labels (or colors) may be necessary. When an edge is inserted which is not connected to the subgraph already built, a new pair of labels is created. When an edge joins two previously disconnected components, the relevant labels are merged.

Depending on the shape of the container, different approximation qualities for the underlying k-center problem can be guaranteed.

Parallelotopes. If (and only if) C is a parallelotope (e.g. a unit cube, if $\|\cdot\|_C = \|\cdot\|_\infty$) the Helly dimension of C is 1; that is, $R(P, C) = d(P, C)$ for all P [8, 14.3]. This implies that P can be packed into k translates of ρC if and only if $G(\rho)$ is k-colorable [4], [29]. Hence, Algorithm 2 solves the k-center problem for parallelotopes[8] exactly.

Note that solving the 2-center problem in l_∞ via diameter partitioning is not optimal. A faster algorithm is proposed in [5]. It computes a minimal axis-parallel enclosing box for P and determines the position of the two cubes in this box by maximizing consecutively in the directions of the n coordinate axes. However, Algorithm 2 has the advantage of being adaptable to *general* 2-containment problems, whereas the algorithm in [5] is limited to two *identical* parallelotopal containers.

Euclidean Containers. We get the following for Euclidean containers:

Lemma 1. *Algorithm 2 computes a $\sqrt{\frac{2n}{n+1}}$-approximation of $R^k(P, C)$ for any point set P and any ellipsoid C.*

Proof. Surely, $d(P_i, C) \leq R^k(P, C)$ for all P_i when $P_1, \ldots, P_k$ is a partition of P such that every two points joint by an edge in the final distance-graph of Algorithm 2 are in different P_i. If $P_1^*, \ldots, P_k^*$ is an optimal partition, $\max_i R(P_i, C) \geq \max_i R(P_i^*, C) = R^k(P, C)$. Hence, by Jung's inequality [26],

$$\max_i d(P_i, C) \leq R^k(P, C) \leq \max_i R(P_i, C) \leq \max_i \sqrt{\frac{2n}{n+1}} d(P_i, C).$$

In computations, an incomplete partition can be extended in a greedy manner upon all points in P. Besides the lower bound output ρ of Algorithm 2, an upper bound $\bar{\rho} = \max_i R(P_i, C)$ is obtained. Surely, this upper bound is often much smaller than $\sqrt{2n/(n+1)}\rho$ in practice (compare Table 2).

General, Identical Containers. For general containers C, the bounds are weaker, but only slightly when C is (almost) symmetric.

Lemma 2. *Algorithm 2 computes an $\frac{n}{n+1}(1 + \frac{1}{s_C})$-approximation of the optimal radius $R^k(P, C)$ for any point set $P \subset \mathbb{R}^n$ and any container $C \subset \mathbb{R}^n$.*

Proof. Following the proof of Lemma 1 it suffices to show that $R(P, C) \leq \frac{n}{n+1}(1 + \frac{1}{s_C})d(P, C)$ for any point set P. Suppose $d(P, C) = 1$, i.e. every two points in P can be covered by a translate of C. It easily follows that every two points of $P - P$ can be covered by $C - C$, and since both $P - P$ and $C - C$ are symmetric $R(P - P, C - C) = d(P - P, C - C) = 1$ [16]. Since $(1 + s_P)P$ can be covered by a translate of $P - P$ and $C - C$ by a translate of $(1 + \frac{1}{s_C})C$, we conclude with $s_P \geq \frac{1}{n}$ that P is contained in a translate of $\frac{n}{n+1}(1 + \frac{1}{s_C})C$.

---

[8] When the parallelotope is given in $\mathcal{H}$-representation $C = \bigcap_i \{x : a_i^T x \leq 1\}$, and especially for l_∞-containment, $R(\{p, q\}, C) = \max_i a_i^T(p - q)$ can easily be computed.

Remark 1. a) If C is symmetric, a well known inequality about the ratio between the outer radius and the diameter of convex sets (or point sets) in arbitrary Minkowski spaces [7] can be obtained as a corollary of Lemma 2:

$$\frac{R(P,C)}{d(P,C)} \leq \frac{2n}{n+1}.$$

b) If $\mathcal{C}^n$ is a small subset of the set of convex bodies in $\mathbb{R}^n$, like the parallelotopes or ellipsoids (or – maybe even non-symmetric – sets close to these shapes) in Sects. 5.1 and 5.1, the approximation error may be much better than predicted by Lemma 2.

c) If a better guarantee on lower bounds on the Minkowski symmetry of the input point set P can be given, the bounds in Lemma 2 can be improved.

5.2   Different Containers

Regarding the general k-$\mathrm{MCP}^{\mathcal{P}}_{Hom}$, two points p and q which are far apart in the (non-symmetric) norm induced by one container may be close in the norm induced by another. Definition 1 has to be adapted.

Definition 2. *Let $G = (V, E_1, \ldots, E_k)$ be an (edge-colored) multigraph with vertex set V and edge sets $E_1, \ldots, E_k$. A generalized k-coloring of G is a partition $V_1, \ldots, V_k$ of the vertices V such that for any $\{v, w\} \in E_i$ it follows $\{v, w\} \not\subset V_i$, $i = 1, \ldots, k$.*

Again, we can define the ρ-distance graph:

Definition 3. *For every $\rho > 0$ the ρ-distance graph of $(P, C_1, \ldots, C_k)$ is the edge-colored multigraph $G(\rho) = (P, E_1, \ldots, E_k)$ with edges in E_i for every pair $\{p, q\}$ with $R(\{p, q\}, C_i) > \rho$.*

Now a solution of the generalized k-coloring problem for the ρ-distance graph $G(\rho)$ implies again that ρ is a lower bound for $R(P, C_1, \ldots, C_k)$.

Algorithm 3
> *for all l combinations of pairs $\{p, q\}$ of points in P and $i \in \{1, \ldots, k\}$:*
> > *compute $\rho_j = R(\{p, q\}, C_i)$, $1 \leq j \leq l$*
>
> *label such that $\rho_1 \geq \ldots \geq \rho_l$*
> *for $j = 1, \ldots, l$:*
> > *if $G(\rho_j)$ has no valid generalized k-coloring*
> > > *break*
> >
> > *set $\rho = \rho_j$*
>
> *return ρ*

Respecting the edge colors seems to make generalized k-coloring more difficult than usual coloring. Yet, if $k = 2$, the problem can still be solved efficiently:

Lemma 3. *The generalized 2-coloring problem can be reduced to 2-SAT.*

Proof. By assigning boolean variables z_i, where $z_i \Leftrightarrow (v_i \in V_i)$, the generalized 2-coloring instance (V, E_1, E_2) is equivalent to the following instance of 2-SAT:

$$\bigwedge_{\substack{(p_i, p_j) \\ E_1\text{-edges}}} (\neg z_i \vee \neg z_j) \quad \wedge \quad \bigwedge_{\substack{(p_i, p_j) \\ E_2\text{-edges}}} (z_i \vee z_j).$$

A valid assignment of a 2-SAT instance (or evidence that no valid assignment exists) can be found in linear time (in the size of $G(\rho)$), see, e.g., [34].

Any valid assignment of the variables z_i in the corresponding 2-SAT formula yields a partition into two sets P_1 and P_2 with the following property: $R(\{p, q\}, C_i) \leq \rho$, $i = 1, 2$ for any pair of points $p, q \in S_i$.

Lemma 4. *Algorithm 3 computes*

a) *an $\frac{n}{n+1}(\max_{1 \leq i \leq k}(\frac{1}{s_{C_i}}) + 1)$-approximation for the general k-MCP$^{\mathcal{P}}_{Hom}$.*

b) *a $\frac{2n}{n+1}$-approximation for k-MCP$^{\mathcal{P}}_{Hom}$ if all containers are 0-symmetric.*

c) *a $\sqrt{\frac{2n}{n+1}}$-approximation for k-MCP$^{\mathcal{P}}_{Hom}$ if all containers are ellipsoids or parallelotopes.*

d) *an exact solution of k-MCP$^{\mathcal{P}}_{Hom}$ if all containers are parallelotopes (compare Fig. 2).*

Proof. This follows directly from Lemma 3 and Sect. 5.1.

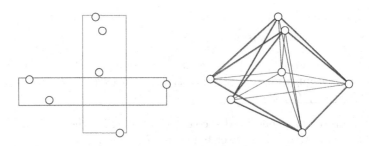

Fig. 2. An example of an optimal containment with two boxes as containers and the corresponding edges in the final ρ-distance graph. For parallelotopal containers, Algorithm 3 computes the exact solution.

5.3   Partitioning Procedures

Algorithms 2 and 3 approximate the 2-containment problem within the bounds given in Lemmas 1, 2, and 4.[9] For better approximations, we rely on the B&B procedure. If $|P|$ is not too big, the super-quadratic[10] running time of the diameter partitioning is not too expensive and it is even possible to combine Algorithms 1 and 2 (resp. 3) to compute an (almost) exact solution of the underlying 2-center problem (compare Table 2).

---

[9] E.g., the error is at most 0.225 if both C_i are parallelotopes or ellipsoids and $d \leq 3$.
[10] Since the edges have to be sorted.

Table 2. Test results for diameter partitioning where the containers are either two Euclidean balls or two arbitrarily, independently rotated unit cubes. The "rand. box" data sets refer to 100 equally distributed points within boxes with randomly scaled axes. The "norm. dist." data sets refer to 100 $(0, 1)$-normally distributed points. Due to the page limit, we report only the mean running times (in seconds) and the approximation quality after the diameter partitioning step (DP) over samples of 20 here. The accuracy is $\varepsilon = 0.01$ for all tests. See Table 1 for details on the environment used.

			pure B&B	B&B with diameter partitioning			
containers	data set	n	time	DP error	DP time	B&B time	overall time
Euclidean	rand. box	10	7.2	0.06	0.5	2.6	3.1
Euclidean	rand. box	20	26.2	0.11	0.5	14.1	14.6
Euclidean	rand. box	30	88.3	0.15	0.8	67.3	68.2
Euclidean	norm. dist.	10	10.9	0.12	0.5	7.1	7.6
Euclidean	norm. dist.	20	56.1	0.15	0.6	32.5	33.1
Euclidean	norm. dist.	30	135.5	0.17	0.8	89.0	89.9
rot. cubes	rand. box	10	12.8	$< \varepsilon$	2.1	-	2.1
rot. cubes	rand. box	20	103.2	$< \varepsilon$	2.4	-	2.4
rot. cubes	rand. box	30	639.9	$< \varepsilon$	3.1	-	3.1
rot. cubes	norm. dist.	10	21.5	$< \varepsilon$	1.1	-	1.1
rot. cubes	norm. dist.	20	210.9	$< \varepsilon$	2.1	-	2.1
rot. cubes	norm. dist.	30	1153.1	$< \varepsilon$	2.7	-	2.7

Combining the two algorithms is accomplished as follows. First, consider identical containers. A good upper bound obtained by Algorithm 2 decreases the running time as many branches need not be considered. Secondly, it provides a-priori information about point pairs not fitting in the same container: If $R(\{p, q\}, C) > \bar{\rho}$ for two points in P, assigning one of them to P_1 forces the other one into P_2. Since all pairwise distances have been computed and sorted, all such pairs of points can easily be identified and assigned to different partition sets. This is equivalent to building the distance graph $G(\bar{\rho})$ and 2-coloring it. As all possible 2-colorings have to be considered and the resulting bipartite graph is not connected in general, this leads to (usually several) disjoint subset pairs of P. During the B&B routine, each of those subset pairs can be considered as a whole and requires only one node in the B&B tree. For instance, when we choose such a point as first core-set point, we can assign all the points from the corresponding pair of colored subsets to the right cluster – even before the branching has started. The same can be done for 2-containment problems with different containers using Algorithm 3. Yet, here, each color yields a distinct set of subset pairs which has to be taken into account in the B&B procedure.

As one can conclude from the experiments in Table 2, Euclidean 2-center problems can be approximated to a good level of accuracy even in higher dimensions. One should especially recognize the quality of the bounds computed by the diameter partitioning before starting the B&B. Of course, the approximation quality achieved by the diameter partitioning is even better for some classes of non-Euclidean containers where nothing is known about the existence of small core-sets or even fast algorithms to compute those. This becomes clear when looking at the results in Table 2 for applying an implementation of Algorithm 3 on 2-containment problems with rotated cubes.

Algorithms 2 and 3 get slow on large point sets. However, it is not necessary to abandon their advantages. We restrict the algorithms to small subsets (e.g. random samples) of the input data, perform the diameter partitioning, compute an upper bound $\bar{\rho}$ (for the complete point set) and apply the pre-partitioning only to the sample. Any time the B&B algorithm picks a new core-set point, we test whether this point supplies additional information and if applicable add more points from the sample to the core-set. Surely, there is no guarantee for the quality of the computed bounds. But even this simple strategy improves the running times in experiments. Further reductions should be possible by more advanced strategies to avoid the evaluation of the complete graph over P.

6   An Application of Non-euclidean Container Shapes

In the case of non-Euclidean containers, a typical setting for instance in facility location is covering a 2D (point) set with several objects. However, different from the problems addressed before, rotations of the containers in addition to homothetics are of interest.

Fig. 3. Solution of a 4-containment problem with 18512 data points allowing rotations of the containers (accuracy 2%)

Figure 3 depicts the solution of such a 4-containment problem with identical 2-dimensional containers being conical sections of circles. These 'pie slice' shapes arise in applications when points should be within the sight of cameras, in the transmission range of oriented senders, or reachable by robot arms with joint limits [18]. A discretization of the possible space of rotations is considered, and included in the B&B algorithm. Note that the computational effort increases severely since the rotations of the four containers have to be addressed independently of each other. Still, the full computation takes less than 5 hours.

References

1. Agarwal, P.K., Procopiuc, C.M.: Exact and approximation algorithms for clustering. In: Proc. 9th ACM-SIAM Symp. Discrete Alg., pp. 658–667 (1998)
2. Agarwal, P.K., Sharir, M.: Efficient algorithms for geometric optimization. ACM Comput. Surv. 30(4), 412–458 (1998)
3. Anderberg, M.R.: Cluster analysis for applications. Probability and mathematical statistics. Academic Press, London (1973)
4. Avis, D.: Diameter partitioning. Discrete Comput. Geom. 1, 265–276 (1986)
5. Bespamyatnikh, S., Kirkpatrick, D.: Rectilinear 2-center problems. In: Proc. 11th Canad. Conf. Comp. Geom., pp. 68–71 (1999)
6. Bespamyatnikh, S., Segal, M.: Covering a set of points by two axis-parallel boxes. Inf. Process. Lett. 75(3), 95–100 (2000)
7. Bohnenblust, H.F.: Convex regions and projections in Minkowski spaces. Ann. Math. 39, 301–308 (1938)
8. Boltyanski, V., Martini, H., Soltan, P.S.: Excursions into Combinatorial Geometry. Springer, Heidelberg (1997)
9. Brandenberg, R., Gerken, T., Gritzmann, P., Roth, L.: Modeling and optimization of correction measures for human extremities. In: Jäger, W., Krebs, H.-J. (eds.) Mathematics – Key Technology for the Future. Joint Projects between Universities and Industry 2004-2007, pp. 131–148. Springer, Heidelberg (2008)
10. Brandenberg, R., Roth, L.: Optimal containment under homothetics, a practical approach (submitted, 2007)
11. Bădoiu, M., Har-Peled, S., Indyk, P.: Approximate clustering via core-sets. In: Proc. 34th Annu. ACM Symp. Theor. Comput., pp. 250–257 (2002)
12. Bunschoten, R.: A fully vectorized function that computes the Euclidean distance matrix between two sets of vectors (1999), http://www.mathworks.com/matlabcentral/fileexchange/loadFile.do?objectId=71
13. Chan, T.M.: More planar two-center algorithms. Comp. Geom. Theor. Appl. 13(3), 189–198 (1999)
14. Eppstein, D.: Faster construction of planar two-centers. In: Proc. 8th ACM-SIAM Symp. Discrete Alg., pp. 131–138 (1997)
15. Gonzalez, T.F.: Clustering to minimize the maximum intercluster distance. Theor. Comput. Sci. 38, 293–306 (1985)
16. Gritzmann, P., Klee, V.: Inner and outer j-radii of convex bodies in finite-dimensional normed spaces. Discrete Comput. Geom. 7, 255–280 (1992)
17. Gritzmann, P., Klee, V.: On the complexity of some basic problems in computational convexity I: Containment problems. Discrete Math. 136, 129–174 (1994)
18. Halperin, D., Sharir, M., Goldberg, K.: The 2-center problem with obstacles. J. Alg. 42(1), 109–134 (2002)
19. Hartigan, J.A.: Clustering algorithms. Wiley series in probability and mathematical statistics. John Wiley and Sons, New York (1975)
20. Helly, E.: Über Mengen konvexer Körper mit gemeinschaftlichen Punkten. Jahresbericht Deutsch. Math. Verein 32, 175–176 (1923)
21. Hershberger, J.: A faster algorithm for the two-center decision problem. Inf. Process. Lett. 47(1), 23–29 (1993)
22. Hoffmann, M.: A simple linear algorithm for computing rectilinear 3-centers. Comput. Geom. Theor. Appl. 31(3), 150–165 (2005)
23. Jain, A.K., Dubes, R.C.: Algorithms for clustering data. Prentice Hall, Englewood Cliffs (1988)

24. Jaromczyk, J.W., Kowaluk, M.: An efficient algorithm for the Euclidean two-center problem. In: Symp. Comp. Geom, pp. 303–311 (1994)
25. John, F.: Extremum problems with inequalities as subsidiary conditions. In: Courant Anniversary Volume, pp. 187–204. Interscience (1948)
26. Jung, H.W.E.: Über die kleinste Kugel, die eine räumliche Figur einschließt. J. Reine Angew. Math. 123, 241–257 (1901)
27. Kumar, P.: Clustering and reconstructing large data sets. PhD thesis, Department of Computer Science, Stony Brook University (2004)
28. Mangasarian, O.L., Setiono, R., Wolberg, W.H.: Pattern recognition via linear programming: theory and application to medical diagnosis. In: Coleman, T.F., Li, Y. (eds.) Large-Scale Numerical Optimization, pp. 22–31. SIAM, Philadelphia (1990); Computer Sciences TR 878 (1989)
29. Megiddo, N.: On the complexity of some geometric problems in unbounded dimension. J. Symb. Comput. 10(3/4), 327–334 (1990)
30. Pólik, I.: Addendum to the sedumi user guide version 1.1. Technical report, Advanced Optimization Laboratory, McMaster University (2005)
31. Procopiuc, C.M.: Clustering problems and their applications: A survey. Department of Computer Science, Duke University (1997)
32. Sharir, M.: A near-linear algorithm for the planar 2-center problem. In: Proc. Symp. Comp. Geom., pp. 106–112 (1996)
33. Sturm, J.F.: Using SEDUMI 1.02, a MATLAB toolbox for optimization over symmetric cones. Optim. Method. Softw. 11-12, 625–653 (1999)
34. del Val, A.: On 2-SAT and renamable Horn. In: Proc. 17th Nat. Conf. on Artif. Intel. AAAI / MIT Press (2000)
35. Wei, H., Murray, A.T., Xiao, N.: Solving the continuous space p-centre problem: planning application issues. IMA J. Management Math. 17, 413–425 (2006)

Separating Sublinear Time Computations by Approximate Diameter

Bin Fu[1] and Zhiyu Zhao[2]

[1] Dept. of Computer Science, University of Texas - Pan American
TX 78539, USA
binfu@cs.panam.edu

[2] Department of Computer Science, University of New Orleans, New Orleans, LA
70148, USA
zzha2@cs.uno.edu

Abstract. We study sublinear time complexity and algorithm to approximate the diameter for a sequence $S = p_1 p_2 \cdots p_n$ of points in a metric space, in which every pair of two consecutive points p_i and p_{i+1} in the sequence S has the same distance. The diameter of S is the largest distance between two points p_i and p_j in S. The approximate diameter problem is investigated under deterministic, zero error randomized, and bounded error randomized models. We obtain a class of separations about the sublinear time computations using various versions of the approximate diameter problem based on the restriction about the format of input data.

1 Introduction

Sublinear time computation is an active area of computer science in the recent years. A sublinear time algorithm has a sequence of elements $a_1, a_2, \cdots, a_n$ as input and can only access a part of the elements. Many sublinear time algorithms have been developed in the recent years. We give an incomplete list of sublinear time algorithms such as approximating matrix product [7], checking the polygon intersection [2], approximating the average degree in graph [8,14], estimating the cost of minimum spanning tree [3,5,6], finding the geometric separators [10], computing the basis of abelian groups [4], property testing [16,13], and facility location [1]. Initially, the main research of sublinear time algorithms has been in the property testing with surveys in [9,11,12,15,17]. People tend to believe that there will be more and more sublinear time algorithms to emerge in the future. Therefore, it is important to study the power and limitation of sublinear time computations in both deterministic and randomized computation models.

A sublinear time algorithm usually uses a randomized method to access the input since it does not have enough time to see the entire input data. Most of the sublinear time algorithms developed in the recent years are randomized. A recent interesting derandomization approach by Zimand [19] showed that for some $\alpha > 0$, randomized algorithms of time complexity $T(n) < n^\alpha$ can be

B. Yang, D.-Z. Du, and C.A. Wang (Eds.): COCOA 2008, LNCS 5165, pp. 79–88, 2008.
© Springer-Verlag Berlin Heidelberg 2008

simulated by deterministic algorithms of time poly($T(n)$) except on at most an $\exp(-\Omega(T(n)\log T(m)))$ fraction of the instances.

In this paper we study the number of queries about the input sequence. In order to separate the power of sublinear time computations with different query complexity bounds, we select the problem to compute the diameter for a sequence of points in a metric space. We realized this problem and its connection to sublinear time computation from our research on the protein backbone alignment [18]. From this approximate diameter problem, we show the existence of sublinear time algorithms at three different models, which are deterministic, bounded error randomized, and zero-error randomized. We study the complexity of the sublinear time algorithms to approximate the diameter of a sequence of points. The separations of sublinear time computations under various complexity bounds and models in this paper are based on the several versions of the diameter problem.

Three sublinear time computing models including deterministic, bounded error randomized, and zero error randomized models are studied in this paper. We obtain a class of separations about the power of sublinear time computations using several versions of the approximate diameter problem. We derive a dense sublinear time hierarchy for each of the three models. For every $0 < r < 1$ and $0 < \epsilon < r$, we show that the sublinear time deterministic computation with $O(n^r)$ queries to the input sequence is more powerful than sublinear time deterministic computation with $O(n^{r-\epsilon})$ queries and also the sublinear time deterministic computation with $O(n^r)$ queries to the input sequence cannot be simulated by sublinear time randomized computation with $O(n^{r-\epsilon})$ queries. We show that those separations by the number of queries imply similar dense time separations among sublinear time computations.

It is an interesting problem to identify what computational problems have the sublinear time algorithms. Our results show that the existence of sublinear time algorithms and their computational time depend on the restrictions on the format of input points in the metric space. We will show how those restrictions affect the existence of a sublinear time algorithm and its complexity. We identify the parameters to control the diameter length and the permutation of the input points, and we also show how the sublinear time model and the time complexity for computing an approximate diameter depend on those parameters.

We also show that the zero-error randomized sublinear time computation is more powerful than the deterministic sublinear time algorithm with similar time complexity and the bounded-error randomized sublinear time computation is more powerful than the zero-error randomized sublinear time algorithm with similar time complexity. We show that the bounded error randomized sublinear time algorithms in time $O(n^r)$ cannot be simulated by a zero-error randomized sublinear time algorithm in $o(n)$ time or queries, where r is an arbitrary parameter in $(0, 1)$. We also show that zero-error randomized sublinear time algorithms in time $O(n^r)$ cannot be simulated by a deterministic sublinear time algorithm in $o(n)$ time or queries, where r is an arbitrary parameter in $(0, 1)$.

2   Notations

A metric space S has a distance function dist$(.,.)$ that satisfies the following conditions: 1) dist$(p,p) = 0$ for every point $p \in S$; 2) dist$(p_1,p_2) = $ dist(p_2,p_1) for any two points $p_1, p_2 \in S$; and 3) dist$(p_1,p_3) \leq $ dist$(p_1,p_2) + $ dist(p_2,p_3) for any three points $p_1, p_2, p_3 \in S$.

For an integer $d \geq 1$, R^d is the d-dimensinal Euclidean space, which is clearly a metric space. Let $A = a_1, \cdots, a_n$ be a sequence of n points in a metric space. We often use $|A| = n$ to represent the number of points in A. Let $A = a_1, \cdots, a_n$ be a sequence of n points in a metric space. If for every pair of two consecutive points a_i and a_{i+1}, dist$(a_i, a_{i+1}) = t$, then the sequence A is called a t-sequence.

Definition 1.   – *Let $A = a_1, \cdots, a_n$ be a sequence of n points in a metric space. For every pair of two consecutive points a_i and a_{i+1}, if $t_1 \leq$ dist$(a_i, a_{i+1}) \leq t_2$, then the sequence A is called a (t_1, t_2)-sequence. Define* minInterDist$(A) = \min_{1 \leq i \leq n-1}($dist$(a_i, a_{i+1}))$ *and* maxInterDist$(A) = \max_{1 \leq i \leq n-1}($dist$(a_i, a_{i+1}))$.

– *For a sequence of points A in a metric space,* diameter(A) *is the largest distance between two points of A.*

– *A real number d is $(1 - \epsilon)$-approximate to the diameter of S of a sequence of points, if $(1 - \epsilon)$*diameter$(S) \leq d \leq$ diameter(S).

– *A path of a randomized computation C of $r(n)$ random bits with the input sequence S is determined by a binary sequence B of length $r(n)$. Its output in the path B is denoted by $C(S, B)$.*

– *A deterministic $(1 - \epsilon)$-approximate algorithm C with query complexity $q(n)$ for the diameter of sequence satisfies that 1) $C(S)$ is a $(1 - \epsilon)$ approximation to* diameter(S); *and 2) C makes at most $q(n)$ queries to the points in S, where input S is a sequence of n points. Its query complexity is defined by a function $q(n)$ that for every input of length n points, the algorithm makes at most $q(n)$ queries. Its time complexity is defined by a function $t(n)$ that for every input of length n points, the algorithm stops in $t(n)$ steps.*

– *A randomized $(1 - \epsilon)$-approximate algorithm C with $r(n)$ random bits for the diameter of sequence satisfies that 1) $C(S, B)$ is a $(1 - \epsilon)$ approximation to* diameter(S) *with probability at least $\frac{3}{4}$; and 2) each path of C makes at most $q(n)$ queries to the points in S, where input S is a sequence of n points and B is a random binary sequence of length $r(n)$. A randomized algorithm can be also called bounded error randomized algorithm.*

– *A zero-error randomized $(1 - \epsilon)$-approximate algorithm C with $r(n)$ random bits for the diameter of sequence satisfies that 1) $C(S, B)$ is a $(1 - \epsilon)$ approximation to* diameter(S) *with probability at least $\frac{3}{4}$; 2) no path gives a result that is not an $(1 - \epsilon)$ approximation to* diameter(S).

– *A randomized $(1 - \epsilon)$-approximate algorithm C (either bounded error or zero-error) with $r(n)$ random bits and time complexity $t(n)$ for the diameter of sequence satisfies that $C(S, B)$ stops in $t(n)$ steps, where input S is an arbitrary sequence of n points and B is a random binary sequence of length $r(n)$.*

- A randomized $(1-\epsilon)$-approximate algorithm C (either bounded error or zero-error) with $r(n)$ random bits and query complexity $q(n)$ for the diameter of sequence satisfies that $C(S, B)$ makes at most $q(n)$ queries, where input S is an arbitrary sequence of n points and B is a random binary sequence of length $r(n)$.

3 Tight Separations among Sublinear Time Computations

We separate sublinear time computable functions with time complexity n^r from those with time complexity $n^{r-\epsilon}$ for any $0 < r < 1$ and any small $\epsilon > 0$. The separation is achieved in both deterministic and randomized computation models.

Definition 2. Let r be an integer ≥ 0 and $S = p_1 p_2 \cdots p_n$ be a (t_1, t_2)-sequence. The sequence $S' = p'_1 p'_2 \cdots p'_n$ is r-reliable rearrangement of S if S' is a permutation of $p_1 p_2 \cdots p_n$ and for each p_i, $p_i = p_{i'}$ for some i' with $1 \leq i' \leq n$ and $|i - i'| \leq r$.

Let M be a metric space, r, m, and n be non-negative integers, and c be an real number at least 1. Define $\Phi_M(c, r, m, n)$ to be the set of all sequences $H = q_1 q_2 \cdots q_n$ of n points in M such that H is an r-reliable rearrangement for a (t_1, t_2)-sequence S for some $0 < t_1 \leq t_2$ with $\frac{t_2}{t_1} \leq c$ and diameter$(S) \geq mt_1$. In particular, $\Phi_M(c, 0, m, n)$ is the set of all (t_1, t_2)-sequence S of length n in M with $\frac{t_2}{t_1} \leq c$ and diameter$(S) \geq mt_1$. Sequence S is called a $\Phi_M(c, r, m, n)$-sequence if $S \in \Phi_M(c, r, m, n)$.

We first present a deterministic sublinear time approximate algorithm to compute the diameter of a t-sequence in a metric space. Its computational time is reversely propositional to the length of the diameter. The algorithm is described in a more generalized format by the following theorem.

Theorem 1. Assume that c is a positive constant, and α, μ and ϵ are constants in $(0, 1)$. Assume that M is a metric space with a $(1 - \mu)$-factor approximate algorithm App_M of time complexity $C(k)$ for the diameter of k points in M for some nondecreasing function $C(k) : N \to N$. Then there exists a deterministic algorithm such that given a $\Phi_M(c, \frac{\epsilon(1-\alpha)}{2c} m, m, n)$-sequence B, it makes at most $O(\frac{n}{m})$ non-adaptive queries to the points of B and outputs a number x with $(1 - \epsilon)(1 - \mu) \cdot \text{diameter}(B) \leq x \leq \text{diameter}(B)$ in total time $O(\frac{n}{m}) + C(O(\frac{n}{m}))$.

Proof. Our algorithm selects an $O(\frac{n}{m})$ points set Q from the input sequence B and uses the diameter of Q to approximate the diameter of B. Select $\delta = \frac{\epsilon\alpha}{2c}$ and $\beta = \frac{\epsilon(1-\alpha)}{2c} m$. Assume that $A = p_1 p_2 \cdots p_n$ is a (t_1, t_2)-sequence such that B is a β-reliable rearrangement of A with $0 < t_1 \leq t_2$, $\frac{t_2}{t_1} \leq c$, and diameter$(A) \geq mt_1$. By the condition of the theorem, let $t_1 = \text{minInterDist}(A)$ and $t_2 = \text{maxInterDist}(A)$ be two positive real numbers with $t_1 \leq t_2$ and $\frac{t_2}{t_1} \leq c$. Our algorithm is described as follows:

Algorithm

Input: $B = p'_1, p'_2, \cdots, p'_n$ that is β-reliable-rearrangement of a (t_1, t_2)-sequence $A = p_1, p_2, \cdots, p_n$.

Output: an approximation x to diameter(A).

    let $h = \lfloor \delta m \rfloor$;

    select $q_i = p'_{h \cdot i}$ for $i = 1, \cdots, k = \lceil \frac{n}{h} \rceil$;

    let Q be the sequence $q_1 \cdots q_k$;

    output $x = App_M(Q)$;

End of Algorithm

Now we are going to prove that for the sequence Q constructed from B in the algorithm, $(1 - \epsilon)$diameter$(A) = (1 - \epsilon)$diameter$(B) \leq$ diameter$(Q) \leq$ diameter$(B) =$ diameter(A). Assume that p_i and p_j are two points in A such that dist$(p_i, p_j) =$ diameter(A). Let i_1 be the number $1 \leq i_1 \leq k$ such that $|i_1 h - i| = \min_{1 \leq i_2 \leq k} |i_2 h - i|$ and j_1 be the number $1 \leq j_1 \leq k$ such that $|j_1 h - j| = \min_{1 \leq j_2 \leq k} |j_2 h - j|$. It is easy to see that $|i_1 h - i| \leq h$ and $|j_1 h - j| \leq h$. Since two consecutive points in A have distance at most t_2, we have

$$\text{dist}(p_i, p_{i_1 h}) \leq h \cdot t_2 \tag{1}$$

$$\text{dist}(p_j, p_{j_1 h}) \leq h \cdot t_2 \tag{2}$$

For each p'_k, it has another p_s such that $p_s = p'_k$ and $|s - k| \leq \beta$ since B is a β-reliable rearrangement of A. Therefore, we have

$$\text{dist}(p_k, p'_k) = \text{dist}(p_k, p_s) \leq \beta t_2. \tag{3}$$

We have the following inequalities:

$$\text{diameter}(A) = \text{diameter}(p_i, p_j) \tag{4}$$

$$\leq \text{dist}(p_i, p_{i_1 h}) + \text{dist}(p_{i_1 h}, p'_{i_1 h}) + \text{dist}(p'_{i_1 h}, p'_{j_1 h}) + \text{dist}(p'_{j_1 h}, p_{j_1 h}) \tag{5}$$

$$+ \text{dist}(p_{j_1 h}, p_j) \tag{6}$$

$$\leq h \cdot t_2 + \beta t_2 + \text{dist}(p'_{i_1 h}, p'_{j_1 h}) + \beta t_2 + h \cdot t_2 \tag{7}$$

$$\leq h \cdot t_2 + \beta t_2 + \text{diameter}(Q) + \beta t_2 + h \cdot t_2 \tag{8}$$

$$\leq 2(h + \beta) t_2 + \text{diameter}(Q) \tag{9}$$

$$\leq 2(\frac{\epsilon \alpha}{2c} + \frac{\epsilon(1 - \alpha)}{2c}) c \cdot m \cdot t_1 + \text{diameter}(Q) \tag{10}$$

$$\leq \epsilon \cdot m t_1 + \text{diameter}(Q) \tag{11}$$

$$\leq \epsilon \cdot \text{diameter}(A) + \text{diameter}(Q). \tag{12}$$

The transition from (4) to (6) is due to the triangle inequality in the metric space. The transition from (6) to (7) is due to inequalities (1), (2), and (3). The transition from (7) to (8) is because $p'_{i_1 h}$ and $p'_{j_1 h}$ are in Q. By (4)-(12), we have $(1 - \epsilon)$diameter$(A) \leq$ diameter(Q). On the other hand, all points in Q are from A. So, diameter$(Q) \leq$ diameter(A). Therefore, $(1 - \epsilon)$diameter$(A) \leq$ diameter$(Q) \leq$ diameter(A). Since App_M gives factor $(1 - \mu)$ approximation for the diameter of set Q, the output x satisfies $(1 - \epsilon)(1 - \mu) \cdot$ diameter$(A) \leq x \leq$ diameter(A). Since B is a permutation of A, we have diameter$(B) =$ diameter(A). Therefore, $(1 - \epsilon)(1 - \mu) \cdot$ diameter$(B) \leq x \leq$ diameter(B).

The number of queries of the algorithm is $|Q| = O(\frac{n}{m})$. The time for generating Q is $O(\frac{n}{m})$ and the time for computing $App_M(Q)$ is $C(O(\frac{n}{m}))$.   □

Corollary 1. *Assume that α is a constant with $0 < \alpha < 1$, and ϵ is a small constant greater than 0. Let t be a positive real number. Then there exists a deterministic $O(\frac{n}{m})$-time algorithm such that given an $\epsilon(1-\alpha)m/2$-reliable-rearrangement sequence B for a t-sequence A of n points in a metric space with diameter at least $m \cdot t$, it outputs a number x with $\frac{1-\epsilon}{2}$diameter$(B) \leq x \leq$ diameter(B).*

Proof. It is known that there exists an $O(k)$ time $\frac{1}{2}$-factor approximation algorithm to compute the diameter of k points in a metric space. The algorithm selects an arbitrary point and finds the point with the largest distance to the other points. It is at least half of the diameter. Apply Theorem 1.   □

Corollary 2. *Assume that α is a constant with $0 < \alpha < 1$, and ϵ is a small constant greater than 0. Let t be a positive real number. Then there exists a deterministic $O(\frac{n}{m})$-time algorithm such that given an $\epsilon(1-\alpha)m/2$-reliable-rearrangement sequence B for a t-sequence A of n points in R^1 with diameter at least $m \cdot t$, it outputs a number x with $(1 - \epsilon)$diameter$(B) \leq x \leq$ diameter(B).*

Proof. In R^1, finding the diameter takes $O(k)$ time for an input of k points.   □

Corollary 3. *Assume that c is a positive constant, d is a fixed dimension number, α is a constant in $(0,1)$, and ϵ is a small constant greater than 0. Let t be a positive real number. Then there exists a deterministic $O(\frac{n}{m} + (\frac{1}{\epsilon^{2d}}))$-time algorithm such that given an $\epsilon(1-\alpha)m/2$-reliable-rearrangement sequence B for a t-sequence A of n points in R^d with diameter at least $m \cdot t$, it outputs a number x with $(1 - \epsilon)$diameter$(A) \leq x \leq$ diameter(A).*

Proof. We just need to prove that for any constant $\delta \in (0,1)$, there exists an $O(k + (\frac{1}{\delta^{2d}}))$ time $(1 - \delta)$-factor approximate algorithm App_{R^d} to compute the diameter of k points set H in R^d. Let d be a fixed dimensional number. Find a $\frac{1}{2}$-factor approximate diameter D of H (see the proof of Corollary 1). The approximate diameter D can be found in time $O(k)$ as described in the proof of Corollary 1. There exists a $(4D)^d$ cube region G that contains all points in H. Partition G into small cubes of size $(\frac{\delta D}{2\sqrt{d}})^d$. For each cube C that contains points in H, select one point from $H \cap C$ and put it into set Q. The number of small cubes of size $(\frac{\delta D}{2\sqrt{d}})^d$ in G is at most $O((\frac{1}{\delta})^d)$ since d is fixed. We have $|Q| = O((\frac{1}{\delta})^d)$. Compute the diameter of Q by brute force method in time $O(|Q|^2)$.   □

Lemma 1. *For any even number n and two numbers $p_1 < p_2$ in R^1, there exists a* dist(p_2, p_1)-*sequence $S = p_1 q_1 q_2 \cdots q_{n-2} p_2$ in R^1 such that $p_1 < q_i$ for $i = 1, \cdots, n - 2$ and diameter$(S) \geq \frac{n \cdot \text{dist}(p_1,p_2)}{2}$. The sequence S is denoted as* unfolding$_{R^1}(p_1, p_2, n)$.

Proof. Let $n = 2h$ and $t = \text{dist}(p_1, p_2)$. We construct a t-sequence of n points as follows: Let 1) $q_1 = p_1 + t$, 2) $q_s = q_{s-1} + t$ for $s = 2, \cdots, h$, and 3) $q_s = q_{s-1} - t$ for $s = h + 1, h + 2, \cdots, 2h - 2$. It is easy to see that $S = p_1 q_1 q_2 \cdots q_{2h-2} p_2$ is a t-sequence of $n = 2h$ points in R^1 and $\text{diameter}(S) = ht = \frac{nt}{2}$.   □

Theorem 2 gives a lower bound about the randomized sub-linear time algorithms and matches the upper bound of Theorem 1.

Theorem 2. *Assume that ϵ is a constant in $(0, 1)$ and $m = o(n)$. Then there is no randomized algorithm such that given a $\Phi_{R^1}(1, 0, m, n)$-sequence S, the algorithm makes at most $o(\frac{n}{m})$ adaptive queries and outputs $(1 - \epsilon)$-approximate diameter for S.*

Proof. Assume that C is a randomized $(1 - \epsilon)$ approximate algorithm with $o(\frac{n}{m})$ adaptive queries for computing the approximate diameter for all of the t-sequences of diameter at least $m \cdot t$. Let $h = 2(\lceil \frac{\epsilon m}{1-\epsilon} \rceil + 2)$, $g = 2h$ and $n = m + kg$, where k is a parameter that is flexible. Since $m = o(n)$, we always assume that $1 \le m < \frac{n}{2}$. We have $k = \frac{n-m}{g} = \frac{n-m}{4(\lceil \frac{\epsilon m}{1-\epsilon} \rceil + 2)} \le \frac{n}{4(\lceil \epsilon m \rceil)}$. On the other hand, $k \ge$

$$\frac{(n-m)}{4(\lceil \frac{\epsilon m}{1-\epsilon} \rceil + 2)} > \frac{(n-m)}{4(\frac{\epsilon m}{1-\epsilon} + 3)} \ge \frac{(n-m)}{4(\frac{\epsilon m + 3(1-\epsilon)}{1-\epsilon})} \ge \frac{(1-\epsilon)(n-m)}{4(\epsilon + 3(1-\epsilon))m} \ge \frac{(1-\epsilon)(n-m)}{4(3-2\epsilon)m} \ge \frac{(1-\epsilon)n}{8(3-2\epsilon)m}.$$

Let constant $c_0 = 0.09 \cdot \frac{(1-\epsilon)}{8(3-2\epsilon)}$. Let t be a constant greater than 0.

Since each path queries $o(\frac{n}{m})$ points, we assume that every path of C queries at most $\frac{c_0 n}{m}$ points in every t-sequence A. Let A be the t-sequence of points $q_1, q_2, \cdots, q_{m+1}, p_1, p_2, \cdots, p_{n-m}$, where $q_i = (i - 1)t$ for $i = 1, 2, \cdots, m + 1$, $p_i = (m - 1)t$ for odd number $i = 1, 3, \cdots$, and $p_i = mt$ for even number $i = 2, 4, \cdots$. Clearly, A is a t-sequence in one dimensional axis of diameter $m \cdot t$.

Partition the points $p_1 p_2 \cdots p_{n-m}$ sequentially into $P_1 P_2 \cdots P_k$ with $|P_i| = g$. In the next phase, we will show that there exists some P_i such that no more than $10\% G$ paths of C query the points in P_i, where G is the number of total paths in C. Assume that for every P_i, there are at least $10\% G$ paths of C to query the points in P_i. Thus, the total number of queries is at least $k \cdot 10\% G > \frac{c_0 n}{m} G$ among all paths. On the other hand, since every path of C queries at most $\frac{c_0 n}{m}$ points, the total number of queries by all paths of C is at most $\frac{c_0 n}{m} G$. This is a contradiction. Therefore, we have a P_i that no more than 10% paths of C query the points in P_i.

We can arrange the points in P_i so that it has greatly different diameters. Since P_i has at least $2h$ points, we can make $\text{diameter}(P_i)$ as large as ht and as small as t without changing the positions of first and last points of P_i. Formally, assume that P_i has the sequence of points $p_u, p_{u+1}, \cdots, p_{u+g-1}$.

Clearly, $\text{dist}(p_u, p_{u+g-1}) = t$ and $p_u < p_{u+g-1}$ by the definition of A. We replace $p_{u+1}, \cdots, p_{u+g-2}$ by $p'_{u+1}, \cdots, p'_{u+g-2}$, where unfolding $_{R^1}(p_u, p_{u+g-1}, g) = p_u p'_{u+1} p'_{u+2} \cdots, p'_{u+g-2} p_{u+g-1}$.

If the sequence A' is derived from A that P_i is replaced by $P'_i = p_u p'_{u+1} p'_{u+2} \cdots, p'_{u+g-2} p_{u+g-1}$. $C(A, B)$ and $C(A', B)$ will be the same at 90% paths B. On the other hand, the diameter of A is $m \cdot t$ and the diameter of A' is at least $mt + ht - t > \frac{1}{(1-\epsilon)} mt$ by Lemma 1. Thus, C is not an $(1 - \epsilon)$-approximation to the diameter of a t-sequence of n points in R^1 with diameter at least mt. A contradiction.   □

Corollary 2 and Theorem 2 imply the following dense separation for the sublinear time computations.

Corollary 4. *Assume that ϵ is a constant in $(0,1)$. Then for every constant r in $(0,1)$ and constant δ in $(0,r)$, there is a function that can be $(1-\epsilon)$-approximated by n^r sublinear time deterministic algorithm, but there is no $n^{r-\delta}$ sublinear time $(1-\epsilon)$-approximate randomized algorithm.*

4   Randomized and Deterministic Computations

In this section, we show that randomized algorithms are more powerful than deterministic algorithms with the same computational time. We first present a randomized algorithm, then show that similar computation cannot be done in the deterministic algorithm with the similar complexity.

Theorem 3. *Assume that c is a positive constant, and α, μ and ϵ are constants in $(0,1)$. Assume that M is a metric space with a $(1-\mu)$-factor approximate algorithm App_M of complexity $C(k)$ for the diameter of k points in M for some nondecreasing function $C(k) : N \to N$. Then there exists a randomized algorithm such that given a $\Phi_M(c,\infty,m,n)$-sequence B, it makes at most $O(\frac{n}{\epsilon m})$ non-adaptive queries to the points of B and outputs a number x with $(1-\epsilon)(1-\mu) \cdot$ diameter$(B) \le x \le$ diameter(B) in total time $O(\frac{n}{\epsilon m}) + C(\frac{n}{\epsilon m})$, where $m = o(n)$.*

Corollary 5. *Assume that c is a positive constant, α, μ and ϵ are constants in $(0,1)$. Then there exists a randomized algorithm such that given a $\Phi_{R^1}(c,\infty,m,n)$-sequence B, it makes at most $O(\frac{n}{\epsilon m})$ non-adaptive queries to the points of B and outputs a number x with $(1-\epsilon) \cdot$ diameter$(B) \le x \le$ diameter(B) in total time $O(\frac{n}{\epsilon m})$.*

Theroem 4 gives a lower bound for the deterministic algorithms for computing the approximate diameter problem. Corollary 5 and Theroem 4 give the separation between randomized and deterministic computations.

Theorem 4. *Let ϵ be a constant in $(0,1)$ and $m = o(n)$. Then there is no deterministic algorithm that given a $\Phi_{R^1}(1, 8(\lceil \epsilon m \rceil + 2), m, n)$ sequence B, it makes no more than $(n - m - 1)/2$ adaptive queries to the input points and outputs a $(1-\epsilon)$-approximation to the diameter of B.*

5   Zero-Error Randomized Algorithm and Its Complexity

In this section, we show a zero-error randomized algorithm. We also derive a lower bound for the deterministic algorithms. This shows that zero-error randomized algorithms are more powerful than deterministic algorithms.

Definition 3. *Let M be a metric space.*

– *Let $S' = q_1, q_2, \cdots, q_n$ be a rearrangement of a sequence of points $S = p_1 p_2, \cdots, p_n$. A point q_i is called a still point if $q_i = p_i$.*
– *A function $f(x) \to N$ can be c-approximated by a $FZ[n^r]$ computation algorithm if the algorithm makes at most n^r queries, gives output with probability at least $\frac{2}{3}$, and each output y has $cf(x) \le y \le f(x)$.*
– *Let $S' = q_1, q_2, \cdots, q_n$ be a rearrangement of a sequence of points $S = p_1 p_2, \cdots, p_n$. A point q_i in S' is called v-stable if $q_i = p_j$ with $|i - j| \le v$.*
– *Let $S' = q_1, q_2, \cdots, q_n$ be a rearrangement of a sequence of points $S = p_1 p_2, \cdots, p_n$. S' is called (u, v, α)-stable if for every u consecutive points set Q from S', Q has at least αu v-stable points.*
– *For a sequence $S = q_1 q_2 \cdots q_n$ of points in M, the sequence $S^* = (q'_1, i_1)(q'_2, i_2) \cdots (q'_n, i_n)$ is called a marked sequence of S, where $(q'_1, i_1)(q'_2, i_2) \cdots (q'_n, i_n)$ is a permutation of $(q_1, 1)(q_2, 2) \cdots (q_n, n)$. Define $E(S^*) = S$.*
– *Let $\Lambda_M(c, m_1, m_2, r, m, n)$ be the set of all marked sequences $(q_1, a_1)(q_2, a_2) \cdots (q_n, a_n)$ such that 1) $S' = q_1 q_2 \cdots q_n$ is a permutation of a (t_1, t_2)-sequence $S = p_1 p_2 \cdots p_n$ of n points in M for some $0 < t_1 < t_2$ with $\frac{t_2}{t_1} \le c$; 2) every m_1 consecutive points in S' have at least m_2 points q_i which are r-stable between S' and S; 3) the diameter of S is at least $m \cdot t_1$. and 4) $(q_1, a_1)(q_2, a_1) \cdots (q_n, a_n)$ is a permutation of $(p_1, 1)(p_2, 2) \cdots (p_n, n)$*
– *Let Γ be a class of marked sequences. A zero-error randomized $(1 - \epsilon)$-approximate algorithm C with $r(n)$ random bits for the diameter of sequence in Gamma if for every input $S \in \Gamma$, we have 1) at least $\frac{3}{4}$ paths of C has non-empty output; and 2) each non-empty output in a path is a $(1 - \epsilon)$ approximation to diameter(S). Its time complexity and query complexity are defined similarly as that in Definition 1.*

Theorem 5 shows a zero-error randomized algorithm to approximate the diameter of a marked sequence.

Theorem 5. *Assume that M is a metric space with a $(1-\mu)$-factor approximate algorithm App_M of time complexity $C(k)$ for the diameter of k points in M for some nondecreasing function $C(k) : N \to N$. Then for every constant $\epsilon \in (0, 1)$, there exist positive constants β_1, β_2, and $\alpha < \beta_1$, and a zero-error randomized $(1 - \epsilon)$-approximate algorithm such that given a $\Lambda_M(c, \beta_1 m, \alpha m, \beta_2 m, m, n)$-sequence $S' = (q_1, a_1) \cdots (q_n, a_n)$, the algorithm makes at most $O(\frac{n}{m} \log \frac{n}{m})$ non-adaptive queries to the items of S' and outputs a number x with $(1 - \epsilon)(1 - \mu) \cdot$ diameter$(E(S')) \le x \le$ diameter$(E(S'))$ in total time $O(\frac{n}{m}) + C(O(\frac{n}{m}))$, where $m = o(n)$.*

We have the following theorem to separate the sublinear time zero-error randomized computations from sublinear time deterministic computations.

Theorem 6. *Assume that c is a positive constant, ϵ is a constant in $(0, 1)$, β is a constant in $(0, c)$, and $m = o(n)$. Then there is no deterministic algorithm such that given a $\Lambda_{R^1}(1, cm, \beta m, 0, m, n)$-sequence S' it makes $o(n)$ adaptive queries to the input and outputs a $(1 - \epsilon)$ approximation to the diameter of $E(S')$.*

References

1. Badoiu, M., Czumaj, A., Indyk, P., Sohler, C.: Facility location in sublinear time. In: Proceedings of 32nd Annual International Colloquium on Automata, Languages and Programming, pp. 866–877 (2005)
2. Chazelle, B., Liu, D., Magen, A.: Sublinear geometric algorithms. SIAM Journal on Computing 35, 627–646 (2005)
3. Chazelle, B., Rubfinfeld, R., Trevisan, L.: Approximating the minimum spanning tree weight in sublinear time. SIAM Journal on computing 34, 1370–1379 (2005)
4. Chen, L., Fu, B.: Linear and sublinear time algorithms for the basis of abelian groups. Electronic Colloquium on Computational Complexity, TR07-052 (2007)
5. Czumaj, A., Ergun, F., Fortnow, L., Magen, I.N.A., Rubinfeld, R., Sohler, C.: Sublinear approximation of euclidean minimum spanning tree. SIAM Journal on Computing 35, 91–109 (2005)
6. Czumaj, A., Sohler, C.: Estimating the weight of metric minimum spanning trees in sublinear-time. In: Proceedings of the 36th Annual ACM Symposium on Theory of Computing, pp. 175–183 (2004)
7. Drineas, P., Kannan, R.: Fast monte-carlo algorithms for approximate matrix multiplication. In: Proceedings of the 42nd IEEE Symposium on Foundations of Computer Science, pp. 452–459 (2001)
8. Feige, U.: On sumes of independent random variables with unbounded variance and estimating the average degree in a graph. SIAM Journal on Computing 35, 964–984 (2006)
9. Fischer, E.: The art of uninformed decision: A primer to property testing. Bulletin of the EATCS 75, 97–126 (2001)
10. Fu, B., Chen, Z.: Sublinear-time algorithms for width-bounded geometric separators and their applications to protein side-chain packing problems. Journal of Combinatorial Optimization 15, 387–407 (2008)
11. Goldreich, O.: Combinatorial proterty testing (a survey). In: Pardalos, P., Rajasekaran, S., Rolim, J. (eds.) Proceedings of the DIMACS workshop on radnomziation methods in algorithm design, vol. 43, pp. 45–59 (1997)
12. Goldreich, O.: Property testing in massive graphs. In: Abello, J., Pardalos, P.M., Resende, M. (eds.) Handbook of massive data sets, pp. 123–147 (2002)
13. Goldreich, O., Ron, D.: On testing expansion in bounded-degree graphs. Technical Report 00-20, Electronic Colloquium on Computational Complexity, http://www.eccc.uni-trier.de/eccc/ (2000)
14. Goldreich, O., Ron, D.: Approximating average parameters of graphs. Technical Report 05-73, Electronic Colloquium on Computational Complexity (2005), http://www.eccc.uni-trier.de/eccc/
15. Kumar, R., Rubinfeld, R.: Sublinear time algorithms. SIGACT News 34, 57–67 (2003)
16. Goldreich, S.G.O., Ron, D.: Property testing and its connection to learning and approximation. J. ACM 45, 653–750 (1998)
17. Ron, D.: Handbook of randomzied algorithm. Bulletin of the EATCS II, 597–649 (2001)
18. Zhao, Z., Fu, B.: A flexible algorithm for pairwise protein structure alignment. In: Proceedings International Conference on Bioinformatics and Computational Biology 2007 (2007)
19. Zimand, M.: On derandomizing probabilistic sublinear-time algorithms. In: Proceedings of the 22nd IEEE conference on computational complexity, pp. 1–9 (2007)

Computational Study on Dominating Set Problem of Planar Graphs

Marjan Marzban[1], Qian-Ping Gu[1], and Xiaohua Jia[2]

[1] School of Computing Science, Simon Fraser University, Burnaby BC Canada
{mmarzban,qgu}@cs.sfu.ca
[2] Department of Computer Science, City University of Hong Kong
csjia@cityu.edu.hk

Abstract. Recently, there has been significant theoretical progress towards fixed-parameter algorithms for the DOMINATING SET problem of planar graphs. It is known that the problem on a planar graph with n vertices and dominating number k can be solved in $O(c^{\sqrt{k}}n)$ time using tree/branch-decomposition based algorithms, where c is some constant. However there has been no computational study report on the practical performances of the $O(c^{\sqrt{k}}n)$ time algorithms. In this paper, we report computational results of Fomin and Thilikos algorithm which uses the branch-decomposition based approach. The computational results show that the algorithm can solve the DOMINATING SET problem of large planar graphs in a practical time for the class of graphs with small branchwidth. For the class of graphs with large branchwidth, the size of instances that can be solved by the algorithm in a practical time is limited to a few hundreds edges. The practical performances of the algorithm coincide with the theoretical analysis of the algorithm. The results of this paper suggest that the branch-decomposition based algorithms can be practical for some applications on planar graphs.

Keywords: PLANAR DOMINATING SET, branch-decomposition, fixed-parameter algorithms, data reduction, computational study.

1   Introduction

Given an undirected graph $G(V, E)$, a *k-dominating set* D of G is a subset of k vertices of G such that for every vertex $v \in V(G)$, either $v \in D$ or v is adjacent to a vertex $u \in D$. The *dominating number* of G, denoted by $\gamma(G)$, is the minimum k such that G has a k-dominating set. Given G and an integer k, The DOMINATING SET problem is to decide if $\gamma(G) \leq k$. The optimization version of the problem is to find a dominating set D with $|D| = \gamma(G)$. The DOMINATING SET problem is a core NP-complete problem in combinatorial optimization and graph theory [17]. It also has wide practical applications such as resource allocations [21], domination problems in electric networks [19], and wireless ad hoc networks [33]. The books of Haynes et al. give a survey on the rich literature of algorithms and complexity of the DOMINATING SET

B. Yang, D.-Z. Du, and C.A. Wang (Eds.): COCOA 2008, LNCS 5165, pp. 89–102, 2008.
© Springer-Verlag Berlin Heidelberg 2008

problem [20,21]. A recent experimental study on the heuristic algorithms for the DOMINATING SET problem can be found in [30].

The DOMINATING SET problem is NP-hard. Approximation algorithms and exact fixed-parameter algorithms have been extensively studied to tackle the intractability of the problem. A minimization problem P of size n is α-approximable if there is an algorithm which runs in polynomial time in n and produces a solution of P with value at most αOPT, where OPT is the value of the optimal solution of P and $\alpha \geq 1$. If P is $(1 + \epsilon)$-approximable for every fixed $\epsilon > 0$, P is polynomial time approximable (i.e., has a PTAS). Problem P is fixed-parameter tractable if given a parameter k, OPT can be computed in $O(f(k)n^{O(1)})$ time, where $f(k)$ may be an exponentially fast (or faster) growing function in k. For arbitrary undirected graph G of n vertices, the DOMINATING SET problem is known $(1 + \log n)$-approximable [22], but not approximable within a factor of $(1 - \epsilon) \ln n$ for any $\epsilon > 0$ unless $NP \subseteq DTIME(n^{\log \log n})$ [15]. The problem is also known fixed-parameter intractable unless the parameterized complexity classes collapse [13,14]. If the problem is restricted to planar graphs, it is known as the PLANAR DOMINATING SET problem which is still NP-hard [17]. But the PLANAR DOMINATING SET problem is known polynomial time approximable [7] and fixed-parameter tractable [13].

In recent years, there have been significant improvements on the fixed-parameter algorithms for the PLANAR DOMINATING SET problem. Algorithms with running time $O(11^k n)$ [13] and $O(8^k n)$ [5] are known for graphs with $\gamma(G) = k$. The running time is further reduced and $O(c^{\sqrt{k}} n)$ time algorithms are known for a constant c [4,16,23]. Most of the sublinear exponent algorithms use a tree-decomposition based approach: First a tree decomposition of the given graph is computed and then a dynamic programming algorithm based on the tree-decomposition is used to compute a minimum dominating set. For a planar graph G with $\gamma(G) = k$, a tree decomposition of width $b\sqrt{k}$, b is a constant, can be computed and the dynamic programming part runs in $O(2^{2b\sqrt{k}} n)$ time [4]. One problem with those algorithms is that the constant $c = 2^{2b}$ is too large for solving the PLANAR DOMINATING SET problem in practice. In relation to treewidth and tree decompositions [27,28], Robertson and Seymour introduce branchwidth and branch decompositions [29]. Instead of a tree decomposition, a branch decomposition can be used in the above dynamic programming algorithms for the PLANAR DOMINATING SET problem. Fomin and Thilikos give such an algorithm (called FT Algorithm in what follows) which reduces the constant c to $2^{15.13}$ [16]. Dorn proposes an approach of applying the distance product of matrices to the dynamic programming step in branch/tree-decomposition based algorithms for the problem [11]. If the distance product of matrices is realized by the $O(n^\omega)$ ($\omega < 2.376$) time fast matrix multiplication method [10], the constant c in is improved to $2^{11.98}$. However the constant hidden in the Big-Oh may be huge. Dorn also proposes a tree-decomposition based algorithm for the problem [12]. Although expressed in terms of treewidth tw of G, the algorithm has time complexity $O(3^{tw} n^{O(1)})$, it has actually the same running time as that of FT Algorithm. An encouraging fact on branch decomposition is that an optimal branch decomposition of a planar graph can be computed in polynomial time [18,32]. This makes the branch-decomposition based algorithms receiving increasing attention for the problems on planar graphs.

Another important progress on the algorithmic tractability of the PLANAR DOMINATING SET problem is that the problem is shown having a linear size kernel [6]. More specifically, Alber et al. give an $O(n^3)$ time algorithm which, given a planar graph G with $\gamma(G) = k$, produces a reduced graph H (kernel) such that H has $O(k)$ vertices, $\gamma(H) = k' \leq k$, and a minimum dominating set of G can be constructed from a minimum dominating set of H in linear time [6]. In general, H and k' are smaller than G and k, respectively, since in the reduction process, a number of vertices in a minimum dominating set of H have been decided. This reduction process reduces the sublinear exponent from $c^{\sqrt{k}}$ to $c^{\sqrt{k'}}$ and thus improves the running time of the fixed-parameter algorithms for the PLANAR DOMINATING SET problem. This result is used in FT Algorithm which has three major steps [16]: Step I computes a kernel H of G by the data reduction process of [6] in $O(n^3)$ time. Step II finds an optimal branch decomposition of H with width $bw(H)$. This can be done by algorithms of [9,18] in $O(k^3)$ time. Step III computes a minimum dominating set D' of H using the dynamic programming method based on the branch decomposition in $O(2^{3 \log_4 3bw(H)} k)$ time and constructs a minimum dominating set D of G from D' in linear time. It is proved in [16] that the branchwidth $bw(H) \leq 3\sqrt{4.5k'}$ and FT Algorithm has time complexity $O(2^{15.13k'} k + n^3)$. Alber et al. report that the data reduction computes a much smaller kernel in practice for a class of planar graphs [3,6]. Very recently, Bian et al. report that an optimal branch decomposition of a planar graph can be computed efficiently in practice [8,9]. These results provide the base for testing the practical efficiency of FT Algorithm for the PLANAR DOMINATING SET problem.

Although significant theoretical progresses have been made towards the fixed-parameter algorithms for the PLANAR DOMINATING SET problem, the authors are not aware of any report on the practical performances of these algorithms. In this paper, we report the computational study on FT Algorithm for the PLANAR DOMINATING SET problem. In our implementation of FT Algorithm, in addition to the data reduction rules of [3,6], we introduce new data reduction rules and use the recent works on planar branch decompositions. The new data reduction rules further reduce the kernel size and improve the running time of FT Algorithm. We have tested our implementation of FT Algorithm on several classes of planar graphs, including the maximal planar graphs and their subgraphs from LEDA [2,25], Delaunay triangulations of point sets taken from TSPLIB [26], triangulations and intersection graphs of segments from LEDA, Gabriel graphs, and planar graphs from PIGALE library [1]. The computational results show that the size of instances that can be solved in a practical time mainly depends on the branchwidth of the kernels. For example, the maximal planar graphs and their subgraphs have branchwidth at most four. This class of graphs are used as the test instances for the data reduction in previous studies [3,6]. Step I reduces the problem size significantly (often finds the solution already) and the PLANAR DOMINATING SET problem can be solved efficiently for very large instances in this class. On the other hand, for Delaunay triangulation and Gabriel graphs, because the branchwidth of kernels increases fast in instance size, the size of instances that can be solved in a practical time is limited to a few hundreds edges. For triangulation graphs, intersection graphs, and graphs from PIGALE library, the branchwidth of kernels increases slowly or does not grow in instance size, instances of size up to about ten thousands edges can be solved in a practical time. These results coincide with the theoretical analysis of FT Algorithm [16]: it runs exponentially in the branchwidth of the kernel and $k \geq b(bw(G))^2$ for some constant b. Because the kernel of G has $O(k)$ vertices, the analysis suggests that a large branchwidth of the instance implies a large kernel, Step I may not reduce the problem size much, and the kernel may have a large branchwidth. For a kernel H with large branchwidth, FT Algorithm is not practical because Step III of the algorithm runs exponentially in the branchwidth of H.

The results of this paper give a concrete example on using branch-decomposition based algorithms for solving important hard problems in planar graphs and show that the PLANAR DOMINATING SET problem can be solved in practice for some applications. This work may bring the theory of branch decomposition closer to practice.

The rest of the paper is organized as follows. In the next section, we review FT Algorithm. We introduce the data reduction rules in Section 3. Computational results of FT Algorithm are reported in Section 4. The final section concludes the paper.

2   Fomin and Thilikos Algorithm

We first introduce some definitions and terminology. Readers may refer to a textbook on graph theory (e.g., the one by West [34]) for basic definitions and terminology on graphs. In this paper, graphs are undirected unless otherwise stated. Let G be a graph with vertex set $V(G)$ and edge set $E(G)$. A *branch decomposition* of G is a tree T_B such that the set of leaves of T_B is $E(G)$ and each internal node of T_B has node degree three. For each link e of T_B, removing e separates T_B into two subtrees. Let E' and E'' be the sets of leaves of the subtrees. Let S_e be the set of vertices of G incident to both an edge of E' and an edge of E''. The width of e is $|S_e|$ and the width of T_B is the maximum width of all links of T_B. The *branchwidth* $bw(G)$ of G is the minimum width of all branch-decompositions. We call a link $e = \{x, y\}$ a leaf link if one of x and y is a leaf node of T_B, otherwise an internal link. Notice that S_e is a set which separates G into two subgraphs induced by edges of E' and E'', respectively.

We say a vertex u is dominated by a vertex v if u and v are adjacent. A vertex set U is dominated by a vertex set V if for every vertex $u \in U$ there is a vertex $v \in V$ such that u and v are adjacent or $u \in V$. Given two graphs G and H, we say $size(H) \leq size(G)$ if $|V(H)| \leq |V(G)|$ and $|E(H)| \leq |E(G)|$. In the rest of the paper, the PLANAR DOMINATING SET problem is used for the optimization version of the problem unless otherwise stated.

Now we briefly review FT Algorithm. Readers may refer to [16] for more details. FT Algorithm solves the PLANAR DOMINATING SET problem of G in three steps. Step I computes a kernel H of G by the data reduction process such that $size(H) \leq size(G)$, $\gamma(H) \leq \gamma(G)$, and a minimum dominating set D of G can be computed from a minimum dominating set D' of H in linear time. Step II finds an optimal branch decomposition T_B of H. Step III computes a minimum dominating set D' of H using the dynamic programming method based on T_B and constructs a minimum dominating set D of G from D'.

In Step I, the principle of data reduction introduced in [6] is that based on some rules we check the vertices of G to decide if some vertices can be included into D or excluded for computing D. More specifically, each vertex v of G is colored by black or grey as follows. Initially, every vertex v is colored grey, meaning that whether v should be included in D or not has not been decided. If v has been decided to be included in D, v is colored black. If v has been decided to be excluded for computing D in the future, v is removed from G. After the reduction process, we get a kernel $H(B \cup C, E)$, where B and C are the sets of black and grey vertices, respectively. The specific reduction rules will be introduced in the next section.

To compute an optimal branch decomposition T_B of H, either the edge-contraction algorithms [18,32] or the divide-and-conquer algorithms [9] can be used. The divide-and-conquer algorithms are faster for large graphs in practice.

In Step III, given a kernel $H = (B \cup C, E)$, we find a minimum $D' \subseteq (B \cup C)$ such that $D' \supseteq B$ and D' dominates all vertices of C. As shown later, a minimum dominating set D of G can be constructed from D' in linear time. To compute D', first the branch decomposition T_B of H is converted into a rooted binary tree by replacing a link $\{x, y\}$ of T_B by three links $\{x, z\}, \{z, y\}$, and $\{z, r\}$, where z and r are new nodes to T_B, r is the root, and $\{z, r\}$ is an internal link. For every internal link e of T_B, e has two children links incident to e. For every link e of T_B, let T_e be the subtree of T_B consisting of all descendant links of e. Let H_e be the subgraph of H induced by the edges at leave nodes of T_e. To compute a minimum dominating set D' of H, we find all dominating sets (solutions) of H_e from which D' may be constructed for every link e of T_B by a dynamic programming method: the solutions of H_e for each leaf link e is computed by enumeration and the solutions for an internal link e is computed by merging the solutions for the children links of e. To find a solution of H_e, each vertex of S_e is colored by one of the following colors.

Black. denoted by 1, meaning that the vertex is included in the dominating set.

White. denoted by 0, meaning that the vertex is dominated at the current step of the algorithm and is not in the dominating set.

Grey. denoted by $\hat{0}$, meaning that we have not decided to color the vertex into black or white yet at the current step.

A solution of H_e subject to a coloring $\lambda \in \{0, \hat{0}, 1\}^{|S_e|}$ is a minimum set $D_e(\lambda)$ satisfying

- for $u \in B \cap S_e$, $\lambda(u)$ is black;
- every vertex of $V(H_e) \setminus S_e$ is dominated by a vertex of $D_e(\lambda)$; and
- for every vertex $u \in S_e$ if $\lambda(u)$ is black then $u \in D_e(\lambda)$, if $\lambda(u)$ is white then $u \notin D_e(\lambda)$ and u is dominated by a vertex of $D_e(\lambda)$.

Intuitively, $D_e(\lambda)$ is a minimum set to dominate the vertices of H_e with grey vertices removed, subject to the condition that the vertices of S_e are colored by λ.

For a leaf link e, colorings λ and sets $D_e(\lambda)$ are computed by enumeration. An internal link e has children edges e_1 and e_2 in T_B. The colorings λ of S_e and sets $D_e(\lambda)$ are computed from the colorings λ_1 of S_{e_1}, sets $D_{e_1}(\lambda_1)$, colorings λ_2 of S_{e_2}, and sets $D_{e_2}(\lambda_2)$. A coloring λ of S_e is formed from λ_1 and λ_2 if:

- For $u \in S_e \setminus S_{e_2}$, $\lambda(u) = \lambda_1(u)$.
- For $u \in S_e \setminus S_{e_1}$, $\lambda(u) = \lambda_2(u)$.
- For $u \in S_e \cap S_{e_1} \cap S_{e_2}$, if $\lambda_1(u) = \lambda_2(u) = 1$ then $\lambda(u) = 1$; if $\lambda_1(u) = \lambda_2(u) = \hat{0}$ then $\lambda(u) = \hat{0}$; and if $\lambda_1(u) = 0$ and $\lambda_2(u) = \hat{0}$, or $\lambda_1(u) = \hat{0}$ and $\lambda_2(u) = 0$ then $\lambda(u) = 0$.
- For $u \in (S_{e_1} \cup S_{e_2}) \setminus S_e$, $\lambda_1(u) = \lambda_2(u) = 1$, or $\lambda_1(u) = 0$ and $\lambda_2(u) = \hat{0}$, or $\lambda_1(u) = \hat{0}$ and $\lambda_2(u) = 0$.

For a coloring λ of S_e formed from λ_1 and λ_2, the minimum dominating set $D_e(\lambda)$ is the minimum set among the sets of $D_{e_1}(\lambda_1) \cup D_{e_2}(\lambda_2)$. For $e = \{z, r\}$, a minimum set $D_e(\lambda)$ among all colorings λ of S_e is a minimum dominating set of H.

3   Data Reduction

In this section, we introduce the data reduction rules used in our implementation of FT Algorithm for Step I. All reduction rules of [3,6] are used. To enhance the data reduction effect, we also propose some new reduction rules. Following the convention of FT Algorithm, we color each vertex of G by black or grey, and may remove some vertices from G by those reduction rules. After the data reduction step, we get a kernel $H(B \cup C, E)$, recall that B and C are the sets of black and grey vertices, respectively. For a vertex v, let $N(v) = \{u | \{u, v\} \in E(G)\}$, $N[v] = N(v) \cup \{v\}$, $B(v) = B \cap N(v)$, and $C(v) = C \cap N(u)$. For a set U of vertices, let $N(U) = \cup_{v \in U} N(v)$. For a vertex u, if there is a black vertex $v \in N[u]$, we mark u *dominated*. Initially, every vertex of G is unmarked. In the data reduction step, some vertices are marked. Let X be the set of marked vertices and Y be the set of unmarked vertices. For $v \in V(G)$, the following is introduced in [6]:

$$N_1(v) = B(v) \cup \{u | u \in C(v), N(u) \setminus N[v] \neq \emptyset\},$$
$$N_2(v) = \{u | u \in N(v) \setminus N_1(v), N(u) \cap N_1(v) \neq \emptyset\}, \text{ and}$$
$$N_3(v) = N(v) \setminus (N_1(v) \cup N_2(v)).$$

Rule 1 [6]. For $v \in V(G)$, if $N_3(v) \cap Y \neq \emptyset$ then remove $N_2(v)$ and $N_3(v)$ from G, color v black, and mark $N[v]$ dominated.

For a pair of vertices $v, w \in V(G)$, let $N(v, w) = N(v) \cup N(w) \setminus \{v, w\}$, $B(v, w) = B \cap N(v, w)$, $C(v, w) = C \cap N(v, w)$, and $N[v, w] = N[v] \cup N[w]$. The following is introduced in [6]:

$$N_1(v, w) = B(v, w) \cup \{u | u \in C(v, w), N(u) \setminus N[v, w] \neq \emptyset\},$$
$$N_2(v, w) = \{u | u \in N(v, w) \setminus N_1(v, w), N(u) \cap N_1(v, w) \neq \emptyset\},$$
$$N_3(v, w) = N(v, w) \setminus (N_1(v, w) \cup N_2(v, w)).$$

Rule 2 [6]. For $v, w \in V(G)$ with both v and w grey, assume that $|N_3(v, w) \cap Y| \geq 2$ and $N_3(v, w) \cap Y$ can not be dominated by a single vertex of $N_2(v, w) \cup N_3(v, w)$.

Case 1: $N_3(v, w) \cap Y$ can be dominated by a single vertex of $\{v, w\}$.

- (1.1) If $N_3(v, w) \cap Y \subseteq N(v)$ and $N_3(v, w) \cap Y \subseteq N(w)$ then remove $N_3(v, w)$ and $N_2(v, w) \cap N(v) \cap N(w)$ from G and add new gadget vertices z and z' with edges $\{v, z\}, \{w, z\}, \{v, z'\}$, and $\{w, z'\}$ to G.
- (1.2) If $N_3(v, w) \cap Y \subseteq N(v)$ but $N_3(v, w) \cap Y \not\subseteq N(w)$ then remove $N_3(v, w)$ and $N_2(v, w) \cap N(v)$ from G, color v black, and mark $N[v]$ dominated.
- (1.3) If $N_3(v, w) \cap Y \subseteq N(w)$ but $N_3(v, w) \cap Y \not\subseteq N(v)$ then remove $N_3(v, w)$ and $N_2(v, w) \cap N(w)$ from G, color w black, and mark $N[w]$ dominated.

Case 2: If $N_3(v, w) \cap Y$ can not be dominated by a single vertex of $\{v, w\}$ then remove $N_2(v, w)$ and $N_3(v, w)$ from G, mark v and w black, and mark $N[v, w]$ dominated.

In Rule 1 and Rule 2 (Cases 1.2, 1.3, and 2) of [6], gadget vertices are used to guarantee some vertices to be included in the solution set. In [3] the rules are implemented in a way that the vertices to be included in the solution set are removed. Our descriptions are slightly different from the previous ones: we do not use gadget vertices nor remove the vertices to be included to the solution set but color them black. Our descriptions allow us to have new reduction rules given below that may further reduce the size of the kernel.

Rule 3

3.1: For $v, w \in V(G)$ with v black and w grey, if $(N_3(v, w) \cap Y) \setminus N(v) \neq \emptyset$ then remove $N_2(v, w) \cup N_3(v, w)$, color w black, and mark $N[w]$ dominated; otherwise remove $(N_2(v, w) \cup N_3(v, w)) \cap N(v)$.

3.2: For $v, w \in V(G)$ with v grey and w black, if $(N_3(v, w) \cap Y) \setminus N(w) \neq \emptyset$ then remove $N_2(v, w) \cup N_3(v, w)$, color v black, and mark $N[w]$ dominated; otherwise remove $(N_2(v, w) \cup N_3(v, w)) \cap N(w)$.

3.3: For $v, w \in V(G)$ with both v and w black, remove $N_2(v, w) \cup N_3(v, w)$.

Lemma 1. *Given a graph G, let G' be the graph obtained by applying Rule 3 for $v, w \in V(G)$. Then $size(G') \leq size(G)$, $\gamma(G') \leq \gamma(G)$, and a minimum dominating set D' of G' that contains all black vertices of G' is a minimum dominating set of G that contains all black vertices of G.*

Proof: For $v, w \in V(G)$ with v black and w grey, assume that $(N_3(v, w) \cap Y) \setminus N(v) \neq \emptyset$. For $u \in (N_3(v, w) \cap Y) \setminus N(v)$ and x which dominates u, $x \in \{w\} \cup N_2(v, w) \cup N_3(v, w)$. Since $N(N_2(v, w) \cup N_3(v, w)) \subseteq N[v] \cup N[w]$, we should include w into D to dominate $(N_3(v, w) \cap Y) \setminus N(v)$. Therefore, we can remove $N_2(v, w) \cup N_3(v, w)$ from G. Assume that $(N_3(v, w) \cap Y) \setminus N(v) = \emptyset$. For $u \in (N_2(v, w) \cup N_3(v, w)) \cap N(v)$, u is dominated by v and $N(v) \cup N(u) \subseteq N(v) \cup N(w)$. This implies that we can at least include w rather than u to get D. At this point, we can not decide if we should include w into D or not because there might be a vertex x with $N(w) \subseteq N(x)$ that should be included in D. But we can exclude $(N_2(v, w) \cup N_3(v, w)) \cap N(v)$ from D. Since $(N_2(v, w) \cup N_3(v, w)) \cap N(v)$ is dominated by v, we can remove $(N_2(v, w) \cup N_3(v, w)) \cap N(v)$ from G. This completes the proof for (3.1).

The proof for (3.2) is a symmetric argument of that for (3.1).

For $v, w \in V(G)$ with both v and w black, since $N(N_2(v, w) \cup N_3(v, w)) \subseteq N[v] \cup N[w]$, we can remove $N_2(v, w) \cup N_3(v, w)$ from G. $\qquad\square$

Rule 4 [3]

4.1: Delete edges between vertices of X (vertices marked dominated).

4.2 If $u \in X$ has $|C(u)| \leq 1$ then remove u.

4.3 For $u \in X$ with $C(u) \cap Y = \{u_1, u_2\}$, if u_1 and u_2 are connected by a path of length at most 2 then remove u.

4.4 For $u \in X$ with $C(u) \cap Y = \{u_1, u_2, u_3\}$, if $\{u_1, u_2\}, \{u_2, u_3\} \in E(G)$ then remove u.

To perform the data reduction, we first apply Rule 1 for every vertex of G. Next for every pair of vertices v and w of G, we apply either Rule 2 or Rule 3 depending on the colors of v and w. Then we apply Rule 4. We repeat the above until Rules 1-4 do not change the graph. From the results of [6,3] on Rules 1,2, and 4, and Lemma 1, we have the following result.

Theorem 1. *Given a planar graph G, let $H(B \cup C, E)$ be the kernel obtained by applying the reduction rules described above and D' be a minimum vertex set of $H(B \cup C, E)$ such that $D' \supseteq B$ and D' dominates C. Then a minimum dominating set D of G can be constructed from D' in linear time.*

Given a planar graph G, let $H(B \cup C, E)$ be the kernel obtained from Step I, T_B be an optimal branch decomposition of H, and $l(H) = \max\{|C \cap S_e|, e \in E(T_B)\}$. It is shown in [6] that $H(B \cup C, E)$ can be computed in $O(n^3)$ time. T_B can be computed by either the edge-contraction algorithm [18] or a divide-and-conquer algorithm [9] in $O(|E(H)|^3)$ time. It is shown in [16] that Step III has time complexity $O(2^{3 \log_4 3l(H)}|E(H)|)$. Therefore, FT Algorithm takes $O(2^{3 \log_4 3l(H)}|E(H)| + n^3)$ time to solve the PLANAR DOMINATING SET problem. In what follows, we use $l(H)$ for the branchwidth of kernel H.

4   Computational Results

We implemented FT Algorithm and tested our implementation on six classes of planar graphs from some libraries including LEDA [2,25] and PIGALE [1]. LEDA generates two types of planar graphs. One type of graphs are the random maximal planar graphs and their subgraphs and the other type of graphs are the planar graphs based on some geometric properties, including the Delaunay triangulations and triangulations of points, and the intersection graphs of segments, uniformly distributed in a two-dimensional plane. Instances of Class (1) are the random maximal graphs and their subgraphs generated by LEDA. This class of instances have been used by Alber et al. in their study on the data reduction rules used in Step I [3,6]. Instances of Class (2) are Delaunay triangulations of point sets taken from TSPLIB [26]. Instances of Classes (3) and (4) are the triangulations and intersection graphs generated by LEDA, respectively. Instances of Class (5) are Gabriel graphs of the points uniformly distributed in a two-dimensional plane. Instances of Classes (2)-(5) are graphs based some geometric properties. The DOMINATING SET problem on those graphs has important applications such as the virtual backbone design of wireless networks [24]. Instances of Class (6) are random planar graphs generated by the PIGALE library [1]. PIGALE provides a number of planar graph generators. We used a function in the PIGALE library that randomly generates one of all possible 2-connected planar graphs with a given number of edges based on the algorithms of [31].

Step I of FT Algorithm is implemented as described in the previous section. To compute an optimal branch decomposition T_B, we use the divide-and-conquer algorithm [9]. In Step III, to save memory, we compute the colorings λ and sets $D_e(\lambda)$ for each link e of T_B in the postorder. Once the colorings λ and sets $D_e(\lambda)$ are computed for a link e, the solutions for the children links of e are discarded. We sort the tables for the colorings to have an efficient implementation of Step III. The computer used for testing has an AMD Athlon(tm) 64 X2 Dual Core Processor 4600+ (2.4GHz) and 4Gbyte memory. The operating system is SUSE Linux 10.2 and the programming language used is C++.

We report the computational results of FT Algorithm in Table 1. For Step I, we give the number $|B|$ of vertices of an optimal dominating set decided in the data reduction and the running time of the step. For Step II, we give the size $|E(H)|$ and branchwidth $l(H) = \max\{|C \cap S_e|, e \in E(T_B)\}$ of kernel H, and the running time of the step. For Step III, we give the dominating number $\gamma(G)$ obtained by FT Algorithm and the running time of the step. The running time is in seconds, and Steps I, II, and III have time complexities $O(|E(G)|^3)$, $O(|E(H)|^3)$, and $O(2^{3\log_4 3l(H)}|E(H)|)$, respectively. We use the number of edges to express the size of an instance or a kernel.

It is easy to show that the instances of Class (1) have branchwidth at most four. These instances have small kernels and Step I is very effective. For the instances included in the table, $|B|$ is very close to $\gamma(G)$ (i.e., Step I finds most vertices in an optimal dominating set) and the kernels are much smaller than the original instances. For some smaller instances not reported in the table, Step I already finds optimal dominating sets. Because the kernels have small size and branchwidth, FT Algorithm is efficient for the instances in this class, for example, an optimal dominating set can be computed for large instances of size up to about 40,000 edges in about 20 minutes.

Table 1. Computational results of FT Algorithm for instances of Classes (1)-(6)

Class	Graph G	$|E(G)|$	$bw(G)$	Step I		Step II			Step III		total
				$|B|$	time	$|E(H)|$	$l(H)$	time	$\gamma(G)$	time	time
(1)	max1500	4047	4	209	4	23	2	< 1	211	< 1	4
	max6000	7480	4	2214	55	32	2	< 1	2219	< 1	55
	max8000	13395	4	2186	336	194	3	< 1	2211	< 1	337
	max11000	28537	4	1679	815	208	4	1	1695	< 1	816
	max13500	38067	4	1758	1203	302	3	1	1779	< 1	1204
(2)	pr144	393	9	2	< 1	291	6	1	20	1	3
	ch130	377	10	0	< 1	377	10	1	21	12734	12735
	kroB150	436	10	0	< 1	436	10	1	23	43094	43095
	pr226	586	7	12	1	126	6	< 1	21	< 1	2
	pr299	864	11	1	1	824	11	1	47	392931	392933
(3)	tri1000	2980	7	69	10	1657	7	4	163	26	40
	tri2000	5977	8	136	56	3192	7	146	321	120	322
	tri3000	8976	8	209	87	4805	7	379	489	190	656
	tri4000	11969	9	252	251	6888	7	1667	653	413	2331
	tri5000	14969	8	384	285	7271	8	1547	804	915	2747
(4)	rand2000	3247	8	371	8	1219	7	1	548	14	23
	rand3000	4943	10	514	19	2093	8	3	806	173	195
	rand4000	6676	11	678	35	2956	8	4	1068	217	256
	rand5000	8451	11	755	57	4177	8	13	1315	363	433
	rand6000	10293	11	839	93	5598	9	25	1563	2933	3051
(5)	Gab100	182	7	3	< 1	162	7	< 1	24	5	6
	Gab200	366	8	3	< 1	344	8	1	47	192	193
	Gab300	552	10	5	< 1	516	10	32	70	28014	28046
(6)	p1277	2128	9	116	8	1353	9	14	323	1953	1975
	p2518	4266	9	329	31	1876	5	26	621	3	60
	p4206	7101	6	596	75	2901	5	7	1039	2	84
	p5995	10092	7	708	181	5142	5	20	1504	6	207
	p7595	12691	6	998	259	5702	5	16	1893	7	272

For Class (2) and (5), the branchwidth of instance increases fast in instance size (e.g., Class (2) instances rd400 of 1,183 edges and u2152 of 6,312 edges have branchwidth 17 and 31, respectively, Class (5) instances Gab500 of 932 edges and Gab2000 of 3,911 edges have branchwidth 12 and 26, respectively). For the instances tested, the kernel H of an instance G has the same branchwidth as that of G ($l(H) = bw(G)$) and has the same size as or only slightly smaller than that of G. The size of those instances for which the PLANAR DOMINATING SET problem can be solved in a practical time is limited to a few instances of size up to only a few hundreds edges. The computation time for Instances ch130, kroB150, and pr299 in Class (2) is significantly larger than that for Instances pr144 and pr226 in the same class. As shown in the table, this huge difference comes from the difference between the branchwidthes of kernels (Step III), the kernels of Instances ch130 and kro150 have branchwidth 10 while those of pr144 and pr226 have branchwidth 6. This coincides with the theoretical time complexity of FT

Algorithm which runs exponentially in $l(H)$. Similar difference is observed for Instances Gab100 and Gab300 of Class (5) as well.

For Classes (3), (4), and (6), the branchwidth of instance increases slowly or does not grow in instance size. The data reduction is effective for instances in these classes. For most instances, the kernel size is at most half of the instance size and the branchwidth of the kernel is usually smaller than that of the instance as well. Our data show that the PLANAR DOMINATING SET problem can be solved for instances in these classes of size up to about 10,000 edges in a practical time. For large instances, the size $|E(H)|$ of kernel H is also important to the running time of Step III. For example, FT Algorithm takes more time to solve Instance rand6000 than that for rand2000. The time difference comes from the differences of both $l(H)$ and $|E(H)|$.

Due to the page limit, Table 1 only contains the instances well scaled within some size ranges. We have tested FT Algorithm on more instances. The results are similar to those in Table 1, the running time mainly depends on $l(H)$ and then $|E(H)|$. For a kernel H with large $l(H)$, Step III is time consuming, because this step runs exponentially in $l(H)$. Our computational results suggest that it may not be practical to use FT Algorithm to solve the PLANAR DOMINATING SET problem of instances with $l(H) > 10$ on a PC with a CPU of about 3GHz (e.g., it takes more than 100 hours to solve the instance pr299 with 864 edges and $l(H) = 11$).

Both the theoretical analysis and computational study suggest that computing a kernel H with smaller $l(H)$ and $|E(H)|$ is a most effective way to improve the efficiency of FT Algorithm. For this purpose, we proposed new reduction rules (Rule 3). Recall that H is the kernel obtained by new reduction rules (Rules 1,2,3, and 4) and let H' be the kernel obtained by applying only the previous known reduction rules (Rules 1,2, and 4). Since all nodes colored black (resp. nodes deleted) by previous rules are also colored (resp. deleted) by new rules, $l(H) \leq l(H')$ and $|E(H)| \leq |E(H')|$. For Classes (2) and (5), $l(H) = l(H') = bw(G)$ for all instances testes and $|E(H)| = |E(H')| = |E(G)|$ for most instances, that is, the effect of data reduction is very limited. However, for instances in other classes, data reduction is effective and our new rules improve the efficiency of FT Algorithm. For instances of Classes (1),(3),(4), and (6), Table 2 shows the computational results of FT Algorithm when previous rules and new rules are used. In the table, t_{old} and t_{new} (resp. $|B'|$ and $|B|$) are the total running times (resp. the numbers of vertices in an optimal dominating set decided in Step I) when previous rules and new rules are used, respectively. The data show that $l(H) = l(H')$ and $|E(H)| < |E(H')|$ for most instances. The total running time is improved when new rules are used: $t_{new} < t_{old}$ for all instances in the table. The improvement is instance dependent and t_{new}/t_{old} varies from 48% to 97%. The average of t_{new}/t_{old} over the five instances of Class (1) is about 90%. Similarly, the averages of t_{new}/t_{old} for Classes (3),(4), and (6) are about 70%, 85%, and 90%, respectively. The improvement of the total running time is obtained mainly from Step III. The running time of Step I when new rules are used is about the

Table 2. The results of using new data reduction rules and without using the new rules in Step I

Class	Graph G	$|E(G)|$	$bw(G)$	Results without new rules				Results with new rules			
				$|B'|$	$|E(H')|$	$l(H')$	time	$|B|$	$|E(H)|$	$l(H)$	time
(1)	max1500	4047	4	209	23	2	5	209	23	2	4
	max6000	7480	4	2212	41	2	58	2214	32	2	55
	max8000	13395	4	2183	218	3	357	2186	194	3	337
	max11000	28537	4	1671	287	4	893	1679	208	4	816
	max13500	38067	4	1752	362	4	1294	1758	302	3	1204
(3)	tri1000	2980	7	63	1752	7	84	69	1657	7	40
	tri2000	5977	8	102	3787	7	490	136	3192	7	322
	tri3000	8976	8	175	5442	7	877	209	4805	7	656
	tri4000	11969	9	214	7541	7	2499	252	6888	7	2331
	tri5000	14969	8	333	8201	8	4118	384	7271	8	2747
(4)	rand2000	3247	8	361	1293	7	25	371	1219	7	23
	rand3000	4943	10	512	2120	8	216	514	2093	8	195
	rand4000	6676	11	669	3043	8	263	678	2956	8	256
	rand5000	8451	11	748	4254	8	474	755	4177	8	433
	rand6000	10293	11	832	5675	9	5586	839	5598	9	3051
(6)	p1277	2128	9	112	1371	9	2134	116	1353	9	1975
	p2518	4266	9	291	2139	5	67	329	1876	5	60
	p4206	7101	6	555	3189	5	91	596	2901	5	84
	p5995	10092	7	652	5508	5	297	708	5142	5	207
	p7595	12691	6	925	6159	5	281	998	5702	5	272

same as that when previous rules are used (instance dependent) and we omit the details here due to the page limit.

5   Concluding Remarks

We tested the practical performances of FT Algorithm on a wide range of planar graphs. The computational results coincide with the theoretical analysis of the algorithm, it is efficient for graphs with small branchwidth but may not be practical for graphs with large branchwidth. By a PC with a CPU of about 3GHz, it is possible to solve the PLANAR DOMINATING SET problem for graphs with the branchwidth of their kernels at most 10 in a few hours. Since FT Algorithm runs exponentially in the branchwidth $l(H)$ of a kernel H for a given graph, it is worth to develop more powerful data reduction rules to reduce $l(H)$. Another research direction is to develop heuristics to reduce $l(H)$ to compute approximate solutions for the PLANAR DOMINATING SET problem by branch-decomposition based algorithms. Those heuristics should provide solutions very close to the optima but runs faster than FT Algorithm for graphs with large branchwidth. It is also interesting to find heuristics which are efficient in practice and have guaranteed performance for the Planar Dominating Set problem.

Acknowledgement

The authors thank anonymous reviewers for constructive comments. The work was partially supported by NSERC Research Grant of Canada and Research Grant Council of Hong Kong (Project No. CityU 114307).

References

1. Public Implementation of a Graph Algorithm Library and Editor (2008), http://pigale.sourceforge.net/
2. The LEDA User Manual, Algorithmic Solutions, Version 4.2.1 (2008), http://www.mpi-inf.mpg.de/LEDA/MANUAL/MANUAL.html
3. Alber, J., Betzler, N., Niedermeier, R.: Experiments on data reduction for optimal domination in networks. In: Proc. of the International Network Optimization Conference (INOC 2003), pp. 1–6 (2003)
4. Alber, J., Bodlaender, H.L., Fernau, H., Kloks, T., Niedermeier, R.: Fixed parameter algorithms for dominating set and related problems on planar graphs. Algorithmica 33, 461–493 (2002)
5. Alber, J., Fan, H., Fellows, M., Fernau, H., Niedermeier, R.: Refined search tree technique for dominating set on planar graphs. In: Sgall, J., Pultr, A., Kolman, P. (eds.) MFCS 2001. LNCS, vol. 2136, pp. 111–122. Springer, Heidelberg (2001)
6. Alber, J., Fellows, M.R., Niedermeier, R.: Polynomial time data reduction for dominating set. Journal of the ACM 51(3), 363–384 (2004)
7. Baker, B.S.: Approximation algorithms for np-complete problems on planar graphs. Journal of ACM 41, 153–180 (1994)
8. Bian, Z., Gu, Q., Marzban, M., Tamaki, H., Yoshitake, Y.: Empirical study on branchwidth and branch decomposition of planar graphs. In: Proc. of the 9th SIAM Workshop on Algorithm Engineering and Experiments (ALENEX 2008), pp. 152–165 (2008)
9. Bian, Z., Gu, Q.-P.: Computing branch decomposition of large planar graphs. Technical report, SFU-CMPT-TR 2008-04, School of Computing Science, Simon Fraser University (2008)
10. Coppersmith, D., Winograd, S.: Matrix multiplication via arithmetic progressions. Journal of Symbolic Computation 9, 251–280 (1990)
11. Dorn, F.: Dynamic programming and fast matrix multiplication. In: Azar, Y., Erlebach, T. (eds.) ESA 2006. LNCS, vol. 4168, pp. 280–291. Springer, Heidelberg (2006)
12. Dorn, F.: How to use planarity efficiently: new tree-decomposition based algorithms. In: Proc. of the 33rd International Workshop on Graph-Theoretic Concepts in Computer Science (WG 2007). LNCS, vol. 4769, pp. 280–291 (2007)
13. Downey, R.G., Fellow, M.R.: Parameterized complexity. In: Monographs in Computer Science. Springer, Heidelberg (1999)
14. Downey, R.G., Fellows, M.R.: Fixed parameter tractability and completeness. Cong. Num. 87, 161–187 (1992)
15. Fiege, U.: A threshold of $\ln n$ for approximating set cover. Journal of ACM 45, 634–652 (1998)
16. Fomin, F.V., Thilikos, D.M.: Dominating sets in planar graphs: branch-width and exponential speed-up. SIAM Journal on Computing 36(2), 281–309 (2006)
17. Garey, M.R., Johnson, D.S.: Computers and Intractability, a Guide to the Theory of NP-Completeness. Freeman, New York (1979)

18. Gu, Q.P., Tamaki, H.: Optimal branch decomposition of planar graphs in $O(n^3)$ time. In: Caires, L., Italiano, G.F., Monteiro, L., Palamidessi, C., Yung, M. (eds.) ICALP 2005. LNCS, vol. 3580, pp. 373–384. Springer, Heidelberg (2005)
19. Haynes, T.W., Hedetniemi, S.M., Hedetniemi, S.T., Henning, M.A.: Domination in graphs applied to electronic power networks. SIAM J. on Discrete Mathematics 15(4), 519–529 (2002)
20. Haynes, T.W., Hedetniemi, S.T., Slater, P.J.: Domination in graphs. In: Monographs and Textbooks in Pure and Applied Mathematics, vol. 209. Marcel Dekker, New York (1998)
21. Haynes, T.W., Hedetniemi, S.T., Slater, P.J.: Fundamentals of domination in graphs. In: Monographs and Textbooks in Pure and Applied Mathematics, vol. 208. Marcel Dekker, New York (1998)
22. Johnson, D.S.: Approximation algorithms for combinatorial problems. Journal of Computer and System Sciences 9, 256–278 (1974)
23. Kanj, I.A., Perkovic, L.: Improved parameterized algorithms for planar dominating set. In: Diks, K., Rytter, W. (eds.) MFCS 2002. LNCS, vol. 2420, pp. 399–410. Springer, Heidelberg (2002)
24. Li, X.-Y.: Algorithmic, geometric and graphs issues in wireless networks. Journal of Wireless Communications and Mobile Computing (WCMC) 6(2), 119–140 (2003)
25. Mehlhorn, K., Näher, S.: LEDA: A Platform for Combinatorial and Geometric Computing. Cambridge University Press, New York (1999)
26. Reinelt, G.: TSPLIB-A traveling salesman library. ORSA J. on Computing 3, 376–384 (1991)
27. Robertson, N., Seymour, P.D.: Graph minors I. Excluding a forest. Journal of Combinatorial Theory Series B 35, 39–61 (1983)
28. Robertson, N., Seymour, P.D.: Graph minors II. Algorithmic aspects of tree-width. Journal of Algorithms 7, 309–322 (1986)
29. Robertson, N., Seymour, P.D.: Graph minors X. Obstructions to tree decomposition. J. of Combinatorial Theory Series B 52, 153–190 (1991)
30. Sanchis, L.A.: Experimental analysis of heuristic algorithms for the dominating set problem. Algorithmica 33, 3–18 (2002)
31. Schaeffer, G.: Random sampling of large planar maps and convex polyhedra. In: Proc. of the 31st Annual ACM Symposium on the Theory of Computing (STOC 1999), pp. 760–769 (1999)
32. Seymour, P.D., Thomas, R.: Call routing and the ratcatcher. Combinatorica 14(2), 217–241 (1994)
33. Wan, P.J., Alzoubi, K.M., Frieder, O.: A simple heuristic for minimum connected dominating set in graphs. International Journal of Found. Comput. Sci. 14(2), 323–333 (2003)
34. West, D.B.: Introduction to Graph Theory. Prentice Hall Inc., Upper Saddle River (1996)

Optimal Movement of Mobile Sensors for Barrier Coverage of a Planar Region

(Extended Abstract)

B. Bhattacharya[1,*], B. Burmester[2], Y. Hu[1], E. Kranakis[3,*],
Q. Shi[1], and A. Wiese[4]

[1] School of Computing Science, Simon Fraser University, Burnaby, Canada
{binay,yhu1,qshi1}@cs.sfu.ca
[2] Department of Computer Science, Florida State University,
Talahassee, Florida, USA
burmester@cs.fsu.edu
[3] School of Computer Science, Carleton University, Ottawa, Ontario, Canada
kranakis@scs.carleton.ca
[4] Institut für Mathematik, Technische Universität Berlin, Berlin, Germany
hopeneverdies@web.de

Abstract. Intrusion detection, area coverage and border surveillance
are important applications of wireless sensor networks today. They can
be (and are being) used to monitor large unprotected areas so as to detect
intruders as they cross a border or as they penetrate a protected area.
We consider the problem of how to optimally move mobile sensors to the
fence (perimeter) of a region delimited by a simple polygon in order to
detect intruders from either entering its interior or exiting from it. We
discuss several related issues and problems, propose two models, provide
algorithms and analyze their optimal mobility behavior.

1 Introduction

Monitoring and surveillance are two of the main applications of wireless sensor
networks today. Typically, one is interested in monitoring a given geographic region either for measuring and surveying purposes or for reporting various types
of activities and events. Another important application concerns critical security
and safety monitoring systems. One is interested in detecting intruders (or move-
ments thereof) around *critical* infrastructure facilities and geographic delimiters
(chemical plants, forests, etc). As a matter of fact, since the information security
level of the monitoring system might change rapidly because of hostile attacks
targeted at it, research efforts are currently underway to extend the scalability
of wireless sensor networks so that they can be used to monitor international
borders as well. For example, [11] reports the possibility of using wireless sensor
networks for replacing traditional barriers (more than a kilometer long) at both
the building and estate level. Also, "Project 28" concerns the construction of

---

* Research was partially supported by MITACS and NSERC.

B. Yang, D.-Z. Du, and C.A. Wang (Eds.): COCOA 2008, LNCS 5165, pp. 103–115, 2008.
© Springer-Verlag Berlin Heidelberg 2008

a virtual fence as a way to complement a physical fence that will include 370 miles of pedestrian fencing and 300 miles of vehicle barrier (see [8] which reports delays in its deployment along the U.S.-Mexico border).

To begin, we say that a point is *covered* by a sensor if it is within its range. In this paper we will use the concept of *barrier coverage* as used in [11] and which differs from the more traditional concept of *full coverage*. In the latter case one is interested in covering the entire region by the deployment of sensors, while in the former all crossing paths through the region are covered by sensors. Thus, one is not interested in covering the entire deployment region but rather to detect potential intruders by guaranteeing that there is no path through this region that can be traversed undetected by an intruder as it traverses the border. Clearly, barrier coverage is an appropriate model of movement detection that is more efficient than full coverage since it requires less sensors for detecting intruders (this is the case, for example, when the width of the deployment region is three times the range of the sensors).

In [3] the authors consider the problem of how individual sensors can determine barrier coverage *locally*. In particular, they prove that it is possible for individual sensors to locally determine the existence of barrier coverage, even when the region of deployment is arbitrarily curved. Although local barrier coverage does not always guarantee global barrier coverage, they show that for thin belt regions, local barrier coverage almost always provides global barrier coverage. They also consider the concept of *L-local barrier coverage* whereby if the bounding box that contains the entire trajectory of a crossing path has length at most L then this crossing path is guaranteed to be detected by at least one sensor.

Motivation, model and problem statement. Motivated from the works of [3] and [11], in this paper we go beyond by asking a different question not examined by any of these papers. More precisely, given that the mobile sensors have detected the existence of a crossing path (e.g., using any of the above algorithms) how do they reposition themselves *most efficiently* within a specified region so as to repair the existing *security hole* and thereby prevent intruders.

Further, we stipulate the existence of a geometric planar region (the critical region to be protected) delimited by a simple polygon and mobile sensors are lying in the interior of this polygon. We consider a set of mobile sensors (or robots) lying within a region that can move autonomously in the plane. Each sensor has knowledge of the region to be barrier-covered, of its geographic location and can move from its starting position p to a new position p' on the perimeter of this polygon. For each sensor, we look at the distance $d(p, p')$ between the starting and final positions of the sensors, respectively, and investigate how to move the sensors within this region so as to optimize either the *minimum sum* or the *minimum maximum* of the distances covered by the respective sensors. In the sequel we investigate the complexity of this problem for various types of regions and types of movement of the mobile sensors.

Related work. An interesting research article is by [1] which surveys the different kinds of holes that can form in geographically correlated problem areas of wireless sensor networks. The authors discuss relative strengths and short-comings of existing solutions for combating different kinds of holes such as coverage holes, routing holes, jamming holes, sink/black holes, worm holes, etc. [2] looks at critical density estimates for coverage and connectivity of thin strips (or annuli) of sensors. In addition, [5] and [6] design a distributed self deployment algorithm for coverage calculations in mobile sensor networks and consider various performance metrics, like coverage, uniformity, time and distance traveled till the algorithm converges. Related is also the research on art gallery theorems (see [14]) which is concerned with finding the minimal number of positions for guards or cameras so that every point in a gallery is observed by at least one guard or camera.

In addition to the research on barrier coverage already mentioned there is extensive literature on detection and tracking in sensor networks. [12] considers the problem of event tracking and sensor resource management in sensor networks and transforms the detection problem into finding and tracking the cell that contains the point in an arrangement of lines. [9] addresses the problem of tracking multiple targets using a network of communicating robots and stationary sensors by introducing a region-based approach for controlling robot deployment. [16] considers the problem of accurate mobile robot localization and mapping with uncertainty using visual landmarks. Finally, related to the problem of detecting a path through a region that can be traversed undetected by an intruder is the paper [15] which gives necessary and sufficient conditions for the existence of vertex disjoint simple curves homotopic to certain closed curves in a graph embedded on a compact surface.

Outline and results of the paper. Section 2 gives the formal model on a circle and defines the min-max (minimizing the maximum) and min-sum (minimizing the sum) problems for a set of sensors within a circle or a simple polygon. Section 3 looks at the simpler one dimensional case and derives simple optimal algorithms for the case the sensors either all lie on a line or on the perimeter of circle. Section 4 and Section 5 are the core of the paper and provide algorithms solving the min-sum and min-max problems, respectively. That is, in Section 4, an $O(n^{3.5} \log n)$-time algorithm for the min-max problem on a circle and an $O(mn^{3.5} \log n)$-time algorithm for the min-max problem on a simple polygon are proposed (m is the number of edges of the simple polygon). Our approximation algorithms for min-sum problems on a circle or a simple polygon are presented in Section 5. Finally, Section 6 gives the conclusion.

2   Preliminaries and Formal Model

First we describe the formal model on a circle and provide the basic definitions and preliminary concepts.

2.1   Optimization on a Circle

The simpler scenario we envision concerns n mobile sensors which are located in the interior of a unit-radius circular region. A set of n sensors are located inside the disk. Further, assume that the sensors are location aware (i.e., they know their geometric coordinates) and also know the location of the center of the disk. We would like to move all the sensors from their initial positions to the perimeter of the circle so as to 1) form a regular n-gon, and 2) minimize the total/maximum distance covered.

The motivation for placing the sensors on the perimeter is because it provides the most efficient way to protect the disk from intruders. Observe that when all n sensors lie equidistant on the vertices of a regular n-gon, they each need to cover a circular arc of size $2\pi/n$ so as to be able to monitor the entire perimeter. Using elementary trigonometry, it follows easily that the transmission range of each sensor must be equal to $r = \sin(\pi/n)$.

More formally, for n given sensors in positions $A_1, A_2, \ldots, A_n$, respectively, which move to new positions $A'_1, A'_2, \ldots, A'_n$ at the corners of a regular n-gon the total distance covered is $\sum_{i=1}^{n} d(A_i, A'_i)$. Every sensor moves from its current position A_i to a new position A'_i. It is clear that the sum is minimized when each sensor moves to its new position in a straight line.

The reason for having the sensors at the corners of a regular n-gon is because this is evidently the optimal final arrangement that will enable them to detect intruders (i.e., by being equidistant on the perimeter). Thus, since the final position $A'_1 A'_2 \cdots A'_n$ of the sensors forms a regular n-gon it is clear that all possible solutions can be parametrized by using a single angle $0 \le \theta \le 2\pi$. However, a difficulty arises in view of the fact that we must also specify a permutation $\sigma : \{1, 2, \ldots, n\} \to \{1, 2, \ldots, n\}$ of the sensors such that the i-th sensor moves from position $A_{\sigma(i)}$ to the new position A'_i.

Let the n sensors have coordinates (a_i, b_i), for $i = 1, 2, \ldots, n$. Let us parametrize the regular polygon with respect to the angle of rotation say θ. The n vertices of the regular n-gon that lie on the perimeter of the disk can be described by

$$(a_i(\theta), b_i, (\theta)) = \left(\cos\left(\theta + \frac{(i-1)2\pi}{n}\right), \sin\left(\theta + \frac{(i-1)2\pi}{n}\right)\right), \text{ (for } i = 1, 2, \ldots, n),$$

$$(1)$$

respectively, where $(a_i(\theta), b_i(\theta))$ are the vertices of the regular n-gon when the angle of rotation is θ.

Minimizing the sum. The optimization problem is $\min_\theta S_n(\theta)$, where the function $S_n(\theta)$ is defined by $S_n(\theta) := \sum_{i=1}^{n} \sqrt{(a_i - a_i(\theta))^2 + (b_i - b_i(\theta))^2}$, as a function of the angle θ. This of course assumes that the i-th sensor is assigned to position $(\cos(\theta + (i-1)2\pi/n), \sin(\theta + (i-1)2\pi/n))$ on the perimeter. In general, we have to determine the minimum over all possible permutations σ of the sensors. If for a given angle θ and permutation σ we define $S_n(\sigma, \theta) := \sum_{i=1}^{n} \sqrt{(a_{\sigma(i)} - a_i(\theta))^2 + (b_{\sigma(i)} - b_i(\theta))^2}$ then the more general optimization problem is $\min_{\sigma, \theta} S_n(\sigma, \theta)$.

Minimizing the maximum. The previous problem was concerned with minimizing the sum of the distances of the robots to their final destinations. In view of the fact that the robots are moving simultaneously it makes sense to ask for minimizing the maximum of the distances of the robots to their final destinations $\max_{1 \leq i \leq n} d(A_i, A_i')$. The optimization problem is $\min_\theta M_n(\theta)$, where $M_n(\theta) := \max_{1 \leq i \leq n} \sqrt{(a_i - a_i(\theta))^2 + (b_i - b_i(\theta))^2}$, as a function of the angle θ. This of course assumes that the i-th sensor is assigned to position $(\cos(\theta + (i - 1)2\pi/n), \sin(\theta + (i-1)2\pi/n))$ on the perimeter. In general, we have to determine the minimum over all possible permutations σ. If for a given permutation σ we define the following maximum $M_n(\sigma, \theta) := \max_{1 \leq i \leq n} \sqrt{(a_{\sigma(i)} - a_i(\theta))^2 + (b_{\sigma(i)} - b_i(\theta))^2}$ then the general optimization problem is $\min_{\sigma, \theta} M_n(\sigma, \theta)$.

2.2   Optimization on a Simple Polygon

We similarly define the problem of minimizing the sum and minimizing the maximum on a simple polygon as follows.[1] Let P be a simple polygon. (From now on, a polygon is always assumed to be simple.) We denote the boundary of P by ∂P. We assume that ∂P is oriented in the clockwise (also called positive) direction. For any two points $a, c \in \partial P$, we write $\hat{\pi}_P(a, c)$ to denote the set of all points $b \in \partial P$ such that when starting after a in positive direction along ∂P, b is reached before c. Let $p_0, p_1, \ldots, p_{m-1}$ denote the vertices on P ordered in the positive direction. The edges of ∂P are $e_0, e_1, \ldots, e_{m-1}$, where edge e_i has endpoints p_i and p_{i+l}, where $0 \leq i < m$ (i.e., the indices are computed modulo m; e.g., $p_0 = p_m$). We denote by $l(e_i)$ the length of edge $e_i, 0 \leq i < m$, and by $\hat{d}_P(a, b)$ the length of $\hat{\pi}_P(a, b)$ for any two points a and b on ∂P (called *polygonal distance* between a and b). Let $L(P) = \sum_{i=0}^{m-1} l(e_i)$.

We are given n mobile sensors which are located in the interior of P. Each sensor has the knowledge of its geometric coordinates and the simple polygon (i.e., the geometric coordinates of all vertices $p_i, 0 \leq i < n$ and the clockwise ordering of these vertices). The objective is to move all the sensors from their initial positions to ∂P such that 1) the polygonal distance between any two consecutive sensors on the polygon is $L(P)/n$, and 2) minimize the total/maximum distance covered. We postulate that if n given sensors are located at positions $A_1, A_2, \ldots, A_n$, and the destination positions are $A_1, A_2, \ldots, A_n$, respectively, then $\hat{d}_P(A_i', A_{i+1}') = L(P)/n, 0 \leq i < n$.

3   Mobile Sensors in One Dimension

In this section we look at the one dimensional problem and provide efficient algorithmic solutions. In particular, since optimization for the minimum maximum is similar (and simpler than the two dimensional analogue) we provide algorithms only for the minimum sum.

---

[1] Although the approach proposed later (parametric search) will also work for arbitrary simple curves, we refrain from such a generalization so as to avoid unnecessary complications.

3.1   Sensors on a Line Segment

In this model we suppose that the sensors can move on a line segment. Further, instead of protecting a circular range the sensor can now protect an interval of a given size centered at the sensor. Consider the minimum sum optimization problem for the case of n sensors on a line. Without loss of generality assume the segment has length 1 and let the n sensors be at the initial locations $x_0 < x_1 < \cdots < x_{n-1}$, respectively. The destination locations are $\frac{i}{n-1}$, for $i = 0, 1, \ldots, n-1$.

Theorem 1. *The optimal arrangement is obtained by moving point x_i to position $\frac{i}{n-1}$, for $i = 0, 1, \ldots, n-1$, respectively.*

3.2   Sensors on the Perimeter of a Circle

In this model we suppose that the sensors can move on the perimeter of a circle. Further, instead of protecting a circular range the sensor can now protect an arc on the perimeter of a given size centered at the sensor. The same idea as for a line segment should work for the case of a unit circle when the sensors lie on the perimeter of the circle. The main difficulty here is that we no longer have a unique destination. Instead, we can parametrize all possible destinations of the n points by $\phi + \frac{2j\pi}{n}$, for $j = 0, 1, \ldots, n-1$, using a fixed angle $0 \le \phi < \frac{2\pi}{n}$.

Theorem 2. *There is an algorithm that computes an optimal cost arrangement of the sensors.*

When the sensors' movement is in the interior of the circle. In this model we suppose that the sensors and their destination positions are located on the perimeter of a circle and the sensors can move to their destination positions along a straight line. The following theorem is based on the fact there is an optimal solution in which one sensor does not move at all.

Theorem 3. *There is a linear time algorithm that computes an optimal cost arrangement of the sensors.*

4   Min-Max Problem in 2D

In this section we study the problem of minimizing the maximum (min-max problem) on a unit circle and a simple polygon, and provide efficient algorithmic solutions.

4.1   On a Circle

Let $\lambda^*_{m,C}$ be the optimal value of the min-max problem on a circle C, i.e., $\lambda^*_{m,C} = \min_{\sigma,\theta} M_n(\sigma, \theta)$. It is easy to see that $\lambda^*_{m,C}$ is no more than the diameter of the circle $\mathcal{C}$, i.e., $\lambda_{m,C}^* \le 2$. In this section we propose a parametric-searching approach [13] to compute $\lambda_{m,C}^*$.

A non-negative value λ is *feasible* in the min-max problem if all the sensors can move from their initial positions to the perimeter of the circle such that the new positions form a regular n-gon and the maximum moving distance is no more than λ, otherwhise λ is *infeasible*. Clearly, the min-max problem is to compute the minimum feasible value, which is equal to $\lambda_{m,C}^*$.

The remaining part of this section is organized as follows. We first show that a feasibility test of a given value $\lambda(0 \le \lambda \le 2)$ can be performed in time $O(n^{3.5})$. Then, a parametric-searching approach for the min-max problem is presented, which runs in $O(n^{3.5} \log n)$ time.

Algorithm to check the feasibility test of λ. For each $i, 1 \le i \le n$, we construct a circle of radius λ centered at position A_i, denoted by C_i. If a circle C_i for some i is contained in $\mathcal{C}$, then λ is infeasible since sensor A_i cannot move to the perimeter of $\mathcal{C}$ within distance λ. We therefore assume that for each $i, 1 \le i \le n$, either circle C_i contains $\mathcal{C}$ or C_i intersects with $\mathcal{C}$.

For each $i, 1 \le i \le n$, we denote by Q_i the arc of $\mathcal{C}$ that lies in C_i. Let $q_{i(1)}, q_{i(2)}$ be the angles of two endpoints of arc Q_i in clockwise order, $i = 1, \ldots, n$. We let $q_{i(1)} = 0$ and $q_{i(2)} = 2\pi$ if C_i contains $\mathcal{C}$.

The following property is important to our algorithm for the feasibility test of λ. Its proof is omitted here.

Lemma 1. *A given non-negative value λ is feasible if and only if there exists a regular n-gon on the perimeter of $\mathcal{C}$ such that one of its corner points is an endpoint of arc Q_i for some $i(1 \le i \le n)$.*

The algorithm (Algorithm **Check**) to check the feasibility of λ is is formally described below.

Algorithm Check

1. The first step is to sort the angles of endpoints of arcs $Q_i, 1 \le i \le n$. Let $q_1', \ldots, q_{2n}'$ be the angles in increasing order. These angles partition the interval $[0, 2\pi]$ into at most $2n + 1$ pairwise disjoint intervals, denoted by $I_1, \ldots, I_{2n+1}$.
2. For each interval $I_j, 1 \le j \le 2n+1$, we determine the set of sensors, denoted by S_j, that lie within distance λ to its corresponding arc on $\mathcal{C}$.
3. In the third step, we do the following for a regular n-gon with rotation $q_j', i = 1, \ldots, 2n$.
   (a) It is easy to see that the angles of corner points of such regular n-gon are $q_j', (q_j' + \frac{2\pi}{n}) \bmod 2\pi, \ldots, (q_j' + (n-1)\frac{2\pi}{n}) \bmod 2\pi$. We compute the intervals where these angles lie. Let $B_i, i = 1, 2, \ldots, n$ be the corner points.
   (b) Construct a bipartite graph between the set of corner points of the regular n-gon and the set of sensors. An edge is linked between corner point B_k and sensor A_i if $d(A_i, B_k) \le \lambda$ ($1 \le i, k \le n$). The bipartite graph can be obtained from the steps 2 and 3(a) .

(c) Check if there exists a perfect matching. If it is so, terminate the process and return "Feasible".

4. Return "Infeasible" .

It is easy to see that the sorting in the first step can be done in $O(n \log n)$ and the computation of $S_j, j = 1, \ldots, 2n+1$ can be done in $O(n^2)$. In the third step, the process might try all $O(n)$ regular n-gons. For each regular n-gon, it takes $O(n^{2.5})$ time (see [7]). Therefore, we have the following lemma.

Lemma 2. *Whether a given positive value λ is feasible in the min-max problem can be determined in $O(n^{3.5})$ time.*

A parametric-searching approach. Our approach for the solution to the min-max problem is to run Algorithm *Check* parametrically, which has a single parameter λ, without specifying the value of $\lambda^*_{m,C}$ a priori. Note that for a fixed value of the parameter, the algorithm is executed in $O(n^{3.5})$ steps. Imagine that we start the algorithm without specifying a value of the parameter λ. The parameter is restricted to some interval which is known to contain the optimal value $\lambda^*_{m,C}$. (Initially, we may start with the interval $[0, 2]$.) As we go along, at each step of the algorithm we update and shrink the size of the interval, ensuring that it includes the optimal value $\lambda^*_{m,C}$. The final interval contains $\lambda^*_{m,C}$ and any value in it is feasible. Therefore, the minimum value of the final interval is the optimal value $\lambda^*_{m,C}$.

Theorem 4. *The min-max problem on a circle can be solved in $O(n^{3.5} \log n)$ time.*

Note that our algorithm can be easily extended to the model in which all sensors are arbitrarily located on the plane (not restricted to the interior of the circle C).

4.2   On a Simple Polygon

The parametric-searching approach for a circle (described in section 4.1) should work for the case of a polygon where the destination positions of all sensors lie on the perimeter of the polygon. The main difficulty here is that to check the feasibility of a positive value λ, there might be $O(m)$ isolated polygonal chains of ∂P within the circle C_i (of radius λ centered at position A_i) for each sensor A_i. In other words, for a given positive value of λ each sensor will contribute $O(m)$ candidate sets of n destination positions on P instead of at most two candidate sets on a circle. Hence, whether a given positive value λ is feasible in the min-max problem on a simple polygon can be determined by solving $O(mn)$ matching problems of size n. Therefore, the feasibility test of the min-max problem on a simple polygon can be solved in $O(mn^{3.5})$ time.

Theorem 5. *The min-max problem on a simple polygon can be solved in $O(mn^{3.5} \log n)$ time where m is the size of the simple polygon.*

5 Approximation Algorithms for the Min-Sum Problem in 2D

In this section we discuss the problem of minimizing the sum (min-sum problem) on a circle and a simple polygon, and provide approximation solutions for them.

5.1 On a Circle

Let $\lambda_{s,C}^*$ be the optimal value of the min-sum problem on a circle, i.e., $\lambda_{s,C}^* = \min_{\sigma,\theta} S_n(\sigma, \theta)$. We present two approximation algorithms for the min-sum problem. One algorithm (labeled as the *first approach*) has an approximation ratio $\pi + 1$ (section 5.2). The other one (labeled as the *second approach*) uses the first approach as a subroutine to obtain lower and upper bounds of $\lambda_{s,C}^*$ and has an approximation ratio $1 + \epsilon$, where ϵ is an arbitrary constant (Section 5.3).

More notations are introduced as follows. Let $\hat{d}_C(x, y)$ denote the arc distance between two points x and y on the boundary if the cycle C and let $\hat{\pi}_C(x, y)$ denote the arc of length $\hat{d}_C(x, y)$ between x and y. For a point x on C, we denote by $\hat{Q}_x(r)$ the arc consisting of all points y on C such that $\hat{d}_C(x, y) \leq r$.

For each $i = 1, \ldots, n$, let ω_i be the smallest distance between A_i and a point on the cycle C, and we denote by B_i the point on C such that the distance $d(A_i, B_i) = \omega_i$. We note that for each $i = 1, \ldots, n$, B_i is unique if A_i is not located at the center of C. In the case when A_i is located at the center of C, an arbitrary point on C is selected to be B_i. Let $\Omega = \sum_{i=1}^{n} \omega_i$. Obviously, we have the following lemma.

Lemma 3. $\Omega \leq \lambda_{s,C}^*$.

5.2 The First Approach

The first approach (called Algorithm 1) consists of three steps.

Step 1. For each sensor $A_i, 1 \leq i \leq n$, compute B_i.
Step 2. Compute a destination regular n-gon for the set of n points $B_1, \ldots, B_n$, and find the optimal arrangement of the n points to the vertices of the n-gon, by using the algorithm for sensors on the perimeter of a circle described in Section 3.2. Let B_i' be the destination vertex of $B_i, 1 \leq i \leq n$.
Step 3. Move A_i to $B_i', 1 \leq i \leq n$, and compute $S_n^1 = \sum_{i=1}^{n} d(A_i, B_i')$.

In section 3.2 we showed that step 2 of Algorithm 1 can be implemented in $O(n^2)$ time. Thus the above algorithm can be solved in $O(n^2)$ time.

Approximation bound of Algorithm 1. In this section, we show that S_n^1 computed by the first approach is bounded by $(\pi + 1) \times \lambda_{s,C}^*$. Suppose that A_i' is the destination of sensor $A_i, i = 1, \ldots, n$, in an optimal solution. Clearly, $A_1', \ldots, A_n'$ lie on C and form a regular n-gon. Obviously, $\sum_{i=1}^{n} \hat{d}_C(B_i, B_i') \leq \sum_{i=1}^{n} \hat{d}_C(B_i, A_i')$ since $\{B_1', \ldots, B_n'\}$ is an optimal solution for the one dimensional min-sum problem with the input $\{B_1, \ldots, B_n\}$. The following lemma is easy to show.

Lemma 4. *For any two points x, y on C, $\hat{d}_C(x, y) \leq \frac{\pi}{2} \times d(x, y)$.*

Theorem 6. *Algorithm 1 can be implemented in $O(n^2)$ time and its approximation ratio is no more than $\pi + 1$.*

5.3   The Second Approach

The second approach designs a PTAS pproximation algorithm whic is described below.

Algorithm 2

Step 1. Using Algorithm 1, compute S_n^1 defined above.

Step 2. For each $i = 1, \ldots, n$, find the arc $\hat{Q}_{B_i}(\frac{\pi}{2} \times \frac{S_n^1}{n})$ and compute a set of points that partitions the arc into $\lceil \frac{1}{\epsilon'} \rceil$ pieces of equal length where $\epsilon' = \frac{2\epsilon}{\pi(\pi+1)}$.

Step 3. Clearly, there are $n \times (\lceil \frac{1}{\epsilon'} \rceil + 1)$ points in total. For each point x, construct a regular n-gon P_x such that one of the corners of P_x is located at x, and find the optimal arrangement of the n sensors $(A_1, \ldots, A_n)$ to the vertices of the n-gon by solving a weighted bipartite matching problem. (The Hungarian method to solve the weighted matching problem in a complete bipartite graph of size n takes $O(n^3)$ time (see [10])).

Step 4. Among all $n \times (\lceil \frac{1}{\epsilon'} \rceil + 1)$ regular n-gons thus constructed, find the one with the minimum cost (denoted by S_n^2) and output the optimal arrangement of the n sensors to the vertices of the n-gon.

The following lemma is crucial for the second approach.

Lemma 5. *In an optimal solution, there exists at least one sensor $A_i (1 \leq i \leq n)$ such that its destination A_i' on C is on the arc $\hat{Q}_{B_i}(\frac{\pi}{2} \times \frac{S_n^1}{n})$.*

Proof. It is clear that $S_n^1 \geq \lambda_{s,C}^*$. Let A_i' be the destination of sensor A_i in an optimal solution, $i = 1, \ldots, n$. Then there is at least one sensor, say $A_k (1 \leq k \leq n)$, such that the distance $d(A_k, A_k')$ is no more than $\frac{S_n^1}{n}$. According to Lemma 4, all points on C with the distance to A_k of no more than $\frac{S_n^1}{n}$ lie on the arc $\hat{Q}_{B_k}(\frac{\pi}{2} \times \frac{S_n^1}{n})$ (recall that B_k is the point on C closest to A_k), which completes the proof of Lemma 5.

Analysis of the second approach. First, it is evident that the running time of the second approach is determined by the time needed for solving $n \times (\lceil \frac{1}{\epsilon'} \rceil + 1) \in O(\frac{n}{\epsilon})$ bipartite matching problems.

According to Lemma 5, there exists an optimal solution in which one of the corners of the corresponding regular n-gon is located at a point on the arc $\hat{Q}_{B_k}(\frac{\pi}{2} \times \frac{S_n^1}{n})$ for some k, $1 \leq k \leq n$. In Step 2, the arc $\hat{Q}_{B_k}(\frac{\pi}{2} \times \frac{S_n^1}{n})$ is partitioned into $\lceil \frac{1}{\epsilon'} \rceil$ pieces, and therefore, the length of each piece is no more than $\frac{\pi S_n^1 \epsilon'}{n}$

(note that the length of $Q_{B_k}'(\frac{\pi}{2} \times \frac{S_n^1}{n})$ is $\frac{\pi S_n^1}{n}$). Since all possible values of k are considered, the difference between S_n^2 (computed by the second approach) and $\lambda_{s,C}^*$ (the optimal cost) is no more than $n \times \frac{1}{2} \times \frac{\pi S_n^1 \epsilon'}{n} = \frac{\pi S_n^1 \epsilon'}{2} = \frac{S_n^1 \epsilon}{\pi+1} \le \epsilon \lambda_{s,C}^*$, by Theorem 6). Therefore, we have the following theorem.

Theorem 7. *The approximation ratio of Algorithm 2 is no more than $1 + \epsilon$ for a given constant ϵ, and the running time of the second approach is $O(\frac{1}{\epsilon}n^4)$.*

5.4   On a Simple Polygon

Let $\lambda_{s,P}^*$ be the optimal value of the min-sum problem on a polygon P. In this subsection we present an approximation algorithm for the min-sum problem on P, which has an approximation ratio $1 + \epsilon$ (ϵ is an arbitrary constant).

Our algorithm for a simple polygon is very similar to the second approach for a circle. In the second approach for a circle, we use Algorithm 1 as a subroutine to obtain lower and upper bounds of $\lambda_{s,C}^*$. However, our approximation algorithm for a simple polygon will use the solution for the min-max problem on the polygon to obtain lower and upper bounds of $\lambda_{s,P}^*$. Let $\lambda_{m,P}^*$ be the optimal value of the min-max problem on P. It is easy to see that $\lambda_{m,P}^* \le \lambda_{s,P}^* \le n \times \lambda_{m,P}^*$. Our algorithm for a simple polygon P is described below.

Min-Sum Algorithm on a Simple Polygon

Step 1. Using the approach for the min-max problem on P, compute $\lambda_{m,P}^*$ described above.

Step 2. For each i, j where $1 \le i \le n$ and $0 \le j < n$, find the sub-edge $e_{i,j}'$ of edge e_j that is within the circle of radius $\lambda_{m,P}^*$ centered at position A_i, and compute a set of points that partitions the the sub-edge into $\lceil \frac{n}{\epsilon} \rceil$ pieces of equal length.

Step 3. Clearly, there are $mn \times (\lceil \frac{n}{\epsilon} \rceil + 1) \in O(\frac{mn^2}{\epsilon})$ points in total. For each point x, construct a set of n positions on P such that one of them is located at x and the polygonal distance between any two consecutive positions is $L(P)/n$, and find the optimal arrangement of the n sensors $(A_1, \ldots, A_n)$ to the set of n positions by using the algorithm [10].

Step 4. Among all $O(\frac{mn^2}{\epsilon})$ candidate sets of n positions thus constructed, find the one with the minimum cost.

Theorem 8. *The approximation ratio of the approach for a simple polygon is no more than $1 + \epsilon$ for a given constant ϵ, and the running time of the second approach is $O(\frac{1}{\epsilon}mn^5)$.*

Proof. (**Theorem 8**) The reason why the approximation ratio of the above approach is bounded by $1 + \epsilon$, is as follows. Since $\lambda_{s,P}^* \le n \times \lambda_{m,P}^*$, there is at least one sensor whose moving distance to its destination is no more than $\lambda_{m,P}^*$ in an optimal solution. Let A_i be one such sensor and its destination position lies on edge e_j in that optimal solution. In Step 2, the sub-edge $e_{i,j}'$ is partitioned into

$\lceil \frac{n}{\epsilon} \rceil$ pieces, and therefore, the length of each piece is no more than $\frac{2\epsilon\lambda^*_{m,P}}{n}$. Since all possible values of i and j are considered, the difference between the value computed by the above approach and $\lambda^*_{s,P}$ (the optimal cost) is no more than

$$n \times \frac{1}{2} \times \frac{2\lambda^*_{m,P}}{n} = \epsilon\lambda^*_{m,P} \le \epsilon\lambda^*_{s,P}.$$

It is evident that the running time of the above approach is determined by the time needed for solving $O(\frac{mn^2}{\epsilon})$ weighted bipartite matching problems.

6   Conclusion and Open Problems

In this paper we gave an algorithm for solving the min-max problem and a PTAS (Polynomial Time Approximation Scheme) for the min-sum problem in both one and two dimensions. Although it is unknown whether the min-sum problem is NP-hard, we conjecture that it can be solved in polynomial time. Evidence for this also comes from experimental results on finding the number of different counter-clockwise orderings of n sensors on the perimeter of a circle when we sweep a regular n-gon along the perimeter (shown in the full version of this paper.). In addition, several other variants of the problem on simple polygons and regions are of interest for further investigation, including k-barrier coverage, regions with holes, and various types of sensor placements and motions. Thus, in Subsection 2.2, in order to minimize the number of sensors used when scanning the perimeter one should take into account sections already scanned. For example, this is the case if the polygon is a narrow rectangle of height less than the range of a sensor; this in itself is an interesting optimization problem which is worth of further investigation. Also of interest is to refine the sensor motion model, the network model, and the communication model in order to enable effective intrusion detection and barrier coverage. For example, the communication model becomes crucial when assuming the sensors either do not have knowledge of the region or do not know their coordinates.

References

1. Ahmed, N., Kanhere, S.S., Jha, S.: The holes problem in wireless sensor networks: a survey. ACM SIGMOBILE Mobile Computing and Communications Review 9(2), 4–18 (2005)
2. Balister, P., Bollobas, B., Sarkar, A., Kumar, S.: Reliable density estimates for coverage and connectivity in thin strips of finite length. In: Proceedings of the 13th annual ACM international conference on Mobile computing and networking, pp. 75–86 (2007)
3. Chen, A., Kumar, S., Lai, T.H.: Designing localized algorithms for barrier coverage. In: Proceedings of the 13th annual ACM international conference on Mobile computing and networking, pp. 63–74 (2007)
4. Cole, R.: Slowing down sorting networks to obtain faster sorting algorithms. Journal of the ACM (JACM) 34(1), 200–208 (1987)

5. Heo, N., Varshney, P.K.: A distributed self spreading algorithm for mobile wireless sensor networks. In: Wireless Communications and Networking, 2003. WCNC 2003. 2003 IEEE, vol. 3 (2003)

6. Heo, N., Varshney, P.K.: Energy-efficient deployment of Intelligent Mobile sensor networks. Systems, Man and Cybernetics, Part A, IEEE Transactions on 35(1), 78–92 (2005)

7. Hopcroft, J.E., Karp, R.M.: An $n^{2.5}$ algorithm for maximum matchings in bipartite graphs. SIAM Journal on Computing 2(4), 225–231 (1973)

8. Hu, S.S.: 'Virtual Fence' along border to be delayed. Washington Post, Thursday (February 28, 2008)

9. Jung, B., Sukhatme, G.S.: Tracking Targets Using Multiple Robots: The Effect of Environment Occlusion. Autonomous Robots 13(3), 191–205 (2002)

10. Kuhn, H.W.: The Hungarian Method for the assignment problem. Naval Research Logistics Quarterly 2, 83–97 (1955)

11. Kumar, S., Lai, T.H., Arora, A.: Barrier coverage with wireless sensors. Wireless Networks 13(6), 817–834 (2007)

12. Liu, J., Cheung, P., Zhao, F., Guibas, L.: A dual-space approach to tracking and sensor management in wireless sensor networks. In: Proceedings of the 1st ACM international workshop on Wireless sensor networks and applications, pp. 131–139 (2002)

13. Megiddo, N.: Applying Parallel Computation Algorithms in the Design of Serial Algorithms. Journal of the ACM (JACM) 30(4), 852–865 (1983)

14. O'Rourke, J.: Art gallery theorems and algorithms. Oxford University Press, Inc., New York (1987)

15. Schrijver, A.: Disjoint circuits of prescribed homotopies in a graph on a compact surface. Journal of Combinatorial Theory Series B 51(1), 127–159 (1991)

16. Se, S., Lowe, D., Little, J.: Mobile Robot Localization and Mapping with Uncertainty using Scale-Invariant Visual Landmarks. The International Journal of Robotics Research 21(8), 735 (2002)

Parameterized Algorithms for Generalized Domination[*]

Venkatesh Raman[1], Saket Saurabh[2], and Sriganesh Srihari[3]

[1] The Institute of Mathematical Sciences, C.I.T. Campus, Chennai 600 113
vraman@imsc.res.in
[2] Department of Informatics, University of Bergen, Bergen, Norway
saket.saurabh@ii.uib.no
[3] School of Computing, National University of Singapore, Singapore 117590
srigsri@comp.nus.edu.sg

Abstract. We study the parameterized complexity of a generalization of DOMINATING SET problem, namely, the VECTOR DOMINATING SET problem. Here, given an undirected graph $G = (V, E)$, with $V = \{v_1, \cdots, v_n\}$, a vector $l = (l(v_1), \cdots, l(v_n))$ and an integer parameter k, the goal is to determine whether there exists a subset D of at most k vertices such that for every vertex $v \in V \setminus D$, at least $l(v)$ of its neighbors are in D. This problem encompasses the well studied problems – VERTEX COVER (when $l(v) = d(v)$ for all $v \in V$, where $d(v)$ is the degree of vertex v) and DOMINATING SET (when $l(v) = 1$ for all $v \in V$). While VERTEX COVER is known to be fixed parameter tractable, DOMINATING SET is known to be $W[2]$-complete. In this paper, we identify vectors based on several measures for which this generalized problem is fixed parameter tractable and W-hard. We also show that the VECTOR DOMINATING SET is fixed parameter tractable for graphs of bounded degeneracy and for graphs excluding cycles of length four.

1 Introduction

DOMINATING SET is among the most fundamental problems in graph theory, algorithms and combinatorial optimization. DOMINATING SET asks for a minimum set of vertices such that every vertex of the graph not in this set has a neighbor in it. This is a classical NP-hard problem and is well-studied from the point of view of approximation algorithms [8,10,14] and parameterized complexity [7,11,15,16]. DOMINATING SET is a "hard" problem, in the sense that it is known to be W[2]-complete in parameterized complexity [7] and $(1 - o(1)) \ln n$ is a threshold below which it can not be approximated efficiently (unless NP has slightly super-polynomial time algorithm [10]).

In [16], the authors identified "easy" instances for DOMINATING SET by giving fixed parameter tractable algorithms (FPT) for it in graphs with no short cycles. Here, instead of making any assumptions about the input graph, we study

---

[*] The work was done when the second and the third authors were at The Institute of Mathematical Sciences.

B. Yang, D.-Z. Du, and C.A. Wang (Eds.): COCOA 2008, LNCS 5165, pp. 116–126, 2008.

variations of DOMINATING SET which are FPT in general graphs. The VECTOR DOMINATING SET problem [13] generalizes the DOMINATING SET problem and encompasses many fixed parameter tractable (FPT) variations of it. In particular, this includes VERTEX COVER, which asks for a minimum set of vertices that covers all the edges of the graph. Let $G = (V, E)$ be a graph on vertex set $V = \{v_1, \cdots, v_n\}$ and let $d(v_i)$ be the degree of a vertex v_i (the number of edges incident on the vertex v_i). Given an integral vector $\boldsymbol{l} = (l(v_1), \cdots, l(v_n))$ with $1 \leq l(v_i) \leq d(v_i)$, for $1 \leq i \leq n$, a set $D \subseteq V$ is called $\boldsymbol{l}$-dominating if for all $v \in V \setminus D$, $|N(v) \cap D| \geq l(v)$, where $N(v)$ is the set of neighbors of v. We also call D as *vector dominating set*. Now we define the problem formally.

VECTOR DOMINATING SET (VDS): Given an undirected graph $G = (V, E)$, an integral vector $\boldsymbol{l} = (l(v_1), \cdots, l(v_n))$ with $1 \leq l(v_i) \leq d(v_i)$ for $1 \leq i \leq n$ and a positive integer k, does there exist an $\boldsymbol{l}$-dominating set D with $|D| \leq k$?

When $l(v) = 1$ for all $v \in V$, the VDS problem is the well known DOMINATING SET problem, while $l(v) = d(v)$ for all $v \in V$ corresponds to the well known VERTEX COVER problem. Since both these problems are NP-complete [12], it follows that the VDS problem is NP-complete.

The VDS problem was first introduced in [13] and two NP-Completeness classification results were given. The authors show that (a) the VDS problem remains NP-complete for bipartite graphs even when $\lceil cn \rceil$ coordinates of the input vector $\boldsymbol{l}$ is greater than 1 for any $0 < c < 1/2$, and (b) the VDS problem is polynomial time solvable for bipartite graphs when the number of edges in the graph induced by vertices with $l(v) < d(v)$ is at most $c \log n$ for some $0 < c < 1$.

Unlike classical complexity where VERTEX COVER, INDEPENDENT SET and DOMINATING SET are just NP-complete, parameterized complexity does a finer classification. While INDEPENDENT SET and DOMINATING SET are hard for different levels of W-hierarchy, VERTEX COVER is a celebrated fixed parameter tractable problem where the problem is well solved for parameter size up to 100 [7]. Thus we observe different complexities at both ends of the 'vector hierarchy' for the VDS problem from the view point of parameterized complexity. This makes one curious about the complexities of VDS for other vectors. Hence, apart from generalizing the results of [13], our purpose is two fold:

(a) to study the parameterized complexity of the generalized version of DOMINATING SET, under various scenarios set by the vectors; this would help address the parameterized complexity of problems that can be modeled as these generalized dominating sets;

(b) explain the differences between the parameterized complexity of VERTEX COVER and DOMINATING SET.

In this paper, we identify vectors based on several measures for which VDS is fixed parameter tractable and W-hard. We also show that the VDS is fixed parameter tractable for graphs of bounded degeneracy and for graphs excluding cycles of length four.

Organization of the rest of the Paper. In section 2, we set up notations used in the paper, define FPT and state a result which is used extensively to argue the running time of our algorithms. In section 3 we show that VDS remains W[2]-complete even for some restricted l vectors. In section 4, we identify certain (a) vectors and (b) graph classes for which the VDS problem is FPT. Finally, we conclude with some remarks and discussions in Section 5.

2   Preliminaries

We assume that all our graphs are simple and undirected. Given a graph $G = (V, E)$, n represents the number of vertices, and m represents the number of edges. For a subset $V' \subseteq V$, by $G[V']$ we mean the subgraph of G induced on V'. By $N(u)$ we mean all vertices that are adjacent to u, and by $N[u]$, we refer to $N(u) \cup \{u\}$. Similarly, for a subset $D \subseteq V$, we define $N(D) = \cup_{v \in D} N(v) \setminus D$ and $N[D] = \cup_{v \in D} N[v]$. A vertex is said to *dominate* all its neighbors.

Parameterized complexity is a two-dimensional framework for studying the computational complexity of problems [7,11,15]. One dimension is the input size n and the other one is *parameter* k. A problem is called *fixed-parameter tractable (FPT)* if it can be solved in time $f(k) \cdot n^{O(1)}$, where f is a computable function depending only on k. We refer to [7,11,15] for definition of W[t] and the notion of parameterized reduction. Next, we give a lemma which is used extensively in the later sections to show that certain variations of VDS are fixed parameter tractable.

Lemma 1 ([2]). *For every $f(n) \in n^{o(1)}$ there exists a function $g(k)$ such that for all n and k, $(f(n))^k$ is $O(g(k)n^{O(1)})$. In other words if a parameterized problem, say A, can be solved in time $O((n^{o(1)})^k n^{O(1)})$ then A is FPT.*

3   VDS Is W[2]-Complete

In this section we show that VDS is W[2] complete. VDS is W[2]-hard (in general) follows from the fact that when $l(v) = 1$ for all v, it corresponds to W[2]-complete DOMINATING SET problem [7]. Here we show that it remains W[2]-hard even when we slightly "relax" the condition of $l(v) = 1$ for all v. We call this relaxed problem where at most a constant number of the l-vector coordinates are unbounded while the rest are bounded by a fixed constant c as ALMOST c-BVDS.

Theorem 1. [⋆][1] ALMOST c-BVDS *is W[2]-complete for bipartite graphs.*

Finally, we show that VDS is W[2]-complete by showing it to be in W[2]. Before we prove our result, we state a simple but useful lemma.

Lemma 2. *Let $G = (V, E)$ be an undirected graph and l be an l-vector. If G has a vector dominating set of size at most k then every vertex v such that $l(v) > k$ is part of every vector dominating set of size at most k.*

---

[1] Proofs of results labeled with [⋆] will appear in the long version of the paper.

Theorem 2. *VDS is* W[2]-*complete.*

Proof. Let (G, l, k) be an instance of VDS. Lemma 2 allows us to assume that for every vertex v, we have $1 \leq l(v) \leq k$. (Otherwise, v has to be in the solution). For $v \in V$, let x_v be a boolean variable associated with it. Now consider the following boolean formula

$$\bigwedge_{v \in V} \left(x_v \bigvee_{S \in Y} \left(\bigwedge_{u \in S} x_u \right) \right),$$

where Y is the collection of all subsets of size $l(v)$ of $N(v)$. Hence $|Y|$ is $\binom{|N(v)|}{l(v)}$. Suppose this formula has a satisfying assignment with weight at most k (where the weight is the number of true variables in the assignment). Then the subset Z of V defined by $Z = \{u \in V | x_u = 1\}$ is of cardinality k and forms a vector dominating set of size at most k for G. For the other direction, let D be a vector dominating set of size at most k for G. A satisfying assignment of weight at most k for the above boolean formula is formed by taking $x_u = 1$ for $u \in D$ and 0 otherwise. In the above boolean formula, we have two levels of alterations of unbounded fan-in and and one level of alteration of at most k fan-in (since $l(v) \leq k$ for all $v \in V$). This implies that VDS is in $W^*[2] = W[2]$ [6,11]. □

4   FPT Algorithms for Some Natural Instances of VDS

In this section we give FPT algorithms for some natural special instances of VDS problem by (a) looking at the vector values and (b) by looking at the structure of the input graph.

We need a (generalized) "colored" version of VDS problem as an intermediate problem for some of our FPT algorithms. In order to define the problem we introduce the notion of colored graph. A graph $G = (V, E)$ is called a *colored graph* if its vertices are either colored Black or White. Let B and W be the disjoint set of vertices colored black and white respectively and $V = B \cup W$. The problem is defined as follows.

COLORED VECTOR DOMINATING SET (CVDS): Let $G = (V, E)$ be a colored graph, together with an integral vector $l = (l(v_1), \cdots, l(v_n))$ with $1 \leq l(v) \leq d(v)$ for each $v \in B$ while $l(v) = 0$ for each $v \in W$ and a positive integer k. Does G have a set D of at most k vertices such that for every $v \in B$, either v is in D or $|N(v) \cap D| \geq l(v)$?

To solve the general VDS problem, simply color all vertices black and solve the resulting CVDS problem.

4.1   Vectors for Which VDS Is FPT

Bounded Differential Slack. We first introduce a term *differential slack vector* and define it as follows: Let $G = (V, E)$ be a given graph, $l = (l(v_1), \cdots, l(v_n))$ be an l-vector corresponding to the VDS problem and $D(G) = (d(v_1), \cdots, d(v_n))$.

Then $S = D(G) - l = (d(v_1) - l(v_1), \cdots, d(v_n) - l(v_n)) = (s(v_1), \cdots, s(v_n))$ is defined as the differential slack vector and $s(v_i)$ is the differential slack corresponding to the vertex v_i. As discussed in the introduction, the VERTEX COVER problem, for which the maximum differential slack value is 0, is known to be FPT. Generalizing this, we show

Theorem 3. *Let $G = (V, E)$ be a graph, l be an l-vector and S be the corresponding differential slack vector. If $M = max_{v \in V}\{s(v)\}$ then VDS problem can be solved in time $O((M + 2)^k n^{O(1)})$. In particular, if $M \leq n^{o(1)}$, then the VDS problem is FPT.*

Proof. Our algorithm is based on the following observation: If D is a l-dominating set of G, then for any vertex $v \in V \setminus D$, at least $l(v)$ neighbors are in D and at most $s(v) = d(v) - l(v)$ of its neighbors are outside D. Hence if we select a set of $s(v) + 1$ neighbors of v, then it must intersect D or v must be in D.

Now we design an algorithm based on the above observation as follows. Start with $B = V$ and $D = W = \emptyset$ and the parameter k. At any stage of the algorithm we choose a vertex v in the set B and select a set X of $s(v) + 1$ neighbors of v and then recursively solve the problem by selecting a vertex $u \in (X \cup \{v\})$ in D. If any of the recursive branches returns YES, we say YES and return the corresponding vector dominating set D else we return NO (meaning there does not exist any vector dominating set of size at most k). When we select a vertex $u \in (X \cup \{v\})$ in D, we update the information as follows:

- $D := D \cup \{u\}$; For every vertex $w \in B$ do: If $|N(w) \cap D| \geq l(w)$ recolor w to white, i. e. $W := W \cup \{w\}$ and $B := B \setminus \{w\}$, and set $l(w) = 0$ else set $l(w) := l(w) - |N(w) \cap D|$; (Here $N(w)$ refers to the neighborhood of w in the current graph and hence $|N(w) \cap D|$ is either u or $\emptyset$.) set $k := k - 1$ and $G := G \setminus \{u\}$.

If at any recursive step $k = 0$ and the current B is non-empty then we return NO. Else, if $k \geq 0$ and B is empty, we return YES and the set D.

The correctness of the algorithm is clear from the description. For the time complexity, observe that the depth of recursion tree is at most k and number of recursive calls made at any stage (number of children of any node in the recursion tree) is at most $max_{v \in V} s(v) + 2 \leq M + 2$. Hence the total number of nodes in the recursive tree is bounded by $(M + 2)^k$. Since at each node of the recursion tree, we spend polynomial time, the overall time complexity of our algorithm is bounded by $O((M + 2)^k n^{O(1)})$. Now by Lemma 1, if M is at most $n^{o(1)}$ the problem is FPT. This completes the proof.   □

Now we show that the problem becomes FPT if there are only a few vertices with non zero slack (even if the slack values are unbounded).

Theorem 4. *Let $G = (V, E)$ be a graph, l be an l-vector, S be the corresponding slack vector and S be the set of vertices with non-zero slack. Then the VDS can be solved in time $O((|S|^k + 1.2738^k)n^{O(1)})$. In particular, if $|S| \leq n^{o(1)}$, then the VDS problem is FPT.*

Proof. We first introduce a notion of valid partition for the set $S \subseteq V$. A partition of the set S as (S_1, S_2) is called *valid* if the following holds: (a) $|S_1| \leq k$; (b) let $X = (S_1 \cup N(S_2))$; $|X| \leq k$ (here $N(S_2)$ is in G, not in S); (c) for all $v \in S_2$, $|N(v) \cap X| \geq l(v)$. The motivation is that S_1 denotes the intersection of S with the desired vector dominating set. Hence all the neighbors of $S_2 = S \setminus S_1$ in $V \setminus S$ must be in D. This is because for all $v \in V \setminus S$, either $v \in D$ or $N(v) \subseteq D$ as the slack value of these vertices is zero. For our algorithm, we proceed as follows:

– We enumerate all the valid partitions (S_1, S_2) of S. If there is no valid partition then return No.
– For each of the valid partitions check whether $G[V \setminus (X \cup S)]$ has a vertex cover of size at most $k - |X|$. If there is a vertex cover V' of size at most $k - |X|$ then return $D = X \cup V'$ as the desired vector dominating set else return No.

The correctness of the algorithm follows from the fact that $D \cap (V \setminus S)$ is a vertex cover of $G[V \setminus S]$, as all vertices of $V \setminus S$ have slack value zero.

For the time complexity, we first note that a vertex cover of size at most k in G can be found in time $O(1.2738^k + kn)$ [3]. Hence, the time complexity of our algorithm is bounded by $O\left(\left(\binom{|S|}{k} + 1.2738^k\right)n^{O(1)}\right)$. If $|S| \leq n^{o(1)}$ then the time complexity is bounded by $O(((n^{o(1)})^k + 1.2738^k)n^{O(1)})$, which makes the problem FPT, by Lemma 1.   □

In Theorems 3 and 4 , we needed an upper bound of $n^{o(1)}$ for the maximum slack value or the number of l-values with non-zero slack. We can show that this bound is asymptotically tight.

Theorem 5. [⋆] *Given an $0 < \epsilon < 1$, a constant, the VDS problem is W[2]-complete when at least $\lceil n^\epsilon \rceil$ vertices have non zero slack. VDS problem also remains W[2]-complete if the maximum of differential slack is at least $\lceil n^\epsilon \rceil$. Here n is the number of vertices of the input graph.*

Bounded Fractional Slack. In the last subsection we gave FPT algorithms for the VDS problem with respect to bounds related to differential slack. In this section we study the VDS problem in terms of fractional slack of vertices. For a graph G and a vector l, the *fractional slack* of a vertex v is defined as $d(v)/l(v)$.

Theorem 6. *Let $G = (V, E)$ be a graph, l be an l-vector and k be a positive integer. If $p = \max_{v \in V}\{d(v)/l(v)\}$ then we can solve the VDS problem in $O\left(\binom{k^3p^2 + 2k^2p + k}{k}n^{O(1)}\right)$ time. If $p \leq n^{o(1)}$ then the VDS problem is FPT.*

Proof. Let $S = \{v \in V : l(v) > k\}$. By Lemma 2, we know that every vertex of the set S is part of every vector dominating set of size at most k. If $|S| > k$ then G does not have any vector dominating set of size at most k and we return No. So we assume that $|S| \leq k$. Thus, we obtain an equivalent instance for CVDS in the following way:

- For every vertex $w \in (V \setminus S)$ do: If $|N(w) \cap S| \geq l(w)$ color w to white, i. e. $W :=$ $W \cup \{w\}$ and $B := B \setminus \{w\}$, and set $l(w) = 0$ else set $l(w) := l(w) - |N(w) \cap S|$ and color w black; set $k' := k - |S|$ and $G := G \setminus S$.

Let B and W be the set of black and white vertices of $V \setminus S$ respectively. Now we need a set $S' \subseteq (B \cup W)$ of at most k' vertices, such that every vertex $v \in (B \setminus S')$ has at least $l(v)$ neighbors in S'. But for vertices in $B \cup W$, $k \geq l(v_i) \geq d(v_i)/p$ which implies that $d(v_i) \leq kp$. So the degree of every vertex in $B \cup W$ is bounded by kp. Since a vector dominating set S', of the above defined instance, is also a *dominating set* for B with vertex degree bounded by kp, it can only dominate $(|S'|kp)$ vertices outside S'. Note that $|S'| \leq k'$. So if $|B| > k'(kp+1)$ then we return No.

Now we bound $|W|$ as follows: If there is a vertex $v \in W$ such that it has no neighbor in B then remove this vertex from the graph as this can not be part of any minimal S'. So the remaining vertices of W have at least one neighbor in B. But then B is a dominating set for W and hence $|W|$ is bounded by $|B| \times (\text{ max degree of the vertex in } B) \leq (kp)(k')(kp+1) \leq k^3p^2 + k^2p$.

So now we can solve the problem by enumerating all subsets of size at most k' as the required S' from the vertices of $B \cup W$. Since $|B \cup W| \leq k^3p^2 + 2k^2p + k$, this takes $O\left(\binom{k^3p^2+2k^2p+k}{k'} n^{O(1)}\right) \leq O\left(\binom{k^3p^2+2k^2p+k}{k} n^{O(1)}\right)$ time. Notice that if $p \leq n^{o(1)}$ then by Lemma 1, VDS problem is FPT. This completes the proof of the theorem. $\square$

We can prove the following theorem for fractional slack, analogous to Theorem 5. We call an instance of VDS as p-ratio-VDS, if $p \geq \max_{v \in V} d(v)/l(v)$.

Theorem 7. [⋆] *Given an $0 < \epsilon < 1$, the n^ϵ-ratio-VDS is W[2]-complete.*

Restriction on vector values. Finally we show that if the number of values in the l-vector that are unbounded is at least some function of n then the problem becomes FPT, complementing Theorem 1.

Theorem 8. *Let $G = (V, E)$ be a graph, l be an l-vector and k be a positive integer. Furthermore let $f(n)$ be increasing function of n (that is, for all $x < y$, $x, y \in \mathbb{R}^+$, $f(x) < f(y)$) and let $|S = \{v \mid l(v) \geq f(n)\}| \geq g(n)$, for some increasing function $g(n)$. In addition if f and g are invertible functions then the VDS problem is FPT.*

Proof. Given an instance of VDS, our algorithm will either show that the given instance is a No instance or bound the size of G as a function of k. Once the size of G is bounded as a function of k, we can find the desired vector dominating set of size at most k by enumerating all subsets D of size at most k of V and then checking whether D is a vector dominating set.

If $k \geq g(n)$ then $n \leq g^{-1}(k)$ and the number of vertices is bounded by $g^{-1}(k)$. Else, $k < g(n)$ and hence not all vertices of S can be in a vector dominating set D of size at most k. However each vertex of S, not in D requires at least $f(n)$ neighbors in D. Hence $f(n) \leq k$ as otherwise the answer is No. Thus $n \leq f^{-1}(k)$

and again we have bounded the number vertices of G as a function of k. This completes the proof of the theorem. □

4.2 Graphs with Some Special Structures

FPT algorithm for VDS for Graphs Excluding 4-Cycles. In [16], the authors initiated a parameterized classification of DOMINATING SET problem based on the girth (length of the shortest cycle) of the graph and showed that the DOMINATING SET problem is FPT for graphs with girth at least 5. Our Theorem 1 implies that the VDS problem is W[2]-hard for bipartite graphs and hence for triangle free graphs. We complement this theorem by showing that if the input graph excludes four cycles (it may contain triangles) then the VDS problem is FPT.

Our algorithm is based on the following observation which is formulated in the next lemma.

Lemma 3. *Let $G = (V, E)$ be a graph which does not contain cycles of length four, l be an l-vector and k be a positive integer. Then any vector dominating set of size at most k, if there exists one, contains all the vertices v of G such that $d(v) > 2k$.*

Proof. If we do not include a vertex v of degree $> 2k$ in D then we can not have any dominating set of size at most k. This is because any neighbor u of v can dominate at most one another neighbor w of v and hence just to dominate all the neighbors of v, we require more than k vertices. □

Theorem 9. [⋆] *Let $G = (V, E)$ be a graph, l be an l-vector and k be a positive integer. If G excludes cycles of length 4 then the VDS problem is FPT.*

Sparse Graphs – Graphs of Bounded Degeneracy. The results of this section are motivated by a result of Alon and Gutner [1] for DOMINATING SET in graphs of bounded degeneracy. A graph G is d-degenerated if every induced subgraph of G has a vertex of degree at most d. A d-degenerated graph with n vertices has less than dn edges and therefore its average degree is less than $2d$. A d-degenerated graph with every vertex being colored black or white is a colored d-degenerated graph and that B and W respectively are the vertices with color black and white.

Our result is based on the following key lemma from [1].

Lemma 4 ([1]). *Let $G = (V = B \cup W, E)$ be a colored d-degenerated graph. If $|B| > (4d + 2)k$, then there are at most $(4d + 2)k$ vertices in G that dominate at least $|B|/k$ vertices of B.*

Theorem 10. *Let $G = (V, E)$ be a d-degenerated graph, l be an l-vector and k be a positive integer. Then VDS problem can be solved in time $k^{O(dk^2)}n^{O(1)}$.*

Proof. We solve the CVDS starting with $B = V$ and $D = W = \emptyset$ and the parameter k. Our algorithm solves the CVDS problem either by branching on vertices of set $S = \{v \mid v \in (B \cup W), |N(v) \cap B| \geq |B|/k\}$ or solving some $f(d,k)$ number of instances of "matching problem". The branching step is performed when $|B| > (4d+2)k$. In this case Lemma 4 bounds the size of S by $(4d+2)k$. Note that every l-dominating set D' of size at most k contains a vertex which dominates at least $|B|/k$ black vertices and hence $D' \cap S$ is nonempty for any k sized l-dominating set for black vertices. Hence, when the size of S is bounded, we solve the problem by including every $v \in S$ in the desired dominating set D of size at most k and recursively solving following instance:

- $D := D \cup \{v\}$; For $u \in (N(v) \cap B)$ do: If $l(u) = 1$, color u white, i. e. $W := W \cup \{u\}$ and $B := B \setminus \{u\}$, and set $l(u) = 0$ else $l(u) = l(u) - 1$; set $k := k - 1$ and $G := G - \{v\}$.

In the other case when the size of B is bounded by $(4d+2)k$, we find the desired vector dominating set for black vertices as follows. Here we assume that for all $v \in B$, $l(v) \leq k$ otherwise we include $v \in D$ deterministically and re-color the vertices of graph appropriately. We make a multiset M from B by having $l(u)$ copies for each vertex $u \in B$. Clearly the size of $|M|$ is bounded above by $(4d+2)k^2$. A partition $\mathscr{P} = \{P_1, P_2, \cdots, P_\alpha\}$ of M is called *valid* if (a) there exists a subset $S \subseteq B \cup W$ forming a *system of distinct representatives*; that is for all $1 \leq i \leq \alpha$, there exists a *distinct* $u_i \in S$ such that $P_i \subseteq N(u_i)$ and (b) each P_i contains at most one copy of any vertex of B. The set S is a *witness set*. So to find the desired vector dominating set in $B \cup W$ we proceed as follows. For all partitions $\mathscr{P}$ of M in at most k parts, say $\mathscr{P} = \{P_1, P_2, \cdots, P_\alpha\}$, $1 \leq \alpha \leq k$, we check whether $\mathscr{P}$ is a valid partition. If any partition $\mathscr{P}$ is valid then return YES with the corresponding witness set else return NO.

For a fixed partition $\mathscr{P} = \{P_1, P_2, \cdots, P_\alpha\}$, we can do the validity testing and find a corresponding witness set in polynomial time as follows. Testing for duplicate copies in P_i's are easy. For the other part we first define the set $I_i = \{u \in (B \cup W) \mid P_i \subseteq N(u)\}$. Now we make the bipartite incidence graph for the sets $\{I_1, \cdots, I_\alpha\}$, that is a bipartite graph $G^* = (X \cup Y, E'')$, where X has a vertex x_i for every set I_i and $Y = \cup_{l=1}^{\alpha} I_l$ and there is an edge between (x_i, u) if $u \in I_i$. Now finding a valid system of distinct representatives reduces to finding a maximum bipartite matching in G^* saturating X, for which there is a classical polynomial time algorithm of Edmonds [9].

The total number of partitions $\mathscr{P}$ considered for our case is upper bounded by $\sum_{i=1}^{k} i^{(4d+2)k^2} \leq k^{((4d+2)k^2)+1}$. Also the search tree obtained in the branching step of the algorithm can be of size at most at most $(4d+2)^k k!$. This together with the bounds on the number of partition gives the desired time bound of the algorithm.   □

There exists a constant c such that, for every h, every graph with no K_h minor is $ch\sqrt{\log h}$-degenerated [5]. Hence Theorem 10 implies the following result.

Corollary 1. *VDS problem is FPT on graphs excluding a fixed graph H as a minor. In particular VDS is FPT on planar graphs and graphs of bounded genus.*

We remark that one can obtain sub-exponential time algorithm for VDS in H-minor free graphs using bidimensionality [4], though those results do not carry over to d-degenerated graphs.

5 Conclusions and Discussions

In this paper we initiated the study of the VDS problem, a generalization encompassing important problems like VERTEX COVER and DOMINATING SET, from the view point of parameterized complexity. We identified several measures and studied the VDS problem in terms of these measures in the realm of parameterized complexity. Some of these results can also be generalized to the connected or independent versions of the VDS problem, where one is looking for a l-dominating set which is connected or independent respectively. It would be interesting to know whether there are specific natural problems apart from DOMINATING SET, VERTEX COVER and THRESHOLD DOMINATING SET in the middle of the vector hierarchy. Finally, it would be interesting to investigate whether there are some vectors for which VDS problem is W[1]-hard.

References

1. Alon, N., Gutner, S.: Linear Time Algorithms for Finding a Dominating Set of Fixed Size in Degenerated Graphs. In: Lin, G. (ed.) COCOON. LNCS, vol. 4598, pp. 394–405. Springer, Heidelberg (2007)
2. Cai, L., Juedes, D.: On the Existence of Subexponential Algorithms. Journal of Computer and System Sciences 67(4), 789–807 (2003)
3. Chen, J., Kanj, I.A., Xia, G.: Improved Parameterized Upper Bounds for Vertex Cover. In: Královič, R., Urzyczyn, P. (eds.) MFCS 2006. LNCS, vol. 4162, pp. 238–249. Springer, Heidelberg (2006)
4. Demaine, E.D., Fomin, F.V., Hajiaghayi, M.T., Thilikos, D.M.: Subexponential parameterized algorithms on bounded-genus graphs and H-minor-free graphs. Journal of ACM 52(6), 866–893 (2005)
5. Diestel, R.: Graph Theory. Springer, Heidelberg (1997)
6. Downey, R.G., Fellows, M.R.: Threshold Dominating Sets and an Improved Characterization of W[2]. Theoretical Computer Science 209(1-2), 123–140 (1998)
7. Downey, R.G., Fellows, M.R.: Parameterized Complexity. Springer, Heidelberg (1999)
8. Duh, R., Fürer, M.: Approximation of k-set cover by semi-local optimization. In: The Proceedings of STOC, pp. 256–264 (1997)
9. Edmonds, J.: Paths, trees and flowers. Canadian Journal of Mathematics 17, 449–467 (1965)
10. Feige, U.: A Threshold of $\ln n$ for Approximating Set Cover. Journal of ACM 45(4), 634–652 (1998)
11. Flum, J., Grohe, M.: Parameterized Complexity Theory. Springer, Heidelberg (2006)
12. Garey, M.R., Johnson, D.S.: Computers and Intractability: A Guide to the Theory of NP-Completeness. W.H. Freeman, New York (1979)

13. Gerlach, T., Harant, J.: A Note on Domination in Bipartite Graphs. Discuss. Math. Graph Theory 22, 229–231 (2002)
14. Johnson, D.S.: Approximation algorithms for combinatorial problems. Journal of Computer and System Sciences 9(3), 256–278 (1974)
15. Niedermeier, R.: Invitation to Fixed-Parameter Algorithms. Oxford University Press, Oxford (2006)
16. Raman, V., Saurabh, S.: Short Cycles make W-hard problems hard: FPT algorithms for W-hard Problems in Graphs with no short Cycles (to appear in Algorithmica)

Turán Graphs, Stability Number, and Fibonacci Index

Véronique Bruyère and Hadrien Mélot[*]

Department of Theoretical Computer Science, Université de Mons-Hainaut,
Avenue du Champ de Mars 6, B-7000 Mons, Belgium
{veronique.bruyere,hadrien.melot}@umh.ac.be

Abstract. The Fibonacci index of a graph is the number of its stable sets. This parameter is widely studied and has applications in chemical graph theory. In this paper, we establish tight upper bounds for the Fibonacci index in terms of the stability number and the order of general graphs and connected graphs. Turán graphs frequently appear in extremal graph theory. We show that Turán graphs and a connected variant of them are also extremal for these particular problems.

1 Introduction

The Fibonacci index $F(G)$ of a graph G was introduced in 1982 by Prodinger and Tichy [20] as the number of stable sets in G. In 1989, Merrifield and Simmons [16] introduced independently this parameter in the chemistry literature[1]. They showed that there exist correlations between the boiling point and the Fibonacci index of a molecular graph. Since, the Fibonacci index has been widely studied, especially during the last few years. The majority of these recent results appeared in chemical graph theory [12,13,21,23,24,25] and in extremal graph theory [9,11,17,18,19].

In this literature, several results are bounds for $F(G)$ among graphs in particular classes. Lower and upper bounds inside the classes of general graphs, connected graphs, and trees are well known (see Sect. 2). Several authors give a characterization of trees with maximum Fibonacci index inside the class $\mathcal{T}(n, k)$ of trees with order n and a fixed parameter k. For example, Li et al. [13] determine such trees when k is the diameter; Heuberger and Wagner [9] when k is the maximum degree; and Wang et al. [25] when k is the number of pending vertices. Unicyclic graphs are also investigated in similar ways [17,18,24].

The Fibonacci index and the stability number of a graph are both related to stable sets. Hence, it is natural to use the stability number as a parameter to determine bounds for $F(G)$. Let $\mathcal{G}(n, \alpha)$ and $\mathcal{C}(n, \alpha)$ be the classes of – respectively general and connected – graphs with order n and stability number α. The lower bound for the Fibonacci index is known for graphs in these classes. Indeed,

---

[*] Chargé de Recherches F.R.S.-FNRS.
[1] The Fibonacci index is called the Fibonacci number in [20]. Merrifield and Simmons introduced it as the σ-index [16], also known as the Merrifield-Simmons index.

B. Yang, D.-Z. Du, and C.A. Wang (Eds.): COCOA 2008, LNCS 5165, pp. 127–138, 2008.
© Springer-Verlag Berlin Heidelberg 2008

Pedersen and Vestergaard [18] give a simple proof to show that if $G \in \mathcal{G}(n, \alpha)$ or $G \in \mathcal{C}(n, \alpha)$, then $F(G) \geq 2^{\alpha} + n - \alpha$. Equality occurs if and only if G is a complete split graph (see Sect. 2). In this article, we determine upper bounds for $F(G)$ in the classes $\mathcal{G}(n, \alpha)$ and $\mathcal{C}(n, \alpha)$. In both cases, the bound is tight for every possible value of α and n and the extremal graphs are characterized.

A Turán graph is the union of disjoint balanced cliques. Turán graphs frequently appear in extremal graph theory. For example, the well-known Theorem of Turán [22] states that these graphs have minimum size inside $\mathcal{G}(n, \alpha)$. We show in Sect. 3 that Turán graphs have also maximum Fibonacci index inside $\mathcal{G}(n, \alpha)$. Observe that removing an edge in a graph strictly increases its Fibonacci index. Indeed, all existing stable sets remain and there is at least one more new stable set: the two vertices incident to the deleted edge. Therefore, we might have the intuition that the upper bound for $F(G)$ is a simple consequence of the Theorem of Turán. However, we show that it is not true (see Sects. 2 and 5). The proof uses structural properties of α-critical graphs.

Graphs in $\mathcal{C}(n, \alpha)$ which maximize $F(G)$ are characterized in Sect. 4. We call them Turán-connected graphs since they are a connected variant of Turán graphs. It is interesting to note that these graphs again minimize the size inside $\mathcal{C}(n, \alpha)$. Hence, our results lead to questions about the relations between the Fibonacci index, the stability number, the size and the order of graphs. These questions are summarized in Sect. 5.

2   Basic Properties

In this section, we suppose that the reader is familiar with usual notions of graph theory (we refer to Berge [1] for more details). First, we fix our terminology and notation. We then recall the notion of α-critical graphs and give properties of such graphs. We end with some basic properties of the Fibonacci index.

2.1   Notations

Let $G = (V, E)$ be a simple and undirected graph order $n(G) = |V|$ and size $m(G) = |E|$. For a vertex $v \in V(G)$, we denote by $N(v)$ the neighborhood of v; its closed neighborhood is defined as $N[v] = N(v) \cup \{v\}$. The degree of a vertex v is denoted by $d(v)$ and the maximum degree of G by $\Delta(G)$. We use notation $G \simeq H$ when G and H are isomorphic graphs. The complement of G is denoted by $\overline{G}$. The *stability number* $\alpha(G)$ of a graph G is the number of vertices of a maximum stable set of G. Clearly, $1 \leq \alpha(G) \leq n(G)$, and $1 \leq \alpha(G) \leq n(G) - 1$ when G is connected.

Definition 1. *We denote by G^v the induced subgraph obtained by removing a vertex v from a graph G. Similarly, the graph $G^{N[v]}$ is the induced subgraph obtained by removing the closed neighborhood of v. Finally, the graph obtained by removing an edge e from G is denoted by G^e.*

Classical graphs of order n are used in this article: the complete graph K_n, the path P_n, the cycle C_n, the star S_n (composed by one vertex adjacent to $n-1$ vertices of degree 1) and the complete split graph $CS_{n,\alpha}$ (composed of a stable set of α vertices, a clique of $n - \alpha$ vertices and each vertex of the stable set is adjacent to each vertex of the clique). The graph $CS_{7,3}$ is depicted in Fig. 1.

We also deeply study the two classes of Turán graphs and Turán-connected graphs. A *Turán graph* $T_{n,\alpha}$ is a graph of order n and a stability number α such that $1 \leq \alpha \leq n$, that is defined as follows. It is the union of α disjoint balanced cliques (that is, such that their orders differ from at most one) [22]. These cliques have thus $\lceil \frac{n}{\alpha} \rceil$ or $\lfloor \frac{n}{\alpha} \rfloor$ vertices. We now define a *Turán-connected graph* $TC_{n,\alpha}$ with n vertices and a stability number α where $1 \leq \alpha \leq n-1$. It is constructed from the Turán graph $T_{n,\alpha}$ with $\alpha - 1$ additional edges. Let v be a vertex of one clique of size $\lceil \frac{n}{\alpha} \rceil$, the additional edges link v and one vertex of each remaining cliques. Note that, for each of the two classes of graphs defined above, there is only one graph with given values of n and α, up to isomorphism.

Example 2. Figure 1 shows the Turán graph $T_{7,3}$ and the Turán-connected graph $TC_{7,3}$. When $\alpha = 1$, we observe that $T_{n,1} \simeq TC_{n,1} \simeq CS_{n,1} \simeq K_n$. When $\alpha = n$, we get $T_{n,n} \simeq CS_{n,n} \simeq \overline{K_n}$, and when $\alpha = n-1$, we get $TC_{n,n-1} \simeq CS_{n,n-1} \simeq S_n$.

Fig. 1. The graphs $CS_{7,3}$, $T_{7,3}$ and $TC_{7,3}$

2.2 α-Critical Graphs

We recall the notion of α-critical graphs [6,10,14]. An edge e of a graph G is *α-critical* if $\alpha(G^e) > \alpha(G)$, otherwise it is *α-safe*. A graph is *α-critical* if all its edges are α-critical. By convention, a graph with no edge is α-critical. These graphs play an important role in extremal graph theory [10], and in our proofs.

Example 3. Simple examples of α-critical graphs are complete graphs and odd cycles. Turán graphs are also α-critical. On the contrary, Turán-connected graph are not α-critical, except when $\alpha = 1$.

Lemma 4. *Let G be an α-critical graph. If G is connected, then the graph G^v is connected for all vertices v of G.*

Proof. We use two known results on α-critical graphs (see, e.g., [14, Chap. 12]). If a vertex v of an α-critical graph has degree 1, then v and its neighbor w form a connected component of the graph. Every vertex of degree at least 2 in an α-critical graph is contained in a cycle. Hence, by the first result, the minimum degree of G equals 2, except if $G \simeq K_2$. Clearly G^v is connected by the second result or when $G \simeq K_2$.   □

Lemma 5. *Let G be an α-critical graph. Let v be any vertex of G which is not isolated. Then, $\alpha(G) = \alpha(G^v) = \alpha(G^{N[v]}) + 1$.*

Proof. Let $e = vw$ be an edge of G containing v. Then, there exist in G two maximum stable sets S and S', such that S contains v, but not w, and S' contains w, but not v (see, e.g., [14, Chap. 12]). Thus, $\alpha(G) = \alpha(G^v)$ due to the existence of S'. The set S avoids each vertex of $N(v)$. Hence, $S \setminus \{v\}$ is a stable set of the graph $G^{N[v]}$ of size $\alpha(G) - 1$. Note that this stable set is maximum. $\square$

2.3   Fibonacci Index

Let us now recall the Fibonacci index [16,20]. The *Fibonacci index $F(G)$* of a graph G is the number of all the stable sets in G, including the empty set. The following lemma is well-known (see [8,13,20]).

Lemma 6. *Let G be a graph.*

- *Let e be an edge of G, then $F(G) < F(G^e)$.*
- *Let v be a vertex of G, then $F(G) = F(G^v) + F(G^{N[v]})$.*
- *If G is the union of k disjoint graphs G_i, $1 \leq i \leq k$, then $F(G) = \prod_{i=1}^{k} F(G_i)$.*

Example 7. We have $F(\mathsf{K}_n) = n + 1$, $F(\overline{\mathsf{K}_n}) = 2^n$, $F(\mathsf{S}_n) = 2^{n-1} + 1$ and $F(\mathsf{P}_n) = f_{n+2}$ (recall that the sequence of Fibonacci numbers f_n is $f_0 = 0, f_1 = 1$ and $f_n = f_{n-1} + f_{n-2}$ for $n > 1$).

Prodinger and Tichy [20] give simple lower and upper bounds for the Fibonacci index. We recall these bounds in the next lemma.

Lemma 8. *Let G be a graph of order n.*

- *Then $n + 1 \leq F(G) \leq 2^n$ with equality if and only if $G \simeq \mathsf{K}_n$ (lower bound) and $G \simeq \overline{\mathsf{K}_n}$ (upper bound).*
- *If G is connected, then $n + 1 \leq F(G) \leq 2^{n-1} + 1$ with equality if and only if $G \simeq \mathsf{K}_n$ (lower bound) and $G \simeq \mathsf{S}_n$ (upper bound).*
- *If G is a tree, then $f_{n+2} \leq F(G) \leq 2^{n-1} + 1$ with equality if and only if $G \simeq \mathsf{P}_n$ (lower bound) and $G \simeq \mathsf{S}_n$ (upper bound).*

We denote by $\mathcal{G}(n, \alpha)$ the class of general graphs with order n and stability number α; and by $\mathcal{C}(n, \alpha)$ the class of connected graphs with order n and stability number α. Pedersen and Vestergaard [18] characterize graphs with minimum Fibonacci index as indicated in the following theorem.

Theorem 9. *Let G be a graph inside $\mathcal{G}(n, \alpha)$ or $\mathcal{C}(n, \alpha)$, then $F(G) \geq 2^\alpha + n - \alpha$, with equality if and only if $G \simeq \mathsf{CS}_{n,\alpha}$.*

The aim of this article is the study of graphs with maximum Fibonacci index inside the two classes $\mathcal{G}(n, \alpha)$ and $\mathcal{C}(n, \alpha)$. The system GraPHedron [15] allows a formal framework to conjecture optimal relations among a set of graph invariants. Thanks to this system, graphs with maximum Fibonacci index inside each of the two previous classes have been computed for small values of n [7]. We observe that these graphs are isomorphic to Turán graphs for the class $\mathcal{G}(n,\alpha)$, and to Turán-connected graphs for the class $\mathcal{C}(n,\alpha)$. For the class $\mathcal{C}(n,\alpha)$, there is one exception when $n = 5$ and $\alpha = 2$: both the cycle C_5 and the graph $TC_{5,2}$ have maximum Fibonacci index.

Recall that the classical Theorem of Turán [22] states that Turán graphs $T_{n,\alpha}$ have minimum size inside $\mathcal{G}(n,\alpha)$. We might think that Turán graphs have maximum Fibonacci index inside $\mathcal{G}(n,\alpha)$ as a direct corollary of the Theorem of Turán and Lemma 6. This argument is not correct since removing an α-critical edge increases the stability number. Therefore, Lemma 6 only implies that graphs with maximum Fibonacci index inside $\mathcal{G}(n,\alpha)$ are α-critical graphs. In Sect. 5, we make further observations on the relations between the size and the Fibonacci index inside the classes $\mathcal{G}(n,\alpha)$ and $\mathcal{C}(n,\alpha)$.

There is another interesting property of Turán graphs related to stable sets. Byskov [4] establish that Turán graphs have maximum number of maximal stable sets inside $\mathcal{G}(n,\alpha)$. The Fibonacci index counts not only the maximal stable sets but all the stable sets. Hence, the fact that Turán graphs maximize $F(G)$ cannot be simply derived from the result of Byskov.

3   General Graphs

In this section, we study graphs with maximum Fibonacci index inside the class $\mathcal{G}(n,\alpha)$. These graphs are said to be *extremal*. For fixed values of n and α, we show that there is one extremal graph up to isomorphism, the Turán graph $T_{n,\alpha}$ (see Theorem 12).

Before establishing this result, we need some auxiliary results. We denote by $f_T(n,\alpha)$ the Fibonacci index of the Turán graph $T_{n,\alpha}$. By Lemma 6,

$$f_T(n,\alpha) = \left(\left\lceil \frac{n}{\alpha}\right\rceil + 1\right)^p \left(\left\lfloor \frac{n}{\alpha}\right\rfloor + 1\right)^{\alpha-p},$$

where $p = (n \mod \alpha)$. We have also the following inductive formula.

Lemma 10. *Let n and α be integers such that $1 \leq \alpha \leq n$. Then*

$$f_T(n,\alpha) = \begin{cases} n+1 & \text{if } \alpha = 1, \\ 2^n & \text{if } \alpha = n, \\ f_T(n-1,\alpha) + f_T(n - \left\lceil \frac{n}{\alpha}\right\rceil, \alpha - 1) & \text{if } 2 \leq \alpha \leq n - 1 \ . \end{cases}$$

Proof. The cases $\alpha = 1$ and $\alpha = n$ are trivial (see Example 7). Suppose $2 \leq \alpha \leq n-1$. Let v be a vertex of $T_{n,\alpha}$ with maximum degree. Thus v is in a $\left\lceil \frac{n}{\alpha}\right\rceil$-clique. As $\alpha < n$, the vertex v is not isolated. Therefore $T_{n,\alpha}^v \simeq T_{n-1,\alpha}$. As $\alpha \geq 2$, the graph $T_{n,\alpha}^{N[v]}$ has at least one vertex, and $T_{n,\alpha}^{N[v]} \simeq T_{n-\left\lceil \frac{n}{\alpha}\right\rceil, \alpha-1}$. By Lemma 6, we obtain the inductive formula. $\square$

A consequence of Lemma 10 is that $f_T(n-1,\alpha) < f_T(n,\alpha)$. Indeed, the cases $\alpha = 1$ and $\alpha = n$ are trivial, and the term $f_T(n - \left\lceil \frac{n}{\alpha}\right\rceil, \alpha - 1)$ is always strictly positive when $2 \leq \alpha \leq n - 1$.

Corollary 11. *The function $f_T(n, \alpha)$ is strictly increasing in n when α is fixed.*

We now state the upper bound on $F(G)$ inside the class $\mathcal{G}(n, \alpha)$.

Theorem 12. *Let G be a graph of order n with a stability number α, then $F(G) \le f_T(n, \alpha)$, with equality if and only if $G \simeq T_{n,\alpha}$.*

Proof. The cases $\alpha = 1$ and $\alpha = n$ are straightforward. Indeed $G \simeq T_{n,1}$ when $\alpha = 1$, and $G \simeq T_{n,n}$ when $\alpha = n$. We can assume that $2 \le \alpha \le n - 1$, and thus $n \ge 3$. We now prove by induction on n that if G is extremal, then it is isomorphic to $T_{n,\alpha}$.

The graph G is α-critical. Otherwise, there exists an edge $e \in E(G)$ such that $\alpha(G) = \alpha(G^e)$, and by Lemma 6, $F(G) < F(G^e)$. This is a contradiction with G being extremal.

Let us compute $F(G)$ thanks to Lemma 6. Let $v \in V(G)$ of maximum degree Δ. The vertex v is not isolated since $\alpha < n$. Thus by Lemma 5, $\alpha(G^v) = \alpha$ and $\alpha(G^{N[v]}) = \alpha - 1$. On the other hand, If χ is the chromatic number of G, it is well-known that $n \le \chi \cdot \alpha$ (see, e.g., Berge [1]), and that $\chi \le \Delta + 1$ (see Brooks [3]). It follows that

$$n(G^{N[v]}) = n - \Delta - 1 \le n - \left\lceil \frac{n}{\alpha} \right\rceil . \tag{1}$$

Note that $n(G^{N[v]}) \ge 1$ since $\alpha \ge 2$.

We can apply the induction hypothesis on the graphs G^v and $G^{N[v]}$. We obtain

$$
\begin{aligned}
f_T(n, \alpha) \le F(G) & \qquad \text{as } G \text{ is extremal,} \\
= F(G^v) + F(G^{N[v]}), & \qquad \text{by Lemma 6,} \\
\le f_T(n(G^v), \alpha(G^v)) + f_T(n(G^{N[v]}), \alpha(G^{N[v]})), & \qquad \text{by induction,} \\
= f_T(n - 1, \alpha) + f_T(n - \Delta - 1, \alpha - 1), & \\
\le f_T(n - 1, \alpha) + f_T(n - \lceil \tfrac{n}{\alpha} \rceil, \alpha - 1), & \qquad \text{by (1) and Corollary 11,} \\
= f_T(n, \alpha) & \qquad \text{by Lemma 10 .}
\end{aligned}
$$

Hence equality holds everywhere. In particular, by induction, the graphs G^v, $G^{N[v]}$ are extremal, and $G^v \simeq T_{n-1,\alpha}$, $G^{N[v]} \simeq T_{n-\lceil \frac{n}{\alpha} \rceil, \alpha-1}$. Coming back to G from G^v and $G^{N[v]}$ and recalling that v has maximum degree, it follows that $G \simeq T_{n,\alpha}$. □

Corollary 11 states that $f_T(n, \alpha)$ is increasing in n. It was an easy consequence of Lemma 10. The function $f_T(n, \alpha)$ is also increasing in α. Theorem 12 can be used to prove this fact easily as shown now.

Corollary 13. *The function $f_T(n, \alpha)$ is strictly increasing in α when n is fixed.*

Proof. Suppose $2 \le \alpha \le n-1$. By Lemma 8 it is clear that $f_T(n, 1) < f_T(n, \alpha) < f_T(n, n)$. Now, let e be an edge of $T_{n,\alpha}$. Clearly $\alpha(T_{n,\alpha}^e) = \alpha + 1$. Moreover, by Lemma 6 and Theorem 12, $F(T_{n,\alpha}) < F(T_{n,\alpha}^e) < F(T_{n,\alpha+1})$. Therefore, $f_T(n, \alpha) < f_T(n, \alpha + 1)$. □

4   Connected Graphs

We now consider graphs with maximum Fibonacci index inside the class $\mathcal{C}(n, \alpha)$. Such graphs are called *extremal*. If G is connected, the bound of Theorem 12 is clearly not tight, except when $\alpha = 1$, that is, when G is a complete graph. We are going to prove that there is one extremal graph up to isomorphism, the Turán-connected graph $\mathsf{TC}_{n,\alpha}$, with the exception of the cycle C_5 (see Theorem 17). First, we need preliminary results and definitions to prove this theorem.

We denote by $f_{\mathsf{TC}}(n, \alpha)$ the Fibonacci index of the Turán-connected graph $\mathsf{TC}_{n,\alpha}$. An inductive formula for its value is given in the next lemma.

Lemma 14. *Let n and α be integers such that $1 \le \alpha \le n - 1$. Then*

$$f_{\mathsf{TC}}(n, \alpha) = \begin{cases} n + 1 & \text{if } \alpha = 1, \\ 2^{n-1} + 1 & \text{if } \alpha = n - 1, \\ f_{\mathsf{T}}(n - 1, \alpha) + f_{\mathsf{T}}(n', \alpha') & \text{if } 2 \le \alpha \le n - 2, \end{cases}$$

where $n' = n - \left\lceil \frac{n}{\alpha} \right\rceil - \alpha + 1$ and $\alpha' = \min(n', \alpha - 1)$.

Proof. The cases $\alpha = 1$ and $\alpha = n - 1$ are trivial by Lemma 8. Suppose now that $2 \le \alpha \le n - 2$. Let v be a vertex of maximum degree in $\mathsf{TC}_{n,\alpha}$. We apply Lemma 6 to compute $F(\mathsf{TC}_{n,\alpha})$. Observe that the graphs $\mathsf{TC}^v_{n,\alpha}$ and $\mathsf{TC}^{N[v]}_{n,\alpha}$ are both Turán graphs when $2 \le \alpha \le n - 2$.

The graph $\mathsf{TC}^v_{n,\alpha}$ is isomorphic to $\mathsf{T}_{n-1,\alpha}$. Let us show that $\mathsf{TC}^{N[v]}_{n,\alpha}$ is isomorphic to $\mathsf{T}_{n',\alpha'}$. By definition of a Turán-connected graph, $d(v)$ is equal to $\left\lceil \frac{n}{\alpha} \right\rceil + \alpha - 2$. Thus $n(\mathsf{TC}^{N[v]}_{n,\alpha}) = n - d(v) - 1 = n'$.

If $\alpha < \frac{n}{2}$, then $\mathsf{TC}_{n,\alpha}$ has a clique of order at least 3 and $\alpha(\mathsf{TC}^{N[v]}_{n,\alpha}) = \alpha - 1 \le n'$. Otherwise, $\mathsf{TC}^{N[v]}_{n,\alpha} \simeq \overline{\mathsf{K}_{n'}}$ and $\alpha(\mathsf{TC}^{N[v]}_{n,\alpha}) = n' \le \alpha - 1$. Therefore $\alpha(\mathsf{TC}^{N[v]}_{n,\alpha}) = \min(n', \alpha - 1)$ in both cases. By Lemma 6, these observations leads to the inductive formula. □

Definition 15. *A bridge in a connected graph G is an edge $e \in E(G)$ such that the graph G^e is no more connected. To a bridge $e = v_1 v_2$ of G which is α-safe, we associate a decomposition $\mathcal{D}(G_1, v_1, G_2, v_2)$ such that $v_1 \in V(G_1)$, $v_2 \in V(G_2)$, and G_1, G_2 are the two connected components of G^e. A decomposition is said to be α-critical if G_1 is α-critical.*

Lemma 16. *Let G be a connected graph. If G is extremal, then either G is α-critical or G has an α-critical decomposition.*

Proof. We suppose that G is not α-critical and we show that it must contain an α-critical decomposition. Let e be an α-safe edge of G. Then e must be a bridge. Otherwise, the graph G^e is connected, has the same order and stability number as G and satisfies $F(G^e) > F(G)$ by Lemma 6. This is a contradiction with G being extremal. Therefore G contains at least one α-safe bridge defining a decomposition of G.

Let us choose a decomposition $\mathcal{D}(G_1, v_1, G_2, v_2)$ such that G_1 is of minimum order. Then, G_1 is α-critical. Otherwise, G_1 contains an α-safe bridge $e' = w_1 w_2$, since the edges of G are α-critical or α-safe bridges by the first part of the proof. Let $\mathcal{D}(H_1, w_1, H_2, w_2)$ be the decomposition of G defined by e', such that $v_1 \in V(H_2)$. Then $n(H_1) < n(G_1)$, which is a contradiction. Hence the decomposition $\mathcal{D}(G_1, v_1, G_2, v_2)$ is α-critical. $\square$

Theorem 17. *Let G be a connected graph of order n with a stability number α, then $F(G) \leq f_{\mathsf{TC}}(n, \alpha)$, with equality if and only if $G \simeq \mathsf{TC}_{n,\alpha}$ when $(n, \alpha) \neq (5,2)$, and $G \simeq \mathsf{TC}_{5,2}$ or $G \simeq \mathsf{C}_5$ when $(n, \alpha) = (5,2)$.*

Proof. We prove by induction on n that if G is extremal, then it is isomorphic to $\mathsf{TC}_{n,\alpha}$ or C_5. To handle more easily the general case of the induction (in a way to avoid the extremal graph C_5), we consider all connected graphs with up to 6 vertices as the basis of the induction. For these basic cases, we refer to the report of an exhaustive automated verification [7]. We thus suppose that $n \geq 7$.

We know by Lemma 16 that either G has an α-critical decomposition or G is α-critical. We consider now these two situations.

1) G has an α-critical decomposition. We prove in three steps that $G \simeq \mathsf{TC}_{n,\alpha}$: (*i*) We establish that for every decomposition $\mathcal{D}(G_1, v_1, G_2, v_2)$, the graph G_i is extremal and is isomorphic to a Turán-connected graph such that $d(v_i) = \Delta(G_i)$, for $i = 1, 2$. (*ii*) We show that if such a decomposition is α-critical, then G_1 is a clique. (*iii*) We prove that G is itself isomorphic to a Turán-connected graph.

(*i*) For the first step, let $\mathcal{D}(G_1, v_1, G_2, v_2)$ be a decomposition of G, n_1 be the order of G_1, and α_1 its stability number. We prove that $G_1 \simeq \mathsf{TC}_{n_1,\alpha_1}$ such that $d(v_1) = \Delta(G_1)$. The argument is identical for G_2. By Lemma 6, we have $F(G) = F(G_1)F(G_2^{v_2}) + F(G_1^{v_1})F(G_2^{N[v_2]})$.

By the induction hypothesis, $F(G_1) \leq f_{\mathsf{TC}}(n_1, \alpha_1)$. The graph $G_1^{v_1}$ has an order $n_1 - 1$ and a stability number $\leq \alpha_1$. Hence by Theorem 12 and Corollary 13, $F(G_1^{v_1}) \leq f_{\mathsf{T}}(n_1 - 1, \alpha_1)$. It follows that

$$F(G) \leq f_{\mathsf{TC}}(n_1, \alpha_1)F(G_2^{v_2}) + f_{\mathsf{T}}(n_1 - 1, \alpha_1)F(G_2^{N[v_2]}) \ . \tag{2}$$

As G is supposed to be extremal, equality occurs. It means that $G_1^{v_1} \simeq \mathsf{T}_{n_1-1,\alpha_1}$ and G_1 is extremal. If G_1 is isomorphic to C_5, then $n_1 = 5$, $\alpha_1 = 2$ and $F(G_1) = f_{\mathsf{TC}}(5, 2)$. However, $F(G_1^{v_1}) = F(\mathsf{P}_4) < f_{\mathsf{T}}(4, 2)$. By (2), this leads to a contradiction with G being extremal. Thus, G_1 must be isomorphic to $\mathsf{TC}_{n_1,\alpha_1}$. Moreover, v_1 is a vertex of maximum degree of G_1. Otherwise, $G_1^{v_1}$ cannot be isomorphic to the graph $\mathsf{T}_{n_1-1,\alpha_1}$.

(*ii*) The second step is easy. Let $\mathcal{D}(G_1, v_1, G_2, v_2)$ be an α-critical decomposition of G, that is, G_1 is α-critical. By (*i*), G_1 is isomorphic to a Turán-connected graph. The complete graph is the only Turán-connected graph which is α-critical. Therefore, G_1 is a clique.

(*iii*) We now suppose that G has an α-critical decomposition $\mathcal{D}(G_1, v_1, G_2, v_2)$ and we show that $G \simeq \mathsf{TC}_{n,\alpha}$. Let n_1 be the order of G_1 and α_1 its stability number. As $v_1 v_2$ is an α-safe bridge, it is clear that $n(G_2) = n - n_1$ and $\alpha(G_2) =$

$\alpha - \alpha_1$. By (i) and (ii), G_1 is a clique (and thus $\alpha_1 = 1$), $G_2 \simeq \mathsf{TC}_{n-n_1,\alpha-1}$, and v_2 is a vertex of maximum degree in G_2. If $\alpha = 2$, then G_2 is also a clique in G. By Lemma 6 and the fact that $F(\mathsf{K}_n) = n + 1$ we have,

$$F(G) = F(G^{v_1}) + F(G^{N[v_1]}) = n_1(n - n_1 + 1) + (n - n_1) = n + n\,n_1 - n_1^2 \ .$$

When n is fixed, this function is maximized when $n_1 = \frac{n}{2}$. That is, when G_1 and G_2 are balanced cliques. This appears if and only if $G \simeq \mathsf{TC}_{n,2}$.

Thus we suppose that $\alpha \geq 3$. In other words, G contains at least three cliques: the clique G_1 of order n_1; the clique H containing v_2 and a clique H' in G_2 linked to H by an α-safe bridge v_2v_3. Let $k = \frac{n-n_1}{\alpha-1}$, then the order of H is $\lceil k \rceil$ and the order of H' is $\lceil k \rceil$ or $\lfloor k \rfloor$ (recall that $G_2 \simeq \mathsf{TC}_{n-n_1,\alpha-1}$). These cliques are represented in Fig. 2.

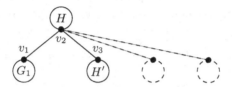

Fig. 2. Cliques in the graph G

To prove that G is isomorphic to a Turán-connected graph, it remains to show that the clique G_1 is balanced with the cliques H and H'. We consider the decomposition defined by the α-safe bridge v_2v_3. By (i), G_1 and H are cliques of a Turán-connected graph, and H is a clique with maximum order in this graph (recall that v_2 is a vertex of maximum degree in G_2). Therefore $\lceil k \rceil - 1 \leq n_1 \leq \lceil k \rceil$, showing that G_1 is balanced with H and H'.

2) G is α-critical. Under this hypothesis, we prove that G is a complete graph, and thus is isomorphic to a Turán-connected graph. Suppose that G is not complete. Let v be a vertex of G with a maximum degree $d(v) = \Delta$. As G is connected and α-critical, the graph G^v is connected by Lemma 4. By Lemma 5, $\alpha(G^v) = \alpha$ and $\alpha(G^{N[v]}) = \alpha - 1$. Moreover, $n(G^v) = n - 1$ and $n(G^{N[v]}) = n - \Delta - 1$.

By the induction hypothesis and Theorem 12, we get

$$F(G) = F(G^v) + F(G^{N[v]}) \leq f_{\mathsf{TC}}(n - 1, \alpha) + f_{\mathsf{T}}(n - \Delta - 1, \alpha - 1) \ .$$

Therefore, G is extremal if and only if $G^{N[v]} \simeq \mathsf{T}_{n-\Delta-1,\alpha-1}$ and G^v is extremal. However, G^v is not isomorphic to C_5 as $n \geq 7$. Thus $G^v \simeq \mathsf{TC}_{n-1,\alpha}$.

So, the graph G is composed by the graph $G^v \simeq \mathsf{TC}_{n-1,\alpha}$ and an additional vertex v connected to $\mathsf{TC}_{n-1,\alpha}$ by Δ edges. There must be an edge between v and a vertex v' of maximum degree in G^v, otherwise $G^{N[v]}$ is not isomorphic to a Turán graph. The vertex v' is adjacent to $\lceil \frac{n-1}{\alpha} \rceil + \alpha - 2$ vertices in G^v and it is adjacent to v, that is, $d(v') = \lceil \frac{n-1}{\alpha} \rceil + \alpha - 1$. Since G is not a complete graph, we have

$$\Delta \geq d(v') > \left\lceil \frac{n-1}{\alpha} \right\rceil, \tag{3}$$

On the other hand, v is adjacent to each vertex of some clique H of G^v since $G^{N[v]}$ has a stability number $\alpha - 1$. As this clique has order at most $\lceil \frac{n-1}{\alpha} \rceil$, v must be adjacent to a vertex $w \notin H$ by (3). We observe that the edge vw is α-safe. This is impossible as G is α-critical. It follows that G is a complete graph and the proof is completed.   □

The study of the maximum Fibonacci index inside the class $\mathcal{T}(n, \alpha)$ of trees with order n and stability number α is strongly related to the study done in this section for the class $\mathcal{C}(n, \alpha)$. Indeed, due to the fact that trees are bipartite, a tree in $\mathcal{T}(n, \alpha)$ has always a stability number $\alpha \geq \frac{n}{2}$. Moreover, the Turán-connected graph $\mathsf{TC}_{n,\alpha}$ is a tree when $\alpha \geq \frac{n}{2}$. Therefore, the upper bound on the Fibonacci index for connected graphs is also valid for trees. We thus get the next corollary with in addition the exact value of $f_{\mathsf{TC}}(n, \alpha)$.

Corollary 18. *Let G be a tree of order n with a stability number α, then*

$$F(G) \leq 3^{n-\alpha-1}2^{2\alpha-n+1} + 2^{n-\alpha-1},$$

with equality if and only if $G \simeq \mathsf{TC}_{n,\alpha}$.

Proof. It remains to compute the exact value of $f_{\mathsf{TC}}(n, \alpha)$. When $\alpha \geq \frac{n}{2}$, the graph $\mathsf{TC}_{n,\alpha}$ is composed by one central vertex v of degree α and α pending paths of length 1 or 2 attached to v. An extremity of a pending path of length 2 is a vertex w such that $w \notin N[v]$. Thus there are $x = n - \alpha - 1$ pending paths of length 2 since $N[v]$ has size $\alpha + 1$, and $y = \alpha - x = 2\alpha - n + 1$ pending paths of length 1. We apply Lemma 6 on v to get $f_{\mathsf{TC}}(n, \alpha) = 3^x 2^y + 2^x$.   □

5   Observations

Turán graphs $\mathsf{T}_{n,\alpha}$ have minimum size inside $\mathcal{G}(n, \alpha)$ by the Theorem of Turán [22]. Christophe et al. [5] give a tight lower bound for the connected case of this theorem, and Bougard and Joret [2] characterized the extremal graphs, which happen to contain the $\mathsf{TC}_{n,\alpha}$ graphs as a subclass.

By these results and Theorems 12 and 17, we can observe the following relations between graphs with minimum size and maximum Fibonacci index. The graphs inside $\mathcal{G}(n, \alpha)$ minimizing $m(G)$ are exactly those which maximize $F(G)$. This is also true for the graphs inside $\mathcal{C}(n, \alpha)$, except that there exist other graphs with minimum size than the Turán-connected graphs.

However, these observations are not a trivial consequence of the fact that $F(G) < F(G^e)$ where e is any edge of a graph G. As shown in our proofs, the latter property only implies that a graph maximizing $F(G)$ contains only α-critical edges (and α-safe bridges for the connected case). Our proofs use a deep study of the structure of the extremal graphs to get Theorems 12 and 17.

We now give additional examples showing that the intuition that more edges imply fewer stable sets is wrong. Pedersen and Vestergaard [18] give the following example. Let r be an integer such that $r \geq 3$, G_1 be the Turán graph $\mathsf{T}_{2r,r}$ and G_2 be the star S_{2r}. The graphs G_1 and G_2 have the same order but G_1 has less edges (r) than G_2 ($2r-1$). Nevertheless, observe that $F(G_1) = 3^r < F(G_2) = 2^{2r-1}+1$. However, we note that $\alpha(G_1) < \alpha(G_2)$.

We propose a similar example of pairs of graphs with the same order and the same stability number (see the graphs G_3 and G_4 on Fig. 3). These two graphs are inside the class $\mathcal{G}(6,4)$, however $m(G_3) < m(G_4)$ and $F(G_3) < F(G_4)$. Notice that we can get such examples inside $\mathcal{G}(n,\alpha)$ with n arbitrarily large, by considering the union of several disjoint copies of G_3 and G_4.

These remarks and our results suggest some questions about the relations between the size, the stability number and the Fibonacci index of graphs. What are the lower and upper bounds for the Fibonacci index inside the class $\mathcal{G}(n,m)$ of graphs order n and size m; or inside the class $\mathcal{G}(n,m,\alpha)$ of graphs order n, size m and stability number α? Are there classes of graphs for which more edges always imply fewer stable sets? We think that these questions deserve to be studied.

Fig. 3. Graphs with same order and stability number

Acknowledgments

The authors thank Gwenaël Joret for helpful suggestions.

References

1. Berge, C.: The Theory of Graphs. Dover Publications, New York (2001)
2. Bougard, N., Joret, G.: Turán Theorem and k-connected graphs. J. Graph Theory 58, 1–13 (2008)
3. Brooks, R.-L.: On colouring the nodes of a network. Proc. Cambridge Philos. Soc. 37, 194–197 (1941)
4. Byskov, J.M.: Enumerating maximal independent sets with applications to graph colouring. Oper. Res. Lett. 32, 547–556 (2004)
5. Christophe, J., Dewez, S., Doignon, J.-P., Elloumi, S., Fasbender, G., Grégoire, P., Huygens, D., Labbé, M., Mélot, H., Yaman, H.: Linear inequalities among graph invariants: using GraPHedron to uncover optimal relationships, 24 pages (Accepted for publication in Networks, 2008)
6. Erdös, P., Gallai, T.: On the minimal number of vertices representing the edges of a graph. Magyar Tud. Akad. Mat. Kutató Int. Közl. 6, 181–203 (1961)
7. GraPHedron: Reports on the study of the Fibonacci index and the stability number of graphs and connected graphs,
   `www.graphedron.net/index.php?page=viewBib\&bib=7`

8. Gutman, I., Polansky, O.E.: Mathematical Concepts in Organic Chemistry. Springer, Berlin (1986)
9. Heuberger, C., Wagner, S.: Maximizing the number of independent subsets over trees with bounded degree. J. Graph Theory 58, 49–68 (2008)
10. Joret, G.: Entropy and Stability in Graphs. PhD thesis, Université Libre de Bruxelles, Belgium (2007)
11. Knopfmacher, A., Tichy, R.F., Wagner, S., Ziegler, V.: Graphs, partitions and Fibonacci numbers. Discrete Appl. Math. 155, 1175–1187 (2007)
12. Li, X., Li, Z., Wang, L.: The Inverse Problems for Some Topological Indices in Combinatorial Chemistry. J. Comput. Biol. 10(1), 47–55 (2003)
13. Li, X., Zhao, H., Gutman, I.: On the Merrifield-Simmons Index of Trees. MATCH Comm. Math. Comp. Chem. 54, 389–402 (2005)
14. Lovász, L., Plummer, M.D.: Matching Theory. Akadémiai Kiadó. North-Holland, Budapest (1986)
15. Mélot, H.: Facet defining inequalities among graph invariants: the system GraPHedron. Discrete Appl. Math., 17 pages (Accepted for publication, 2007)
16. Merrifield, R.E., Simmons, H.E.: Topological Methods in Chemistry. Wiley, New York (1989)
17. Pedersen, A.S., Vestergaard, P.D.: The number of independent sets in unicyclic graphs. Discrete Appl. Math. 152, 246–256 (2005)
18. Pedersen, A.S., Vestergaard, P.D.: Bounds on the Number of Vertex Independent Sets in a Graph. Taiwanese J. Math. 10(6), 1575–1587 (2006)
19. Pedersen, A.S., Vestergaard, P.D.: An Upper Bound on the Number of Independent Sets in a Tree. Ars Combin. 84, 85–96 (2007)
20. Prodinger, H., Tichy, R.F.: Fibonacci numbers of graphs. Fibonacci Quart. 20(1), 16–21 (1982)
21. Tichy, R.F., Wagner, S.: Extremal Problems for Topological Indices in Combinatorial Chemistry. J. Comput. Biol. 12(7), 1004–1013 (2005)
22. Turán, P.: Eine Extremalaufgabe aus der Graphentheorie. Mat. Fiz. Lapok 48, 436–452 (1941)
23. Wagner, S.: Extremal trees with respect to Hosoya Index and Merrifield-Simmons Index. MATCH Comm. Math. Comp. Chem. 57, 221–233 (2007)
24. Wang, H., Hua, H.: Unicycle graphs with extremal Merrifield-Simmons Index. J. Math. Chem. 43(1), 202–209 (2008)
25. Wang, M., Hua, H., Wang, D.: The first and second largest Merrifield-Simmons indices of trees with prescribed pendent vertices. J. Math. Chem. 43(2), 727–736 (2008)

Vertex-Uncertainty in Graph-Problems
(Extended Abstract)

Cécile Murat and Vangelis Th. Paschos

LAMSADE, CNRS UMR 7024 and Université Paris-Dauphine
Place du Maréchal de Lattre de Tassigny, 75775 Paris Cedex 16, France
{murat,paschos}@lamsade.dauphine.fr

Abstract. We study a probabilistic model for graph-problems under vertex-uncertainty. We assume that any vertex v_i of the input-graph G has only a probability p_i to be present in the final graph to be optimized (i.e., the final instance for the problem tackled will be only a sub-graph of the initial graph). Under this model, the original "deterministic" problem gives rise to a new (deterministic) problem on the same input-graph G, having the same set of feasible solutions as the former one, but its objective function can be very different from the original one, the set of its optimal solutions too. Moreover, this objective function is a sum of $2^{|V|}$ terms, where V is the vertex-set of G; hence, its computation is not immediately polynomial. We give sufficient conditions for large classes of graph-problems under which objective functions of the probabilistic counterparts are polynomially computable and optimal solutions are well-characterized.

1 Introduction

Very often people has to make decisions under several degrees of uncertainty, i.e., when only probabilistic information about the future is available. We deal with the following probabilistic model under data uncertainty. Consider a generic instance I of a combinatorial optimization problem Π. Assume that Π is not to be necessarily solved on the whole I, but rather on a (unknown a priori) sub-instance $I' \subset I$. Suppose that any datum d_i in the data-set describing I has a probability p_i, indicating how d_i is likely to be present in the final sub-instance I'. Consider finally that once I' is specified, the solver has no opportunity to solve it directly (for example she/he has to react quasi-immediately, so no sufficient time is given her/him). Concrete examples of such situations dealing with satellite shots planning or with timetabling are given in [1,2].

What can the solver do in this case? A natural way to proceed is to compute an *anticipatory solution* S for Π, i.e., a solution for the entire instance I, and once I' becomes known, to modify S in order to get a solution S' fitting I'. The objective is to determine an initial solution S for I such that, for any sub-instance $I' \subseteq I$ presented for optimization, the solution S' respects some pre-defined quality criterion (for example, optimal for I', or achieving, say, constant approximation ratio, or ...).

B. Yang, D.-Z. Du, and C.A. Wang (Eds.): COCOA 2008, LNCS 5165, pp. 139–148, 2008.

In what follows we restrict ourselves in graph-problems and consider the following very simple and quick modification[1] of S: *take the restriction of S in the present sub-instance of G*. Consider a graph-problem Π, a graph $G(V,E)$ of order n, instance of Π, and an n-vector $\mathbf{Pr} = (p_1, \ldots, p_n)$ of vertex-probabilities any of them, say p_i, measuring how likely is for vertex $v_i \in V$, $i = 1, \ldots, n$ to be present in the final subgraph $G' \subseteq G$, on which the problem will be really solved. For any solution S for Π in G and for any $V' \subseteq V$, denote by S' the restriction of S in V', i.e., the set resulting from S after removal of the vertices that do not belong to V'. As we have mentioned, S', can or cannot (depending of the definition of Π) be a feasible solution of Π in the subgraph G' of G induced by V'. Whenever S' is feasible, denote by $m(G', S')$ the objective value of S' in G'. Then, the value of S for G, denoted by $E(G, S)$ (and frequently called *functional*), is the expectation of $m(G', S')$, over all the possible $G' \subseteq G$. Formally, given S, the functional $E(G, S)$ of S is defined by: $E(G, S) = \sum_{V' \subseteq V} \Pr[V'] m(G', S')$, where $\Pr[V']$ is the probability that V' will be finally the real instance to be optimized and is defined by: $\Pr[V'] = \prod_{v_i \in V'} p_i \prod_{v_i \in V \setminus V'} (1 - p_i)$. Obviously, $E(G, S)$ depends also on $\mathbf{Pr}$ but, for simplicity, this dependency will be omitted. Quantity, $E(G, S)$ can be seen as the objective function of a new combinatorial problem, derived from Π and denoted by PROBABILISTIC Π in what follows, where we are given an instance G of Π and a probability vector $\mathbf{Pr}$ on the vertices of G and the objective is to determine a solution S^* in G optimizing $E(G, S)$ (optimal anticipatory solution). The optimization goal of PROBABILISTIC Π is the same as the one of Π.

This way to tackle uncertainty in combinatorial optimization is sometimes called *a priori optimization* (this term has been introduced by [3]). In a priori optimization, the combinatorial problem to be solved, being subject to hazards or to inaccuracies, is not defined on a static and clearly specified instance, since the instance to be effectively optimized is not known with absolute certainty from the beginning. Probabilistic requirements are due to the fact that uncertainty in the presence of data makes that it is not possible to assign unique values to some parameters of the problems. In such an approach, the goal is to compute solutions that behave "well" under any assignment of values to these parameters. Under this model, restrictive versions of routing and network-design probabilistic minimization problems (in complete graphs) have been studied in [4,5,6,7,3,8,9,10,11]. Recently, in [12], the analysis of the probabilistic minimum travelling salesman problem, originally performed in [5,8], has been revisited and refined. In [13,1,14] the minimum vertex covering and the minimum coloring are tackled, while in [15,16] probabilistic maximization problems, namely, the longest path and the maximum independent set, are studied. Finally, in [17], the Steiner forest problem (a generalization of the well-known Steiner tree problem) is tackled. An early survey about a priori optimization can be found in [18] while, a more recent one appears in [19].

Our goal is to go beyond study of probabilistic versions of particular combinatorial problems and to propose a structural study of uncertainty for the

---

[1] As we will see later such a modification strategy does not always produce feasible solutions; in such a case some more work is needed.

a priori optimization paradigm. Here, the main mathematical issues (assuming that, given an anticipatory solution S, its restriction to $G[V']$, the subgraph of G induced by V', is feasible) are: (i) the complexity of the computation of $E(G, S)$ which, carrying over 2^n additive terms, is non-trivially polynomial; (ii) compact combinatorial characterization (based upon the form of $E(G, S)$) of S^* as optimal solution of PROBABILISTIC Π; (iii) the complexity of computing S^*, at least for particular classes of subproblems of the initial problem.

Notice that, for any problem Π, its combinatorial counterpart PROBABILISTIC Π contains Π as subproblem (just consider probability vector $(1, \ldots, 1)$ for Π). Hence, from a complexity point of view, PROBABILISTIC Π is at least as hard as Π, that is, if Π is **NP**-hard, then PROBABILISTIC Π is also **NP**-hard, while if Π is polynomial, then no immediate indication can be provided for the complexity of PROBABILISTIC Π, until this latter problem is explicitly studied.

In what follows, we tackle four categories of graph-problems exhausting a very large part of the most known ones. For any of these categories, we give sufficient conditions under which functionals are analytically expressible and polynomially computable and anticipatory solutions are well-characterized. These structural results immediately apply to several well-known problems producing so particular results interesting per se. Furthermore, the scope of our results is even larger than for graph-problems, as problems not originally defined on graphs (e.g., MAX SET PACKING or MIN SET COVER), are also captured. So, this work can provide a framework for a systematic classification of a great number of probabilistic derivatives of well-known graph-problems.

Given a combinatorial problem $\Pi \in$ **NPO**, we denote by PROBABILISTIC Π, its probabilistic counterpart defined as described previously and assume that the vertex-probabilities are independent.

The results in this extended abstract are given without detailed proofs which can be found in [20]. Also, definitions of the particular optimization problems discussed can be found in [21].

2   Solutions Are Subsets of the Initial Vertex-Set

In this section, we deal with graph-problems whose solutions are subsets of the vertex-set of the input-graph and where, given such a solution S and a set $V' \subseteq V$, the restriction of S in V', i.e., the set $S' = S \cap V'$ is feasible for $G[V']$. The main result of this section is stated in Theorem 1.

Theorem 1. *Consider a graph-problem Π verifying the following assumptions: (i) an instance of Π is a vertex-weighted graph $G(V, E, \boldsymbol{w})$; (ii) solutions of Π are subsets of V; (iii) for any solution S and any subset $V' \subseteq V$, $S' = S \cap V'$ is feasible for $G' = G[V']$; (iv) the value of any solution $S \subseteq V$ is defined by: $m(G, S) = w(S) = \sum_{v_i \in S} w_i$, where w_i is the weight of $v_i \in V$. Then, the functional of PROBABILISTIC Π is expressed as: $E(G, S) = \sum_{v_i \in S} w_i p_i$ and can be computed in polynomial time. Furthermore, the complexity of PROBABILISTIC Π is the same as the one of Π.*

Although computation of the functional is, as we have mentioned, a priori exponential (since it carries over 2^n subgraphs of G), assumptions (i) through (iv) in Theorem 1 allow polynomial computation of its value. This is due to the fact that, under these assumptions, given a subgraph G' induced by a subset $V' \subseteq V$, the value of the solution for G' is the sum of the weights of the vertices in $S \cap V'$. Furthermore, a vertex not in S will never make part of any solution in any sub-graph of G. Consequently, computation of the functional amounts to determining, for any G', which vertices make part of $S \cap V'$. This is equivalent with the specification, for any $v_i \in S$, of all the subgraphs to which v_i belongs and with a summation of the presence-probabilities of these subgraphs. This sum is equal to p_i (the probability of v_i). This simplification is the main reason that renders functional's computation polynomial, despite of the exponential number of terms in its generic expression.

Notice that Theorem 1 can also be used for getting generic approximation results for PROBABILISTIC Π. Indeed, since this problem is a particular weighted version of Π, one immediately concludes that *if Π is approximable within approximation ratio ρ, so is* PROBABILISTIC Π.

Corollary 1. *Under the hypotheses of Theorem 1, whenever Π and* PROBABILISTIC Π *are* **NP**-*hard, they are equi-approximable.*

Theorem 1 has also the following immediate corollary dealing with the case of probabilistic versions of unweighted combinatorial optimization problems.

Corollary 2. *Consider a probabilistic combinatorial optimization problem* PROBABILISTIC Π *verifying assumptions (i) to (iv) of Theorem 1 with $w = 1$. Then, the functional of* PROBABILISTIC Π, *is expressed as:* $E(G, S) = \sum_{v_i \in S} p_i$ *and can be computed in polynomial time. Furthermore,* PROBABILISTIC Π *is equivalent to a weighted version of Π where vertex-weights are the vertex-probabilities.*

Corollary 2 is somewhat weaker than Theorem 1 since it does not establish the equivalence between Π and PROBABILISTIC Π but rather a kind of reduction from Π to PROBABILISTIC Π stating that the latter is a priori harder than the former one. As a consequence, whenever Π is **NP**-hard, so is PROBABILISTIC Π whereas if Π is polynomial, the status of PROBABILISTIC Π remains unclear by Corollary 2.

Let us note that Theorem 1 can be applied to a broad class of problems that fit its four conditions, as PROBABILISTIC MAX INDEPENDENT SET ([16]), PROBABILISTIC MIN VERTEX COVERING ([13]), PROBABILISTIC MAX INDUCED SUBGRAPH WITH PROPERTY π and PROBABILISTIC MIN FEEDBACK VERTEX-SET ([20]), etc.

Theorem 1 can be used to capture problems that are not originally defined on graphs as PROBABILISTIC MAX SET PACKING and PROBABILISTIC MIN SET COVER. The probabilistic versions dealt for both of them consist of considering presence probabilities $p_1, \ldots, p_n$ for the corresponding elements of the ground set C. A set $S_i \in \mathcal{S}$ (the set-system defined on C) is present if at least one of its elements is present. So, denoting by $\Pr[S_i]$ the presence probability of the set $S_i = \{c_{i_1}, c_{i_2}, \ldots, c_{i_k}\}$, we get: $\Pr[S_i] = 1 - \prod_{j=1}^{k}(1 - p_{i_j})$. PROBABILISTIC

MAX SET PACKING can be transformed into a kind of PROBABILISTIC MAX INDEPENDENT SET on a particular graph derived from the initial instance where sets in $\mathcal{S}$ are transformed into vertices. PROBABILISTIC MIN SET COVER can be transformed into a particular kind of PROBABILISTIC DOMINATING SET on a graph where, once more sets in $\mathcal{S}$ are transformed into vertices. Both of the derived problems fit the assumptions of Theorem 1 considering that a vertex v_i representing $S_i \in \mathcal{S}$ has occurrence probability $\Pr[S_i]$ (detailed proofs are given in [20]).

3   Solutions Are Collections of Subsets of the Initial Vertex-Set

We now deal with problems the feasible solutions of which are collections of subsets of the initial vertex-set. Consider a graph $G(V, E)$ and a combinatorial optimization graph-problem Π whose solutions are collections of subsets of V verifying some specified non-trivial hereditary property[2] (e.g., independent set, clique, etc.). The following theorem characterizes functionals and optimal anticipatory solutions for such problems.

Theorem 2. *Consider a graph- problem Π verifying the following assumptions: (i) an instance of Π is a graph $G(V, E)$; (ii) a solution of Π on an instance G is a collection $S = (V_1, \ldots, V_k)$ of subsets of V any of them satisfying some specified non-trivial hereditary property; (iii) for any solution S and any subset $V' \subseteq V$, the restriction S' of S in V', i.e., $S' = (V_1 \cap V', \ldots, V_k \cap V')$, is feasible for $G' = G[V']$; (iv) the value of any solution $S \subseteq V$ of Π is defined by: $m(G, S) = |S| = k$. Then, $E(G, S) = \sum_{j=1}^{k} (1 - \prod_{v_i \in V_j}(1 - p_i))$ and can be computed in polynomial time. PROBABILISTIC Π amounts to a particular weighted version of Π, where the weight of any vertex $v_i \in V$ is $1 - p_i$, the weight $w(V_j)$ of a subset $V_j \subseteq V$ is defined by $w(V_j) = 1 - \prod_{v_i \in V_j}(1 - p_i)$ and the objective function to be optimized is equal to $\sum_{V_j \in C} w(V_j)$.*

What does play a central role for yielding result of Theorem 2, is the fact that property satisfied by the sets of the collection is hereditary. This allows to preserve sets $V_1, \ldots, V_k$ in the solution returned by $S \cap V_i$, $i = 1, \ldots, k$, unless they are empty.

Assume that $p_i = 1$, for any $v_i \in V$. Then, $E(G, S) = k$ and PROBABILISTIC Π coincides in this case with Π.

Corollary 3. *If Π is **NP**-hard, then PROBABILISTIC Π is also **NP**-hard.*

As for Corollary 2, Corollary 3 settles complexity only for the case where Π is **NP**-hard, leaving unclear the status of PROBABILISTIC Π when $\Pi \in \mathbf{P}$.

Theorem 2 has also application for numerous combinatorial optimization problems, as PROBABILISTIC MIN COLORING ([1]) PROBABILISTIC MIN COMPLETE BIPARTITE SUBGRAPH COVER, PROBABILISTIC MIN CUT COVER ([20]), etc.

---

[2] A property π is *hereditary* if, whenever is satisfied by a graph G, it is satisfied by any subgraph of G.

As in Section 2, application of Theorem 2 can go beyond graphs. In fact, Theorem 2 provides an equivalent way for analyzing MIN SET COVER always assuming that uncertainty carries over ground elements and a set is present if at least one of its element is so. Then, an instance $(\mathcal{S}, C)$ of MIN SET COVER can be transformed into a particular graph whose nodes correspond to the elements of C. A solution for MIN SET COVER is a particular vertex cover by cliques. Seen in such terms, MIN SET COVER perfectly fits conditions of Theorem 2 ([20]).

We now extend an approximation result of [1] to capture the whole of problems meeting the conditions of Theorem 2. Consider such a problem Π, an instance $G(V, E)$ of Π, set $n = |V|$ and consider a solution $S = (V_1, \ldots, V_k)$ of Π on G. Denote by $p_{\min}$ and $p_{\max}$ the minimum and maximum vertex-probabilities, respectively. Then, the following bounds hold for $E(G, S)$ ([20]):

$$\max \left\{ \sum_{i=1}^{n} p_i - \sum_{i=1}^{n} \sum_{j=i+1}^{n} p_i p_j, k p_{\min} \right\} \leqslant E(G, S) \leqslant \min \left\{ \sum_{i=1}^{n} p_i \leqslant n p_{\max}, k \right\}$$

Fix a vertex-probability p', assume that there exists a ρ-approximation polynomial time algorithm A for Π, and run the following algorithm RA for PROBABILISTIC Π: (i) partition the vertices of G into three subsets: V_1 including the vertices with probabilities at most $1/n$, V_2, including the vertices with probabilities in the interval $[1/n, p']$ and V_3, including the vertices with probabilities greater than p'; (ii) feasibly solve Π in $G[V_1]$ and $G[V_2]$ separately, run A in $G[V_3]$ and take the union of the solutions computed as solution for G.

Theorem 3. *If A achieves approximation ratio ρ for Π, then RA approximately solves in polynomial time the probabilistic version of Π within ratio $O(\sqrt{\rho n})$.*

4   Solutions Are Subsets of the Initial Edge-Set

We deal in this section with problems for which solutions are sets of edges. Notice that whenever a vertex is absent from some subset $V' \subseteq V$, the edges incident to it are also absent from $G[V']$. So, our assumption is that, given a solution (in terms of a set of edges) S, and a set $V' \subseteq V$ inducing a subgraph $G[V'] = G'(V', E')$ of G, the set $S \cap E'$ is feasible for Π in G'. The main result for this case, is the following theorem.

Theorem 4. *Consider a graph-problem Π verifying the following assumptions: (1) an instance of Π is an edge- (or arc-) valued graph $G(V, E, \boldsymbol{\ell})$; (2) any solution of Π on any instance G is a subset of E; (3) for any solution S and any subset $V' \subseteq V$, denoting by $G'(V', E')$ the subgraph of G induced by V', the set $S \cap E'$ is feasible; (4) the value of any solution $S \subseteq E$ of Π is defined by: $m(G, S) = w(S) = \sum_{(v_i, v_j) \in S} \ell(v_i, v_j)$, where $\ell(v_i, v_j)$ is the valuation of the edge (or arc) (v_i, v_j) of G. Then, the functional of PROBABILISTIC Π is expressed as: $E(G, S) = \sum_{(v_i, v_j) \in S} \ell(v_i, v_j) p_i p_j$ and can be computed in polynomial time. Furthermore, dealing with their respective computational complexities, PROBABILISTIC Π and Π are equivalent.*

The reasons for which the functional derived in Theorem 4 becomes polynomial are quite analogous to the ones in Theorem 1.

Let us note that, as in Section 2, Theorem 4 can be used for getting generic approximation results for PROBABILISTIC Π. Since this problem is a particular weighted version of Π, one immediately concludes that *if Π is approximable within approximation ratio ρ, so is* PROBABILISTIC Π.

Corollary 4. *Under the hypotheses of Theorem 4, whenever Π and* PROBA-BILISTIC *Π are **NP**-hard, they are equi-approximable.*

Corollary 5. *Consider a probabilistic combinatorial optimization problem Π verifying assumptions (1) through (4) of Theorem 4 with $\ell = 1$. Then, the functional of* PROBABILISTIC Π *is expressed as: $E(G, S) = \sum_{(v_i,v_j) \in S} p_i p_j$ and can be computed in polynomial time and* PROBABILISTIC Π *is equivalent to an edge- (or arc-) valued version of Π where the values of an edge is the product of the probabilities of its endpoints.*

Theorem 4 has immediate applications to the study of probabilistic versions of many well-known combinatorial optimization problems like PROBABILISTIC MAX MATCHING, PROBABILISTIC MAX CUT ([20]), etc.

5   When Things Become Complicated

In this section we tackle edge-weighted graph-problems where feasible solutions are connected sets of edges (for example, paths, trees, cycles, etc.) but we assume that, given a solution S and a set $V' \subseteq V$ inducing a subgraph $G[V'] = G'(V', E')$ of G, the set $S \cap E'$ is not always feasible for G'.

Formally, consider a problem Π where a feasible solution is a connected set S of edges. Consider also that vertices in S are ordered in some appropriate order. Assume that $S \cap E'$ is a set of $k = k(G')$ (in other words, k depends on the present graph G') connected subsets $C_1, C_2, \ldots, C_k$ of S but that $S'' = \cup_{i=1}^{k} C_i$ is not connected (i.e., S'' does not constitute a feasible solution for Π). Assume also that connected subsets $C_1, C_2, \ldots, C_k$ are also ranged in this order (always following some appropriate ordering implied by the one of S).

We consider a kind of "completion" of S'' by additional edges linking, for $i = 1, \ldots, k-1$, the last vertex (in the ordering considered for S) of C_i with the first vertex of C_{i+1}. In other words, given S (representing a connected set of edges), we apply the following algorithm, denoted by A in the sequel: (1) range the vertices of S following some appropriate order; (2) compute $S \cap E'$; let $C_1, C_2, \ldots, C_k$ be the resulting connected components of $S \cap E'$; (3) for $i = 1, \ldots, k-1$, use an edge to link the last vertex of C_i with the first vertex of C_{i+1}; (4) output S' the solution so computed.

Obviously, in order that step (3) of A is able to link components C_i and C_{i+1}, an edge must exist between the vertices implied; otherwise, A is definitely unfeasible. So, in order to assure feasibility, we make, for the rest of the section the basic assumption that the input graph for the problems tackled is complete.

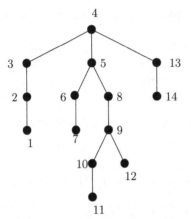

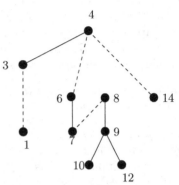

(a) The ordering of the nodes of an anticipatory solution T

(b) The solution T' derived from application of Algorithm A on T

Fig. 1. When anticipatory solution is a tree

In what follows, denote by $V[S']$ the set of vertices in S' and by $G''(V[S'], E'')$ the graph $G[V[S']]$. Also, denote by $[v_i, v_j]$ the set $\{v_{i+1}, v_{i+2}, \ldots, v_{j-1}\}$ ($i < j$ in the ordering assumed for S) such that: (a) for any $\ell = i, i+1, \ldots, j-1$, $(v_\ell, v_{\ell+1}) \in S$ (i.e., $[v_i, v_j]$ is the set of vertices in the path linking v_i to v_j in S, where v_i and v_j themselves are not encountered[3]) and (b) v_i and v_j belong to consecutive[4] connected subsets C_m and C_{m+1}, for some $m < k$. By symmetry, always for $i < j$ in the ordering assumed for S, we denote by $[v_j, v_i]$ the set $\{v_{j+1}, v_{j+2}, \ldots, v_n, v_1, \ldots, v_{i-1}\}$. Obviously, $[v_i, v_j]$ and $[v_j, v_i]$ are both nonempty if S is, say, a cycle, or more generally, it verifies some cyclic property. On the other hand, if S is, say, a path or a tree, then $[v_j, v_i]$ is empty.

Theorem 5. *Consider a probabilistic combinatorial optimization problem* PROBABILISTIC Π *verifying the following assumptions: (i) instances of Π are edge-valued complete graphs $(K_n, \ell) = G(V, E, \ell)$; furthermore, in the probabilistic version of Π any vertex $v_i \in V$ has a presence-probability p_i; (ii) a solution of Π is a subset S of E verifying some connectivity property; (iii) given an anticipatory solution S (the vertices of which are ranged in some appropriate order), Algorithm A computes a feasible solution S', for any subgraph $G'(V', E', \ell) = G[V']$ of G (obviously, G' is complete); (iv) $m(G, S) = \sum_{(v_i, v_j) \in S} \ell(v_i, v_j)$. Then, $E(G, S)$ is computable in polynomial time and is expressed by:*

$$E(G, S) = \sum_{(v_i, v_j) \in S} \ell(v_i, v_j)\, p_i p_j + \sum_{(v_i, v_j) \in E'' \setminus S} \ell(v_i, v_j)\, p_i p_j \prod_{v_l \in [v_i, v_j]} (1 - p_l)$$

$$+ \sum_{(v_i, v_j) \in E'' \setminus S} \ell(v_i, v_j)\, p_i p_j \prod_{v_l \in [v_j, v_i]} (1 - p_l)$$

---

[3] It is assumed that if $[v_i, v_j] = \emptyset$, then $\prod_{v_l \in [v_i, v_j]}(1 - p_l) = 0$.

[4] With respect to the order $C_1, \ldots, C_k$.

Unfortunately, in the opposite of Theorems 1 and 4, Theorem 5 does not derive a "good" characterization for the optimal anticipatory solutions of the problems meeting the assumptions (i) to (iv). In particular, the form of the functional does not imply solution of some well-defined weighted version of Π (the deterministic support of PROBABILISTIC Π). Indeed due to the second term of the expression for $E(G, S)$ in Theorem 5, the "costs" assigned to the edges depend on the structure of the anticipatory solution chosen.

Consider PROBABILISTIC MIN SPANNING TREE. Starting from a tree T we first number its vertices following a kind of depth-first-search. For instance, for the tree of Figure 1(a) starting from the leftmost leaf (numbered by 1), the numbering of the rest of its vertices is as shown in the figure. Assuming that vertices 2, 5, 11 and 13 are absent, the tree T', derived from application of Algorithm A on T is shown in Figure 1(b), where slotted edges represent edges added during execution of step (3) of A.

6   Final Remarks

We have drawn a framework for the classification of probabilistic problems under the a priori optimization paradigm. What seems to be of interest in this classification is that when restriction of the initial solution to the "present" subgraph is feasible, then the complexity of determining the optimal anticipatory solution for the problems tackled, amounts to the complexity of solving some weighted version of the deterministic problem, where the weights depend on the vertex-probabilities. These weights do not depend on particular characteristics of the anticipatory solution considered, thing that allows a compact characterization of optimal anticipatory solution. On the contrary, when more-than-one-stage algorithms are needed for building solutions, then the observation above is no more valid. In this case, one also recovers some weighted version of the original problem, but the weights on the data cannot be assigned independently of the structure of a particular anticipatory solution.

References

1. Murat, C., Paschos, V.T.: On the probabilistic minimum coloring and minimum k-coloring. Discrete Appl. Math. 154, 564–586 (2006)
2. Murat, C., Paschos, V.T.: Probabilistic combinatorial optimization on graphs. ISTE and Hermès Science Publishing, London (2006)
3. Bertsimas, D.J., Jaillet, P., Odoni, A.: A priori optimization. Oper. Res. 38, 1019–1033 (1990)
4. Averbakh, I., Berman, O., Simchi-Levi, D.: Probabilistic a priori routing-location problems. Naval Res. Logistics 41, 973–989 (1994)
5. Bertsimas, D.J.: Probabilistic combinatorial optimization problems. Phd thesis, Operations Research Center. MIT, Cambridge Mass., USA (1988)
6. Bertsimas, D.J.: On probabilistic traveling salesman facility location problems. Transportation Sci. 3, 184–191 (1989)

7. Bertsimas, D.J.: The probabilistic minimum spanning tree problem. Networks 20, 245–275 (1990)
8. Jaillet, P.: Probabilistic traveling salesman problem. Technical Report 185, Operations Research Center. MIT, Cambridge Mass., USA (1985)
9. Jaillet, P.: A priori solution of a traveling salesman problem in which a random subset of the customers are visited. Oper. Res. 36, 929–936 (1988)
10. Jaillet, P.: Shortest path problems with node failures. Networks 22, 589–605 (1992)
11. Jaillet, P., Odoni, A.: The probabilistic vehicle routing problem. In: Golden, B.L., Assad, A.A. (eds.) Vehicle routing: methods and studies. North Holland, Amsterdam (1988)
12. Bianchi, L., Knowles, J., Bowler, N.: Local search for the probabilistic traveling salesman problem: correction to the 2-p-opt and 1-shift algorithms. European J. Oper. Res. 161, 206–219 (2005)
13. Murat, C., Paschos, V.T.: The probabilistic minimum vertex-covering problem. Int. Trans. Opl Res. 9, 19–32 (2002), Preliminary version, http://www.lamsade.dauphine.fr/~paschos/documents/c170.pdf
14. Bourgeois, N., Della Croce, F., Escoffier, B., Murat, C., Paschos, V.T.: Probabilistic coloring of bipartite and split graphs. J. Comb. Optimization (to appear)
15. Murat, C., Paschos, V.T.: The probabilistic longest path problem. Networks 33, 207–219 (1999)
16. Murat, C., Paschos, V.T.: A priori optimization for the probabilistic maximum independent set problem. Theoret. Comput. Sci. 270, 561–590 (2002), http://www.lamsade.dauphine.fr/~paschos/documents/c166.pdf
17. Paschos, V.T., Telelis, O.A., Zissimopoulos, V.: Steiner forests on stochastic metric graphs. In: Dress, A., Xu, Y., Zhu, B. (eds.) COCOA. LNCS, vol. 4616, pp. 112–123. Springer, Heidelberg (2007)
18. Bellalouna, M., Murat, C., Paschos, V.T.: Probabilistic combinatorial optimization problems: a new domain in operational research. European J. Oper. Res. 87, 693–706 (1995)
19. Murat, C., Paschos, V.T.: L'optimisation combinatoire probabiliste. In: Paschos, V.T. (ed.) Optimisation combinatoire : concepts avancés, pp. 221–247. Hermès Science, Paris (2005)
20. Murat, C., Paschos, V.T.: What about future? Robustness under vertex-uncertainty in graph-problems. Cahier du LAMSADE 236, LAMSADE, Université Paris-Dauphine (2006), http://www.lamsade.dauphine.fr/~paschos/documents/pcopclassifcah.pdf
21. Ausiello, G., Crescenzi, P., Gambosi, G., Kann, V., Marchetti-Spaccamela, A., Protasi, M.: Complexity and approximation. In: Combinatorial optimization problems and their approximability properties. Springer, Berlin (1999)

Protean Graphs with a Variety of Ranking Schemes

Paweł Prałat*

Department of Mathematics and Statistics, Dalhousie University, Halifax, NS,
Canada B3H 3J5
pralat@mathstat.dal.ca

Abstract. The World Wide Web may be viewed as a graph each of
whose vertices corresponds to a static HTML web page, and each of
whose edges corresponds to a hyperlink from one web page to another.
Recently there has been considerable interest in using random graphs to
model complex real-world networks to gain an insight into their properties. In this paper, we propose a generalized version of the protean graph
(a random model of the web graph) in which the degree of a vertex depends on its age. Classic protean graphs can be seen as a special case
of the rank-based approach where vertices are ranked according to age.
Here, we investigate graph generation models based on other ranking
schemes and show that these models lead to graphs with a power law
degree distribution.

1 Introduction

Recently many new random graphs models have been introduced and analyzed
by certain common features observed in many large-scale real-world networks
such as the 'web graph' (see, for instance, the survey [1]). The web may be
viewed as a directed graph whose nodes correspond to static pages on the web,
and whose arcs correspond to links between these pages.

One of the most characteristic features of this graph is its degree sequence.
Broder et al. [2] noticed that the distribution of degrees follows a power law: the
fraction of vertices with degree k is proportional to $k^{-\gamma}$, where γ is a constant
independent of the size of the network (more precisely, $\gamma \approx 2.1$ for in-degrees,
$\gamma \approx 2.7$ for out-degrees). These observations suggest that the web is not well
modeled by traditional random graph models such as $G_{n,p}$ (see, for instance [4]).

Łuczak and the author of this paper introduced in [6] another random graph
model of the undirected 'web graphs', the protean graph $\mathcal{P}_n(d, \eta)$, which is con-
trolled by two additional parameters ($d \in \mathbb{N}$ and $0 < \eta < 1$). The major feature
of this model is that older vertices are preferred when joining a new vertex into
the graph. In [6] it is proved that the degrees of the $\mathcal{P}_n(d, \eta)$ are distributed
according to the power law and the behaviour near the connectivity threshold

---

* The author is supported by MITACS and NSERC. This work is part of the MITACS
project Modelling and Mining of Networked Information Spaces (MoMiNIS).

B. Yang, D.-Z. Du, and C.A. Wang (Eds.): COCOA 2008, LNCS 5165, pp. 149–159, 2008.
© Springer-Verlag Berlin Heidelberg 2008

is studied. The author of this paper showed also in [8] that the protean graph $\mathcal{P}_n(d, \eta)$ asymptotically almost surely (a.a.s.) has one giant component, containing a positive fraction of all vertices, whose diameter is equal to $\Theta(\log n)$. (See also [9] where the growing protean graphs are studied.)

Classic protean graphs can be viewed as a special case of the rank-based approach where vertices are ranked according to age. The general approach was first proposed by Fortunato, Flammini and Menczer in [3], and the occurrence of a power law was postulated based on simulations (Janssen and the author of this paper provided rigorous proofs in [5]). In this approach, the vertices are ranked from 1 to n according to some ranking scheme (so the vertex with highest degree has rank 1, etc.), and the link probability of a given vertex is proportional to its rank, raised to the power $-\eta$ for some $\eta \in (0, 1)$; we will refer to η as the *attachment strength*. (Negative powers are chosen since a low value for rank should result in a higher link probability.)

As we will show, protean graphs with rank-based attachment leads to power law graphs for a variety of different ranking schemes. One obvious ranking scheme is to rank vertices by age (*the old get richer*); as we already mentioned, this model was studied in [6,8] and this leads to a power law with the exponent $1 + 1/\eta$. In this paper, we study a ranking scheme where an external prestige label for each vertex is given and vertices are ranked according to their prestige label. In order to allow for a different distribution of "prestige" over the vertices, we considered also a *random ranking* scheme. Here, each vertex is assigned an initial rank according to a given distribution. We consider distributions of the following form. Let R_i be the initial rank of a vertex born at time i. Then $\mathbb{P}(R_i \leq k) = (k/n)^s$. First we show that, if $s = 1$, then the situation is similar to the one described previously, and vertices with initial rank R_i exhibit behaviour as if they had received fitness R_i/n. We also consider the case where $s > 1$, so the rank of new vertices is biased towards the higher ranks.

These results suggest an explanation for the power law degree distribution often observed in real-life networks such as the web graph, protein interaction networks, and social networks. The growth of such networks can be seen as governed by a rank-based attachment scheme, based on a ranking scheme that can be derived from a number of different factors such as age, degree, or fitness. The exponent of the power law is independent of these factors, but is rather a consequence of the attachment strength. In addition, rank-based attachment accentuates the difference between higher ranked vertices: the difference in link probability between the vertices ranked 1 and 2 is much larger than that between the vertices ranked 100 and 101. This again corresponds to our intuition of what constitutes a credible mechanism for link attachment.

In order to establish the right attachment strength to model a given real-life network we should consider the following. In a graph in which the number of vertices of degree k decreases roughly as $k^{-\gamma}$ the fraction of vertices of degree at least k changes roughly as $\sum_{\ell \geq k} O(\ell^{-\gamma}) = O(k^{1-\gamma})$. Thus, in order to imitate this distribution the attachment strength η should be set to $\eta \sim 1/(\gamma - 1)$.

2   Definitions

In this section, we formally define the graph generation model based on rank-based attachment. The model produces a sequence $\{G_t\}_{t=0}^{\infty} = \{(V_t, E_t)\}_{t=0}^{\infty}$ of undirected graphs on n vertices, where t denotes time. Our model has two fixed parameters: initial degree $d \in N$, and attachment strength $\eta \in (0,1)$. At each time t, each vertex $v \in V_t$ has rank $r(v,t) \in [n]$ (we use $[n]$ to denote the set $\{1, 2, \ldots, n\}$). In order to obtain a proper ranking, the rank function $r(\cdot, t) : V_t \to [n]$ is a bijection for all t, so every vertex has a unique rank. In agreement with the common use of the word "rank", high rank refers to a vertex v for which $r(v,t)$ is small: the highest ranked vertex is ranked number one, so has rank equal to 1; the lowest ranked vertex has rank n. The initialization and update of the ranking is done according to a *ranking scheme*. Various ranking schemes can be considered; we first give the general model, and then list the ranking schemes.

Let $G_0 = (V_0, E_0)$ be any graph on n vertices and $r_0 = r(\cdot, 0) : V_0 \to [n]$ any initial rank function. (For random labeling scheme we take any function $l : V_0 \to (0,1)$ and the initial rank function is a function of l; for degree scheme $r_0 = r_0(G_0)$.) For $t \geq 1$ we form G_t from G_{t-1} according to the following rules:

- Choose uniformly at random a vertex $u \in V_{t-1}$, delete u together with all edges incident to it.
- Add a new vertex v_t together with d edges from v_t to existing vertices chosen randomly with weighted probabilities. The edges are added in d substeps. In each substep, one edge is added, and the probability that v is chosen as its endpoint (the link probability), equals

$$\frac{r(v, t-1)^{-\eta}}{\sum_{i=1}^{n} i^{-\eta}} = \frac{1-\eta}{n^{1-\eta} + O(1)} r(v, t-1)^{-\eta}.$$

- Update the ranking function $r(\cdot, t) : V_t \to [n]$ according to the ranking scheme.

Our model allows for loops and multiple edges; there seems no reason to exclude them. However, there will not in general be very many of these, so excluding them can be shown not to affect our conclusions in any significant way.

We now define the different ranking schemes.

- **Ranking by age**: The vertex added at time t obtains an initial rank n; its rank decreases by one each time a vertex with smaller rank is removed.
- **Ranking by inverse age**: The vertex added at time t obtains an initial rank 1; its rank increases by one each time a vertex with higher rank is removed.
- **Ranking by random labeling**: The vertex added at time t obtains a label $l(v_t) \in (0,1)$ chosen uniformly at random. Vertices are ranked according to their labels: if $l(v_i) < l(v_j)$, then $r(v_i, t) < r(v_j, t)$.

- **Random ranking**: The vertex added at time t obtains an initial rank R_t which is randomly chosen from $[n]$ according to a prescribed distribution. Formally, let $F : [0,1] \rightarrow [0,1]$ be any cumulative distribution function. Then for all $k \in [t]$, $\mathbb{P}(R_t \leq k) = F(k/t)$.
- **Ranking by degree**: After each time step t, vertices are ranked according to their degrees in G_t, and ties are broken by age. Precisely, if $\deg(v_i, t) < \deg(v_j, t)$ then $r(v_i, t) < r(v_j, t)$, and if $\deg(v_i, t) = \deg(v_j, t)$ then $r(v_i, t) < r(v_j, t)$ if $i < j$.

In this paper, due to the space limitations, we focus on ranking by random labeling and random ranking with $F(x) = x^s$ for $s \geq 1$. The other ranking schemes will be studied in a journal version of this paper. In particular, it is interesting and non-trivial task to investigate the ranking by degree scheme; in this case, it is not even clear how long we have to wait to obtain a stationary distribution. For the other schemes (except the random labeling case), it is enough to wait L steps for all vertices to be 'renewed' (for the random labeling case we have to wait two times longer: the first round is needed to have labels distributed uniformly at random, during the second one the process 'forgets' about the initial graph) and from that time the protean process is the Markov chain that is in the stationary distribution (that is, the distribution determined by G_t on the set of all ordered graphs on n vertices is identical for all t.) By the coupon collector problem, a.a.s. $L = n(\log n + O(\omega(n)))$ where $\omega(n)$ is any function tending to infinity with n (for random labeling scheme, clearly $L = 2n(\log n + O(\omega(n)))$ a.a.s.). Furthermore, this distribution does not depend on the choice of G_0 and r_0. The random graph G_L corresponding to this distribution is called a protean graph $\mathcal{P}_n(d, \eta)$.

In the rest of the paper, $\{G_t\}_{t=1}^{\infty}$ is assumed to be a graph sequence generated by the rank-based attachment model, with ranking scheme as defined in each particular section, and d and η are assumed to be the initial degree and attachment strength parameters of the model as defined above. The results are generally about the degree distribution in G_L, where the asymptotics are based on n tending to infinity.

We will use the stronger notion of *wep* in favour of the more commonly used a.a.s., since it simplifies some of our proofs. We say that an event holds *with extreme probability* (*wep*), if it holds with probability at least $1 - \exp(-\Theta(\log^2 n))$ as $n \rightarrow \infty$. Thus, if we consider a polynomial number of events that each holds *wep*, then *wep* all events hold. To combine this notion with asymptotic notations such as $O()$ and $o()$, we follow the conventions in [10].

3 Ranking by Random Labeling

In this scheme, each new vertex v_t obtains a label $l(v_t) \in (0,1)$ chosen uniformly at random. (Note that the probability that two vertices receive the same label is zero.) Vertices are ranked by their labels: if $l(v_i) < l(v_j)$, then $r(v_i, t) < r(v_j, t)$.

First we note that the process of choosing a label *uar* from $(0,1)$ does not imply loss of generality. Namely, suppose that the labels are chosen from $\mathbb{R}$

according to any probability distribution with a strictly increasing *cumulative distribution function* F. Since F is an increasing function, labels $F(l(v_i))$ lead to exactly the same ranking as labels $l(v_i)$. But $\mathbb{P}(F(l(v_i)) \leq x) = \mathbb{P}(l(v_i) \leq F^{-1}(x)) = F(F^{-1}(x)) = x$, so the values of labels $F(l(v_i))$ are chosen from $(0,1)$ according to the uniform distribution.

First we investigate the expected degree of a vertex v at time L with a given age-rank and a label. We use $a(\cdot, t)$ for a ranking by age and stay with $r(\cdot, t)$ for a ranking by random labeling.

Theorem 1. *Let* $0 < \eta < 1$, $d \in \mathbb{N}$, $i = i(n) \in [n]$, *and* $0 < l(v_i) = l(v_i)(n) < 1$. *If* $n \cdot l(v_i) > \log^3 n$, *then the expected degree of a vertex* v_i *with an age-rank* $a(v_i, L) = i$ *that obtained a label* $l(v_i)$, *is given by*

$$\mathbb{E} \deg(v_i, L) = d\frac{i-1}{n-1} + (1 + O(\log^{-1/2} n))d(1-\eta)l(v_i)^{-\eta}(1 - i/n),$$

and wep

$$\deg(v_i, L) = \mathbb{E} \deg(v_i, L) + O(\sqrt{\mathbb{E} \deg(v_i, L)} \log n).$$

Proof. It is clear that the expected rank of v_i is equal to $l(v_i)n$ at each step of the process. Moreover, we can use the fact that a sum of independent random variables with large enough expected value is not too far from its mean (see, for example, Theorem 2.8 in [4]). From this it follows that, if $\varepsilon \leq 3/2$, then the following inequality, known as a Chernoff bound, holds

$$\mathbb{P}\left(|r(v_i, t) - \mathbb{E}r(v_i, t)| \geq \varepsilon \mathbb{E}r(v_i, t)\right) \leq 2 \exp\left(-\frac{\varepsilon^2}{3}\mathbb{E}r(v_i, t)\right).$$

Therefore, *wep* $r(v_i, t) = l(v_i)n(1 + O(\log^{-1/2} n))$ during the whole period (since $L = O(n \log n)$).

Let $X(t, j)$ be a random indicator variable for an event that vertex v_t (for which $a(v_t, L) = t$) joins v_i at substep j of step when v_t was born ($i < t \leq n$, $j \in [d]$). It is clear that

$$\mathbb{P}(X(t, j) = 1) = 1 - \mathbb{P}(X(t, j) = 0) = \frac{\left(l(v_i)n(1 + O(\log^{-1/2} n))\right)^{-\eta}}{n^{1-\eta}/(1-\eta) + O(1)}$$

$$= (1 + O(\log^{-1/2} n))(1 - \eta)l(v_i)^{-\eta}/n.$$

The number of neighbours v_t of v_i such that $t > i$ is a random variable and can be expressed as a sum $\sum_{t=i+1}^{n} \sum_{j=1}^{d} X(t, j)$ of independent random variables. Note also that vertex v_i generated exactly d edges at the time it was born but only i vertices (including v_i) have not been 'renewed' since then. Thus,

$$\mathbb{E} \deg(v_i, L) = d\frac{i-1}{n-1} + d(n-i)\mathbb{E}X(t, j)$$

$$= d\frac{i-1}{n-1} + (1 + O(\log^{-1/2} n))d(1-\eta)l(v_i)^{-\eta}(1 - i/n).$$

Finally, since $\deg(v_i, L)$ is expressed as a sum of independent random variables, we can use the Chernoff bound to show the concentration result.

Let $Z_k = Z_k(n, d, \eta)$ denote the number of vertices of degree k and $Z_{\geq k} = \sum_{l \geq k} Z_l$. The following theorem shows that the $Z_{\geq k}$'s follow a power law with exponent $1/\eta$. Since the $Z_{\geq k}$'s represent the cumulative degree distribution, this implies that the degree distribution follows a power law with exponent $1 + 1/\eta$.

Theorem 2. *Let $0 < \eta < 1$ and $d \in \mathbb{N}$, $\log^4 n \leq k \leq n^\eta / \log^{4\eta} n$. Then wep*

$$Z_{\geq k} = \left(1 - O(\log^{-1/3} n)\right) \frac{\eta}{1 + \eta} \left(\frac{d(1 - \eta)}{k}\right)^{1/\eta} n.$$

Proof. This theorem is a simple consequence of Theorem 1. One can show that wep each vertex v_i such that $l(v_i) \geq \left(1 + \log^{-1/3} n\right) \left(\frac{d(1-\eta)(1-i/n)}{k}\right)^{1/\eta}$ has fewer than k neighbours, and each vertex v_i for which $l(v_i) \leq \left(1 - \log^{-1/3} n\right) \left(\frac{d(1-\eta)(1-i/n)}{k}\right)^{1/\eta}$ has more than k neighbours. Thus,

$$\mathbb{E}Z_{\geq k} = \sum_{i=1}^{n} \left(1 - O(\log^{-1/3} n)\right) \left(\frac{d(1-\eta)(1 - i/n)}{k}\right)^{1/\eta}$$

$$= \left(1 - O(\log^{-1/3} n)\right) \left(\frac{d(1 - \eta)}{k}\right)^{1/\eta} n \int_0^1 (1 - x)^{1/\eta}$$

$$= \left(1 - O(\log^{-1/3} n)\right) \frac{\eta}{1 + \eta} \left(\frac{d(1 - \eta)}{k}\right)^{1/\eta} n$$

and the assertion follows from the Chernoff bound since $\mathbb{E}Z_{\geq k} = \Omega(\log^4 n)$.

4   Randomly Chosen Initial Rank

Next, we consider the case where the rank of the new vertex v_i, $R_i = r(v_i, i)$, is chosen at random from $[n]$. As described earlier, the ranks of existing vertices are adjusted accordingly. In contrast to the previous scheme, in this case it does matter according to which distribution R_i is chosen. We make the assumption that all initial ranks are chosen according to a similar distribution. In particular, we fix a continuous bijective function $F : [0, 1] \rightarrow [0, 1]$, and for all integers $1 \leq k \leq n$, we let $\mathbb{P}(R_i \leq k) = F\left(\frac{k}{n}\right)$.

Thus, F represents the limit, for n going to infinity, of the cumulative distribution functions of the variables R_i. To simplify the calculations while exploring a wide array of possibilities for F, we assume F to be of the form $F(x) = x^s$, where $s \geq 1$. (The case $0 < s < 1$ will be studied in the journal version of this paper.)

We start from a special case $s = 1$, where the distribution of each R_i is uniform. We will show that this case is similar to the random labeling case with a label equal to R_i/n. Hence, our aim is to show that the random variable $r(v_i, t)$ is sharply concentrated around R_i. In fact, $r(v_i, t) - r(v_i, i)$ is the sum of the differences $r(v_i, j) - r(v_i, j - 1) = X_j$, $i + 1 \leq j \leq t$. If the differences are independent, then the Chernoff bounds are very useful. When the differences are not independent but there is a large degree of independence, results can be often obtained by using large deviation inequalities for corresponding martingales. It is exactly the case here.

Our proofs use the supermartingale method of Pittel et al. [7], as described in [11, Corollary 4.1]. We need the following lemma.

Lemma 1. *Let $G_0, G_1, \ldots, G_n$ be a random process and X_t a random variable determined by $G_0, G_1, \ldots, G_t$, $0 \le t \le n$. Suppose that for some real β and constants γ_t, $\mathbb{E}(X_t - X_{t-1} \mid G_0, G_1, \ldots, G_{t-1}) < \beta$ and $|X_t - X_{t-1} - \beta| \le \gamma_t$ for $1 \le t \le n$. Then for all $\alpha > 0$,*

$$\mathbb{P}\left(\text{For some } t \text{ with } 0 \le t \le n : X_t - X_0 \ge t\beta + \alpha\right) \le \exp\left(-\frac{\alpha^2}{2\sum_{j=1}^{n} \gamma_t^2}\right).$$

Lemma 2. *Suppose that vertex v obtained an initial rank $R \ge \sqrt{n} \log^2 n$. Then, wep $r(v, t) = R(1 + O(\log^{-1/2} n))$ to the end of its life.*

Proof. Note that $r(v, t+1) - r(v, t) = -1$ (conditionally on the fact that v is not deleted at time $t+1$) with probability $(r(v, t) - 1)(n - r(v, t))/(n - 1)n$ and $r(v, t+1) - r(v, t) = 1$ with probability $(n - r(v, t))r(v, t)/(n - 1)n$. Thus,

$$\beta = \mathbb{E}(r(v, t+1) - r(v, t) \mid r(v, t)) = O(1/n).$$

Clearly, the rank can change by at most one ($\gamma_t = 1$) so we can use Lemma 1 with $\alpha = \sqrt{n} \log^{3/2} n$ to get that *wep* $r(v, t) = R(1 + O(\log^{-1/2} n))$ during the whole life of that vertex (note that *wep* v will be deleted after $O(n \log n)$ steps and $R \ge \sqrt{n} \log^2 n$). $\qquad\blacksquare$

From the previous lemma it follows that the random ranking case for $s = 1$ is very similar to the random labeling case. The proof of the following theorem is the same as the proof of the Theorem 1 so it is omitted. (Note that the range for k is slightly different due to the stronger condition for the initial rank.)

Theorem 3. *Let $0 < \eta < 1$ and $d \in \mathbb{N}$, $\log^4 n \le k \le n^{\eta/2}/\log^{3\eta} n$. Then wep*

$$Z_{\ge k} = \left(1 - O(\log^{-1/3} n)\right)\frac{\eta}{1 + \eta}\left(\frac{d(1 - \eta)}{k}\right)^{1/\eta} n.$$

Next, we consider the case where $s > 1$, but before we move to investigating the rank of vertex v after t steps of the process, we study its age-rank. In other words, we would like to know how many vertices have not been 'renewed' after t steps of the process. For this, we use the differential equations method [11]. Without loss of generality, we can assume that the vertex was born at time 0. It is clear that $a(v, 0) = n$ and $a(v, t)$, $t > 0$, is a random variable, which in time step $t + 1$ decreases by one precisely when vertex u for which $a(u, t) < a(v, t)$ is deleted. So, working in the conditional space under consideration, we obtain

$$\mathbb{E}(a(v, t+1) - a(v, t) \mid G_t) = \frac{a(v, t) - 1}{n - 1}.$$

Defining a real function $z(x)$ to model the behaviour of $a(v, xn)/n$, the above relation implies the following differential equation

$$z'(x) = -z(x) \tag{1}$$

with the initial condition $z(0) = 1$.

The general solution is $z(x) = \exp(-x+C)$, $C \in \mathbb{R}$ and the particular solution is $z(x) = \exp(-x)$. This *suggests* that a random variable $a(v, t)$ should be close to a deterministic function $n \exp(-t/n)$. We will show that it represents the "shape" of a typical process.

Theorem 4. *Let $a(v, t)$ be defined as above. Then wep, for every t in the range $0 \le t \le t_f = \frac{1}{2} n \log n - 2n \log \log n$, we have*

$$a(v, t) = n \exp(-t/n)(1 + O(\log^{-1/2} n)) \tag{2}$$

conditional upon the vertex v surviving until time t_f.

Proof. We transform $a(v, t)$ into something close to a martingale. Consider the following real-valued function

$$H(a(v, t), t) = \log a(v, t) + t/n \tag{3}$$

and the stopping time

$$T = \min\{t \ge 0 : a(v, t) < \sqrt{n} \log^2 n/2 \vee t = t_f\} .$$

(A stopping time is any random variable T with values in $\{0, 1, \dots\} \cup \{\infty\}$ for which it is determined whether $T = \hat{t}$ for any time $\hat{t}$ from knowledge of the process up to and including time $\hat{t}$.)

Let $\mathbf{w}_t = (a(v, t), t)$, and consider the sequence of random variables $(H(\mathbf{w}_t) : 0 \le t \le t_f)$. Note that the second-order partial derivatives of H with respect to $a(v, t)$ and t are $O(1/a(v, t)^2) = O(1/n \log^4 n)$, provided $T > t$. Therefore, with $i \wedge T$ denoting $\min\{i, T\}$, we have

$$H(\mathbf{w}_{(t+1) \wedge T}) - H(\mathbf{w}_{t \wedge T})$$
$$= (\mathbf{w}_{(t+1) \wedge T} - \mathbf{w}_{t \wedge T}) \cdot \operatorname{grad} H(\mathbf{w}_{t \wedge T}) + O(1/n \log^4 n) . \tag{4}$$

Observe also that,

$$\mathbb{E}(\mathbf{w}_{t+1} - \mathbf{w}_t \mid G_t) \cdot \operatorname{grad} H(\mathbf{w}_t)$$
$$= \left(-\frac{a(v, t) - 1}{n - 1}, 1\right) \cdot \operatorname{grad} H(\mathbf{w}_t) = O(1/a(v, t)n) = O(1/n^{3/2} \log^2 n),$$

provided $T > t$, since H was chosen so that $H(\mathbf{w})$ is close to a constant along every trajectory $\mathbf{w}$ of the differential equation (1).

Taking the expectation of (4) conditional on $G_{t \wedge T}$, we obtain that

$$\mathbb{E}(H(\mathbf{w}_{(t+1) \wedge T}) - H(\mathbf{w}_{t \wedge T}) | G_{t \wedge T}) = O(1/n \log^4 n) .$$

From (4), noting that grad $H(\mathbf{w}_t) = (O(1/a(v,t)), 1/n)$, and using the fact that the rank changes by at most one in each step,

$$|H(\mathbf{w}_{(t+1)\wedge T}) - H(\mathbf{w}_{t\wedge T})| = O(1/a(v, t\wedge T)) + O(1/n) + O(1/n\log^4 n) = O(1/\sqrt{n}\log^2 n).$$

Now we may apply Lemma 1 to the sequence $(H(\mathbf{w}_{t\wedge T}) : 0 \le t \le t_f)$, and symmetrically to $(-H(\mathbf{w}_{t\wedge T}) : 0 \le t \le t_f)$, with $\alpha = 1/\log^{1/2} n$, $\beta = O(1/n\log^4 n)$, and $\gamma_t = O(1/\sqrt{n}\log^2 n)$ to show that wep

$$|H(\mathbf{w}_{t\wedge T}) - H(\mathbf{w}_{t_0})| = O(\log^{-1/2} n).$$

As $H(\mathbf{w}_0) = \log n$, this implies from the definition (3) of the function H, that wep equation (2) holds for every $0 \le t \le T$.

To complete the proof we need to show that wep, $T = t_f$. The events asserted by (2) hold with this probability up until time T, as shown above. Thus, in particular, wep $a(v, T) = (1 + o(1))n \exp(-T/n) > (1 + o(1))\sqrt{n}\log^2 n$ which implies that $T = t_f$ wep.

Exactly the same approach can be used to study the rank of vertex after t steps of the process, given that its initial rank is equal to R. We present a sketch of the proof only.

Theorem 5. *Suppose that a vertex v obtained an initial rank $r(v, 0) = R < (1 - 1/\sqrt{n}\log^2 n)n$ at time 0. Then wep, for every $t > 0$ conditional upon the vertex v surviving until time t*

$$r(v, t) = n\left(\left(\left(\frac{R}{n}\right)^{1-s} - 1\right)e^{(s-1)t/n} + 1\right)^{\frac{1}{1-s}}(1 + O(\log^{-1/2} n))$$

provided

$$n\left(\left(\left(\frac{R}{n}\right)^{1-s} - 1\right)e^{(s-1)t/n} + 1\right)^{\frac{1}{1-s}} \ge \sqrt{n}\log^2 n.$$

Proof. Defining a real function $z(x)$ to model the behaviour of $r(v, xn)/n$, we get $z'(x) = -z(x) + z(x)^s$ with the initial condition $z(0) = R/n$. The general solution is $z(x) = (Ce^{(s-1)x} + 1)^{1/(1-s)}$, $C \in \mathbb{R}$ and the particular solution is

$$z(x) = \left(\left(\left(\frac{R}{n}\right)^{1-s} - 1\right)e^{(s-1)x} + 1\right)^{\frac{1}{1-s}}.$$

Now we are ready to state the main theorem in this section. The proof is rather straightforward but again we omit the details in this extended abstract.

Theorem 6. *Let $0 < \eta < 1$ and $d \in \mathbb{N}$, $\log^4 n \le k \le n^{\eta/2}/\log^{3\eta} n$. Then wep*

$$Z_{\ge k} = (1 + o(1))\left(\frac{d(1 - \eta)}{k(1 + \eta)}\right)^{1/\eta} n.$$

Proof. Consider vertices v_i $(i = xn)$ and v_j $(j = yn)$ with the age-ranks $a(v_i, L) = i$ and $a(v_j, L) = j$, respectively. Suppose that v_i obtained an initial rank of R. By Theorem 4, *wep* vertices v_i and v_j were born $(1+o(1))n \log(1/x)$ and, respectively, $(1+o(1))n \log(1/y)$ steps ago. By Theorem 5, *wep* v_i had the following rank at that time

$$n \left(\left(\left(\frac{R}{n} \right)^{1-s} - 1 \right) \left(\frac{y}{x} \right)^{s-1} + 1 \right)^{\frac{1}{1-s}} (1 + O(\log^{-1/2} n)).$$

Thus,

$$\mathbb{E}\deg(v_i, L) = O(d) + (1 + O(\log^{-1/2} n))d(1-\eta)\int_x^1 \left(\left(\left(\frac{R}{n} \right)^{1-s} - 1 \right) \left(\frac{y}{x} \right)^{s-1} + 1 \right)^{\frac{-\eta}{1-s}} dy.$$

If $x + R/n = \Omega(1)$, then the expected degree is a constant and the degree is smaller than $\log n$ *wep*. Otherwise it simplifies to

$$\mathbb{E}\deg(v_i, L) = (1 + O(\log^{-1/2} n))d(1-\eta) \left(\left(\frac{R}{n} \right)^{1-s} - 1 \right)^{\frac{-\eta}{1-s}} x^{-\eta} \int_x^1 y^\eta dy$$

$$= (1 + O(\log^{-1/2} n))\frac{d(1-\eta)}{1+\eta} \left(\left(\frac{R}{n} \right)^{1-s} - 1 \right)^{\frac{-\eta}{1-s}} (x^{-\eta} - x).$$

Therefore, we get a threshold $R_0 = R_0(k, x)$ on the initial rank for heaving degree at least $k \geq \log^4 n$, namely,

$$R_0(k, x) = n \left(\frac{d(1-\eta)}{k(1+\eta)} (x^{-\eta} - x)^{\frac{1-s}{\eta}} + 1 \right)^{\frac{1}{1-s}}.$$

Finally, one can show that the expected number of vertices of degree at least k is asymptotic to

$$\sum_{i=1}^n \left(\frac{R_0(k, i/n)}{n} \right)^s = (1 + o(1))n \int_0^1 \left(\frac{d(1-\eta)}{k(1+\eta)} (x^{-\eta} - x)^{\frac{1-s}{\eta}} + 1 \right)^{\frac{s}{1-s}} dx$$

$$= (1 + o(1)) \left(\frac{d(1-\eta)}{k(1+\eta)} \right)^{1/\eta} n \int_0^\infty (x^{s-1} + 1)^{\frac{s}{1-s}} dx$$

$$= (1 + o(1)) \left(\frac{d(1-\eta)}{k(1+\eta)} \right)^{1/\eta} n.$$

(The antiderivative of $(x^{s-1} + 1)^{\frac{s}{1-s}}$ is $x(x^{s-1} + 1)^{\frac{1}{1-s}}$.) The assertion follows from the Chernoff bound.

References

1. Bonato, A.: A survey of web graph models. In: Proceedings of Combinatorial and Algorithm Aspects of Networking (2004)
2. Broder, A., Kumar, R., Maghoul, F., Rahaghavan, P., Rajagopalan, S., State, R., Tomkins, A., Wiener, J.: Graph structure in the web. In: Proc. 9th International World-Wide Web Conference (WWW), pp. 309–320 (2000)
3. Fortunato, S., Flammini, A., Menczer, F.: Scale-free network growth by ranking. Phys. Rev. Lett. 96(21), 218701 (2006)
4. Janson, S., Łuczak, T., Ruciński, A.: Random Graphs. Wiley, Chichester (2000)
5. Janssen, J., Prałat, P.: Rank-based attachment leads to power law graphs (preprint)
6. Łuczak, T., Prałat, P.: Protean graphs. Internet Mathematics 3, 21–40 (2006)
7. Pittel, B., Spencer, J., Wormald, N.: Sudden emergence of a giant k-core in a random graph. J. Combinatorial Theory Series B 67, 111–151 (1996)
8. Prałat, P.: A note on the diameter of protean graphs. Discrete Mathematics 308, 3399–3406 (2008)
9. Prałat, P., Wormald, N.: Growing protean graphs. Internet Mathematics, 13 (accepted)
10. Wormald, N.C.: Random graphs and asymptotics. Section 8.2. In: Gross, J.L., Yellen, J. (eds.) Handbook of Graph Theory, pp. 817–836. CRC, Boca Raton (2004)
11. Wormald, N.: The differential equation method for random graph processes and greedy algorithms. In: Karoński, M., Prömel, H.J. (eds.) Lectures on Approximation and Randomized Algorithms, pp. 73–155. PWN, Warsaw (1999)

Simplicial Powers of Graphs

Andreas Brandstädt and Van Bang Le

Institut für Informatik, Universität Rostock, D-18051 Rostock, Germany
{ab,le}@informatik.uni-rostock.de

Abstract. In a finite simple undirected graph, a vertex is simplicial if its neighborhood is a clique. We say that, for $k \geq 2$, a graph $G = (V_G, E_G)$ is the *k-simplicial power* of a graph $H = (V_H, E_H)$ (H a *root graph* of G) if V_G is the set of all simplicial vertices of H, and for all distinct vertices x and y in V_G, $xy \in E_G$ if and only if the distance in H between x and y is at most k. This concept generalizes *k-leaf powers* introduced by Nishimura, Ragde and Thilikos which were motivated by the search for underlying phylogenetic trees; *k*-leaf powers are the *k*-simplicial powers of trees. Recently, a lot of work has been done on *k*-leaf powers and their roots as well as on their variants phylogenetic roots and Steiner roots. For $k \in \{3, 4, 5\}$, *k*-leaf powers can be recognized in linear time, and for $k \in \{3, 4\}$, structural characterizations are known. For all other k, recognition and structural characterization of *k*-leaf powers is open.

Since trees and block graphs (i.e., connected graphs whose blocks are cliques) have very similar metric properties, it is natural to study *k*-simplicial powers of block graphs. We show that leaf powers of trees and simplicial powers of block graphs are closely related, and we study simplicial powers of other graph classes containing all trees such as ptolemaic graphs and strongly chordal graphs.

Keywords: Graph powers, leaf powers, simplicial powers, forbidden induced subgraph characterization, chordal graphs, block graphs, ptolemaic graphs, strongly chordal graphs.

1 Introduction

Motivated by background from phylogenetic trees [3,16,35], Nishimura, Ragde and Thilikos [33] introduced the following notions: For an integer $k \geq 2$, a finite undirected graph $G = (V_G, E_G)$ is a *k-leaf power* if there is a tree T with V_G as its set of leaves such that for all distinct $x, y \in V_G$, $xy \in E_G$ if and only if the distance between x and y in T is at most k. Then T is called a *k-leaf root of G*. In general, G is a *leaf power* if G is a *k*-leaf power for some $k \geq 2$.

Obviously, a graph is a 2-leaf power if and only if it is a disjoint union of cliques or, equivalently, it contains no induced path P_3 with three vertices and two edges. In [33], a (very complicated) $\mathcal{O}(n^3)$ time algorithm for recognizing 3-leaf powers and 4-leaf powers, respectively, and constructing 3-leaf roots and 4-leaf roots, respectively, if they exist, was described. Recently, Chang and Ko [15] gave a linear time recognition algorithm for 5-leaf powers. Despite considerable

B. Yang, D.-Z. Du, and C.A. Wang (Eds.): COCOA 2008, LNCS 5165, pp. 160–170, 2008.
© Springer-Verlag Berlin Heidelberg 2008

effort, for $k \geq 6$, no characterization and no efficient recognition of k-leaf powers is known. See [6,7,9,11,12,19,34] for more information on leaf powers and in particular, for new characterizations of 3- and 4-leaf powers as well as of distance-hereditary 5-leaf powers and related classes.

It is known that for every $k \geq 2$, k-leaf powers are strongly chordal [7] (for the definition of strongly chordal graphs see section 2). In [4], Bibelnieks and Dearing introduced and studied so-called NeST graphs (i.e., neighborhood subtree tolerance graphs); for constant tolerances these are exactly the induced subgraphs of powers of trees [8,23] which are closely related to k-leaf powers (see Proposition 1). In [4], an example of a graph is given which is strongly chordal but no fixed tolerance NeST graph (i.e., no k-leaf power for any k), and in [23] this is slightly generalized; [23] mentions the open problem of characterizing fixed tolerance NeST graphs.

Definition 1 gives the key notion of this paper, namely k-simplicial powers of graphs which generalizes the notion of k-leaf powers of trees in a very natural way and which is also of independent interest. A vertex is *simplicial* if its neighborhood is a clique. Simplicial vertices of degree one are called *leaves*.

Definition 1. *For any integer $k \geq 1$, graph $G = (V_G, E_G)$ is the k-simplicial power of graph $H = (V_H, E_H)$ if $V_G \subseteq V_H$ is the set of all simplicial vertices in H and for all distinct vertices $x, y \in V_G$, $xy \in E_G$ if and only if the distance in H between x and y is at most k. Such a graph H is a k-simplicial root of G. If G is the k-simplicial power of H and if, in addition, V_G consists of exactly the degree 1 vertices, i.e., leaves of H, then we also say that G is the k-leaf power of H.*

Since trees and block graphs (i.e., those graphs whose 2-connected components are cliques) have very similar metric properties (see Theorem 2), it is natural to study k-simplicial powers of block graphs. In particular, the main motivation of this paper comes from Theorem 6 which claims that for any $k \geq 2$, a graph is the k-leaf power of a tree if and only if it is the $(k-1)$-simplicial power of a claw-free block graph. Thus, our focus is on simplicial powers of block graphs but we also consider simplicial powers of other graph classes containing all trees such as ptolemaic graphs and strongly chordal graphs. Due to space limitations in this extended abstract, proofs are omitted.

2 Basic Notions and Results

Throughout this paper, let $G = (V_G, E_G)$ denote a finite undirected graph without loops and multiple edges, with vertex set V_G and edge set E_G. Moreover, we assume connectedness unless stated otherwise. For a vertex $v \in V_G$, let $N_G(v) = \{u \mid uv \in E_G\}$ denote the *neighborhood* of v in G, and let $N_G[v] = \{v\} \cup N_G(v)$ denote the *closed neighborhood* of v in G. The *degree $deg_G(v)$* of a vertex v is the number of its neighbors, i.e., $deg_G(v) = |N_G(v)|$. The *complement graph of G* is denoted by $\overline{G}$. A *clique* is a set of mutually adjacent vertices. A *stable set* is a set of mutually non-adjacent vertices.

A *cut vertex* is a vertex whose removal increases the number of connected components. A connected graph is *2-connected* if it has no cut vertex. As usual, the maximal induced 2-connected subgraphs of G are the *blocks* (or *2-connected components*) of G. A block of G which contains at most one cut vertex is an *endblock*. For $U \subseteq V$, let $G[U]$ denote the subgraph of G induced by U. For a set $\mathcal{F}$ of graphs, a graph is $\mathcal{F}$-*free* if none of its induced subgraphs is in $\mathcal{F}$.

Two vertices $x, y \in V$ are *true twins* if $N_G[x] = N_G[y]$. A vertex set $U \subseteq V_G$ is a *module* of G if $U \subseteq N_G(v)$ or $U \cap N_G(v) = \emptyset$ for all $v \in V_G \setminus U$. A *homogeneous set* of G is a module which consists of at least two, but not all vertices of G. A *clique module* in G is a module which is a clique in G. Obviously, true twins form a clique module. *Replacing* a vertex v in a graph G by a graph H (or *substituting* H into v) results in the graph obtained from $G[V_G \setminus \{v\}] \cup H$ by adding all edges between vertices in $N_G(v)$ and vertices in V_H.

For a positive integer $k \geq 1$, let P_k denote the chordless path with k vertices and $k - 1$ edges, and for $k \geq 3$, let C_k denote the chordless cycle with k vertices and k edges. A complete bipartite graph with r vertices in one color class and s vertices in the other color class is denoted by $K_{r,s}$; the $K_{1,3}$ is also called the *claw*. For $k \geq 3$, let S_k denote the *(complete) sun* with $2k$ vertices $u_1, \ldots, u_k$ and $w_1, \ldots, w_k$ such that $u_1, \ldots, u_k$ is a clique, $w_1, \ldots, w_k$ is a stable set and for $i \in \{1, \ldots, k\}$, w_i is adjacent to exactly u_i and u_{i+1} (index arithmetic modulo k). A graph is *sun-free* if it contains no induced S_k for any $k \geq 3$.

A graph is *chordal* if it contains no induced C_k for any $k \geq 4$. A graph is *strongly chordal* if it is chordal and sun-free. It is known that leaf powers are strongly chordal (cf. [7], Proposition 3). A graph is a *split graph* if its vertex set can be partitioned into a clique and a stable set. It is well known that G is a split graph if and only if G and its complement graph $\overline{G}$ are chordal. A graph is *ptolemaic* if it is chordal and gem-free (see Figure 1 for the gem).

A connected graph is a *block graph* if each of its blocks is a clique. Clearly, block graphs are ptolemaic but not vice versa. As block graphs will play a crucial role in this paper, we give here some well-known characterizations of them; the equivalence (i) $\Leftrightarrow$ (ii) in Theorem 1 is Theorem 3.5 in [24], and the equivalence (i) $\Leftrightarrow$ (iii) can be easily seen, e.g., by [11, Observation 3].

Theorem 1. *For every graph G, the following statements are equivalent:*
(i) *G is a block graph.*
(ii) *G is the intersection graph of the blocks of some graph.*
(iii) *G is chordal and diamond-free.*

Let $d_G(x, y)$ denote the distance in G between x and y (i.e., the minimum number of edges of a path in G connecting x and y). A graph G is *distance hereditary* if in every connected induced subgraph H of G, the distance function is the same as in G, i.e., $d_H = d_G|_{V_H}$. In [25] it was shown that a chordal graph is distance hereditary if and only if it is gem-free. In particular, distance-hereditary and chordal graphs, i.e., ptolemaic graphs, are strongly chordal but not vice versa.

Let $G^k = (V_G, E_G^k)$ with $xy \in E_G^k$ if and only if $d_G(x, y) \leq k$ denote the *k-th power of G*. See [1,2,5,10,18,20,21,31,32] for basic properties of powers of strongly chordal graphs (chordal graphs, distance-hereditary graphs, respectively).

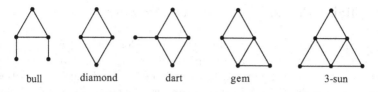

Fig. 1. The bull, diamond, dart, gem, and 3-sun

Buneman's four-point condition (∗) for distances in graphs requires that for every four vertices u, v, x, y, the following inequality holds:

$$(\ast) \quad d_G(u,v) + d_G(x,y) \leq \max\{d_G(u,x) + d_G(v,y), d_G(u,y) + d_G(v,x)\}.$$

The following well-known results show that trees and block graphs have very similar metric properties.

Theorem 2. *Let G be a connected graph.*

(i) [13] *G is a tree if and only if G is triangle-free and fulfills the four-point condition* (∗).

(ii) [26] *G is a block graph if and only if G satisfies* (∗).

Finally, we mention some fundamental but simple properties, among them the following result characterizing 3-leaf powers:

Theorem 3 ([7,19,34]). *For every graph G, the following are equivalent:*

(i) *G is a 3-leaf power.*

(ii) *G is (bull, dart, gem)-free chordal.*

(iii) *G results from substituting cliques into the vertices of a tree.*

In [30], the following notion for $k \geq 1$ is defined: A tree $T = (V_T, E_T)$ is a *k-th Steiner root* of the graph $G = (V_G, E_G)$ if $V_G \subseteq V_T$ and $xy \in E_G$ if and only if $d_T(x,y) \leq k$. In this case, G is a *k-th Steiner power*.

In [11], we say that a graph G is a *basic k-leaf power* if G has a k-leaf root T such that no two leaves of T are attached to the same parent vertex in T (a so-called *basic k-leaf root*). Obviously, for $k \geq 2$, the set of leaves having the same parent node in T form a clique, and G is a k-leaf power if and only if G results from a basic k-leaf power by substituting cliques into its vertices. If T is a basic k-leaf root of G then T minus its leaves is a $(k-2)$-th Steiner root of G. Summarising, the following obvious equivalences hold:

Proposition 1. *For a graph G, the following are equivalent for all $k \geq 2$:*

(i) *G has a k-th Steiner root.*

(ii) *G is an induced subgraph of the k-th power of a tree.*

(iii) *G is a basic $(k+2)$-leaf power.*

3   Simplicial Powers Versus Leaf Powers

Recall that the notion of k-simplicial powers (see Definition 1) is the key notion of this paper. It is easy to see that a graph is the 1-simplicial power of some graph if and only if it is a disjoint union of cliques, i.e., it is P_3-free. As Proposition 2 shows, every graph is the 2-simplicial power of some split graph. Thus, the notion of k-simplicial power is only interesting for some very restricted classes of root graphs.

Proposition 2. *Every graph is*

(i) *the 2-simplicial power of a split graph, and*
(ii) *the 4-leaf power of a bipartite graph.*

Since in the proof of Proposition 2 (i), for given graph G a split graph G' is constructed which might be exponentially larger than G, Proposition 2 (i) suggests the following problem:

2-SIMPLICIAL SPLIT GRAPH ROOT
Instance: A graph $G = (V_G, E_G)$ and an integer N.
Question: Does there exist a split graph $H = (V_H, E_H)$ with $|V_H| \leq N$ such that G is the 2-simplicial power of H?

By reducing the problem INTERSECTION GRAPH BASIS ([22, GT59]) to our problem, we obtain:

Theorem 4. 2-SIMPLICIAL SPLIT GRAPH ROOT *is NP-complete.*

Let $G = (V_G, E_G)$ be a graph. Its *line graph* $L(G)$ has E_G as its vertices, and two edges e, e' are adjacent in $L(G)$ if and only if $e \cap e' \neq \emptyset$.

Theorem 5 ([24], Theorem 8.5). *A graph is the line graph of a tree if and only if it is a claw-free block graph.*

The subsequent Theorem 6 was the main motivation for this paper.

Theorem 6. *For $k \geq 2$, a graph is the k-leaf power of a tree if and only if it is the $(k-1)$-simplicial power of a claw-free block graph.*

Corollary 1. *The class of k-simplicial powers of block graphs contains all t-leaf powers for $t \leq k+1$.*

4   2-Simplicial Powers of Some Subclasses of Chordal Graphs

By Theorem 6, every 3-leaf power is the 2-simplicial power of a claw-free block graph. Theorem 7 characterizes the larger class of 2-simplicial powers of block graphs as the (dart,gem)-free chordal graphs. Note that this graph class appears in other contexts as well:

- In [14], in connection with convexity of graphs, the notion of *contour vertices* is defined, and it is shown that a connected graph G has the property that for all convex sets S in G, the contour vertices of S coincide with the eccentric vertices of S if and only if G is (dart, gem)-free chordal.
- In [28], so-called *strictly chordal graphs* are introduced via rather complicated hypergraph properties, and it is shown that these graphs are leaf powers. It turns out that a graph is strictly chordal if and only if it is (dart,gem)-free chordal [27].
- In [12], the notion of k-leaf root and k-leaf power is modified in the following way: For $k \geq 2$ and $\ell > k$, a tree T is a (k, ℓ)-*leaf root of a graph* $G = (V_G, E_G)$ if V_G is the set of leaves of T, for all edges $xy \in E_G$, $d_T(x, y) \leq k$ and, for all non-edges $xy \notin E_G$, $d_T(x, y) \geq \ell$. A graph G is a (k, ℓ)-*leaf power* if it has a (k, ℓ)-leaf root. Thus, every k-leaf power is a $(k, k+1)$-leaf power. Then, it is shown in [12]: Every block graph is a $(4, 6)$-leaf power, and a $(4, 6)$-leaf root of it can be determined in linear time. Moreover, G is a $(4, 6)$-leaf power if and only if G is (dart,gem)-free chordal.

Recall that Theorem 3 characterizes 3-leaf powers (of trees) as the (bull,dart,gem)-free chordal graphs. Comparing Theorem 7 with Theorem 3 shows how natural the concept of simplicial powers of block graphs fits within the world of leaf powers.

Theorem 7. *For every graph G, the following statements are equivalent:*

  (i) *G is the 2-simplicial power of a block graph.*
 (ii) *G is (dart,gem)-free chordal.*
(iii) *G results from substituting cliques into the vertices of a block graph.*
 (iv) *G is a $(4, 6)$-leaf power.*

Theorem 8. *For every graph G, the following statements are equivalent:*

  (i) *G is the 2-simplicial power of a ptolemaic graph.*
 (ii) *G is the 2-simplicial power of a ptolemaic split graph.*
(iii) *G is ptolemaic.*

An analogous equivalence holds if in Theorem 8, "ptolemaic" is replaced by "strongly chordal" in all three statements.

5   Simplicial Powers of Block Graphs

Theorem 6 indicates the close relationship between leaf powers (of trees) and simplicial powers of (claw-free) block graphs. However, the larger class of simplicial powers of (not necessarily claw-free) block graphs is of independent interest. As already mentioned (see [11] and Proposition 1), leaf powers of trees are exactly those graphs obtainable from an induced subgraph of a tree power by replacing vertices by cliques. A similar statement is true for simplicial powers of block graphs; it is based on the following notion.

Definition 2. *A graph G is a* basic k-simplicial power *of a block graph if G admits a k-simplicial block graph root R in which each block contains at most one simplicial vertex.*

Examples of basic k-simplicial powers of block graphs include block graphs and k-leaf powers. Obviously, every simplicial power of a block graph is obtained from a basic simplicial power of a block graph by replacing vertices by cliques. Moreover, if $G = (V_G, E_G)$ is a basic k-simplicial power, then any (connected) induced subgraph of G is also a basic k-simplicial power of a block graph: If $R = (V_R, E_R)$ is a basic k-simplicial block graph root of G and G' is a subgraph of G induced by $S \subseteq V_G$, then it can be easily seen that the smallest connected subgraph of R containing S is a basic k-simplicial block graph root of G'.

Theorem 9. *Let $k \geq 2$ be an integer. A graph is a basic k-simplicial power of a block graph if and only if it is an induced subgraph of the $(k-1)$-th power of a block graph.*

The proof of Theorem 9 shows directly:

Corollary 2. *Let $k \geq 2$ be an integer. A basic k-simplicial power of a block graph is the $(k-1)$-th power of a block graph if and only if it admits a basic k-simplicial block graph root in which each block contains exactly one simplicial vertex.*

In the rest of this section we will describe the basic 3-simplicial powers of block graphs in more detail.

Definition 3. *A maximal clique Q in a graph $G = (V_G, E_G)$ is* special *if for all $x, y \in V_G - Q$ having a common neighbor in Q, $N(x) \cap Q = N(y) \cap Q$ or $|N(x) \cap Q| = 1$ or $|N(y) \cap Q| = 1$. A vertex v of G is* special *if $N[v]$ is a special clique in G.*

Note that a special vertex is in particular simplicial. It turns out that special vertices play an important role in recognizing 2-connected basic 3-simplicial powers of block graphs.

For a description of 2-connected basic 3-simplicial powers of block graphs, we need the following notion. A *split* of a graph $G = (V_G, E_G)$ is a partition into two disjoint sets V_1 and V_2 such that $|V_1| \geq 2$, $|V_2| \geq 2$ and the set of edges of G between V_1 and V_2 forms a complete bipartite graph. Graphs without split are called *prime*. A *simple split decomposition* of G by the split (V_1, V_2) is the decomposition of G into two graph G_1 and G_2 where G_i is obtained from the subgraph of G induced by V_i and an additional vertex (a so-called *marker*) v by adding all edges between v and those vertices in V_i which have a neighbor in $G - V_i$. Split decomposition can be computed in linear time [17].

We characterize 2-connected basic 3-simplicial powers of block graphs by reducing to smaller ones as follows.

Theorem 10. *A 2-connected graph $G = (V_G, E_G)$ is a basic 3-simplicial power of a block graph if and only if*

(i) G is the square of a block graph, or

(ii) G has a special vertex v such that $N_G(v) = N_G(x) \cap N_G(y)$ for some non-adjacent vertices x and y, and $G - v$ is a 2-connected basic 3-simplicial power of a block graph, or

(iii) G admits a split (V_1, V_2) such that G_1 and G_2 are 2-connected basic 3-simplicial powers of block graphs and the marked vertex is special in both G_1 and G_2.

Theorem 10 gives a recursive procedure that checks in time $\mathcal{O}(n^3)$ whether a 2-connected chordal graph G with n vertices is a basic 3-simplicial power of a block graph: Checking whether G is the square of a block graph can be done in linear time by a result in [29]. If G is not the square of a block graph then check whether G satisfies (ii) or (iii). If yes, recursively check the corresponding 2-connected graphs $G - v$, and G_1 and G_2, respectively. Whether a maximal clique is special can be easily checked in time $\mathcal{O}(n^2)$, the at most n maximal cliques in a chordal graph can be found in linear time, and checking (ii) and (iii) can be done in time $\mathcal{O}(n^3)$.

Observation 1 *Every 2-connected basic 3-simplicial power of a block graph $G = (V_G, E_G)$ admits a basic 3-simplicial block graph root R such that, for all special vertices c of G and all $x \in V_G - c$, $d_R(c, x) \geq 3$.*

Theorem 11. *For every graph G, the following statements are equivalent:*

(i) *G is a basic 3-simplicial power of a block graph.*

(ii) *G is an induced subgraph of the square of a block graph.*

(iii) *Each block of G is a basic 3-simplicial power of a block graph, and each cut vertex v of G is non-special in at most one block containing v.*

Corollary 3. *3-simplicial powers of block graphs can be recognized efficiently.*

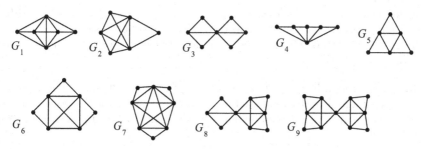

Fig. 2. Forbidden subgraphs $G_1, \ldots, G_9$ characterize induced subgraphs of squares of block graphs

In the full version of [12], induced subgraphs of squares of block graphs (see Theorem 11 (ii)) are also characterized in terms of forbidden subgraphs (see Figure 2), and similarly as for $k = 2$ in Theorem 7, 3-simplicial powers of a block graph are closely related to (6,8)-leaf powers as described in Theorems 11 and 12.

Theorem 12 ([12]). *For every graph G, the following are equivalent:*

(i) G *is a basic* $(6,8)$-*leaf power.*
(ii) G *is an induced subgraph of the square of a block graph.*
(iii) G *is* $(G_1, G_2, \ldots, G_9)$-*free chordal.*

This characterization is inspired by the corresponding results for 4-leaf powers in [11,34]. The graphs $G_1, G_2, G_4, G_5, G_6, G_7$ express separator properties of induced subgraphs of squares of block graphs which are 2-connected, and the graphs G_3, G_8, G_9 express the gluing conditions for the 2-connected components of such graphs.

6   Conclusion

Simplicial powers of block graphs (ptolemaic graphs, strongly chordal graphs, respectively) are a natural generalization of leaf powers. There are close connections between k-leaf powers, $(k, k + 2)$-leaf powers and simplicial powers of block graphs such as described in Theorems 6, 7, 11 and 12.

While every graph is the 2-simplicial power of a split graph and the 4-leaf power of a bipartite graph, 2-simplicial powers of ptolemaic graphs (strongly chordal graphs, respectively) are ptolemaic (strongly chordal, respectively). Since leaf powers are strongly chordal (but not vice versa), our results on simplicial powers of block graphs and of ptolemaic graphs might shed new light on the open problem of characterizing k-leaf powers for $k \geq 5$ and of characterizing leaf powers in general.

We gave various characterizations of classes defined as simplicial powers of certain graph classes. In particular, we obtained the following hierarchy:

- 3-leaf powers (which are exactly the (bull,dart,gem)-free chordal graphs) are a proper subclass of
- 2-simplicial powers of block graphs (which are exactly the (dart,gem)-free chordal graphs), and these are in turn a proper subclass of
- 2-simplicial powers of ptolemaic graphs (which are exactly the gem-free chordal graphs).

The class of 3-simplicial powers of block graphs is an interesting generalization of 4-leaf powers and is characterized in Theorems 11 and 12. We hope that our approach will lead to new insights about the structure of k-leaf powers for $k \geq 5$.

References

1. Bandelt, H.-J., Henkmann, A., Nicolai, F.: Powers of distance-hereditary graphs. Discrete Math. 145, 37–60 (1995)
2. Bandelt, H.-J., Prisner, E.: Clique graphs and Helly graphs. J. Combin. Th. (B) 51, 34–45 (1991)
3. Barthélémy, J.P., Guénoche, A.: Trees and proximity representations. Wiley & Sons, Chichester (1991)

4. Bibelnieks, E., Dearing, P.M.: Neighborhood subtree tolerance graphs. Discrete Applied Math. 43, 13–26 (1993)
5. Brandstädt, A., Dragan, F.F., Chepoi, V.D., Voloshin, V.I.: Dually chordal graphs. SIAM J. Discrete Math. 11, 437–455 (1998)
6. Brandstädt, A., Hundt, C.: Ptolemaic graphs and interval graphs are leaf powers; extended abstract. In: Proceedings of LATIN 2008. LNCS, vol. 4957, pp. 479–491 (2008)
7. Brandstädt, A., Le, V.B.: Structure and linear time recognition of 3-leaf powers. Information Processing Letters 98, 133–138 (2006)
8. Brandstädt, A., Le, V.B., Rautenbach, D.: Exact leaf powers (submitted)
9. Brandstädt, A., Le, V.B., Rautenbach, D.: Distance-hereditary 5-leaf powers (submitted)
10. Brandstädt, A., Le, V.B., Spinrad, J.P.: Graph Classes: A Survey, SIAM Monographs on Discrete Mathematics and Applications, vol. 3. SIAM, Philadelphia (1999)
11. Brandstädt, A., Le, V.B., Sritharan, R.: Structure and linear time recognition of 4-leaf powers. ACM Transactions on Algorithms(accepted)
12. Brandstädt, A., Wagner, P.: On (k, ℓ)-leaf powers; extended abstract. In: Kučera, L., Kučera, A. (eds.) MFCS 2007. LNCS, vol. 4708, pp. 525–535. Springer, Heidelberg (2007) (Full version submitted)
13. Buneman, P.: A note on the metric properties of trees. J. Combin. Th. (B) 1, 48–50 (1974)
14. Cáceres, J., Márquez, A., Oellermann, O.R., Puertas, M.L.: Rebuilding convex sets in graphs. Discrete Math. 293, 26–37 (2005)
15. Chang, M.-S., Ko, T.: The 3-Steiner Root Problem; extended abstract. In: Proceedings 33rd International Workshop on Graph-Theoretic Concepts in Computer Science WG 2007. LNCS, vol. 4769, pp. 109–120 (2007)
16. Chen, Z.-Z., Jiang, T., Lin, G.: Computing phylogenetic roots with bounded degrees and errors. SIAM J. Computing 32, 864–879 (2003)
17. Dahlhaus, E.: Efficient parallel and linear time sequential split decomposition. In: Thiagarajan, P.S. (ed.) FSTTCS 1994. LNCS, vol. 880, pp. 171–180. Springer, Heidelberg (1994)
18. Dahlhaus, E., Duchet, P.: On strongly chordal graphs. Ars Combinatoria 24B, 23–30 (1987)
19. Dom, M., Guo, J., Hüffner, F., Niedermeier, R.: Error compensation in leaf root problems; extended abstract. In: Fleischer, R., Trippen, G. (eds.) ISAAC 2004. LNCS, vol. 3341, pp. 389–401. Springer, Heidelberg (2004); Algorithmica 44, 363–381 (2006)
20. Duchet, P.: Classical perfect graphs. Annals of Discrete Math. 21, 67–96 (1984)
21. Farber, M.: Characterizations of strongly chordal graphs. Discrete Math. 43, 173–189 (1983)
22. Garey, M.R., Johnson, D.S.: Computers and Intractability–A Guide to the Theory of NP-Completeness. Freeman, New York (1979) (twenty-third printing 2002)
23. Hayward, R.B., Kearney, P.E., Malton, A.: NeST graphs. Discrete Applied Math. 121, 139–153 (2002)
24. Harary, F.: Graph Theory. Addison-Wesley, Massachusetts (1972)
25. Howorka, E.: A characterization of distance-hereditary graphs. Quart. J. Math. Oxford, Ser. 2(28), 417–420 (1977)
26. Howorka, E.: On metric properties of certain clique graphs. J. Combin. Th. (B) 27, 67–74 (1979)

27. Kennedy, W.: Strictly chordal graphs and phylogenetic roots, Master Thesis, University of Alberta (2005)
28. Kennedy, W., Lin, G., Yan, G.: Strictly chordal graphs are leaf powers. Journal of Discrete Algorithms 4, 511–525 (2006)
29. Le, V.B., Tuy, N.N.: A good characterization of squares of block graphs (manuscript, 2008)
30. Lin, G.-H., Kearney, P.E., Jiang, T.: Phylogenetic k-root and Steiner k-root. In: Lee, D.T., Teng, S.-H. (eds.) ISAAC 2000. LNCS, vol. 1969, pp. 539–551. Springer, Heidelberg (2000)
31. Lubiw, A.: Γ-free matrices, Master of Science Thesis, Dept. of Combin. and Optim., University of Waterloo (1982)
32. Lubiw, A.: Doubly lexical orderings of matrices. SIAM J. Computing 16, 854–879 (1987)
33. Nishimura, N., Ragde, P., Thilikos, D.: On graph powers for leaf-labeled trees. J. Algorithms 42, 69–108 (2002)
34. Rautenbach, D.: Some remarks about leaf roots. Discrete Math. 306, 1456–1461 (2006)
35. Semple, C., Steel, M.: Phylogenetics. Oxford University Press, Oxford (2003)

On k- Versus $(k+1)$-Leaf Powers

Andreas Brandstädt and Peter Wagner[*]

Institut für Informatik, Universität Rostock, D-18051 Rostock, Germany
{ab,peter.wagner}@informatik.uni-rostock.de

Abstract. For $k \geq 2$ and a finite simple undirected graph $G = (V, E)$, a tree T is a *k-leaf root* of G if V is the set of leaves of T and, for any two distinct $x, y \in V$, $xy \in E$ if and only if the distance between x and y in T is at most k. G is a *k-leaf power* if G has a k-leaf root. Motivated by the search for underlying phylogenetic trees, the concept of k-leaf power was introduced and studied by Nishimura, Ragde and Thilikos and analysed further in many subsequent papers. It is easy to see that for all $k \geq 2$, every k-leaf power is a $(k+2)$-leaf power. However, it was unknown whether every k-leaf power is a $(k+1)$-leaf power. Recently, Fellows, Meister, Rosamond, Sritharan and Telle settled this question by giving an example of a 4-leaf power which is not a 5-leaf power. Motivated by this result, we analyse the inclusion-comparability of k-leaf power classes and show that, for all $k \geq 4$, the k- and $(k+1)$-leaf power classes are incomparable. We also characterise those graphs which are simultaneously 4- and 5-leaf powers.

In the forthcoming full version of this paper, we will show that for all $k \geq 6$ and odd l with $3 \leq l \leq k - 3$, the k- and $(k+l)$-leaf power classes are incomparable. This settles all remaining cases and thus gives the complete inclusion-comparability of k-leaf power classes.

Keywords: k-leaf powers, intersection of leaf power classes, comparability of leaf power classes.

1 Introduction

Nishimura, Ragde and Thilikos [14] introduced the notion of k-leaf power and k-leaf root, motivated by the following: "... a fundamental problem in computational biology is the reconstruction of the *phylogeny*, or evolutionary history, of a set of species or genes, typically represented as a *phylogenetic tree* ...". The species occur as leaves of the phylogenetic tree. Let $k \geq 2$ be an integer and $G = (V, E)$ be a finite simple graph. A tree T is a *k-leaf root* of G, if V is the set of leaves of T and, for any two distinct $x, y \in V$, the distance between x and y in T is at most k if and only if x and y are adjacent in G, i.e., $d_T(x, y) \leq k \iff xy \in E$. We say that G is a *k-leaf power* if G has a k-leaf root. Let $L(k)$ denote the class of all k-leaf powers.

In [5], we introduce the following notion: Let $k \geq 2$ and $\ell > k$ be integers and $G = (V, E)$ be a finite simple graph. A tree T is a *(k, ℓ)-leaf root* of G, if V is

---

[*] Supported by DFG research grant BR 2479/7-1.

the set of leaves of T and, for any two distinct $x, y \in V$, we have $xy \in E \implies d_T(x, y) \leq k$ and $xy \notin E \implies d_T(x, y) \geq \ell$. We say that G is a (k, ℓ)-*leaf power* if G has a (k, ℓ)-leaf root.

Obviously, a graph is a 2-leaf power if and only if it is the disjoint union of cliques, i.e., it contains no induced path P_3 with three vertices. See [10] for the related notions of phylogenetic root and Steiner root and [1,2,3,4,7,8,11,12,13,15] for recent work on leaf powers (including linear time recognition of 3-, 4- and 5-leaf powers and various characterisations). For $k \geq 6$, no characterisation of k-leaf powers and no efficient recognition is known.

By subdividing all edges containing a leaf (i.e., external edges) in a k-leaf root, it is easy to see that, for any $k \geq 2$, we have $L(k) \subset L(k+2)$. It is known [1,4,15] that $L(2) \subset L(3)$ and $L(3) \subset L(4)$, but it was unknown whether, for at least some $k \geq 4$, we have $L(k) \subset L(k+1)$. Fellows, Meister, Rosamond, Sritharan and Telle [9] gave an example of a 4-leaf power on 13 vertices which is not a 5-leaf power, which means that $L(4) \not\subset L(5)$. Motivated by this result, we analyse the inclusion-comparability of k-leaf power classes and show that, for all $k \geq 4$, $L(k)$ and $L(k+1)$ are incomparable. We also give a structural characterisation for $L(4) \cap L(5)$, thereby showing that the example in [9] is not minimal (see Corollary 2).

By subdividing all edges not containing a leaf (i.e., internal edges) in a k-leaf root, it is easy to see that, for any $k \geq 2$, every k-leaf power is a $(2k-2, 2k)$-leaf power (see [5]) and hence a k'-leaf power, for all $k' \geq 2k - 2$. Note that this already implies that, for all $k' \geq 3$, we have $L(2) \subset L(k')$ and, for all $k' \geq 4$, we have $L(3) \subset L(k')$. For $k = 4$ and $k = 5$, we now know, for all $k' > k$, whether $L(k) \subset L(k')$. For $k \geq 6$, the only left open cases are given by all pairs $(k, k+l)$, where l is an odd integer with $3 \leq l \leq k - 3$. In Theorem 7, we settle all these remaining cases by showing that, for all $k \geq 6$ and odd l with $3 \leq l \leq k - 3$, $L(k)$ and $L(k+l)$ are incomparable.

In Theorem 2, we show that for all $2 \leq k < k'$, $L(k') \not\subset L(k)$, which might have been expected but certainly requires a formal proof. We have thus obtained complete information about the inclusion-comparability of all k-leaf power classes, which can, roughly speaking, be summarised as follows: For all $2 \leq k < k'$, the k- and k'-leaf power classes are inclusion-comparable, if and only if every k-leaf root can be transformed into a k'-leaf root by the two simple operations of first possibly subdividing all internal edges exactly once and then possibly subdividing all external edges a fixed number of times.

Due to space limitations, most of the proofs are omitted.

2   Basic Notions and Results

Proposition 1 summarises the immediate inclusions discussed in the previous section.

Proposition 1. *Let $2 \leq k < k'$. If $k' - k$ is even or $k' \geq 2k - 2$, then $L(k) \subset L(k')$.*

Let $d_G(x, y)$ (or $d(x, y)$ for short if G is understood) be the length, i.e., number of edges, of a shortest path in G between x and y. For $k \geq 1$, let $G^k = (V, E^k)$ with $xy \in E^k$ if and only if $d_G(x, y) \leq k$ denote the *k-th power of G*.

A tree is called *basic* if no two of its leaves have the same parent vertex. A k-leaf power is called *basic* if it has a basic k-leaf root.

Two vertices, say x and y, of a graph $G = (V, E)$ are called *true twins* if they have the same set of neighbours in $V \setminus \{x, y\}$ and $xy \in E$.

For $k \geq 2$, let P_k be the chordless path with k vertices $v_0, v_1, \ldots, v_{k-1}$ and $k - 1$ edges $v_0 v_1, \ldots, v_{k-2} v_{k-1}$.

Definition 1. *Let T be a tree. The* first derivative $T^{(1)}$ *of T is the tree obtained from T by deleting its leaves. If T has at most two vertices, then its first derivative is empty. For $k \geq 2$, the kth derivative $T^{(k)}$ of T is the first derivative of $T^{(k-1)}$.*

A well-known fact for distances in trees found by Buneman [6] is the following characterisation in terms of a four-point condition:

Theorem 1. *Let $G = (V, E)$ be a connected graph. G is a tree if and only if G contains no triangles and G satisfies the following four-point condition: For all $u, v, x, y \in V$,*

$$(*) \quad d_G(u, v) + d_G(x, y) \leq \max\{d_G(u, x) + d_G(v, y), d_G(u, y) + d_G(v, x)\}.$$

3 k- and $(k+1)$-Leaf Powers Are Inclusion-Incomparable

Lemma 1 is of central importance in this paper. It is concerned with powers of paths that are subgraphs of powers of trees and states that, roughly speaking, pairs of path vertices that are sufficiently far away from the endvertices (represented in the lemma by the set X) retain a certain path distance property in the tree.

Lemma 1. *Let $p \geq l \geq 2$, let P be the path P_{2p+3-l} and let $X = \{v_{p+1-l}, v_{p+2-l}, \ldots, v_{p+1}\}$. Suppose that P^p is an induced subgraph of T^k, for some tree T and $k \geq 1$. Then, among all unordered pairs of vertices in X, their T-distance is maximal only for $\{v_{p+1-l}, v_{p+1}\}$.*

Proof. Let $a, b \in X$ be two vertices with maximal T-distance; that is, $d_T(a, b) = \max_{x,y \in X} d_T(x, y)$. Suppose that $\{a, b\} \neq \{v_{p+1-l}, v_{p+1}\}$. By symmetry (reversing the vertex labelling swaps v_{p+1-l} and v_{p+1}), we may assume that $v_{p+1-l} \notin \{a, b\}$. Let $x = v_{p+1-l}$ and $y = v_{2p+2-l}$. Then, by Theorem 1, we have $d_T(a, b) + d_T(x, y) \leq \max\{d_T(a, x) + d_T(b, y), d_T(a, y) + d_T(b, x)\}$. Since $x \in X$, we have $d_T(a, x) \leq d_T(a, b)$ and $d_T(b, x) \leq d_T(a, b)$. And since $d_P(a, y) \leq p$ and $d_P(b, y) \leq p$, we have $d_T(a, y) \leq k$ and $d_T(b, y) \leq k$, respectively. The maximum in the inequality is thus bounded above by $d_T(a, b) + k$, implying $d_T(x, y) \leq k$ and hence $d_P(x, y) \leq p$, which contradicts $d_P(x, y) = p + 1$. □

Corollary 1. *Let $p \geq 2$, and let P be the path P_{2p+1}. Suppose that P^p is an induced subgraph of T^k, for some tree T and $k \geq 1$. Then, for all $1 \leq m \leq p$, we have $d_T(v_{p-1}, v_{p-1+m}) \leq k - p + m$.*

Theorem 2, implying that $L(k) \subset L(k')$ can only hold if $k \leq k'$, is an immediate consequence of Corollary 1.

Theorem 2. *For every $k \geq 3$, P_{2k-3}^{k-2} is a k-leaf power which is not a k'-leaf power, for any $2 \leq k' < k$.*

Proof. For $k = 3$, the path P_3 with two edges is an appropriate example. For $k \geq 4$, let P be the path P_{2k-3}. Note that P^{k-2} is a k-leaf power without true twins. Suppose that P^{k-2} is a k'-leaf power, for some $2 \leq k' < k$. Clearly, $k' = 2$ cannot hold, so that we may assume $3 \leq k' < k$. Then there must be a $(k'-2)$-Steiner root T for P^{k-2}; that is, $P^{k-2} \leq T^{k'-2}$. By Corollary 1 (with $m = 1$), we must have $d_T(v_{k-3}, v_{k-2}) \leq (k'-2) - (k-2) + 1 \leq 0$, a contradiction.   □

Let $k \geq 4$. In this section, we are mainly concerned with a special case of Lemma 1 and Corollary 1, as the two involved graph power exponents differ by exactly 1. Roughly speaking, Lemma 2 provides us with valuable structural information about the centre of path powers, which can be used to derive Theorem 3, since trees are spanned by many long paths.

Lemma 2. *Let $k \geq 4$, and let P be the path P_{2k-3}. Suppose that the k-leaf power P^{k-2} without true twins is a $(k+1)$-leaf power, and let T be a $(k-1)$-Steiner root of P^{k-2}. Then the subtree T' of T spanned by the three real vertices corresponding to v_{k-3}, v_{k-2} and v_{k-1}, is obtained from the subpath $v_{k-3}v_{k-2}v_{k-1}$ of P by a subdivision of at most one of its edges by exactly one vertex.*

Theorem 3. *Let $k \geq 4$, and let S be a tree with at least two vertices, such that S^{k-2} has no true twins. Suppose that the k-leaf power S^{k-2} is a $(k+1)$-leaf power, and let T be a $(k-1)$-Steiner root of S^{k-2}; that is, $S^{k-2} \leq T^{k-1}$. Let S' be the $(k-3)^{rd}$ derivative of S. Then the subtree $T[V_{S'}]$ of T spanned by the real vertices corresponding to the vertices of S' is obtained from S' by a subdivision of some of its edges by exactly one vertex.*

The structural information in Theorem 3 is crucial for Theorem 4.

Theorem 4. *For all $k \geq 4$, there is a k-leaf power, which is not a $(k+1)$-leaf power.*

See Figures 1, 2 and 3 for examples. The term $k - 3$ in Figure 3 represents a path with precisely $k - 3$ edges.

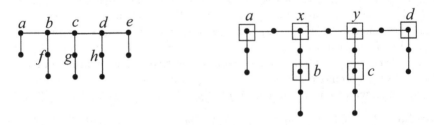

Fig. 1. Trees S_l (left) and S_r (right) with $S_l^2 \in L(4) \setminus L(5)$ and $S_r^3 \in L(5) \setminus L(6)$

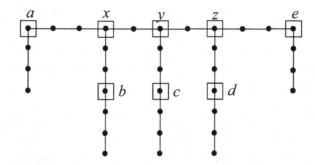

Fig. 2. A tree S with S^4 being a 6-, but not a 7-leaf power

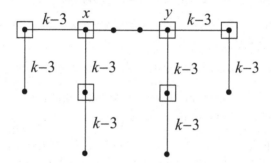

Fig. 3. For $k \geq 7$, a tree S with S^{k-2} being a k-, but not a $(k + 1)$-leaf power

4   The Intersection of 4- and 5-Leaf Powers

We first give a characterisation of 2-connected basic 4-leaf powers that are 5-leaf powers. By [4,15], 2-connected basic 4-leaf powers are precisely squares of basic trees. The square of a tree does not have a pair of true twins (i.e., it is basic) if and only if the tree does not have a pair of leaves with the same parent vertex (i.e., the tree is basic). What we need to do is to find those basic trees whose squares are also 5-leaf powers.

Let S be a basic tree with at least two vertices. Suppose that S^2 is a 5-leaf power, and let T be a 3-Steiner root of S^2. Let S' be the first derivative of S. By Theorem 3, the subtree $T[V_{S'}]$ of T spanned by the real vertices corresponding to the vertices of S' is obtained from S' by a subdivision of some of its edges by exactly one vertex.

This provides useful information about the relationship between S and T. In our special case, we can say more.

Lemma 3. *Let S be a basic tree with at least two vertices. Suppose that S^2 is a 5-leaf power, and let T be a 3-Steiner root of S^2. Let v be a branching vertex of S (i.e., of degree exceeding 2), and let a, b and c be adjacent to v in S. Let C be the claw with vertices a, b, c and v in S. Then the subtree $T[V_C]$ of T spanned* *by the real vertices corresponding to the vertices of C is obtained from C by a subdivision of some (or possibly none) of its edges by exactly one vertex.*

Lemma 3 suggests that branching vertices are helpful when trying to deduce information about T. The following result highlights the consequence of two branching vertices being adjacent in S.

Lemma 4. *Let S be a basic tree with at least two vertices. Suppose that S^2 is a 5-leaf power, and let T be a 3-Steiner root of S^2. Let v and w be two adjacent branching vertices of S, and let a, b, c, d, x and y be six further vertices, such that a and b are adjacent to v, c and d are adjacent to w, x is adjacent to a, and y is adjacent to c in S. Let B be the subtree of S spanned by a, b, c, d, v and w. Then the subtree $T[V_B]$ of T spanned by the real vertices corresponding to the vertices of B is obtained from B by subdividing vw exactly once.*

Corollary 2. *Let S be a basic tree with three consecutive branching vertices. Then S^2 is not a 5-leaf power.*

Note that Corollary 2 improves the example in Figure 1. Simply delete the three leaves adjacent to f, g and h to obtain a tree on ten vertices whose square is a 4-, but not a 5-leaf power. Note further that the obtained tree is a subtree of the tree on 13 vertices given in [9].

Lemma 4 suggests that adjacent pairs of branching vertices in S have, roughly speaking, a significant impact on their neighbourhood. Let a subpath of S be called a *degree-2 path*, if its internal vertices are of degree 2 in S and its endvertices are branching vertices or leaves. It will be important to distinguish between two classes of degree-2 paths, those of lengths 1, 2 and 4 and those of lengths 3, 5 and larger. The following result is a simple consequence of Theorem 3, Lemma 3 and Lemma 4.

Corollary 3. *Let S be a basic tree with at least two pairs of adjacent branching vertices and whose degree-2 paths are exclusively of length 1, 2 or 4. Then S^2 is not a 5-leaf power.*

Conversely, the following holds:

Lemma 5. *Let S be a basic tree with at most one pair of adjacent branching vertices and whose degree-2 paths are exclusively of length 1, 2 or 4. Then there is a 3-Steiner root for S^2, which is obtained by subdividing some of the edges of S exactly once, leaving edges between leaves and branching vertices unaltered.*

The general case with some degree-2 paths of length 3, 5 or larger occurring can be treated by induction on the number of those paths.

Theorem 5. *Let S be a basic tree, for which every subtree with degree-2 paths of length 1, 2 or 4 only contains at most one pair of adjacent branching vertices. Then there is a 3-Steiner root for S^2, which is obtained by subdividing some of the edges of S exactly once, leaving edges between leaves and branching vertices unaltered.*

Fig. 4. A yellow vertex y, a green vertex g and a possible 3-Steiner root

Theorem 6 follows immediately from Corollary 3 and Theorem 5.

Theorem 6. *Let the 2-connected basic 4-leaf power G be the square of the basic tree S. Then G is a 5-leaf power if and only if every subtree of S with degree-2 paths of length $1, 2$ or 4 only contains at most one pair of adjacent branching vertices.*

In order to treat the general connected case, we need to give a description for how the blocks, i.e. the 2-connected components, can be glued together.

As we are dealing with 4-leaf powers, by [4,15], no two non-clique blocks can share a cutvertex. Furthermore, it is clear that any two clique blocks can share a cutvertex and that a trivial clique block (of only two vertices) can be glued to any other block. It remains to discuss the possible gluings of non-clique blocks and non-trivial clique blocks, and, for convenience, we will refer to them as blocks and cliques. Each block B is the square of a tree S. Suppose the block has a pair of true twins. Then, unless S has diameter 3, they arise from being leaves with the same parent node in S. If S has diameter 3, then there is a 3-Steiner root T for it with all distances between real nodes exceeding 1, and B can be glued to any clique. Otherwise, consider one representative for each set of true twins (and temporarily delete the rest) to obtain a new S, and we can apply the above information from the 2-connected case.

In S, mark all vertices of its second derivative with *red*. Furthermore, for each leaf, determine the length of the degree-2 path it lies on. If it has length 1, then mark the leaf *yellow*. If it has length 2 or 4, then mark the leaf *green* and its neighbour *yellow*. If it has length 3, 5 or greater, then mark the leaf and its neighbour *green*. Finally, if a yellow vertex is a vertex of a subtree of S, whose degree-2 paths are of length 1, 2 or 4 only, and which has a pair of adjacent branching vertices, then recolour the yellow vertex red.

Now the block can be glued to any clique at green vertices. When gluing blocks and cliques together we must only prevent an alternating sequence of blocks and cliques from appearing, where we start and end at red vertices in blocks and the intermediate blocks are such that we joined them with cliques at yellow vertices belonging to the same subtree, whose degree-2 paths are of length 1, 2 or 4 only.

A 3-Steiner root realisation for a degree-2 path ending in the yellow vertex y and the green vertex g (a leaf) with a distance of at least 2 from both g and y to every other real vertex is given in Figure 4.

5   The Complete Comparability

So far, we have been comparing $L(k)$ and $L(k + 1)$, for every $k \geq 2$. We have recently shown the following theorem, settling all remaining cases.

Theorem 7. *For all pairs (k, l) of integers $2 \leq k < l$, such that $3 \leq l - k \leq k - 3$ and $l - k$ is not a multiple of 2, we have $L(k) \not\subseteq L(l)$.*

For proving the non-emptiness of the various sets $L(k) \backslash L(l)$, the following notion is needed.

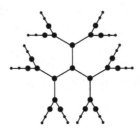

Fig. 5. The tree $C_{3,2}$

Definition 2. *Let C_1 be a claw; that is, a star with three edges. For $i > 1$, let C_i be the tree obtained from C_{i-1} by adding precisely two leaves at every leaf of C_{i-1}. For every $j \geq 1$, let $C_{i,j}$ be the tree obtained from C_i by subdividing every external edge of C_i exactly j times.*

See Figure 5 for an example. Note that, for all $i \geq 1$, every non-leaf of C_i has degree 3. Note further that, for all $1 \leq i' \leq i$ and $1 \leq j' \leq j$, the tree $C_{i',j'}$ is a subtree of $C_{i,j}$. The next two theorems follow immediately from forthcoming work.

Theorem 8. *The 6-leaf power $C_{36,3}^4$ is not a 9-leaf power.*

Theorem 9. *The 7-leaf power $C_{49,4}^5$ is not a 10-leaf power.*

6   Conclusion

In this paper, we compare the classes of k- and l-leaf powers, for all $2 \leq k < l$, in particular, for $l = k+1$. We show that the classes of k- and $(k+1)$-leaf powers are inclusion-incomparable, for all $k \geq 4$. We give a structural characterisation for the intersection of the 4- and 5-leaf powers, extending and improving a result of Fellows, Meister, Rosamond, Sritharan and Telle [9] about an example of a 4-leaf power which is not a 5-leaf power. Finally, by Proposition 1 and Theorem 4, we discuss the remaining interesting case of odd $l - k$ with $3 \leq l - k \leq k - 3$. We have recently shown inclusion-incomparability in that case, the proof of which is to appear in the forthcoming full version of this paper. Characterisations for other interesting intersections, such as the intersection of the 5- and 6-leaf powers, are still open problems.

References

1. Brandstädt, A., Le, V.B.: Structure and linear time recognition of 3-leaf powers. Information Processing Letters 98, 133–138 (2006)
2. Brandstädt, A., Le, V.B., Rautenbach, D.: Exact leaf powers (manuscript) (submitted, 2006)
3. Brandstädt, A., Le, V.B., Rautenbach, D.: Distance-hereditary 5-leaf powers (manuscript) (submitted, 2006)
4. Brandstädt, A., Le, V.B., Sritharan, R.: Structure and linear time recognition of 4-leaf powers. ACM Transactions on Algorithms (manuscript) (accepted, 2006)
5. Brandstädt, A., Wagner, P.: On (k, ℓ)-leaf powers, extended abstract. In: Kučera, L., Kučera, A. (eds.) MFCS 2007. LNCS, vol. 4708, pp. 525–535. Springer, Heidelberg (2007)
6. Buneman, P.: A note on the metric properties of trees. J. Combin. Th. (B) 1, 48–50 (1974)
7. Chang, M.-S., Ko, T.: The 3-Steiner Root Problem, extended abstract. In: Proceedings WG 2007. LNCS, vol. 4769, pp. 109–120 (2007)
8. Dom, M., Guo, J., Hüffner, F., Niedermeier, R.: Error compensation in leaf root problems, extended abstract. In: Fleischer, R., Trippen, G. (eds.) ISAAC 2004. LNCS, vol. 3341, pp. 389–401. Springer, Heidelberg (2004); Algorithmica 44(4), 363-381 (2006)
9. Fellows, M., Meister, D., Rosamond, F., Sritharan, R., Telle, J.A.: On graphs that are k-leaf powers (manuscript, 2007)
10. Jiang, T., Kearney, P.E., Lin, G.-H.: Phylogenetic k-root and Steiner k-root, Extended abstract. In: Lee, D.T., Teng, S.-H. (eds.) ISAAC 2000. LNCS, vol. 1969, pp. 539–551. Springer, Heidelberg (2000)
11. Kennedy, W.: Strictly chordal graphs and phylogenetic roots, Master Thesis, University of Alberta (2005)
12. Kennedy, W., Lin, G.-H.: 5-th phylogenetic root construction for strictly chordal graphs, extended abstract. In: Deng, X., Du, D.-Z. (eds.) ISAAC 2005. LNCS, vol. 3827, pp. 738–747. Springer, Heidelberg (2005)
13. Kennedy, W., Lin, G.-H., Yan, G.: Strictly chordal graphs are leaf powers. J. Discrete Algorithms 4, 511–525 (2006)
14. Nishimura, N., Ragde, P., Thilikos, D.M.: On graph powers for leaf-labeled trees. J. Algorithms 42, 69–108 (2002)
15. Rautenbach, D.: Some remarks about leaf roots. Discrete Math. 306(13), 1456–1461 (2006)

Flows with Unit Path Capacities
and Related Packing and Covering Problems[*]

Maren Martens[1] and Martin Skutella[2]

[1] University of British Columbia, Sauder School of Business,
2053 Main Mall, Vancouver, BC V6T 1Z2, Canada
maren.martens@sauder.ubc.ca
[2] TU Berlin, Institut für Mathematik, MA 5-2, Str. des 17. Juni 136, 10623 Berlin, Germany
skutella@math.tu-berlin.de

Abstract. Since the seminal work of Ford and Fulkerson in the 1950s, network flow theory is one of the most important and most active areas of research in combinatorial optimization. Coming from the classical maximum flow problem, we introduce and study an apparently basic but new flow problem that features a couple of interesting peculiarities. We derive several results on the complexity and approximability of the new problem. On the way we also discover two closely related basic covering and packing problems that are of independent interest.

Starting from an LP formulation of the maximum s-t-flow problem in path variables, we introduce unit upper bounds on the amount of flow being sent along each path. The resulting (fractional) flow problem is NP-hard; its integral version is strongly NP-hard already on very simple classes of graphs. For the fractional problem we present an FPTAS that is based on solving the k shortest paths problem iteratively. We show that the integral problem is hard to approximate and give an interesting $O(\log m)$-approximation algorithm, where m is the number of arcs in the considered graph. For the multicommodity version of the problem there is an $O(\sqrt{m})$-approximation algorithm. We argue that this performance guarantee is best possible, unless P=NP.

1 Introduction

Problem Definition and Notation. The classical maximum s-t-flow problem has been studied from many different points of view. Numerous algorithms are known to solve the problem in polynomial time. Ford and Fulkerson [2] proved already in the 1950s that there always exists an integral optimal solution to the maximum s-t-flow problem provided that all arc capacities are integral. It is also well known that any s-t-flow can be decomposed into flow along paths and cycles. Omitting flow along cycles (which does not contribute to the flow value) yields an alternative LP formulation of the problem in path variables. In this paper we study a new network flow problem in which the flow on any path is bounded by 1, i.e., we add box constraints to the LP formulation of the maximum flow problem in path variables. We call the resulting problem the *maximum one-flow problem.* Our motivation for studying it is mainly academic,

---

[*] This work was partially supported by the DFG Research Center Matheon in Berlin, by the Graduate School of Production Engineering and Logistics, North Rhine-Westphalia, by the DFG Focus Program 1126 and the DFG grants SK 58/4-1 and SK 58/5-3, and by an NSERC Operating Grant. Part of this work was done while the authors were at Universität Dortmund.

B. Yang, D.-Z. Du, and C.A. Wang (Eds.): COCOA 2008, LNCS 5165, pp. 180–189, 2008.
© Springer-Verlag Berlin Heidelberg 2008

but the problem is also well motivated when we think of applications in transportation/communication networks where every single path might be unreliable. In such situations it is reasonable to diversify a commodity/information among several different paths. This can be accomplished by forbidding to send more than a fixed amount of flow along a single path. A more formal definition of the problem is as follows: We are given a network (digraph) $D = (V, A)$ with arc capacities $u : A \rightarrow \mathbb{R}^+$ and two distinguished nodes s, $t \in V$. We assume that $u(a) \geq 1$, for all $a \in A$. If not stated otherwise, $m := |A|$ denotes the number of arcs in the network. Let $\mathcal{P}$ be the set of simple directed s-t-paths in D. Then the *maximum one-flow problem (max-1FP)* and its dual can be formulated as follows, where the path variable x_P denotes the amount of flow sent along path $P \in \mathcal{P}$:

$$\max \sum_{P \in \mathcal{P}} x_P \qquad\qquad \min \sum_{a \in A} u(a) y_a + \sum_{P \in \mathcal{P}} z_P$$

$$\text{s.t.} \sum_{P \ni a} x_P \leq u(a) \quad \forall a \in A \quad (1) \qquad \text{s.t.} \; z_P + \sum_{a \in P} y_a \geq 1 \qquad \forall P \in \mathcal{P}$$

$$0 \leq x_P \leq 1 \qquad \forall P \in \mathcal{P} \quad (2) \qquad z_P, y_a \geq 0 \qquad \forall a \in A, P \in \mathcal{P}$$

Notice that omitting the constraints $x_P \leq 1$ yields the classical maximum s-t-flow problem. An s-t-flow fulfilling (1) and (2) is called a *one-flow*. In an *integral one-flow* each s-t-path sends either 0 or 1 unit of flow. To emphasize that a certain one-flow is not necessarily integral, we sometimes call it *fractional*. Note that in general the encoding size of a maximum one-flow is not polynomial in the input size of the problem since one might want to send flow along exponentially many s-t-paths. Therefore, the best one can expect in terms of complexity are algorithms with running time polynomially bounded in the input plus output size.

The integer version of the dual can be interpreted as a special minimum cut problem, where each s-t-path must be destroyed and this can be done by deleting either a single arc on the path or the path itself. The deletion of an arc a is in general more expensive than that of a whole path ($u(a)$ instead of 1), but can also destroy more than one path simultaneously. The dual separation problem of the classical maximum s-t-flow problem is a shortest path problem. It is not difficult to observe that the dual separation problem of the max-1FP can be solved by computing the k shortest s-t-paths with respect to the dual arc lengths y_a, where k is the number of paths $P \in \mathcal{P}$ with $z_P > 0$ plus 1.

Related Results from the Literature. To the best of our knowledge, the one-flow problem (1FP) is studied for the first time here. However, there is some literature dealing with problems related to it. The problem to compute the number of different (simple) s-t-paths in a network is a special case of the max-1FP. (Consider the case when all arc capacities are infinite.) Valiant [10] shows that this problem is #P-complete under polynomial-time reductions. Similar to the problem of counting paths in a graph is the edge-disjoint paths problem (EDP) that is the same as the integral multicommodity 1FP when we fix all capacities in the considered network to 1. Overviews on the EDP can, e.g., be found in [3,4,5]. For more general packing problems also many results have been obtained. One that is of particular interest for the present paper is obtained by Plotkin, Shmoys, and Tardos [8] and will be introduced later on. In [7] the authors have already considered a certain network flow problem with path capacities, where the flow of any commodity is restricted to at most k paths whose flow values may not exceed given bounds (capacities). The general k shortest paths problem (for a single source and a single sink) is of great interest for the problem considered in this paper. To solve the Lagrange relaxation of the max-1FP that is obtained by penalizing the violation of arc capacities in the objective function, we make use of a result by Lawler [6] who shows how to compute k shortest (simple) paths in a digraph in $O(kn^3)$ time.

Contribution of this Paper. As mentioned above, the problem of computing the number of different simple s-t-paths in a network is #P-complete and a special case of the max-1FP. It therefore follows immediately that computing the maximum one-flow value is NP-hard. This holds for the fractional as well as for the integral 1FP. We prove that the integral max-1FP is strongly NP-hard already on very simple acyclic networks consisting of a chain of parallel arcs. Even worse, the integral max-1FP is APX-hard, even in networks where the number of s-t-paths is polynomially bounded in the size of the network. One interesting consequence of these hardness results is the following: It is NP-hard to decide whether a given integral s-t-flow has an integral path decomposition such that each path carries at most one unit of flow. In Section 2 we establish a close relation between two interesting new combinatorial problems and the special case of the max-1FP on networks consisting of a chain of parallel arcs. The first problem is to cover the edges of a complete graph by cuts of bounded size where the size of a cut is the cardinality of the smaller of the two vertex subsets. The second problem is a packing problem: Consider a set where each element has a given integral weight and find a pre-specified number of different subsets such that the number of subsets containing an element is bounded by the element's weight. The two problems are equivalent and, maybe surprisingly, strongly NP-hard. This also yields the strong NP-hardness of the integral max-1FP on chains of parallel arcs. Moreover, we show that already on a chain of parallel arcs of length 3 the max-1FP has an integrality gap. We also prove that it might happen that each arc in a network carries an integral amount of flow in a maximum one-flow but no maximum one-flow is integral. In Section 3 we show that the approach of Plotkin et al. [8] yields an FPTAS for the fractional max-1FP. The core of the algorithm consists of iteratively solving k shortest paths problems on the given network with varying arc lengths. In Section 4 we derive several approximation algorithms for the integral max-1FP. Our main result is a randomized approximation algorithm with performance ratio $O(\log m)$. Finally, in Section 5 we study multicommodity versions of the 1FP. We show that the FPTAS from Section 3 can be generalized to the fractional multicommodity 1FP. For the integral maximum multicommodity 1FP we present a randomized $O(\sqrt{m})$-approximation algorithm and show that, unless P=NP, no better approximation is possible. Moreover, we present an $O(\log m)$-approximation algorithm for the problem to find an integral multicommodity one-flow with minimum congestion. This extended abstract leaves out most proofs.

2   Interesting Related Problems

In this section we study the max-1FP on a restricted class of networks that are given by chains of parallel arcs. In order to obtain a better understanding of the max-1FP on this particular class of networks we consider two equivalent combinatorial optimization problems, one of which is a covering and the other a packing problem. Although these two problems are easy to formulate and seem quite natural, they have not appeared in the literature before to the best of our knowledge.

We consider networks that consist of $n + 1$ vertices $v_0, v_1, \ldots, v_n$ and $2n$ arcs ($n \in \mathbb{N}$) such that there are two parallel arcs from v_{i-1} to v_i, for $i = 1, \ldots, n$. Vertex v_0 is the source and v_n is the sink. We call one arc of each pair of parallel arcs the *upper* and the other one the *lower arc*. All lower arcs have infinite capacity. The capacity of the ith upper arc is $c_i \in \mathbb{N}$, for $i = 1, \ldots, n$. We call such a capacitated network a *chain of parallel arcs*. An integral one-flow is given by a set of s-t-paths that are pairwise distinct. Notice that two s-t-paths in the considered network are different if and only if there is a pair of arcs where one path uses the upper arc and the other path uses the lower arc. In particular, the ith pair of arcs can distinguish a subset of at most c_i paths from all other paths. This motivates the following problem.

Bounded Cut Cover Problem

GIVEN: k numbers $c_1, \ldots, c_k \in \mathbb{N}$ and a number $q \in \mathbb{N}$.

TASK: Find k subsets $M_1, \ldots, M_k \subseteq \{1, \ldots, q\}$ with $|M_i| \leq c_i$, for $i = 1, \ldots, k$, such that for any pair $j, \ell \in \{1, \ldots, q\}$ with $j \neq \ell$ there is some $i \in \{1, \ldots, k\}$ with $|M_i \cap \{j, \ell\}| = 1$; or decide that no such family of subsets exists.

The name that we choose for this problem stems from the following graph-theoretic interpretation: Consider a complete undirected graph with vertex set $\{1, \ldots, q\}$. The question is whether the edges of the complete graph can be covered by k cuts where, for $i = 1, \ldots, k$, the ith cut partitions the vertex set into two subsets the smaller of which has cardinality at most c_i.

Lemma 1. *The bounded cut cover problem has a solution if and only if there exists an integral one-flow of value q in a chain of parallel arcs of length k where the capacities of the upper arcs are $c_1, \ldots, c_k$.*

Fractional Bounded Cut Cover Problem

GIVEN: k numbers $c_1, \ldots, c_k \in \mathbb{N}$ and a number $q \in \mathbb{R}^+$.

TASK: For some $r \geq q$, find weights $x_1, \ldots, x_r \in [0, 1]$ with $\sum_{j=1}^{r} x_j = q$ and determine k subsets $M_1, \ldots, M_k \subseteq \{1, \ldots, r\}$ with $\sum_{j \in M_i} x_j \leq c_i$, for $i = 1, \ldots, k$, such that for any pair $j, \ell \in \{1, \ldots, r\}$ with $j \neq \ell$ there is some $i \in \{1, \ldots, k\}$ with $|M_i \cap \{j, \ell\}| = 1$; or decide that this is not possible.

There is again a graph-theoretic interpretation of the problem. The task is to find a complete graph with weights on the vertices such that the weight x_i of every vertex i is between 0 and 1 and the weights sum up to q. Moreover, the edges of the complete graph must be covered by k cuts such that the ith cut partitions the vertex set into two subsets the lighter of which has total weight at most c_i. Associating the weighted nodes of the complete graph with s-t-paths of corresponding flow value yields the following observation.

Lemma 2. *The fractional bounded cut cover problem has a solution if and only if there exists a (fractional) one-flow of value q in a chain of parallel arcs of length k where the capacities of the upper arcs are $c_1, \ldots, c_k$.*

It is natural to ask whether the fractional bounded cut cover problem allows for larger values of q with feasible solutions than the non-fractional version. By Lemmas 1 and 2, this is equivalent to the question whether for chains of parallel arcs there always exists a maximum one-flow that is integral. In the fractional bounded cut cover problem the price of the additional degree of freedom given by the possibility to assign fractional weights to the nodes is an increase in the number of nodes (since the node weights still have to sum up to q). On the one hand, a larger number of nodes makes it more difficult to cover all edges of the complete graph. On the other hand, fractional weights on the vertices allow for more balanced cuts that contain more edges. We show below that there exist instances with a larger feasible value of q in the fractional version of the problem than in the integral version. Before we discuss this issue in more detail, we present another equivalent packing problem.

In a chain of parallel arcs, every s-t-path is uniquely determined by the subset of upper arcs contained in the path. Therefore, computing an integral one-flow of value q corresponds to finding a family of q pairwise distinct subsets of $\{1, \ldots, k\}$ such that $i \in \{1, \ldots, k\}$ is contained in at most c_i of these subsets.

Capacitated Set Packing Problem
GIVEN: k numbers $c_1, \ldots, c_k \in \mathbb{N}$ and a number $q \in \mathbb{N}$.
TASK: Find q pairwise distinct subsets of $\{1, \ldots, k\}$ such that element $i \in \{1, \ldots, k\}$ is contained in at most c_i of these subsets, for $i = 1, \ldots, k$; or decide that no such family of subsets exists.

Fractional Capacitated Set Packing Problem
GIVEN: k numbers $c_1, \ldots, c_k \in \mathbb{N}$ and a number $q \in \mathbb{R}^+$.
TASK: For some $r \geq q$, find pairwise distinct subsets $N_1, \ldots, N_r$ of $\{1, \ldots, k\}$ with weights $x_1, \ldots, x_r \in [0, 1]$ such that $\sum_{j=1}^{r} x_j = q$ and $\sum_{j:i \in N_j} x_j \leq c_i$, for $i = 1, \ldots, k$; or decide that this is not possible.

Lemma 3. *The (fractional) capacitated set packing problem has a solution if and only if there exists an integral (fractional) one-flow of value q in a chain of parallel arcs of length k where the capacities of the upper arcs are $c_1, \ldots, c_k$. In particular, the (fractional) capacitated set packing problem is equivalent to the (fractional) bounded cut cover problem.*

The following instance shows that the fractional capacitated set packing problem in general allows for strictly larger values of q with feasible solutions than the non-fractional version. Due to Lemma 3, the same holds for the bounded cut cover problem. Let $k = 3$ and $c_1 = c_2 = c_3 = 2$. It is not difficult to check that $q = 5$ is the largest value of q with a feasible solution to the non-fractional version of the problem: Choose for example the subsets $\emptyset$, $\{1\}$, $\{2\}$, $\{3\}$, and $\{1, 2, 3\}$. But there is a solution to the fractional version of the problem with $q = 5.5$: Choose subsets $\emptyset$, $\{1\}$, $\{2\}$, and $\{3\}$ all with weight 1. In addition choose subsets $\{1, 2\}$, $\{1, 3\}$, and $\{2, 3\}$ all with weight $1/2$.

We can also show that in general there is no integral optimal solution to the dual problem of the max-1FP. Consider the chain of parallel arcs corresponding to the instance of the capacitated set packing problem introduced above. An optimal dual solution destroys the path using all lower arcs and half of each path that uses exactly one upper arc; further, one half of each upper arc is deleted. The following result underlines the discrepancy between the fractional and the integral one-flow problem even more.

Proposition 1. *The existence of a maximum one-flow where the flow value on each arc is integral does in general not imply the existence of an integral path decomposition where each path carries at most one unit of flow.*

In contrast to the classical set packing problem and many similar problems known from the literature, the capacitated set packing problem allows to choose arbitrary subsets that do not have to belong to a given family of subsets. This might make the problem seem to be easier. However, we can prove the following somewhat surprising theorem.

Theorem 1. *The capacitated set packing problem is strongly NP-hard. If q is polynomially bounded in k, the problem is strongly NP-complete.*

One can reduce the strongly NP-hard 3-PARTITION problem to the capacitated set packing problem. Let A with $|A| = 3\ell =: k$ be the ground set of a given 3-PARTITION instance. The weights of elements in A sum up to ℓB for some $B \in \mathbb{N}$. The main idea of the reduction is to identify A with the set $\{1, \ldots, k\}$ and define $c_1, \ldots, c_k$ and q such that all subsets of A with total weight less than B and the subsets of a 3-Partition of A must be chosen to end up with the desired q subsets of A. The following is an immediate implication of this reduction.

Corollary 1. *For an integral one-flow given in arc variables, it is NP-hard to compute an integral path decomposition.*

A reduction to 3-SAT proves that the situation is even worse: The integral max-1FP is APX-hard, even in networks where the number of s-t-paths is polynomially bounded in the size of the network. As an immediate consequence of Theorem 1 we can state the following hardness results.

Theorem 2. *(i) The bounded cut cover problem is strongly NP-hard. If q is polynomially bounded in k, then the problem is strongly NP-complete. (ii) The problem of finding an integral one-flow of maximum value for a chain of parallel arcs is strongly NP-hard, even if the maximum flow value is polynomially bounded in the size of the network (i.e., number of vertices).*

It follows that, in contrast to the problem to count the number of s-t-paths in a digraph, the integral max-1FP is already strongly NP-hard in acyclic networks. Further, the strong NP-hardness of the capacitated set packing problem immediately implies that it is even strongly NP-hard to compute only the value of a maximum integral one-flow in those networks. Note that in the integral 1FP the flow value bounds the number of paths that are used to route a flow. Thus, we can derive the following from Theorem 2 (ii).

Corollary 2. *Unless P = NP, there is no algorithm for the integral max-1FP on chains of parallel arcs whose runtime is pseudo-polynomial in input plus output size.*

Theorem 3. *It is NP-hard to decide whether a given (integral) s-t-flow can be decomposed into integral flows along paths and cycles such that no path carries more than one unit of flow.*

3   An FPTAS for the Fractional Max-1Fp

Theorem 4. *For any $\epsilon > 0$ and any instance of the max-1FP with maximum flow value F^*, it is possible to compute a maximum one-flow of value $(1-\epsilon)F^*$ in time polynomial in the input size, ϵ^{-1}, and F^*.*

First we show how to compute a one-flow of a given value F that does not violate arc capacities by more than a factor $(1 + \epsilon)$, for some $\epsilon > 0$, or decide that no valid flow of value F exists. A flow violating all arc capacities by at most a factor $(1 + \epsilon)$ is called $(1 + \epsilon)$-*approximate*. Plotkin, Shmoys, and Tardos [8] developed an appropriate algorithm for the general fractional set packing problem. In that problem sets of capacitated elements are given and the task is to search for a packing, i.e., a selection of sets such that each element is contained in at most as many sets as its capacity permits. To compute $(1 + \epsilon)$-approximate packings of a given size, Plotkin et al. use a Lagrange relaxation that penalizes the violation of the capacity constraints in the objective function. Iteratively, they choose reasonable Lagrange multipliers, compute a solution to the relaxed problem, and combine this with the current solution. This algorithm runs in time polynomial in the input size and ϵ^{-1}. It can be adapted to the fractional 1FP. For given Lagrange multipliers $\lambda : A \to \mathbb{R}^+$ and $\lambda_P := \sum_{a\in P} \lambda(a)$ for all s-t-paths $P \in \mathcal{P}$, the Lagrange relaxation is $\min\{\sum_{P\in\mathcal{P}} \lambda_P x_P \mid \sum_{P\in\mathcal{P}} x_P \geq F, 0 \leq x_P \leq 1 \forall P \in \mathcal{P}\}$. It can be solved in time polynomial in the input size and $\lceil F \rceil$ by computing the $\lceil F \rceil$ shortest paths according to the length function λ (see, e.g., [6]). The $\lfloor F \rfloor$ shortest paths must carry one unit of flow ($x_P = 1$) and the $(\lfloor F \rfloor + 1)$-shortest path gets a flow value of $F - \lfloor F \rfloor$. To obtain a one-flow that obeys all arc capacities and approximates the maximum flow value F^* within a factor $(1 - \epsilon)$, we embed the algorithm by Plotkin et al. in a binary search. That binary search can be implemented to run in time polynomial in the input size, ϵ^{-1}, and F^*.

4   Approximating the Integral Max-1FP

We assume that all arc capacities are integral and start with the observation that, for the integral max-1FP, the additive integrality gap is at most m. This result follows from basic linear programming theory. With the FPTAS from the previous section we obtain Corollary 3 and Theorem 5 as immediate consequences.

Proposition 2. *The difference of the value F_F^* of a maximum fractional one-flow and the value F_I^* of a maximum integral one-flow is less than m.*

Corollary 3. *For any $\epsilon > 0$, an integral one-flow of value at least $(1 - \epsilon)F_F^* - m$ can be computed in time polynomial in the input size and F_I^*.*

Theorem 5. *There exists a constant factor approximation algorithm for the integral max-1FP whose runtime is polynomial in input plus output size if we restrict to instances whose maximum fractional flow value is larger than some constant $c > 1$ times its number of arcs.*

Subsequently we develop a randomized $O(\log m)$-approximation algorithm for the integral max-1FP that works for arbitrary instances. We start with a simple observation.

Lemma 4. *Applying Raghavan and Thompson's [9] randomized rounding method to a fractional solution computed by the FPTAS from Section 3, we obtain a constant factor approximation algorithm for the integral max-1FP if the minimum arc capacity is at least $\Omega(\log m)$.*

In order to obtain a randomized $O(\log m)$-approximation algorithm for arbitrary instances of the integral max-1FP, we compute an approximate maximum integral one-flow in a modified network by giving a special treatment to arcs whose capacity is less than $\log m$. We call such arcs *thin*, whereas an arc with capacity at least $\log m$ is called *thick*. A path is called *thick* if all its arcs are thick; otherwise it is called *thin*.

For a given instance of the max-1FP, we compute an approximate solution to the fractional problem using the FPTAS from Section 3 for some constant $\epsilon > 0$. If the total flow value along thick paths is at least half of the total flow value (and thus at least a constant fraction of the maximum integral flow value), we can use randomized rounding as explained above in order to obtain a constant factor approximation. Otherwise we can use the algorithm described in the following which computes an $O(\log m)$-approximation from the flow that is routed along thin paths.

The algorithm works as follows. First we delete the flow routed along thick paths. From now on we consider only the part of the underlying graph which is used by thin paths. For each thin arc (v, w) insert a new node $\tilde{v}$, delete the arc (v, w), insert the arcs $(v, \tilde{v})$ and $(\tilde{v}, w)$ and assign the capacity of (v, w) to them. (The flow is adjusted adequately using arcs $(v, \tilde{v})$ and $(\tilde{v}, w)$ instead of (v, w).) The resulting network is denoted by $D = (V, A)$, the set of newly inserted nodes by U. Next, we make a copy $D' = (V', A')$ of D. From each node in U we insert an arc to its copy in D'. The resulting graph is denoted by $\bar{D} = (\bar{V}, \bar{A})$. We define v' to be the clone of $v \in V$ in V' and a' to be the clone of $a \in A$ in A'. An arc connecting a node $u \in U$ with u' is denoted by a_u. For $u \in U$ that was inserted to divide an arc a of the original digraph, the capacity of a_u is the same as that of a. We modify the considered fractional flow by rerouting all its paths from D to D' along the last thin arc at which a rerouting is possible. More precisely, this works as follows. Consider any path P that is used in the original fractional flow and let $(v, w) \in P$ be the last thin arc on P. (This arc does not exist in D.) Then the analogon to P in $\bar{D}$ uses the adjusted path P in D until it reaches v, then uses $(v, \tilde{v})$ and is rerouted to D' along $a_{\tilde{v}}$. In D' the new path uses the arc from $\tilde{v}'$ to w' and then the arcs corresponding to the ones P used in D after (v, w). Let the resulting s-t'-flow be denoted by $\bar{x}$. Note that the value $|\bar{x}|$ of $\bar{x}$ is still only some constant factor smaller than the value of a maximum one-flow in the original network. We choose integral capacities $u(a)$ and $u(a')$ for a thick arc $a \in A$ and its clone a' as follows. If $\lceil \bar{x}(a) \rceil + \lceil \bar{x}(a') \rceil$ is not larger than the original capacity of a, we set $u(a) = \lceil \bar{x}(a) \rceil$ and $u(a') = \lceil \bar{x}(a') \rceil$. Otherwise, we choose the capacities by rounding the smaller flow value up and the larger one down. (The sum of the resulting values is not larger than the original capacity of a, because this was assumed to be integral.) For all thin arcs $a \in \bar{A}$, the capacity $u(a)$ is set to 1. It is easy to prove that $u(a) > \bar{x}(a) / \log m$, for all $a \in \bar{A}$. It follows immediately that a (usual) maximum s-t'-flow in $\bar{D}$ with capacities u has flow value at least $|\bar{x}| / \log m$, because $\bar{x} / \log m$ is a feasible s-t'-flow in that network. By network flow theory, an integral maximum s-t'-flow can be computed in polynomial time. Since the value of this flow is at least $|\bar{x}| / \log m$, it is only by a factor $O(\log m)$ smaller than the value of a maximum one-flow in the original network. If we reroute this flow to D, i.e., do not let it pass over from D to D' and let it use the corresponding arcs in D instead, it is still feasible, because the sum of arc capacities $u(a)$ and $u(a')$ is at most the original capacity of $a \in A$. Further, the resulting flow does not send more than one unit of flow along each path, because each path uses at least one thin arc (otherwise it would not have been able to get from D to D') whose capacity is now 1. This gives us the result in Theorem 6. It also follows immediately from our analysis that the (multiplicative) integrality gap of the max-1FP is $O(\log m)$.

Theorem 6. *There exists a randomized $O(\log m)$-approximation algorithm for the integral max-1FP whose runtime is polynomial in input plus output size.*

5   Multicommodity One-Flows

In this section we consider the multicommodity version of the 1FP in that we have several source-sink-pairs. We still have a digraph $D = (V, A)$ with arc capacities $u : A \to \mathbb{R}^+$. Instead of a single source-sink-pair we now have *requests* $(s_i, t_i) \in V \times V$ for $i = 1, \ldots, K$, where $K \in \mathbb{N}$ denotes the total number of such pairs. Different optimization problems for multicommodity flows have been considered in the literature. Among them are maximization of the total flow sent through a network and minimization of the congestion of a flow that satisfies a given demand for each request. (The congestion measures the relative overload on an arc—a detailed definition follows later.) We consider both such optimization problems in the context of one-flows. We use $\mathcal{P}_i$ to denote the set of s_i-t_i-paths in D, for all $i = 1, \ldots, K$, and $\mathcal{P} := \bigcup_{i=1}^{K} \mathcal{P}_i$.

Maximum Multicommodity One-Flows. With the previous definitions we can describe the *maximum multicommodity one-flow problem (max-mc-1FP)* by the linear program in Section 1. Since solving the min-cost LP from Section 3 can easily be adjusted to the new situation, we can use the given FPTAS for the max-mc-1FP without any changes.

Theorem 7. *For any $\epsilon > 0$ and any instance of the max-mc-1FP with maximum flow value F^*, it is possible to compute a maximum multicommodity one-flow of value $(1 - \epsilon)F^*$ in time polynomial in the input size, ϵ^{-1}, and F^*.*

Proposition 2 still holds in the multicommodity case giving us a result similar to the one stated in Theorem 5. A result by Guruswami et al. [3] shows that, in general, it is NP-hard to approximate the integral max-mc-1FP within a factor $O(m^{1/2-\epsilon})$, for any $\epsilon > 0$. However, randomized rounding yields an $O(\sqrt{m})$-approximation algorithm for this problem. This can be proven using a more detailed analysis of the randomized rounding method given in [1].

Theorem 8. *There exists a randomized $O(\sqrt{m})$-approximation algorithm for the integral max-mc-1FP whose runtime is polynomial in input plus output size.*

Minimizing Congestion. Here, each request (s_i, t_i) $(i = 1, \ldots, K)$ has a corresponding positive demand d_i which has to be satisfied by a one-flow. We look for a solution of minimum congestion. The *congestion* of a flow $(x_P)_{P \in \mathcal{P}}$ is the minimum μ such that $\sum_{P \ni a} x_P \leq \mu u(a)$, for all $a \in A$. The FPTAS from Section 3 can be adapted with the following result.

Theorem 9. *For any $\epsilon > 0$ and any instance of the min-cong-1FP with minimum congestion μ^*, it is possible to compute a multicommodity one-flow of congestion at most $(1 + \epsilon)\mu^*$ in time polynomial in the input size, ϵ^{-1}, and $d_{\max} := \max_i d_i$.*

Using Raghavan and Thompson's [9] randomized rounding method for the integral min-cong-1FP we obtain an $O(\log m)$-approximation algorithm. Since the congestion version of the considered problem does not give any strict restrictions on the arc capacities, the algorithm can be derandomized by the method of conditional probabilities.

Theorem 10. *Applying randomized rounding to a nearly optimal fractional one-flow yields an $O(\log m)$-approximation to the integral min-cong-1FP.*

References

1. Baveja, A., Srinivasan, A.: Approximation algorithms for disjoint paths and related routing and packing problems. Mathematics of Operations Research 25, 255–280 (2000)
2. Ford, L.R., Fulkerson, D.R.: Maximal flow through a network. Canadian J. of Math. 8, 399–404 (1956)
3. Guruswami, V., Khanna, S., Rajaraman, R., Shepherd, B., Yannakakis, M.: Near-optimal hardness results and approximation algorithms for edge-disjoint paths and related problems. In: Proc. 31st Annual ACM Symposium on Theory of Computing, pp. 19–28 (1999)
4. Kleinberg, J.M.: Approximation Algorithms for Disjoint Path Problems. PhD thesis. MIT (May 1996)
5. Kolliopoulos, S.G.: Edge-disjoint paths and unsplittable flow. In: Gonzalez, T.F. (ed.) Handbook of Approximation Algorithms and Metaheuristics, ch. 57. Chapman-Hall/CRC Press (2007)
6. Lawler, E.L.: A procedure for computing the K best solutions to discrete optimization problems and its application to the shortest path problem. Management Science 18, 401–405 (1972)
7. Martens, M., Skutella, M.: Flows on few paths: Algorithms and lower bounds. Networks 48(2), 68–76 (2006)
8. Plotkin, S.A., Shmoys, D.B., Tardos, E.: Fast approximation algorithms for fractional packing and covering problems. Mathematics of Operations Research 20, 257–301 (1995)
9. Raghavan, P., Thompson, C.D.: Randomized rounding: A technique for provably good algorithms and algorithmic proofs. Combinatorica 7, 365–374 (1987)
10. Valiant, L.G.: The complexity of enumeration and reliability problems. SIAM J. on Computing 8(3), 410–421 (1979)

Strong Formulations for 2-Node-Connected Steiner Network Problems

Markus Chimani[1], Maria Kandyba[1,*], Ivana Ljubić[2,**], and Petra Mutzel[1]

[1] Faculty of Computer Science, Dortmund University of Technology, Germany
{markus.chimani,maria.kandyba,petra.mutzel}@cs.uni-dortmund.de
[2] Faculty of Business, Economics and Statistics, University of Vienna, Austria
ivana.ljubic@univie.ac.at

Abstract. We consider a survivable network design problem known as the *2-Node-Connected Steiner Network Problem* (2NCON): we are given a weighted undirected graph with a node partition into two sets of customer nodes and one set of Steiner nodes. We ask for the minimum weight connected subgraph containing all customer nodes, in which the nodes of the second customer set are nodewise 2-connected. This problem class has received lively attention in the past, especially with regard to exact ILP formulations and their polyhedral properties.

In this paper, we present a transformation of 2NCON into a related problem on directed graphs and use this to establish two novel ILP formulations, based on multi-commodity flow and on directed cuts, respectively. We prove the strength of our formulations over the known formulations, and compare our ILPs theoretically and experimentally. This paper thereby consitutes the first experimental study of exact 2NCON algorithms considering more than ∼100 nodes, and shows that graphs with up to 4900 nodes can be solved to provable optimality.

1 Introduction

Various survivable network design problems occur prominently in real-world fiber-optic networks and telecommunication applications, see, e.g., [13,24] for general surveys. We concentrate on the following NP-hard problem class: Given an undirected graph $G = (V, E)$, a cost function $c : E \rightarrow \mathbb{R}^+$ and a vector of connectivity requirements $\varrho \in \{0, 1, \ldots, k\}^{|V|}$ for some constant $k > 0$. A solution of the *k-Node-Connected Steiner Network* problem $(k\text{NCON})$[1] [21] is a subgraph $N = (V_N, E_N)$ of G which contains all nodes $v \in V$ with $\varrho_v > 0$, minimizes $\sum_{e \in E_N} c_e$ and satisfies the following connectivity property: for every

---

* Supported by the German Research Foundation (DFG) through the Collaborative Research Center "Computational Intelligence" (SFB 531).

** Supported by the Hertha-Firnberg Fellowship of the Austrian Science Foundation (FWF).

[1] In the literature there are various names for this problem, with sometimes slightly differing definitions. kNCON is also known as $\{0,\ldots,k\}$-(N)SND (Node Survivable Network Design) and Generalized Steiner Network problem.

B. Yang, D.-Z. Du, and C.A. Wang (Eds.): COCOA 2008, LNCS 5165, pp. 190–200, 2008.
© Springer-Verlag Berlin Heidelberg 2008

pair of nodes $s, t \in V_N$, N contains $\varrho_{st} := \min\{\varrho_s, \varrho_t\}$ node-disjoint paths connecting them. We can relax the problem by replacing the node-disjointness with edge-disjointness, and obtain the *k-Edge-Connected Steiner Network Problem* (*k*ECON). For simplicity, we define $\mathcal{R}_i := \{v \in V \mid \varrho_v = i\}$ for all $0 \leq i \leq k$, and call the set $\mathcal{R} := \bigcup_{i>0} \mathcal{R}_i$ the *customer nodes*. We can assume that $|\mathcal{R}_2| \geq 2$, since otherwise we obtain the traditional Steiner tree problem.

A lot of research has been conducted on this problem, both in the fields of effective heuristics and approximation algorithms, see [13] for an overview. However, these are beyond the scope of this paper, as we will concentrate on the exact ILP formulations. This is based on the fact that recent advances in computational power and ILP solvers, when used in conjunction with strong models, allow to solve real-world instances for other network problems to provable optimality within reasonable time bounds; see, e.g., [1,4,18]. Furthermore, ILP formulations also often form the basis of approximation schemata. For *k*ECON and *k*NCON, Grötschel, Monma and Stoer [9] described cut-based integer linear programs (ILP). Apart from such formulations, the problem can also be formulated in terms of multi-commodity flow, as done by Raghavan [19].

Regarding 2ECON, it has been shown that for both concepts formulations based on *oriented* graphs are stronger than undirected formulations: an *orientation* of an undirected graph G' is a directed graph $\hat{G}'$, which is obtained by transforming each edge of G' into a directed arc. Robbins [20] showed that a graph G' is 2-edge-connected if and only if there exists an orientation $\hat{G}'$ with directed paths $(v \rightarrow w)$ and $(w \rightarrow v)$ for every pair of nodes v, w. This fact has been exploited by Chopra [6] for solving 2ECON via directed graphs, who proved his formulation to be polytope-wise superior to the undirected formulation mentioned above. Goemans [8] and Stoer [21, pp. 31–32] extended this formulation to *k*ECON for the case that all connectivity requirements are 0, 1 or even; later Magnanti and Raghavan [17] extended it for general k.

Yet, it was not possible to extend such orientation techniques to *k*NCON, since Robbins' theorem is not strong enough to exploit it for node-connectivity constraints. It has been an open problem [19, p. 183],[21, pp. 32,134] whether a similar orientation technique can at all be used for *k*NCON-type problems. As a first step, Chimani, Kandyba, and Mutzel [5] showed recently that this is indeed possible for the *2-Root-Connected Steiner Network Problem* (2RSN) and its prize-collecting variant (2RPCSN), where the nodewise 2-connectedness is required only with a special root node $r \in V$; i.e., each node $v \in \mathcal{R}$ has to have ϱ_v node-disjoint paths with r. A certain orientability property for the feasible networks of this problem was shown, and was used to obtain a novel ILP formulation based on directed cuts for this problem.

In this paper, we present how the orientability for the more prominent 2NCON can be established (Section 2) and derive two new ILP formulations based on different concepts: one based on multi-commodity flow (Section 3) and one based on directed cuts, only requiring easily separable constraints (Section 4). We show the theoretical advantages of both new formulations compared to the previously known ILPs. Furthermore, we prove that our approaches are equivalent from the polyhedral point of view (Section 5). Nonetheless, the cut formulation is much stronger in practice, as we show in an experimental study (Section 6). This study is the first time that instances with more than ~100 nodes are considered and solved to provable optimality.

2   Directed 2NCON

The main problem with node-disjointness is that Robbins' theorem can only be exploited for 2ECON. Furthermore, it is in general not possible to orient an undirected 2-connected graph such that for every pair of nodes $v, w \in \mathcal{R}_2$, there are two node-disjoint directed paths, one from v to w and one from w to v. Up until now, this was the main hindrance why there were no orientation-based formulations for 2NCON.

We require the following theorem, shown in [5], as a foundation to show the validity of the formulations below. We rephrase it such that it is more useful in the following:

Theorem 1 ([5]). *Let $G' = (V', E')$ be a graph with some root node $r \in V'$ and the property that each non-trivial (i.e., larger than a single edge) 2-connected component contains r. Then there exists an orientation $\hat{G}'$ such that:*

- *For each node $v \in V' \setminus \{r\}$ which does not share a common non-trivial 2-connected component (block) with r, $\hat{G}'$ contains a directed path $(r \to v)$.*
- *For each node $v \in V' \setminus \{r\}$ which shares a common non-trivial block with r, $\hat{G}'$ contains a directed path $(r \to v)$ and a directed path $(v \to r)$, which are node-disjoint except for r and v.*

The proof of this theorem allows an insight which will be of particular use for our problem:

Lemma 1. *Let the graph G' be defined as in Theorem 1, and let there be only a single non-trivial block. There exists an orientation with the properties of Theorem 1 where the root has in-degree 1.*

Proof. The idea of the proof for Theorem 1 works as follows: we first direct all the edges of the trivial blocks from the node nearer to r to the node farther away. We then consider the non-trivial blocks separately and orient them as follows: We start with a directed cycle through r and label the contained nodes in an increasing fashion. Thereby, the root r obtains the smallest label. Hence there is only a single edge $\hat{e}$ which is directed from a larger label number to the smaller number, i.e., the root. Then the proof proceeds by incrementally choosing a non-oriented path between two labeled nodes: we orient these paths and label their nodes uniquely in such a way that the property of increasing labels remains true. Hence we require only a single edge $\hat{e}$ per block to be directed towards r. If G' only contains a single non-trivial block, we get the lemma's claim.   □

Definition 1. *Let (G', U) be a tuple of an undirected graph $G' = (V', E')$ and $U \subseteq V'$. G' is $\{1, 2\}$-node-connected, if it is connected and all nodes U lie in the same biconnected component.*

Let N be any solution of the *2NCON* problem. We can observe that $(N, \mathcal{R}_2)$ is $\{1, 2\}$-node-connected. This leads to our generalization of Robbins' theorem in the context of 2-node-connectedness:

Theorem 2. *Let (G', U) be a tuple of an undirected graph $G' = (V', E')$ and $U \subseteq V'$. G' is $\{1, 2\}$-connected if and only if for any arbitrarily chosen $r \in U$ there exists an orientation of G' with* in-deg$(r) = 1$ *such that there is a directed path $(r \rightarrow v)$ for each $v \in V'$; if $v \in U$, there furthermore is a directed path $(v \rightarrow r)$ which is node-disjoint from $(r \rightarrow v)$ except for r and v.*

Hence we can reformulate 2NCON as the directed problem *D2NCON*: Let $\bar{G} = (V, A)$ be the bidirected graph obtained from G by replacing every undirected edge $\{u, v\} \in E$ by two directed arcs $(u, v), (v, u) \in A$ with costs $c_{uv} = c_{vu} = c_{\{u,v\}}$. We seek a weight-minimal oriented subgraph $\hat{N} = (V_N, A_N)$ in $\bar{G}$ with $\mathcal{R} \subseteq V_N$ which satisfies Theorem 2 w.r.t. $(\hat{N}, \mathcal{R}_2)^2$.

Corollary 1. *Given an undirected graph $G = (V, E)$, a cost function $c : E \rightarrow \mathbb{R}^+$, and a vector of connectivity requirements $\varrho \in \{0, 1, 2\}^{|V|}$. Choose $r \in V$ with $\varrho_r = 2$ arbitrarily. For this input, any solution of D2NCON can be transformed into an equivalent solution of 2NCON with the same objective value, and vice versa.*

In the following sections, the problem input will always be defined as above. Let $\mathcal{R}'_i := \mathcal{R}_i \setminus \{r\}$, for $0 \leq i \leq 2$, and $\mathcal{R}' := \mathcal{R} \setminus \{r\}$.

3 Multi-commodity Flow for D2NCON

We start with presenting a novel ILP formulation (DFLOW) based on multi-commodity flow for D2NCON, and therefore for 2NCON. As there has been much research on flow-based formulations for the latter problem, we compare our formulation to the currently strongest one, by Raghavan [19, pp. 180–181][3], and show that our formulation is beneficial.

The idea is to consider $\bar{G}$ and send exactly one unit of flow from the root to each $\mathcal{R}'$ node. Furthermore, we send one unit of flow from each $\mathcal{R}'_2$ customer back to the root. Thereby it has to be ensured that the pairs of forward- and backward-flows do not use common nodes and edges except for v and r. We define the set of commodities $\mathcal{C} = \{(r, v) \mid v \in \mathcal{R}'\} \cup \{(v, r) \mid v \in \mathcal{R}'_2\}$; a flow of commodity $\chi \in \mathcal{C}$ on the arc $(i, j) \in A$ is modeled by the variable f_{ij}^χ. Finally,

---

[2] *Remark.* One may try to model node-connectivity by only computing edge-connectivity in a modified underlying graph, by replacing each node by a directed arc. This is not valid in our case as the orientability theorems require bidirectedness of the underlying graphs.

[3] Various aspects of the considered problems were studied in papers by Stoer and Raghavan (with coauthors) [9,10,11,17]. For simplicity and common notations we reference their theses [19,21], including page numbers when suitable.

we introduce the variables x_{ij} which are 1, if the solution network contains the arc $(i, j) \in A$.

$$\text{DFLOW}: \quad \min \sum_{(i,j) \in A} c_{ij} \cdot x_{ij} \tag{1}$$

$$\sum_{(i,v) \in A} f_{iv}^{\chi} - \sum_{(v,i) \in A} f_{vi}^{\chi} = \begin{cases} -1, \text{ if } v = s \\ 1, \text{ if } v = t \\ 0, \text{ else} \end{cases} \quad \forall \chi = (s,t) \in \mathcal{C}, \forall v \in V \tag{2}$$

$$\sum_{(i,w) \in A} \left(f_{iw}^{(v,r)} + f_{iw}^{(r,v)} \right) \leq 1 \qquad \forall v \in \mathcal{R}_2', \forall w \in V \setminus \{r, v\} \tag{3}$$

$$0 \leq f_{ij}^{\chi} \leq x_{ij} \qquad \forall (i,j) \in A, \forall \chi \in \mathcal{C} \tag{4}$$

$$x_{vw} + x_{wv} \leq 1 \qquad \forall \{v, w\} \in E \tag{5}$$

$$\sum_{(i,r) \in A} x_{ir} = 1 \tag{6}$$

$$x_{vw} \in \{0, 1\} \qquad \forall (v, w) \in A \tag{7}$$

Clearly, the flow- and capacity constraints (2) and (4) guarantee the directed paths from r to all customer nodes, and for each $\mathcal{R}_2'$ customer we also have a directed path backwards to r. Due to Robbins' theorem [20] this—together with (5)—ensures that every $\mathcal{R}_2'$ customer belongs to the same edge-biconnected component as r. Theorem 2, (3) and (6) guarantee that this component is 2-node-connected. Hence we obtain:

Theorem 3. *An optimal solution for* DFLOW *gives an optimal solution for the corresponding 2NCON problem.*

Raghavan presented the currently strongest formulation for the 2NCON by computing two multi-commodity flows g and h simultaneously [19, p. 180–181]: g represents directed flow for the induced 2ECON problem, h represents an non-oriented flow with node-disjointness constraints.[4] The two flows are bound to each other only by their common use of the z_e variables, for $e \in E$, which define whether the given undirected edge e is contained in the solution network or not. We denote Raghavan's formulation as MFLOW (mixed flow) and show that our new formulation is superior.

Let $\mathcal{P}_{DF}$ and $\mathcal{P}_{MF}$ be the polyhedra of the feasible solutions of the LP relaxations for DFLOW and MFLOW, respectively. We then consider their projections into the space of z variables, i.e., $\text{proj}_z(\mathcal{P}_{DF}) = \{z \in [0,1]^{|E|} \mid (x,f) \in \mathcal{P}_{DF}, z_{ij} = x_{ij} + x_{ji} \ \forall \{i,j\} \in E\}$ and $\text{proj}_z(\mathcal{P}_{MF}) = \{z \in [0,1]^{|E|} \mid (z,g,h) \in \mathcal{P}_{MF}\}$. We also consider extended projections including the flow variables $f \in [0,1]^{|A| \cdot |\mathcal{C}|}$, i.e., variables not in the objective function. Let $\text{proj}_{z,f}(\mathcal{P}_{DF}) = \{(z,f) \mid (x,f) \in \mathcal{P}_{DF}, z_e = x_{ij} + x_{ji} \ \forall e = \{i,j\} \in E\}$ be the projection of $\mathcal{P}_{DF}$ into the variable space of z and retaining the flow f. Let $\text{proj}_{z,f}(\mathcal{P}_{MF}) = \{(z,f) \mid (z,g,h) \in \mathcal{P}_{MF}, f = g\}$ be the projection of $\mathcal{P}_{MF}$ ignoring the h flow. In other words, we identify the flows f and g.

---

[4] Note that this formulation has been developed for general k, where it is called *improved undirected flow formulation with node-disjointness constraints.*

We show that the lower bounds obtained by the LP relaxations of our new formulation are at least as tight as those of the mixed flow formulation. Therefore note that the flow f is a kind of natural fusion of the flow g and the node-disjointness properties of h. Please refer to [3] for the proof.

Theorem 4. DFLOW *is at least as strong as* MFLOW, *i.e.,* $\text{proj}_z(\mathcal{P}_{DF}) \subseteq \text{proj}_z(\mathcal{P}_{MF})$. *Furthermore we even have* $\text{proj}_{z,f}(\mathcal{P}_{DF}) \subset \text{proj}_{z,f}(\mathcal{P}_{MF})$.

DFLOW requires less variables and constraints than MFLOW, hence:

Observation 1. DFLOW is more compact than MFLOW.

Our formulation answers the question by Raghavan [19, p. 183] whether his flow variables g, h can be bounded together more tightly. Note that Theorem 2 is crucial for the validity of our approach, which explains why this compact formulation could not be used legitimately before.

4   Directed-Cut for D2NCON

We now present an ILP formulation (DCUT) based on directed cuts. Its number of variables is independent of $\mathcal{R}$ as it only requires variables x_e for all $e \in A$. On the other hand, it has an exponential number of constraints. We will see that these are of traditional cut type and therefore easily and polynomially separable within a Branch-and-Cut approach. A main motivation for DCUT is that cut formulations often outperform flow formulations in practice, as e.g. in [5,14].

Let $S \subset V$, then $\delta_G^+(S) := \{(s,t) \in A \mid s \in S, t \in V \setminus S\}$ and $\delta_G^-(S) := \{(s,t) \in A \mid s \in V \setminus S, t \in S\}$ denote the arcs leaving and entering S, respectively. If G is clear from the context, we will omit the subscript. Furthermore, we use the shorthands $G_w := G \setminus \{w\}$, for some $w \in V$, and $x(B) := \sum_{e \in B} x_e$ for some $B \subseteq A$.

$$\text{DCUT}: \quad \min \sum_{(i,j) \in A} c_{ij} \cdot x_{ij} \tag{8}$$

$$x_{vw} + x_{wv} \leq 1 \qquad \forall \{v,w\} \in E \tag{9}$$

$$x(\delta^-(S)) \geq 1 \qquad \forall S \subseteq V \setminus \{r\}, S \cap \mathcal{R}' \neq \emptyset \tag{10}$$

$$x(\delta^+(S)) \geq 1 \qquad \forall S \subseteq V \setminus \{r\}, S \cap \mathcal{R}'_2 \neq \emptyset \tag{11}$$

$$x(\delta_{\bar{G}_w}^-(S_1)) + x(\delta_{\bar{G}_w}^+(S_2)) \geq 1 \qquad \begin{array}{l} \forall w \in V \setminus \{r\}, \forall S_1, S_2 \subseteq V \setminus \{r,w\}, \\ S_1 \cap S_2 \cap \mathcal{R}'_2 \neq \emptyset \end{array} \tag{12}$$

$$\sum_{(i,r) \in A} x_{ir} = 1 \tag{13}$$

$$x_{vw} \in \{0,1\} \qquad \forall (v,w) \in A \tag{14}$$

As before, (9) guarantees the unique orientation of chosen edges. The constraints (10) and (11) ensure the existence of the required paths, and (12) assures the node-disjointness of these paths. Finally, (13) requires exactly one edge being directed towards the root, in order to guarantee a single non-trivial block in the solution network. Based on this description and Corollary 1 we obtain:

Theorem 5. *An optimal solution for* DCUT *gives an optimal solution for the corresponding 2NCON problem.*

We can compare this formulation with the currently best known cut-formulation presented in [21, p. 14], denoted by UCUT: it is based on undirected cuts in the original (undirected) graph, and uses variables z_e for all $e \in E$ which are set to 1 if the corresponding edge is selected into N. For each pair of $\mathcal{R}_i$ ($i = \{1, 2\}$) customers it requires all their cuts to be at least i. For all pairs of $\mathcal{R}_2$ it further requires all cuts to be at least 1, considering all graphs resulting from removing a single node, in order to ensure node-2-connectedness.

We can show that our (rooted, directed) DCUT formulation is stronger than the (unrooted, undirected) UCUT formulation. Let $\mathcal{P}_{DC}$ and $\mathcal{P}_{UC}$ be the polyhedra of the feasible solutions of the LP relaxations for DCUT and UCUT, respectively. We can use the natural projection $x_{vw} + x_{wv} = z_{e'}$ for all $e' = \{v, w\} \in E$ in order to obtain $\mathrm{proj}_z(\mathcal{P}_{DC})$. See [3] for the formal proof.

Theorem 6. DCUT *is strictly stronger than* UCUT, *i.e.,* $\mathrm{proj}_z(\mathcal{P}_{DC}) \subset \mathcal{P}_{UC}$.

5   Analysis of D2NCON Formulations

Let $\mathrm{proj}_x(\mathcal{P}_{DF})$ be the natural projection obtained from $\mathcal{P}_{DF}$ by only considering its x variables.

Theorem 7. DFLOW *and* DCUT *are equally strong, i.e.,* $\mathrm{proj}_x(\mathcal{P}_{DF}) = \mathcal{P}_{DC}$.

We omit the rather long proof of the theorem due to space limitation. It can straight-forwardly be shown that the projection of any fractional feasible DFLOW solution corresponds to a fractional feasible DCUT solution; anyhow, the reverse requires some deeper investigation of the solution structure, see [3] for details.

By dropping the constraints (12) and (13) from DCUT, we obtain the directed cut formulation for 2ECON. For this formulation, we know that it inherently includes the *partition inequalities* [6], the (polynomially separable) *Prodon inequalities* [21, pp. 130–134], and the *odd-hole inequalities* and *combinatorial-design inequalities* [19, pp. 165–180]. Hence we can conclude Proposition 1. In contrast to this, we can show Proposition 2, cf. [3].

Proposition 1. DCUT *and* DFLOW *inherently ensure the validity of the partition, Prodon, odd-hole, and combinatorial-design inequalities.*

Proposition 2. *None of the above formulations induces the undirected node-partition inequalities [21, pp. 91–94], i.e., undirected partition inequalities where one node is removed from the graph. This result constitutes a negative answer for the open question [19, p. 183] whether* MFLOW *would induce this constraint class.*

Algorithmical Remarks. From the point of formulation strength, using DFLOW instead of DCUT might seem like a reasonable choice in general, as both the number of variables and constraints are bound by a polynomial. But in practice, the latter has certain advantages: it requires much less variables, especially when $\mathcal{R}$ is large. Furthermore, its drawback of an exponential number of constraints can turn out to be beneficial, as the actual computation of an optimal solution will in general not require all of these constraints. Therefore, traditional Branch-and-Cut techniques can be expected to be highly efficient, since all constraints in DCut can be easily separated using simple polynomial max-flow algorithms, see, e.g., [25]. The latter property make it possible to solve not only the LP-relaxation of DFlow but also the (equivalent) LP-relaxation of DCut in polynomial time.

6   Experiments

The main purpose of our experimental studies is to obtain an unskewed comparison between our two formulations DCut and DFlow. Therefore, we implemented both formulations using CPLEX 9.0's Branch-and-Bound framework, without any preprocessing or primal heuristics. We turned off all automatic cut-generation etc., usually performed by CPLEX, to be able to compare our formulations in a pure manner. The additionally necessary separation routines for DCut are implemented in C++ using LEDA 5.1.1 and the efficient max-flow algorithm of [2]. All the tests were performed on an Intel Xeon 2.33Ghz CPU with 2GB of RAM per process, and a time limit of 2 hours per problem instance. As a basis for our experiments we took three different benchmark sets, which have, e.g., also been used and thoroughly described in [5]:

ClgS instances. This benchmark set contains 25 instances based on real-world map data of the city district Cologne-Ossendorf. The underlying graph has 190 nodes and 377 edges. The instances differ in the customer nodes, and have 3–6 $\mathcal{R}_1$, and 2–3 $\mathcal{R}_2$ customers. However, such small number of customers is quite unusual in practice. Therefore, we also generated modified instances $ClgS^+$, with 20% (10 % $\mathcal{R}_1$ and 10% $\mathcal{R}_2$) customer nodes.

Grid instances. We consider the artificial instances of [23] based on grid graphs with 100, 400, 900,..., 4900 nodes. For each graph size there are 2×15 instances, using two different cost functions, respectively. They have 5–13 $\mathcal{R}_1$ and 3–8 $\mathcal{R}_2$ customers. As for ClgS we also generated the additional set Grid$^+$, where 20% of all nodes are customers.

PCSTLib$^+$ instances. The PCSTLib benchmark [12], was used in several studies, e.g., [15,16], and contains graphs divided into two groups K and P, where 15%–27% and 34%–50% of the nodes are customers, respectively. The former are similar to street map layouts. We also consider the augmented $PCSTLib^+$ [5,23] benchmark, where roughly 1/3 of the customer nodes are selected to be in $\mathcal{R}_2$. In each group the underlying graphs have 100 and 400 nodes. Note that some of these instances are infeasible for 2NCON as the underlying graph does not allow two node-disjoint paths between certain customers. Hence we report only on the feasible instances.

Comparison of DFlow and DCut. Our experiments show that DCut outperforms DFlow in terms of the running time and computational power on all

Table 1. Average CPU time in seconds. (*) DFLOW solves 67% of the instances. None of the other tested instances could be solved by DFLOW within 2 hours.

	ClgS	ClgS+	G 100	G 400	G 900	G 1600*	G+ 100	K 100	P 100
DFlow	0.3	446.3	7.0	226.0	1505	(6209)	22.4	19.9	1500
DCut	0.1	0.6	0.2	2.0	22.9	74.2	0.3	0.7	0.5

Table 2. Computed via DCUT, times in seconds. **G** denotes the Grid instances. **(left)** Average CPU time for large grids (without branching) (*) 73.3% solved. **(right)** Quality of the LP-relaxation, times in seconds. (**) 50% solved (left column); the right column gives the statistics of the unsolved instances after 2h, using a heuristic upper bound to estimate the gap.

Grid	t_{ILP}
2500	334.3
3600	930.4
4900*	(3216)

	ClgS+	G+ 100	K 100	P 100	G+ 400	K 400**		P 400
gap	0.12%	0.16%	0.06%	0.18%	0.35%	0.12%	(0.53%)	0.37%
t_{LP}	0.4	0.2	0.6	1.1	21.0	32.3	22.5	23.4
t_{ILP}	0.6	0.3	0.7	0.5	170	38.1	—	260.4
req.br.	42.1%	16.7%	22.2%	100%	100%	33.3%	100%	100%
#BB	1.0	2.0	1.9	17.8	101.3	5.33	(641.3)	382.4

instance sets. See Table 1 for the overview of the corresponding running times. Thereby we only report on instances which could— at least in part—be solved to optimality by both DCUT and DFLOW.

We observe that all Grid and Grid+ instances with 100 nodes and all ClgS and ClgS+ instances are solved to optimality by DCUT in less than 2 seconds on average. The only instance set where both algorithms perform comparably well is ClgS, which is due to the fact that the underlying LPs of DFLOW are rather small for this small number of customers, and the overhead of DCUT's cut separation routines is comparably expensive. Note that neither approach requires any branching for ClgS and small Grid instances. Already a slight increase of the customer count is sufficient for DCUT to outperform DFLOW, see, e.g., the results for Grid instances of size 100. This effect is further amplified by the larger underlying graphs, as this results in an even larger increase of variables for DFLOW. While the cut approach is able to solve all Grid instances up to 3600 nodes to optimality within 2 hours, the largest instances which can be completely solved by DFLOW contains 900 nodes. For the PCSTLib+ instances, DCUT is on average 100 times faster for the K group and 3000 times faster for the P group.

LP-relaxations. A common measure to assess and compare ILP formulations is to look at the lower bounds resulting from their LP-relaxations, i.e., the solution at the root node of the branch-and-bound tree (LP_r). In our case, these values are identical as the corresponding polytopes are equivalent, cf. Theorem 7.

We observe that the LP-relaxation of our ILPs usually gives a strong lower bound. For many instances—i.p. all Grid and ClgS instances—the relaxation already gives an integer, and thus optimal, solution. In Table 2 we report on the quality and time (t_{LP}) of the solutions at the root node, i.e., the LP-relaxation, for the instance sets where DCUT requires branching. For each set we compute the average relative **gap** $:= \frac{(\mathrm{OPT}-\mathrm{LP_r})}{\mathrm{OPT}}$ in percent, whereby OPT denotes the optimal objective value of the ILP. Additionally, we give the average total runtime t_{ILP}, the number of instances which require branching (req.br.), and the average average number of Branch-and-Bound nodes (#BB).

DCUT also outperforms DFLOW in terms of running times needed to solve the (equivalent) LP-relaxation. When DFLOW is not able to solve the given instance to optimality within 2 hours, it is due to a large size of the LP and the most part of the computation time is needed to solve the root relaxation. By contrast, when branching is required, DCUT uses only a comparably small percentage of the total running time to solve the root relaxation. For the Grid$^+$ instances with 400 nodes, DFLOW cannot even solve the LP-relaxation within the given time bound. DCUT, on the other hand, requires only 170 seconds on average to solve the ILP, whereby the LP-relaxation is solved within 10–30 seconds. Analogous observations sustain for the other instances like PCSTLib$^+$.

Further investigation of DCUT. The comparison of our formulations shows DCUT to be the clear winner for all practical purposes. We briefly report on further experiments which shows the applicability of DCUT to larger instances in real-world settings, and summarize our findings thereto.

Primal Heuristic. In many large instances, the size of the branch-and-bound tree becomes the bottleneck. We developed an LP-based primal heuristic to obtain strong upper bounds early in our algorithm. Interestingly, although the heuristic often finds the optimal solution very soon, it only rarely leads to pruning B&B-nodes and therefore only seldomly reduces the running time.

Further instances. We furthermore used the TSPLib-based instances [22] which are euclidian Delauney-triangulations on varying graph sizes with 25% (10%) R_1 (R_2) customers. The findings are analogous to the ones reported before, as only DCUT was able to solve instances. It solved all but one instance with up to 300 nodes, and 43% of the larger graphs with up to 700 nodes. Due to a lack of other instance sets, other papers often consider complete graphs, eventhough most real-world applications seem to be based on rather sparse graphs. We considered the complete graphs presented in [7] for a hybrid meta-heuristic to solve 2NCON problems without R_1 customers. While their algorithm finds heuristic solutions requiring 160–200 iterations (using 32–40 seconds per iteration on a presumably old machine), DCUT solves all instances to optimality within an average of 0.29 seconds.

References

1. Applegate, D.L., Bixby, R.E., Chvátal, V., Cook, W.J.: The Traveling Salesman Problem: A Computational Study. Princeton University Press, Princeton (2006)
2. Cherkassky, B.V., Goldberg, A.V.: On implementing push-relabel method for the maximum flow problem. Algorithmica 19, 390–410 (1997)
3. Chimani, M., Kandyba, M., Ljubić, I., Mutzel, P.: Strong formulations for the 2-node-connected steiner network problems (tr). Technical Report TR07-1-008, Chair for Algorithm Engineering, TU Dortmund (November 2007)
4. Chimani, M., Kandyba, M., Ljubić, I., Mutzel, P.: Obtaining optimal k-cardinality trees fast. In: Proc. Siam ALENEX 2008 (2008)

5. Chimani, M., Kandyba, M., Mutzel, P.: A new ILP formulation for 2-root-connected prize-collecting Steiner networks. In: Arge, L., Hoffmann, M., Welzl, E. (eds.) ESA 2007. LNCS, vol. 4698, pp. 681–692. Springer, Heidelberg (2007)
6. Chopra, S.: Polyhedra of the equivalent subgraph problem and some edge connectivity problems. SIAM J. Discrete Math. 5(3), 321–337 (1992)
7. Ghashghai, E., Rardin, R.L.: Using a hybrid of exact and genetic algorithms to design survivable networks. Computers & OR 29(1), 53–66 (2002)
8. Goemans, M.X.: Analysis of linear programming relaxations for a class of connectivity problems. PhD thesis. MIT, Cambridge (1990)
9. Grötschel, M., Monma, C.L., Stoer, M.: Polyhedral Approaches to Network Survivability. In: Reliability of Computer and Communication Networks, Proc. Workshop 1989. Discrete Mathematics and Theoretical Computer Science, vol. 5, pp. 121–141. American Mathematical Society (1991)
10. Grötschel, M., Monma, C.L., Stoer, M.: Computational results with a cutting plane algorithm for designing communication networks with low-connectivity constraints. Operatios Research 40(2), 309–330 (1992)
11. Grötschel, M., Monma, C.L., Stoer, M.: Facets for polyhedra arising in the design of communication networks with low-connectivity constraints. SIAM Journal on Optimization 2(3), 474–504 (1992)
12. Johnson, D.S., Minkoff, M., Phillips, S.: The prize-collecting steiner tree problem: Theory and practice. In: Proceedings of 11th ACM-SIAM Symposium on Distcrete Algorithms, San Fransisco, CA, pp. 760–769 (2000)
13. Kerivin, H., Mahjoub, A.R.: Design of survivable networks: A survey. Networks 46(1), 1–21 (2005)
14. Ljubić, I.: Exact and Memetic Algorithms for Two Network Design Problems. PhD thesis, TU Vienna (2004)
15. Ljubić, I., Weiskircher, R., Pferschy, U., Klau, G., Mutzel, P., Fischetti, M.: An algorithmic framework for the exact solution of the prize-collecting steiner tree problem. Math. Prog. Ser. B 105(2–3), 427–449 (2006)
16. Lucena, A., Resende, M.G.C.: Strong lower bounds for the prize-collecting steiner problem in graphs. Discrete Applied Mathematics 141(1-3), 277–294 (2003)
17. Magnanti, T.L., Raghavan, S.: Strong formulations for network design problems with connectivity requirements. Networks 45(2), 61–79 (2005)
18. Polzin, T., Daneshmand, S.V.: Improved algorithms for the Steiner problem in networks. Discrete Applied Mathematics 112(1-3), 263–300 (2001)
19. Raghavan, S.: Formulations and Algorithms for the Network Design Problems with Connectivity Requirements. PhD thesis. MIT, Cambridge(1995)
20. Robbins, H.E.: A theorem on graphs with an application to a problem of traffic control. American Mathematical Monthly 46, 281–283 (1939)
21. Stoer, M.: Design of Survivable Networks. LNM, vol. 1531. Springer, Heidelberg (1992)
22. Wagner, D.: Generierung und Adaptierung von Testinstanzen für das OPT und SST Problem. Technical Report 03/2007, Carinthia Tech Institute, Klagenfurt, Austria (2007) (in German)
23. Wagner, D., Raidl, G.R., Pferschy, U., Mutzel, P., Bachhiesl, P.: A multicommodity flow approach for the design of the last mile in real-world fiber optic networks. In: Proc. OR 2006, pp. 197–202. Springer, Heidelberg (2006)
24. Winter, P.: Steiner problem in networks: A survey. Networks 17(2), 129–167 (1987)
25. Wolsey, L.A.: Integer Programming. Wiley-Interscience, Chichester (1998)

Algorithms and Implementation for Interconnection Graph Problem

Hongbing Fan[1,*], Christian Hundt[2], Yu-Liang Wu[3,**], and Jason Ernst[4]

[1] Wilfrid Laurier University, Waterloo, ON Canada N2L3C5
hfan@wlu.ca
[2] University of Rostock, Germany
Christian.Hundt@uni-rostock.de
[3] The Chinese University of Hong Kong, Shatin, N.T., Hong Kong
ylw@cse.cuhk.edu.hk
[4] University of Guelph, Guelph, ON Canada N1G2W1

Abstract. The Interconnection Graph Problem (IGP) is to compute for a given hypergraph $H = (V, R)$ a graph $G = (V, E)$ with the minimum number of edges $|E|$ such that for all hyperedges $N \in R$ the subgraph of G induced by N is connected. Computing feasible interconnection graphs is basically motivated by the design of reconfigurable interconnection networks. This paper proves that IGP is NP-complete and hard to approximate even when all hyperedges of H have at most three vertices. Afterwards it presents a search tree based parameterized algorithm showing that the problem is fixed-parameter tractable when the hyperedge size of H is bounded. Moreover, the paper gives a reduction based greedy algorithm and closes with its experimental justification.

1 Introduction

A Reconfigurable Interconnection Network (RIN, for short) consists of terminals and switches. A switch joins a pair of terminals, and when the switch is set to ON (OFF), the two terminals of the switch are connected (disconnected). In the problem of designing a customized RIN, a set of terminals and a set of routing requirements over the terminals are given, where a routing requirement consists of a group of disjoint subsets of the terminals. The RIN design aims at finding a minimal RIN (of the minimum number of switches) satisfying that the network can be reconfigured to fulfill every one of the given routing requirements. That is, for each of the given routing requirements, there is a valid routing: an ON/OFF assignment to all switches, such that only the terminals in the same subset of the routing requirement are connected.

The customized RIN design problem arises from the design of reconfigurable Systems-on-a-Chips (SoC) for multiple applications, where different interconnections of terminals (or ports) of functional modules are required for different

---

* Research partially supported by the NSERC, Canada.
** Research partially supported by RGC Earmarked Grant 2150500 and ITSP Grant 6902308, Hong Kong.

B. Yang, D.-Z. Du, and C.A. Wang (Eds.): COCOA 2008, LNCS 5165, pp. 201–210, 2008.
© Springer-Verlag Berlin Heidelberg 2008

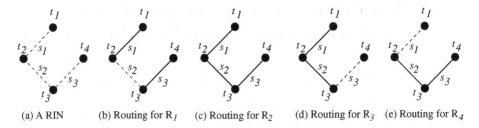

(a) A RIN   (b) Routing for R_1 (c) Routing for R_2 (d) Routing for R_3 (e) Routing for R_4

Fig. 1. An example of reconfigurable interconnection network with valid routings

applications. The general RIN structures such as mesh and trees in Field Programmable Gate Arrays (FPGA) [2,3] tend to use much more resources than necessary for the routing requirements imposed by applications. The goal of the customized RIN design is to minimize the resource usage (the number of switches) of RIN subject to the routing requirement specifications.

Figure 1(a) shows for example four terminals t_1 to t_4. A set of the given routing requirements are represented by sets of disjoint terminal subsets, $R_1 = \{\{t_1, t_2\}, \{t_3, t_4\}\}, R_2 = \{\{t_1, t_2, t_3, t_4\}\}, R_3 = \{\{t_1, t_2, t_3\}, \{t_4\}\}$ and $R_4 = \{\{t_1\}, \{t_2, t_3, t_4\}\}$. A RIN design tries to minimize the number of switches needed to route all the given routing requirements. For that, Figure 1(a) shows by dashed lines a minimum set $E = \{s_1, s_2, s_3\}$ of switches. Then the Figures 1(b) to (e) present valid routings according to $R_1, \ldots, R_4$, where solid switches represent ON and dashed switches represent OFF, respectively.

Hence, RINs can be modeled as graphs $G = (V, E)$ with terminals as vertices V and switches as edges E. The challenge to design a RIN G from a set V of terminals and a number of routing requirements $R_i = \{N_{i,1}, \ldots, N_{i,r_i}\}, i = 1, \ldots, r$, where $N_{i,j} \subseteq V, j \in \{1, \ldots, r_i\}$ and $N_{i,1}, \ldots, N_{i,r_i}$ are mutually disjoint for every $i \in \{1, \ldots, r\}$, can be formulated as the following hypergraph problem.

Let R denote the subsumption of all given routing requirements, i.e., $R = \cup_{i=1}^r R_i = \cup_{i=1}^r \{N_{i,1}, \ldots, N_{i,r_i}\}$. Then $H = (V, R)$ forms a hypergraph. Given a hypergraph $H = (V, R)$, a graph $G = (V, E)$ is said to be an *interconnection graph* of H if the induced subgraphs $G[N]$ of all hyperedge $N \in R$ are connected. We search for a minimal interconnection graph, namely an interconnection graph $G = (V, E)$ of H with the number of edges $|E|$ minimized. Minimal interconnection graph models minimum-switch RIN meeting the given requirements. In particular, for every $R_i = \{N_{i,1}, \ldots, N_{i,r_i}\}$ the routing can be established by computing a spanning tree $T_{i,j}$ for every subgraph $G[N_{i,j}], j \in \{1, \ldots, r_i\}$ and turning on only those switches represented by edges in the trees $T_{i,j}$.

For example, Figure 2(a) shows a hypergraph $H = (V, R)$ of eight vertices t_1 to t_8. We call $N \in R$ am m-hyperedge iff $|N| = m$. The 2-hyperedges N_1 to N_7 are illustrated as solid lines. In turn, the vertices of every triangle made up by one solid and two dashed lines give one of the 3-hyperedges N_7 to N_{16}. For an interconnection graph G of H select a subset of lines such that in every hyperedge all vertices become connected. Every 2-hyperedge $N = \{u, v\}$ enforces the line uv into G. Thus, all solid lines are in G. The connection of each 3-hyperedge

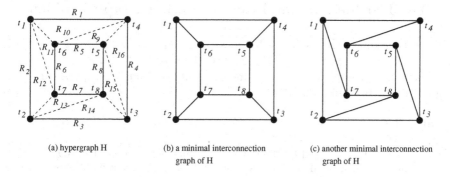

(a) hypergraph H

(b) a minimal interconnection
graph of H

(c) another minimal interconnection
graph of H

Fig. 2. Examples for minimal interconnection graphs

needs at least one dashed line because the solid one is already in G. Since there are four line disjoint triangles, the two minimal interconnection graphs of H have four dashed lines (see Figure 2(b) and (c)),

The combinatorial optimization problem of computing interconnection graphs is called Minimum Interconnection Graph Problem ($\text{IGP}_{\min}$):

Input: A hypergraph $H = (V, R)$.
Output: A graph $G = (V, E)$ such that $G[N]$ is connected for all $N \in R$.
Costs: The number $|E|$ of edges.

The decision version of this problem, denoted by IGP, is to decide whether hypergraph H has an interconnection graph of costs at most some given k. Furthermore, we introduce the problem variants $3\text{IGP}_{\min}$ and 3IGP where all hyperedges of H have at most three vertices, i.e., are at most 3-hyperedges.

To compute any interconnection graph G for $H = (V, R)$ is simple, e.g., by taking the complete graph on vertex set V or choosing the clique graph $C_H = (V, \{uv : u, v \in N, N \in R\})$ of H. We can also derive an interconnection graph by computing spanning trees for all $N \in R$ and then joining them. Moreover, the problem becomes trivial when H contains only 2-hyperedges because then H is itself its own interconnection graph.

However, Section 2 of this paper shows by reductions from the well-known Minimum Vertex Cover ($\text{VC}_{\min}$) optimization problem that the versions $3\text{IGP}_{\min}$ and 3IGP are already computationally hard. The selection of a minimum vertex cover in a graph can be realized by the optimal selection of edges in 3-hyperedge. For the proof we need the decision variant VC of $\text{VC}_{\min}$ and the degree-bounded version $3\text{VC}_{\min}$, where each vertex has degree at most three. Despite this bad news Section 3 presents a search tree based parameterized algorithm for $\text{IGP}_{\min}$. As a result, the problem becomes fixed parameter tractable [4] when the edge size of input hypergraphs is bounded. Finally, in Section 4 we give a reduction based greedy algorithm and justify its performance experimentally.

2   The Hardness of the Interconnection Graph Problem

This section shows that IGP is NP-hard and remains so when restricted to
3IGP. In addition, unless P=NP there is no PTAS for IGP_{min} since the $3IGP_{min}$
is already APX-complete. Used terminology and notations of graph theory and
complexity theory are from [1,6].

Lemma 2.1. *(1) $VC \leq_p 3IGP$, (2) $3VC_{min} \leq_p^A 3IGP_{min}$, (3) $3IGP_{min} \leq_p^A VC_{min}$.*

Proof. For (2) and (3) we present two polynomial algorithms f_1 and f_2 respectively. The first transforms instances x of the first problem to instances $f_1(x)$ of
the second one. Subsequently, if y is a solution to $f_1(x)$ then $f_2(f_1(x), y)$ gives
the solution for x with the property: If y costs c times optimum, then $f_2(f_1(x), y)$
costs c' times optimum. Statement (1) is a byproduct of the proof for (2).

(1,2): Let $G = (V, E)$ be a graph. Algorithm $f_1(G)$ computes the hypergraph
$H = (V', R)$ with $V' = V \cup \{x\}$, where $x \notin V$, and $R = E \cup \{\{u, v, x\} :
uv \in E\}$. Trivially, if G has a vertex cover C then H has the interconnection
graph $G'(C) = (V', E' = E \cup \{vx : v \in C\})$ with $|E'| = |C| + |E|$ edges. For
any solution interconnection graph $G' = (V', E')$ for H the algorithm $f_2(G, G')$
returns the set $C = \{v : vx \in E' \setminus E\}$ of size $|C| = |E'| - |E|$. If C was not a
vertex cover of G, then there must be $uv \in E$ such that $\{u, v\} \cap C = \emptyset$. Therefore
$\{u, v, x\} \in R$ induces an unconnected subgraph in G' which contradicts G' being
interconnection graph for H. The polynomial complexity of f_1 and f_2 is straight
forward. Now (1) has been shown already, since f_1 reduces VC to 3IGP. For (2)
let $G = (V, E)$ have maximum degree three. Obviously, $|E| \leq 3|C_{opt}|$ since in a
minimum vertex cover C_{opt} of G every vertex covers at most three edges. If E''
is the edge set of $G'(C_{opt})$ and $|E'|$ is c times optimal costs then $|E'| \leq c|E''|$.
For $C = f_2(G, G')$ it follows $|C| \leq c|C_{opt}| + (c-1)|E|$. Thus, $|C| \leq c'|C_{opt}|$ with
$c' = 4c - 3$.

(3): Let $H = (V, R)$ be a hypergraph of at most 3-hyperedges. For each 3-
hyperedge $N = \{u, v, w\} \in R$ we let $G_N = (U_N, E_N)$ be the triangle on the
vertex set $U_N = \{uv, uw, vw\}$ and for each 2-hyperedge $N = \{u, v\} \in R$ we let
$G_N = (U_N, E_N)$ be the artificial self-loop (uv, uv). Algorithm $f_1(H)$ composes
a graph $G = (U, E)$ by merging the gadgets G_N for all $N \in R$ in the way
$U = \bigcup_{N \in R} U_N$ and $E = \bigcup_{N \in R} E_N$. If H has an interconnection graph $G' =
(V, C)$ then C is a vertex cover for G. It contains every 2-hyperedge N and for
every 3-hyperedge N at least two edges, thus, covering triangles and loops G_N.
Given any solution vertex cover C for G the function $f_2(H, C)$ returns the graph
$G' = (V, C)$ which must be an interconnection graph for H. The set C has to
cover all triangles G_N by at least two vertices and thus connects N. Moreover,
for all $N = \{u, v\}$ the loop G_N forces the vertex uv into C and thus ensures the
connection of N. Again, f_1 and f_2 are obviously polynomial. Because C and G'
have equal costs it follows that C having c times optimal costs implies the same
for G'.   □

Corollary 2.2. *The problem IGP is NP-complete, even when restricted to 3IGP.*

Proof. IGP (and 3IGP) is in NP by guessing in non-deterministic fashion for given $H = (V, R)$ all subgraphs G of the complete graph with vertex set V and testing in polynomial time the connectivity of the graphs $G[N]$ for all $N \in R$. The hardness of 3IGP (and thus, IGP) in NP is shown in Lemma 2.1.   □

Corollary 2.3. *The optimization problem 3IGP*$_{min}$ *is APX-complete.*

Proof. By Lemma 2.1 3IGP$_{min}$ $\leq_p^A$ VC$_{min}$ and thus, each c-approximation algorithm for VC$_{min}$ approximates 3IGP$_{min}$ with factor c by the given A-reduction. VC$_{min}$ can be 2-approximated [7] and so can 3IGP$_{min}$. Hence, 3IGP$_{min}$ is in APX. Since 3VC$_{min}$ is APX-complete [8] this follows due to Lemma 2.1 for 3IGP$_{min}$ by 3VC$_{min}$ $\leq_p^A$ 3IGP$_{min}$.   □

3   A Search Tree Based Parameterized Algorithm

This section considers the fixed-parameter IGP, where integer k is given as a fixed integer rather than an input. The parameterized complexity theory [4] concerns whether the parameterized problem is Fixed-Parameter Tractable (FPT), i.e., the existence of an algorithm of time complexity $O(f(k)p(n))$, where $p(n)$ is a polynomial function of input size n.

Figure 3 shows a parameterized algorithm for IGP using the well-known search tree approach [4,5]. The algorithm decides whether H has an interconnection graph of cost at most k by setting global variable *exists* to False, then invoking $InterGraph(V, R, \emptyset, k)$, and finally printing the value of variable *exists*.

```
InterGraph(V, R, E, k){
    G = (V, E)
    If R = ∅ and k ≥ 0
        exists = True
    Else if R ≠ ∅ and k > 0
        Choose N ∈ R
        If G[N] is connected
            R = R \ {N}
            InterGraph(V, R, E, k)
        Else
            For every pair {u, v} ⊆ N with uv ∉ E
                E = E ∪ {uv}
                InterGraph(V, R, E, k-1)
            End For
        End If
    End If
}
```

Fig. 3. Search tree based parameterized algorithm

Theorem 3.1. *For hypergraphs $H = (V, R)$ of at most h-hyperedges, the search tree based interconnection graph algorithm answers whether H has an interconnection graph of cost at most k in time $O((h(h-1)/2)^{k+1}n^2m)$, where $n = |V|, m = |R|$.*

Proof. The execution of the algorithm builds a search tree. At each node of the search tree, if $R = \emptyset$, an interconnection graph $G = (V, E)$ is returned; otherwise an N is chosen from R and processed. If $G[N]$ is connected, then no edge is needed for N. If $G[N]$ is not connected, then at least one edge $uv \notin E$ with $\{u, v\} \in N$ has to be chosen for the interconnection graph to make $G[N]$ connected. Since $|N| \leq h$, there are at most $h(h-1)/2$ possible choices for uv, hence there are at most $h(h-1)/2$ children at a node. The height of the search tree is at most k since k decreases by 1 while branching. Hence, there are at most $(h(h-1)/2)^{k+1}$ nodes. At each node, there are at most m hyperedges N to check for branching. For each N, it computes $G[N]$ and tests if $G[N]$ is connected, which can be done in time $O(n^2)$. Therefore, the time spent at each node is bounded by $O(n^2m)$, and the total time of the algorithm is bounded by $O((h(h-1)/2)^{k+1}n^2m)$.   □

By Theorem 3.1 the parameterized IGP is FPT when the edge size of input hyperedges is bounded. In particular, 3IGP can be solved in time $O(3^{k+1}n^2m)$. However, it is not known whether the IGP (i.e., without edge size restriction) is FPT. We leave it as an open problem.

Problem 3.2. Is the parameterized IGP fixed parameter tractable for any input hypergraphs?

4   A Reduction Based Greedy Algorithm

This section presents a reduction based greedy algorithm for IGP$_{min}$. Starting with an empty graph, the algorithm constructs an interconnection graph of hypergraph H by adding edges and removing hyperedges when it reduces a connected subgraph with the currently constructed graph until no hyperedge left. We call an interconnection graph containing an edge set E' minimal if it is of the minimum number of edges among all interconnection graphs containing E'. The algorithm has two modes, reduction mode and greedy mode, switching alternately. In the reduction mode, a reduction procedure is called, which adds new edges under certain conditions. The new edges added during the reduction are in a minimal interconnection graph containing the current edge set. In the greedy mode, the algorithm greedily chooses an edge e^* that reduces the most number of components of all induced subgraphs of hyperedges. But this operation may choose an edge which is not in a minimal interconnection graph containing the current edge set. Let $H' = (V, R')$ be the current hypergraph and $G' = (V, E')$ the current graph constructed. For $u, v \in V$, define

$$d_{H',G'}(uv) = |\{N \in R' : u, v \text{ are not in the same component of } G'[N]\}|$$

The details of the algorithm is shown in Figure 4. Note that the algorithm keeps a current hyperedge set R', an edge set E', and a candidate edge set $\bar{E}'$. The initial setting is $R' = R, E' = \emptyset, \bar{E}' = \{uv : u, v \in N, N \in R\}$. At each iteration, it first invokes the reduction procedure $Reduction(R', E', \bar{E}')$, which returns $R', E', \bar{E}'$. If $R' = \emptyset$, then it outputs $G = (V, E')$, and stops; otherwise, let $H' = (V, R'), G' = (V, E')$ and choose $e^* \in \bar{E}'$ such that $d_{H',G'}(e^*) = max\{d_{H',G'}(e) : e \in \bar{E}'\}$. Add e^* to E' and remove e^* from $\bar{E}'$. Repeat the process until R' becomes empty.

Lemma 4.1. *Suppose that* $Reduction(R', E', \bar{E}')$ *returns* $(R'', E'', \bar{E}'')$. *Then the minimal interconnection graph containing* E'' *has the same number of edges as the minimal interconnection graph containing* E'.

Proof. The reduction done by $Reduction(R', E', \bar{E}')$ consists of a sequence of simple reductions by rules 1 to 4. It is sufficient to show that after a simple reduction, the property holds. Assume that $(R'', E'', \bar{E}'')$ is what returned by a simple reduction from $(R', E', \bar{E}')$. Let $G_1 = (V, E' \cup E_0')$ be a minimal interconnection graph of H containing E', and $G_2 = (V, E'' \cup E_0'')$ a minimal interconnection graph of H containing E'', where $E' \cap E_0' = \emptyset, E'' \cap E_0'' = \emptyset$. We show $|E' \cup E_0'| = |E'' \cup E_0''|$.

Case 1. The reduction is done by Rule 1. Then $E'' = E'$, and G_2 is also a minimal interconnection graph containing E'. Therefore $|E'' \cup E_0''| = |E' \cup E_0'|$.

Case 2. The reduction is done by Rule 2. Then we have $E'' = E' \cup \{uv\}$. We show that G_1 can be transformed to a minimal interconnection graph containing E''. If $uv \in E' \cup E_0'$, then $(V, E' \cup E_0')$ is itself a minimal interconnection graph containing E'', which implies $|E'' \cup E_0''| = |E' \cup E_0'|$. Suppose that $uv \notin E' \cup E_0'$. Then there exists an $N \in R'$ such that u, v are two isolated vertices of $G'[N]$. Since $G_2[N]$ is connected, $G_2[N]$ contains two edges ux, yv such that x and y are in the same component of $G_2[N] - u - v$. Let $T = \{w : wv \in E_0'\}$ and $G_3 = (V, (E' \cup ((E_0' \setminus \{wv : w \in T\}) \cup \{uw : w \in (T \setminus \{y\}) \cup \{v\}\})))$. Then G_3 is a minimal interconnection graph of H containing E_0' with $|E(G_3)| = |E(G_1)|$.

Case 3. The reduction is done by Rule 3. Let $E_2 = \{N : N \in R', |N| = 2\}$. Then $E'' = E' \cup E_2$. Since a minimal interconnection graph containing E' must contain all 2-subset edges in R', G_1 is also a minimal interconnection graph containing E''. We have $|E' \cup E_0'| = |E'' \cup E_0''|$.

Case 4. The reduction is done by Rule 4. Then $E'' = E' \cup \{uw, vw\}$. We show $(V, E' \cup \{uw, wv\} \cup E_0')$ is a minimal interconnection graph containing E'. Suppose that all minimal interconnection graphs containing E' do not contain $\{uw, wv\}$. Let $(V, E' \cup E_0')$ be such a minimal interconnection graph. Then E_0' must contain uv and one of uw and wv, say uw. Since $d_{H',G'}(uv) = 1$, uv is only contained in $N = \{u, v, w\}$. Then $(V, (E' \setminus \{uv\}) \cup \{wv\}$ is a minimal interconnection graph containing $E' \cup \{uw, wv\}$, a contradiction follows.   ▢

Theorem 4.2. *For any hypergraph* $H = (V, R)$, *the reduction based greedy algorithm returns an interconnection graph of* H *in time* $O(n^2 m)$, *where* $n = |V|, m = |R| > n$. *If* $Reduction(R, \emptyset, \bar{E})$ *returns* $(R', E', \bar{E}')$ *and* $R' = \emptyset$, *then* $G = (V, E')$ *is a minimal interconnection graph of* H.

Input: hypergraph $H = (V, R)$
Set $R' = R, E' = \emptyset, \bar{E}' = \{uv : u, v \in N, N \in R\}$.
$H' = (V, R'), G' = (V, E')$
reduction = true
While $R' \neq \emptyset$
    If reduction
        $Reduction(R', E', \bar{E}')$
        reduction = false
    Else
        choose $e^* \in \bar{E}'$ such that $d_{H',G'}(e^*) = max\{d_{H',G'}(e) : e \in \bar{E}'\}$
        $E' = E' \cup \{e^*\}, \bar{E}' = \bar{E}' \setminus \{e^*\}$
        reduction = true
    End If
End While
Output: (V, E')

$Reduction(R', E', \bar{E}')$
    $H' = (V, R'), G' = (V, E')$
    repeat = true
    r1 = r2 = r3 = r4 = true;
    While repeat
        If there is an $N \in R'$ such that $G'[N]$ is connected   **Rule 1**
            remove N from R'
        Else r1 = false
        If $u, v \in V$ satisfy that $\{u, v\} \cap N \neq \emptyset$ implies $\{u, v\} \subseteq N$ and   **Rule 2**
            there is $N \in R'$ such that u, v are two isolated vertices of $G'[N]$
            add uv to E'
            remove v from hyperedges that contain v
            remove edges from $\bar{E}'$ that contain v
            r1 = true
        Else r2 = false
        If there is an $N = \{u, v\} \in R'$ **Rule 3**
            add uv to E' and remove uv from $\bar{E}'$.
            remove N from R'
            r1 = true, r2 = true
        Else r3 = false
        If there is an $N = \{u, v, w\} \in R'$ with $uv \in \bar{E}$ and $d_{H',G'}(uv) = 1$ **Rule 4**
            add $\{uw, vw\} \cap \bar{E}'$ to E' and remove $\{uv, uw, vw\} \cap \bar{E}'$ from $\bar{E}'$.
            remove N from R'.
            r1 = true, r2 = true, r3 = true
        Else r4 = false
        repeat = r1 OR r2 OR r3 OR r4
    End While
    Return $(V', R', E', \bar{E}')$.

Fig. 4. Reduction based greedy algorithm for IGP

Proof. It is obvious that when $R' = \emptyset$, (V, E') is an interconnection graph of H. To derive the time complexity, we need a hyperedge data structure consisting of a list of vertex sets of its connected components in the induced subgraph, and a list of all the hyperedge in increasing order of the size of hyperedges. For each vertex, we use an ordered list of hyperedges that contain the vertex. We use an ordered list of the vertex data structures for all vertices. For each edge $uv \in \bar{E}'$, we use a list of hyperedges such that u, v are in different components of the induced subgraph by the hyperedge. For all edges of $\bar{E}'$, we keep a list of the edge data structures in decreasing order of $d_{H',G'}(uv)$. With this setting, the reduction by rules 1 can be done in time $O(n)$, because a hyperedge N such that $G'[N]$ is connected is be at the top of the hyperedge list, and it takes time $O(n)$ to maintain the data structure after removing N. For rule 2, it takes time $O(n)$ to find a pair u, v with the property, and time $O(m)$ to maintain the hyperedge data structure after adding uv. For rule 3, it takes time $O(1)$ to find an $N = \{u, v\}$, and time $O(m)$ to maintain the hyperedge data structure after adding uv. For rule 4, it takes time $O(1)$ to find N with the property and time $O(m)$ to maintain the data structure after adding edge. When no reduction can be done, it also takes time $O(1)$ find e^* and time $O(m)$ to maintain the data structure after adding e^*. The total number of iterations (i.e., adding edges) is bounded by $O(n^2)$, therefore the time complexity of the algorithm is bounded by $O(n^2 m)$.

When $Reduction(R, \emptyset, \bar{E})$ returns an empty R', then (V, E') is a minimal interconnection graph of H containing empty set by Lemma 4.1, so it is a minimal interconnection graph of H. □

5   Experimental Results

We implemented the reduction based greedy algorithm in C. The main purpose of our implementation is to develop an automation design tool for RIN design. We tested three classes of hypergraphs with the implementation. The first class is hypergraph $H_1(m)$ consisting of vertices $1, \ldots, 2m$ and hyperedges $\{i, i + 1\}, \{i + m, i + m + 1\}, \{i, i + 1, i + m\}, \{i + m, i + m + 1, i + 1\}, i = 1, \ldots, m - 1, \{1, m\}, \{2m, m + 1\}, \{m, 1, 2m\}, \{2m, m + 1, 1\}$. The clique graph of $H_1(m)$ has $4m$ edges and the minimal interconnection graph of $H_1(m)$ has $3m$ edges. The second class is hypergraph $H_2(m)$ consisting of vertices $1, \ldots, 2m + 2$, and hyperedges $\{0, 1, i, i + m\}, i = 2, \ldots, m + 1$, $\{i, i + 1, i + m, i + m + 1\}, i = 2, \ldots, m, \{m + 1, 2, 2m + 1, m + 2\}$. The clique graph of $H_2(m)$ has $9m + 1$ edges and the minimal interconnection graph of $H_2(m)$ has $3m + 1$ edges. The third class is a random hypergraph H consisting of vertices $1, \ldots, n$ and a set of random hyperedges. The number of random hyperedges is set to $5n$. The size of each edge is randomly assigned a value between 2 and 10. There are 10 runs for each n. The program is able to find the minimal interconnection graphs for both $H_1(m)$ and $H_2(m)$. For the random hypergraphs, the program find an interconnection graph with average improvement of over 70% over the initial clique graph. The testing results are shown in Table 1, in which the column name $|V(H)|, |E(H)|, |E(C_H)|$ and $|E(G)|$ denote the number of vertices, the number of hyperedges, the number of edges in the clique graph, and the number of edges in derived interconnection graphs, respectively. The Improvement% column is computed by $(|E(C_H)| - |E(G)|)/|E(C_H)|$. The time column is the running time in second on a Linux machine with 1.8GHz AMD Athlon 64X2 processor.

Table 1. Experimental results of the algorithm

$|V(H)|$	$|E(H)|$	$|E(C_H)|$	$|E(G)|$	Improvement%	Time (s)
Results of test case one: $H_1(m)$					
600	1200	1200	900	25	8.27
800	1600	1600	1200	25	20.6
1000	2000	2000	1500	25	42.28
Results of test case two: $H_2(m)$					
602	600	2701	901	66.6	3.34
802	800	3601	1201	66.6	7.78
1002	1000	4501	1501	66.6	14.79
Results of random hypergraphs with $5n$ hyperedges					
20	100	190	78	59	0.16
40	200	774	238	69	4.17
60	300	1689	456	73	23.31
80	400	2866	703	75	81.99
100	500	4155	989	76	244.39
120	600	5613	1300	77	541.75

Acknowledgments

Thank C.Q. Zhang and J.B. Qian for helpful discussions.

References

1. Bondy, J.A., Murty, U.S.R.: Graph Theory with Applications. Macmillan Press, London (1976)
2. Betz, V., Rose, J., Marquardt, A.: Architecture and CAD for Deep-Submicron FP-GAs. Kluwer-Academic Publisher, Boston (1999)
3. Lemieux, G., Lewis, D.: Design of Interconnection Networks for Programmable Logic. Kluwer-Academic Publisher, Boston (2003)
4. Downey, R.G., Fellows, M.R.: Parameterized Complexity. Springer, Heidelberg (1998)
5. Ellis, J., Fan, H., Fellows, M.: The Dominating Set Problem is Fixed Parameter Tractable for Graphs of Bounded Genus. Journal of Algorithms 52(2), 152–168 (2004)
6. Du, D.-Z., Ko, K.-I.: Theory of Computational Complexity. John Wiley & Sons, Chichester (2000)
7. Monien, B., Speckenmeyer, E.: Ramsey numbers and an approximation algorithm for the vertex cover problem. Acta Inf. 22, 115–123 (1985)
8. Papadimitriou, C.H., Yannakakis, M.: Optimization, approximation, and complexity classes. J. Comput. System Sci. 43, 425–440 (1991)

Algorithms and Experimental Study for the Traveling Salesman Problem of Second Order

Gerold Jäger and Paul Molitor

Computer Science Institute,
University of Halle-Wittenberg,
D-06099 Halle (Saale), Germany
jaegerg@informatik.uni-halle.de,
paul.molitor@informatik.uni-halle.de

Abstract. We introduce a new combinatorial optimization problem, which is a generalization of the Traveling Salesman Problem (TSP) and which we call Traveling Salesman Problem of Second Order (2-TSP). It is motivated by an application in bioinformatics, especially the Permuted Variable Length Markov model. We propose seven elementary heuristics and two exact algorithms for the 2-TSP, some of which are generalizations of similar algorithms for the Asymmetric Traveling Salesman Problem (ATSP), some of which are new ideas. Finally we experimentally compare the algorithms for random instances and real instances from bioinformatics. Our experiments show that for the real instances most heuristics lead to optimum or almost-optimum solutions, and for the random instances the exact algorithms need less time than for the real instances.

Keywords: Traveling Salesman Problem, Assignment Problem, Traveling Salesman Problem of Second Order, Heuristic, Exact Algorithm.

1 Introduction

Gene regulation in higher organisms is accomplished by several cellular processes, one of which is transcription initiation. In order to better understand this process, it would be desirable to have a good understanding of how transcription factors bind to their binding sites. While tremendous progress has been made on the fields of structural biology and bioinformatics, the accuracies of existing models to predict the location and affinity of transcription factor binding sites are not yet satisfactory. The aim is to better understand gene regulation by finding more realistic binding site models. One model that extends the position weight matrix (PWM) model, the weight array matrix (WAM) model, and higher-order Markov models in a natural way is the Permuted Markov (PM) model. Permuted Markov models were proposed by [5] for the recognition of transcription factor binding sites. The class of PM models was further extended by [24] to the class of Permuted Variable Length Markov (PVLM) models, and it was demonstrated that PVLM models can improve the recognition of transcription factor binding sites for many transcription factors. Finding the optimal PM model for a given data set is $\mathcal{NP}$-hard, and

B. Yang, D.-Z. Du, and C.A. Wang (Eds.): COCOA 2008, LNCS 5165, pp. 211–224, 2008.

finding the optimal PVLM model for a given data set is $\mathcal{NP}$-hard, too. Hence, heuristic algorithms for finding the optimal PM model and the optimal PVLM model were proposed in [5] and [24], respectively. Experimental evidence has been accumulated that suggests that the binding sites of many transcription factors fall into distinct classes. Grosse proposed to extend PM models to PM mixture models, to extend PVLM models to PVLM mixture models, and to apply both mixture models to the recognition of transcription factor binding sites [15]. While both the PM mixture model and the PVLM mixture model look appealing from a biological perspective, they pose a computational challenge: the optimal PM mixture model and the optimal PVLM mixture model can be obtained only numerically. One of the commonly used algorithms for finding optimal mixture models is the Expectation Maximization (EM) algorithm. The EM algorithm consists of two steps, the E step and the M step, which are iterated until convergence. Applied to the problem of finding the optimal PM or PVLM mixture model of *order 1*, each M step requires the solution of an ATSP instance. Likewise, applied to the problem of finding the optimal PM or PVLM mixture model of *order 2*, each M step requires the solution of an instance of a generalization of the ATSP, which we call 2-TSP and which is introduced in the following.

For a directed graph $G = (V, E)$ with $n \geq 2$ vertices and a weight function $c : V \times V \to \mathbb{R} \cup \{\infty\}$ with $c(u, u) = \infty$ for all $u \in V$ the *Asymmetric Traveling Salesman Problem* is the problem of finding a complete tour $(v_1, v_2, \ldots, v_n)$ with minimum costs $c(v_n, v_1) + \sum_{j=1}^{n-1} c(v_j, v_{j+1})$. ATSP is $\mathcal{NP}$-hard, which can be shown by a simple polynomial reduction from the *Hamiltonian Cycle Problem* (HCP) and a polynomial reduction of HCP from the $\mathcal{NP}$-complete 3-SAT [20]. The special case that the weight of each arc equals the weight of the corresponding reverse arc is called *Symmetric Traveling Salesman Problem* (STSP).

In this paper we introduce the following problem, which – to the best of our knowledge – has not been considered in literature before. For a directed graph $G = (V, E)$ with $n \geq 3$ vertices and a weight function $c : V \times V \times V \to \mathbb{R} \cup \{\infty\}$ with $c(u, v, w) = \infty$ for $u, v, w \in V$ with $u = v$ or $u = w$ or $v = w$ consider the problem of finding a complete tour $(v_1, v_2, \ldots, v_n)$ with minimum costs $c(v_{n-1}, v_n, v_1) + c(v_n, v_1, v_2) + \sum_{j=1}^{n-2} c(v_j, v_{j+1}, v_{j+2})$. As this problem is a natural generalization of the ATSP, where the costs do not depend on arcs, but on each sequence of three vertices in the tour, we call it *Traveling Salesman Problem of Second Order* (2-TSP).

2-TSP is also $\mathcal{NP}$-hard, which can be seen as follows. Let c be the weight function of an ATSP instance. If you define the three-dimensional weight function $c'(u, v, w) := c(v, w) \; \forall u \in V \setminus \{v, w\}$, then solving this ATSP instance and the corresponding 2-TSP instance are equivalent. Thus ATSP can be reduced in polynomial time to 2-TSP.

In some sense, 2-TSP and ATSP have the same difficulty, as both have $(n-1)!$ feasible tours. In another sense, 2-TSP is much more difficult, as it has a three-dimensional weight function and the ATSP only has a two-dimensional one. As we were not able to find an effective polynomial reduction from 2-TSP to ATSP, 2-TSP has to be considered as a new combinatorial optimization problem.

The purpose of this paper is to develop heuristics and exact algorithms for this problem and to do an experimental study of these algorithms for random and real instances.

The paper is organized as follows. In Section 2 we propose different heuristics and in Section 3 different exact algorithms for the 2-TSP. In Section 4 we give an experimental study for the heuristics and the exact algorithms. Finally we summarize this paper and give suggestions for future research in Section 5.

2 Heuristics for the Traveling Salesman Problem of Second Order

Let in the following for a given tour T and a given vertex $v \in T$, $p(v)$ be the predecessor of v and $s(v)$ be the successor of v.

2.1 Cheapest-Insert Algorithm

The *Cheapest-Insert Algorithm* (CI) is a generalization of an algorithm for the ATSP [23]. We start with an arc (v_1, v_2) as a subtour and choose this arc in such a way that the sum of the costs of a cost minimum predecessor of v_1 and a cost minimum successor of v_2 is minimum. Then step by step, a new vertex is included in the subtour, so that the new subtour is cost minimum. If the tour is complete, we stop this procedure.

2.2 Nearest-Neighbor Algorithm

The *Nearest-Neighbor Algorithm* (NN) is also a generalization of an algorithm for the ATSP [23]. Again we start with one arc (v_1, v_2), now considered as a path. Then step by step, we compute neighbors in the direction $v_1 \rightarrow v_2$ in such a way that the new path becomes cost minimum. We stop, until the path contains n vertices and we receive a tour. As we only walk in one direction, the predecessor of v_1 is chosen in the last step. As in this step only one possibility for the predecessor of v_1 exists, the costs of $(p(p(p(v_1))), p(p(v_1)), p(v_1)))$, $(p(p(v_1)), p(v_1), v_1))$, and $(p(v_1), v_1, s(v_1))$ are irrelevant for this choice. Thus we choose the arc (v_1, v_2) in the first step in such a way that the sum of the costs of a cost minimum successor of v_2 and the average costs of a predecessor of v_1 is cost minimum.

2.3 Two-Directional-Nearest-Neighbor Algorithm

In this section we suggest a variation of the Nearest-Neighbor Algorithm, which we call *Two Directional Nearest-Neighbor Algorithm* (2NN). For this algorithm, we contribute two important ideas. The first idea is to use both directions to find the next neighbor. Thus it is the question, which direction should be chosen in each step. One criteria for this choice is to use the minimum cost neighbor over all new vertices and over both directions. *Our* idea is based on the fact that the tour has to be closed anyway, so that *both* directions have to be used now or at a later step of the algorithm. Thus for a given path $(v_1, \ldots, v_i)$, the cost values $c(v_{i-1}, v_i, x)$ for a cost minimum neighbor vertex x and $c(y, v_1, v_2)$ for a cost minimum neighbor vertex y itself are less important than the difference to the second smallest values in both directions. For both directions, this value can be viewed as an *upper tolerance* of the problem of finding a cost minimum neighbor vertex (for an overview over the theory of tolerances see [12,13]). A similar idea was used for a tolerance based version [11] of the greedy heuristic [7] for the ATSP and a tolerance based version [14] of the contract-or-patch heuristic for the ATSP [7,16]. Thus we choose the direction from which the upper tolerance value is larger, as not using the cost minimum neighbor vertex would cause a larger jump of the costs of the current path.

2.4   Assignment-Patching Algorithm

A well-known technique for the ATSP is the patching technique, which starts from k cycles, where each vertex is visited exactly by one of the cycles, and then – step by step – patches two cycles together, until there is only one cycle, which is the ATSP tour of this heuristic. The most popular version of ATSP patching is based on the Assignment Problem (AP), which is defined as follows. Let a matrix $C = (c_{ij})_{1 \le i,j \le n} \in \mathbb{R}^{n,n}$ be given. Then the AP is to find a node permutation π^* so that $\pi^* = \arg\min\left\{ \sum_{i=1}^n c_{i,\pi(i)} : \pi \in \Pi_n \right\}$, where Π_n is the set of all permutations of $\{1, \ldots, n\}$. There are many efficient algorithms for the AP [2,10,19] (for an experimental comparison of AP algorithms see [4]). The most efficient one is the Hungarian algorithm, which is based on König-Egervary's theorem and has a complexity of $O(n^3)$. In our algorithm we use the implementation of the Hungarian algorithm by Jonker and Volgenant [27].

The corresponding AP instance to an ATSP instance uses the same weight function c with $c_{ii} = \infty$ for $1 \le i \le n$. As the AP solution can be computed efficiently and the solution value is a good lower bound for an optimum ATSP solution value, the AP solution is a good starting point for patching. Karp and Steele suggested for each step to patch the two cycles containing the most number of vertices [21]. For each patching step for cycles C_1 and C_2, two arcs $e_1 \in C_1 = (v_1, w_1)$ and $e_2 = (v_2, w_2) \in C_2$ are replaced by arcs (v_1, w_2) and (v_2, w_1). These arcs are chosen in such a way from both cycles that we receive a minimum cost set of cycles in the next step. For the ATSP this means that the term $c(v_1, w_2) + c(v_2, w_1) - c(v_1, w_1) - c(v_2, w_2)$ is minimum. For the 2-TSP the following term has to be minimized:

$$
\begin{aligned}
&c(p(v_1), v_1, w_2) + c(v_1, w_2, s(w_2)) + c(p(v_2), v_2, w_1) + c(v_2, w_1, s(w_1)) \\
&-c(p(v_1), v_1, w_1) - c(v_1, w_1, s(w_1)) - c(p(v_2), v_2, w_2) - c(v_2, w_2, s(w_2))
\end{aligned}
\tag{1}
$$

As natural extension of the AP we define for a directed graph $G = (V, E)$ with $n \ge 3$ and a weight function $c : V \times V \times V \to \mathbb{R} \cup \{\infty\}$ with $c(u, v, w) = \infty$ for $u, v, w \in V$ with $u = v$ or $u = w$ or $v = w$ the *Assignment Problem of Second*

Order (2-AP) as the problem of finding an one-to-one-mapping $f : V \rightarrow V$ so that the costs $\sum_{i=1}^{n} c(v_i, f(v_i), f(f(v_i)))$ are minimum.

Recently Fischer and Lau [6] have shown by a reduction from SAT that 2-AP is $\mathcal{NP}$-hard. One way to solve it is by integer programming (see Section 3.2), which is not fast enough for an efficient heuristic. Instead we suggest to approximate an optimum solution for 2-AP by a polynomial time solvable heuristic solution. For this purpose define a two-dimensional weight function $c' : V \times V \rightarrow \mathbb{R}$, which depends on the three-dimensional weight function $c : V \times V \times V \rightarrow \mathbb{R}$ as follows: $c'(v, w) = \min_{u \in V \setminus \{v, w\}} c(u, v, w)$ for $v \neq w \in V$.

The solution of the AP for this weight function, which can be computed in $O(n^3)$, is a lower bound for the 2-AP solution, which we call approximated 2-AP solution. We then patch the cycles of this AP solution, and call the approach *Assignment Patching Algorithm* (AK), where "K" stands for "Karp Steele".

2.5   Nearest-Neighbor-Patching Algorithm

One drawback of the NN algorithm is that the number of remaining vertices becomes smaller (by 1) at each step. Thus in average the difference between the weight of the current path after adding one vertex and before should increase at each step. The idea of the following algorithm is to modify the NN algorithm in such a way that it outputs not a tour, but a set of cycles. Then these cycles are patched by the Patching Algorithm.

The main step of the *Nearest Neighbor Patching Algorithm* (NNK) is to stop the NN Algorithm, if closing the current cycle would lead to a "good" subtour. More exactly, we change the path $(v_1, \ldots, v_i)$ to a cycle, if the sum of the two weights $c(v_{i-1}, v_i, v_1)$ and $c(v_i, v_1, v_2)$, which are added by the closing, are smaller than a bound. Experiments have shown that $2 \cdot \sum_{j=1}^{i-2} c(v_j, v_{j+1}, v_{j+2})$ seems to be a good choice for this bound. As all cycles should contain at least 3 vertices and the rest of the graph has also to be divided into cycles, it holds $3 \leq i \leq n-3$. We repeat these steps with the remaining vertices, until each vertex is contained in exactly one cycle.

2.6   Two Directional Nearest-Neighbor-Patching Algorithm

The *Two Directional Nearest Neighbor Patching Algorithm* (2NNK) is exactly the NNK Algorithm with the only difference that for the computation of the cycles instead of the NN Algorithm the 2NN Algorithm is used.

2.7   Greedy Algorithm

The *Greedy Algorithm* (G) is also a generalization of an ATSP algorithm [7] which is based on the contraction procedure.

Let $G = (V, E)$ be a complete graph with $n \geq 3$ vertices and $c : E \rightarrow \mathbb{R}$ a weight function. Furthermore let an arbitrary arc e be given, w.l.o.g. $e = (v_{n-1}, v_n)$. The contraction of e means constructing a new complete graph $G' = (V', E')$ with

$V = \{v'_1, \ldots, v'_{n-1}\}$ and $v'_i = v_i$ for $i = 1, \ldots, n - 2$, $v'_{n-1} = (v_{n-1}, v_n)$ and with weight function $c' : E' \to \mathbb{R}$ defined by

$$c'(v'_i, v'_j) = \begin{cases} c(v_i, v_j) & \text{for } 1 \le i \ne j \le n - 2 \\ c(v_i, v_{n-1}) & \text{for } 1 \le i \le n - 2, j = n - 1 \\ c(v_n, v_j) & \text{for } i = n - 1, 1 \le j \le n - 2 \end{cases}$$

Analogously we define the contraction procedure for a three-dimensional weight function:

$$c'(v'_i, v'_j, v'_k)$$
$$= \begin{cases} c(v_i, v_j, v_k) & \text{for } 1 \le i, j, k \le n - 2, i \ne j, i \ne k, j \ne k \\ c(v_i, v_j, v_{n-1}) & \text{for } 1 \le i \ne j \le n - 2, k = n - 1 \\ c(v_i, v_{n-1}, v_n) + c(v_{n-1}, v_n, v_k) & \text{for } 1 \le i \ne k \le n - 2, j = n - 1 \\ c(v_n, v_j, v_k) & \text{for } 1 \le j \ne k \le n - 2, i = n - 1 \end{cases}$$

The greedy algorithm starts with contracting a "good" arc. We choose such an arc in the same way as in the CI Algorithm. Then we contract this arc, i.e., this arc appears in the final tour, and construct a graph with a vertex less. This step is repeated, until only three vertices remain. For this graph exactly two possible tours exist. We choose the smaller one of those, and finally we re-contract, i.e., all vertices are replaced by the paths which they consist of.

2.8 k-OPT Algorithm

The common characteristic of all previous algorithms is that in different ways they construct a tour. The first tour which is found is also the outputted tour. This is called a *construction heuristic*. In this section we introduce a so called *improvement heuristic*, i.e., it starts with a tour produced by a construction heuristic and improves it. For introducing the k-OPT algorithm [22] we need the following definition. Let a complete graph $G = (V, E)$, $|V| = n$ and a complete tour T be given, and let $k \le n$. Furthermore let a (two-dimensional or three-dimensional) weight function be given. A k-OPT step changes T by omitting k arcs from the tour and adding k arcs not from the tour in such a way that the set of arcs after the change is still a tour. T is called k-*optimum*, if no r-OPT step with $r \le k$ reduces the weight of the tour. Note that in general a k-optimum tour is not unique.

Each tour received by one of the previous construction heuristics can be transformed to a k-optimum tour by doing tour improving k-OPT steps as long, as they exist. As is customary in literature [17] we consider only the case $k \le 5$.

3   Exact Algorithms for the Traveling Salesman Problem of Second Order

3.1   Branch-and-Bound Algorithm

The following Branch-and-Bound Algorithm (BnB) in the worst case visits all possible tours in a lexicographic order and computes the tour with minimum costs.

To avoid that in fact all tours have to be visited, it computes (local) lower bounds and upper bounds by visiting and analyzing subpaths of all possible tours.

First we start with an arbitrary heuristic for the 2-TSP to compute a good upper bound. If at some state of the algorithm we consider a subpath, we compute a lower bound lb for a 2-TSP solution containing this subpath. As lower bound the approximated 2-AP solution (see Section 2.4) is computed. If lb is larger or equal than the current upper bound ub, we can prune this branch. The upper bound is updated, if a whole tour with smaller costs is visited. All tours are started with a fixed vertex v_1, which is chosen in such a way that the sum over all values $c(v_1, x, y)$ with $x \neq v_1, y \neq v_1, x \neq y$ is maximum. This choice is used, because we expect more prunes to appear, as the lower bounds in the first steps should be rather large.

3.2 Integer-Programming Algorithm

The AP can be described as an integer program as follows:

$$\min \sum_{i=1}^{n} \sum_{j=1,j\neq i}^{n} c_{ij}x_{ij} \tag{2}$$

$$\sum_{j=1,j\neq i}^{n} x_{ij} = 1 \ \forall 1 \leq i \leq n \tag{3}$$

$$\sum_{i=1,i\neq j}^{n} x_{ij} = 1 \ \forall 1 \leq j \leq n \tag{4}$$

$$x_{ij} \in \{0;1\} \ \forall 1 \leq i \neq j \leq n \tag{5}$$

where $C = (c_{ij})_{1\leq i\neq j\leq n}$ is the weight matrix of the AP instance. Equation (3) means that each vertex has exactly one out-arc and equation (4) that each vertex has exactly one in-arc. The AP solution consists of all arcs with $x_{ij} = 1$.

Similarly, we suggest to model the 2-AP by the following integer program:

$$\min \sum_{i=1}^{n} \sum_{j=1,j\neq i}^{n} \sum_{k=1,k\neq i,k\neq j}^{n} c_{ijk}x_{ijk} \tag{6}$$

$$\sum_{j=1,j\neq i}^{n} \sum_{k=1,k\neq i,k\neq j}^{n} x_{ijk} = 1 \ \forall 1 \leq i \leq n \tag{7}$$

$$\sum_{i=1,i\neq j}^{n} \sum_{k=1,k\neq i,k\neq j}^{n} x_{ijk} = 1 \ \forall 1 \leq j \leq n \tag{8}$$

$$\sum_{i=1,i\neq k}^{n} \sum_{j=1,j\neq i,j\neq k}^{n} x_{ijk} = 1 \ \forall 1 \leq k \leq n \tag{9}$$

$$\sum_{k=1,k\neq i,k\neq j}^{n} x_{ijk} = \sum_{k=1,k\neq i,k\neq j}^{n} x_{kij} \ \forall 1 \leq i \neq j \leq n \tag{10}$$

$$x_{ijk} \in \{0;1\} \ \forall 1 \leq i,j,k \leq n, \ i \neq j, i \neq k, j \neq k \tag{11}$$

where $C = (c_{ijk})_{1\leq i,j,k\leq n, \ i\neq j,i\neq k,j\neq k}$ is the weight matrix of the 2-AP instance.

Equation (7) means that each vertex appears exactly once as a first vertex of a path of three vertices of the tour, equation (8) that each vertex appears exactly once as a second vertex of a path of three vertices of the tour and equation (9) that each vertex appears exactly once as a third vertex of a path of three vertices of the tour. Equation (10) expresses the condition that, if i is a second vertex and j a third vertex of a path of three vertices of the tour, then there is another path of three vertices of the tour, where i is the first vertex and j is the second vertex.

Like for the AP and the ATSP, a solution of the 2-AP consists of $k \geq 1$ cycles. If $k = 1$, the solution of the 2-AP is an optimum 2-TSP solution. To avoid the possibility $k > 1$, we prevent each cycle $(v_{s_1}, \ldots, v_{s_t})$ by adding the following inequality to the integer program:

$$x_{s_{t-1},s_t,s_1} + x_{s_t,s_1,s_2} + \sum_{i=1}^{t-2} x_{s_i,s_{i+1},s_{i+2}} \leq t - 1 \tag{12}$$

Unfortunately an exponential number of inequalities of that type exists. For this purpose, we solve the integer program and – step by step – add inequalities of that type which are violated.

Furthermore we can use a good upper bound to speed-up the IP Algorithm.

4   Experimental Study

We implemented all algorithms in C++, where as subroutines we used the AP solver implemented by Jonker and Volgenant [27] and the IP solver CPLEX [26]. All experiments were carried out on a PC with an Athlon MP 1900+ CPU with 2GB memory.

As test instances we chose a class of random instances, where each entry c_{ijk} is independently chosen as an integer from $[0, \ldots, 10000]$. For the tests regarding the heuristics we computed the average over 1,000 instances for each dimension $3, 4, \ldots, 44$, and regarding the exact algorithms the average over 10 instances for each dimension $3, 4, \ldots, 24$. Furthermore we considered three real classes *ML*, *BMA*, *MAP* [15], which for each class and each dimension consist of one instance. For the heuristics we used the instances with dimension $3, 4, \ldots, 41$ and for the exact algorithms the instances with dimension $3, 4, \ldots, 17$. We observe that the real instances are much harder to solve than the random instances.

4.1   Comparison of Heuristics

An experimental study of heuristics for the ATSP is given in [18]. In this section we make a similar study for the 2-TSP. In detail, we compare all considered heuristics, which are Cheapest-Insert Algorithm (CI), Nearest-Neighbor Algorithm (NN), Two-Directional Nearest-Neighbor Algorithm (2NN), Assignment-Patching Algorithm (AK), Nearest-Neighbor-Patching Algorithm (NNK),

Two-Directional Nearest-Neighbor-Patching Algorithm (2NNK) and Greedy Algorithm (G). Furthermore we consider for each algorithm a version, where the algorithm is followed by the 5-OPT Algorithm.

All construction algorithms are rather fast, as they have complexity not worse than $O(n^3)$. We observe that the running times of all seven basic algorithms are very similar (except G which is slower) and the running times of all seven basic algorithms plus OPT steps are also very similar. Comparing the algorithms with and without OPT steps, the algorithms with OPT steps are considerably slower. Thus we only compare the upper bounds, where for the random instances we use the average value over 1,000 instances. We use the versions 2NN and 2NN+OPT as a basis for all comparisons and compare the remaining six algorithms against these versions by the difference of upper bounds. The results for the basic versions can be found in Figure 1 and the results for the OPT versions in Figure 2. If some lines do not appear in the diagrams, it means they have y-value 0. To compare the basic and the OPT versions, we only consider 2NN and 2NN+OPT as a representative. These results are shown in Figure 3.

The experiments show that the OPT versions clearly beat the basic versions. Furthermore the results for the random instances are completely different to the results for the real instances. For the real instances, CI and AK are the best algorithms, whereas for the random instances, 2NN and 2NNK are the best algorithms. AK is the worst algorithm for the random instances, and G is rather bad for all instances. Similar results as for the basic versions hold for the OPT versions.

Finally we compare the upper bounds with the optima computed by one of the exact algorithms BnB or IP. Again the results differ for the real and the random instances. Considering real instances, it holds that for 44 of 45 instances at least one (of seven) OPT heuristics finds the optimum. For the single remaining instance as well as for the unsuccessful heuristics we receive upper bounds very close to the optimum. Note that these results lead to many y-values 0 or at least small y-values in the diagrams. In contrast, for the random instances the upper bounds are more far away from the optimum.

4.2   Comparison of Exact Algorithms

In this section we compare the running times of the IP Algorithm and the BnB Algorithm, where all running times are given in seconds. The results can be found in Figure 4.

We observe that the random instances can be solved much faster than the real instances. Furthermore the BnB Algorithm is more efficient than the IP-Algorithm for the real *and* the random instances of our benchmark set. But on the other hand further experiments showed, that the IP-Algorithm was able to solve the MAP instance of dimension 21 in about three days and the BMA instance of dimension 26 in about three weeks, whereas the BnB Algorithm was not able to solve these instances. Thus for larger dimensions the IP-Algorithm becomes more efficient. For the IP Algorithm, the number of iterations of adding equalities of type (12) is the most important criterion for the running time. For

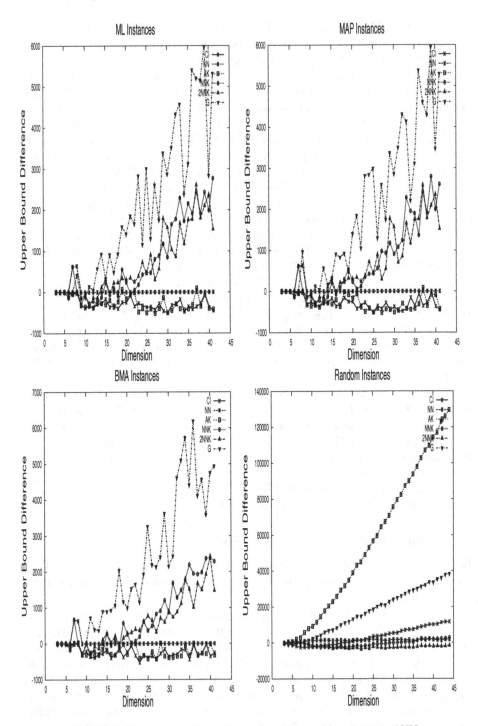

Fig. 1. Quality comparison of basic heuristics with respect to 2NN

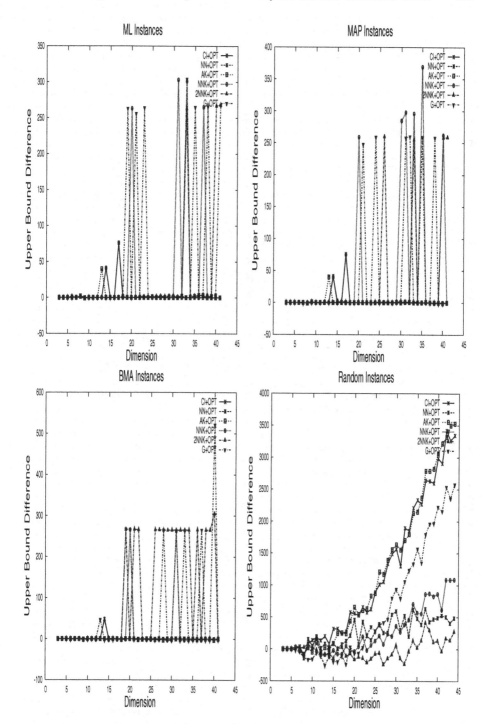

Fig. 2. Quality comparison of OPT heuristics with respect to 2NN+OPT

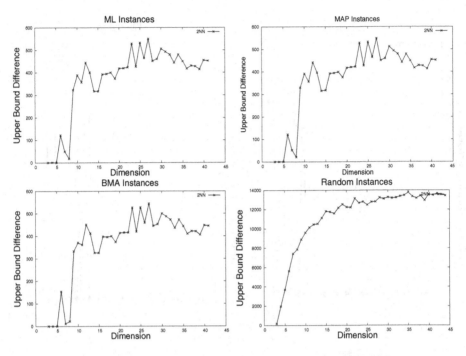

Fig. 3. Quality comparison of 2NN with respect to 2NN+OPT

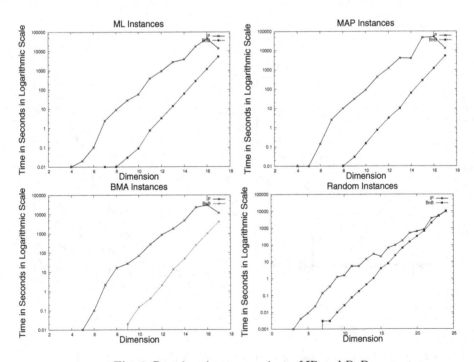

Fig. 4. Running time comparison of IP and BnB

real instances this number is very large: for dimension larger or equal than 10 we have at least 30 iterations. In contrast, the number of iterations is rather small for random instances: not more than 5 iterations for 195 of the 220 examples of dimensions $3, 4, \ldots, 24$. This is the main reason for the running time difference of the IP Algorithm between real and random instances.

5 Summary and Future Research

The purpose of this paper is to introduce a new combinatorial optimization problem with important applications in bioinformatics. We suggest seven heuristics and two exact algorithms for this problem and compare them in an experimental study. For the real instances, the best of our heuristics finds the optima in almost all cases, whereas an exact algorithm is able to compute instances up to dimension 26. For the exact algorithms the running times are considerably smaller for the random instances than for the real instances.

Nevertheless the paper aims to be only a starting point for research in this challenging area. It seems interesting to know, whether completely different heuristics as genetic algorithms [9] or tabu search [8] can be applied for this problem. Also for exact algorithms improvements seem to be possible. For example, the BnB Algorithm might be improved by different branching criteria or better lower bounds. Another possibility is to apply a branch-and-cut approach, as successfully done by the program package Concorde [1,25] for the STSP. Furthermore integer programming can be improved by cut-and-solve, as shown in [3] for the special case of the ATSP.

Acknowledgement

This work was supported by German Research Foundation (DFG) under grant number MO 645/7-3.

References

1. Applegate, D.L., Bixby, R.E., Chvátal, V., Cook, W.J.: The Traveling Salesman Problem. A Computational Study. Princeton University Press, Princeton (2006)
2. Bertsekas, D.P.: A New Algorithm for the Assignment Problem. Math. Program. 21, 152–171 (1981)
3. Climer, S., Zhang, W.: Cut-and-Solve: An Iterative Search Strategy for Combinatorial Optimization Problems. Artificial Intelligence 170(8), 714–738 (2006)
4. Dell'Amico, M., Toth, P.: Algorithms and Codes for Dense Assignment Problems: the State of the Art. Discrete Appl. Math. 100(1-2), 17–48 (2000)
5. Ellrott, K., Yang, C., Sladek, F.M., Jiang, T.: Identifying Transcription Factor Binding Sites Through Markov Chain Optimization. Bioinformatics 18, 100–109 (2002)
6. Fischer, F., Lau, A.: University of Chemnitz. Private Communication
7. Glover, F., Gutin, G., Yeo, A., Zverovich, A.: Construction Heuristics for the Asymmetric TSP. European J. Oper. Res. 129, 555–568 (2001)

8. Glover, F., Laguna, M.: Tabu Search. Kluwer, Dordrecht (1997)
9. Goldberg, D.E.: Genetic Algorithms in Search, Optimization, and Machine Learning. Addison-Wesley, Bonn (1989)
10. Goldberg, A.V., Kennedy, R.: An Efficient Cost Scaling Algorithm for the Assignment Problem. Math. Program. 71, 153–177 (1995)
11. Goldengorin, B., Jäger, G.: How To Make a Greedy Heuristic for the Asymmetric Traveling Salesman Competitive. SOM Research Report 05A11, University of Groningen, The Netherlands (2005)
12. Goldengorin, B., Jäger, G., Molitor, P.: Some Basics on Tolerances. In: Cheng, S.-W., Poon, C.K. (eds.) AAIM 2006. LNCS, vol. 4041, pp. 194–206. Springer, Heidelberg (2006)
13. Goldengorin, B., Jäger, G., Molitor, P.: Tolerances Applied in Combinatorial Optimization. J. Comput. Sci. 2(9), 716–734 (2006)
14. Goldengorin, B., Jäger, G., Molitor, P.: Tolerance Based Contract-or-Patch Heuristic for the Asymmetric TSP. In: Erlebach, T. (ed.) CAAN 2006. LNCS, vol. 4235, pp. 86–97. Springer, Heidelberg (2006)
15. Grosse, I.: University of Halle-Wittenberg, Chair for Bioinformatics. Private Communication
16. Gutin, G., Zverovich, A.: Evaluation of the Contract-or-Patch Heuristic for the Asymmetric TSP. INFOR 43(1), 23–31 (2005)
17. Helsgaun, K.: An Effective Implementation of the Lin-Kernighan Traveling Salesman Heuristic. European J. Oper. Res. 126(1), 106–130 (2000)
18. Johnson, D.S., Gutin, G., McGeoch, L.A., Yeo, A., Zhang, W., Zverovich, A.: Experimental Analysis of Heuristics for the ATSP. In: Gutin, G., Punnen, A.P. (eds.) The Traveling Salesman Problem and Its Variations, ch. 10, pp. 445–489. Kluwer, Dordrecht (2002)
19. Jonker, R., Volgenant, A.: A Shortest Augmenting Path Algorithm for Dense and Sparse Linear Assignment Problems. Computing 38, 325–340 (1987)
20. Karp, R.M.: Reducibility Among Combinatorial Problems. In: Miller, R.E., Thatcher, J.W. (eds.) Complexity of Computer Computations, pp. 85–103. Plenum, New York (1972)
21. Karp, R.M., Steele, J.M.: Probabilistic Analysis of Heuristics. In: Lawler, E.L., Lenstra, J.K., Rinnooy Kan, A.H.G., Shmoys, D.B. (eds.) The Traveling Salesman Problem, ch. 6, pp. 181–205. John Wiley & Sons, Chichester (1985)
22. Lin, S., Kernighan, B.W.: An Effective Heuristic Algorithm for the Traveling-Salesman Problem. Oper. Res. 21, 498–516 (1973)
23. Rosenkrantz, D.J., Stearns, R.E., Lewis, P.M.: An Analysis of Several Heuristics for the Traveling Salesman Problem. SIAM J. Comput. 6, 563–581 (1977)
24. Zhao, X., Huang, H., Speed, T.P.: Finding Short DNA Motifs Using Permuted Markov Models. Journal of Computational Biology 12, 894–906 (2005)
25. Source code of [1] (Concorde), http://www.tsp.gatech.edu/concorde.html
26. Homepage of Cplex, http://www.ilog.com/products/optimization/archive.cfm
27. Source code of [19], http://www.magiclogic.com/assignment.html

Fast Computation of Point-to-Point Paths on Time-Dependent Road Networks

Giacomo Nannicini[1,2], Philippe Baptiste[1], Daniel Krob[1], and Leo Liberti[1]

[1] LIX, École Polytechnique, F-91128 Palaiseau, France
{giacomon,baptiste,dk,liberti}@lix.polytechnique.fr
[2] Mediamobile, 10 rue d'Oradour sur Glane, Paris, France

Abstract. We propose an algorithm for the point-to-point time-dependent shortest path problem, using which good solutions may be found in a short time; our method provides an upper bound to the number of settled nodes for each shortest path computation, which is highly desirable in some industrial applications. In particular, we address a typical server scenario, where we have to compute point-to-point shortest paths in road networks where arc costs (travelling times) are time-dependent, and where each request has to be provided within an allotted time frame.

1 Introduction

Consider a weighted directed graph $G = (V, A)$, a set T of time instants, and a cost function $c : A \times T \to \mathbb{R}_+$: in our case, G represents a road network evaluated by travelling times, so the graph may not be Euclidean, and the travel time $c((u, v), \tau)$ on an arc (u, v) depends on the departure time τ from node u; this allows us to model situations such as "rush hours", where there are congestions at particular times during the day. We assume that G is strongly connected.

Our study is motivated by the following industrial application. We have a server machine that collects traffic information on a road network $G = (V, A)$, and has speed profiles based on historical data that models traffic situation on each arc for each time of the day. We want to provide users (e.g. GPS devices, web services) with point-to-point time-dependent shortest paths on G; for several reasons (secrecy, bandwidth, etc.) these speed profiles cannot be directly provided to the users, so computations have to be carried out by the server machine. This implies that each shortest path query has to be answered in a short time. Since the road network in question is covered by traffic sensors that provide the real-time and historical data used to compute the speed profiles, typically its size will not be huge, but we assume that it is large enough so that an application of Dijkstra's algorithm [1] is too slow for our needs. Assuming that the average number of shortest paths queries that have to be answered in a given time interval is known, we would like to guarantee that each computation can be carried out in the allotted time frame. Our method has a preprocessing phase that provides an upper bound on the number of nodes that have to be explored during a shortest path computation; this can be translated into an upper bound

B. Yang, D.-Z. Du, and C.A. Wang (Eds.): COCOA 2008, LNCS 5165, pp. 225–234, 2008.
© Springer-Verlag Berlin Heidelberg 2008

to the maximum computational time, using an upper bound on the time spent per node and on the time for each priority queue operation. By increasing the value of the approximation constant that is used throughout the whole method, one is able to decrease this upper bound on the number of explored nodes (up to a certain degree), so that the desired time requirements can be met.

In [2], the problem of finding the fastest path between two nodes on a time-dependent graph is addressed: let $\tau \in T$ be the time of arrival at node u; assuming that $c((u,v),\tau)$ is known for each $(u,v) \in A$ and for all possible $\tau \in T$, and that the network has the FIFO property (also called *non-overtaking property*), this problem can be polynomially solved [3] with Dijkstra's algorithm [1], in the same way as if arc costs were not time-dependent. If the FIFO property is not respected, then the problem is **NP**-hard (see [4]). We focus on the FIFO variant of this problem. For many industrial applications, Dijkstra's algorithm is not fast enough if applied on a large graph: a single application may require several seconds on a graph with millions of nodes. If one removes the constraints on the time-dependency of c, some practically efficient algorithms are [5,6]; the ALT algorithm (A^* with landmarks) has been tested in a time-dependent scenario as well (see [7]). The ALT algorithm is more efficient than Dijkstra's algorithm, but even if time requirements for each shortest path computation are small on average, they have a high variance, and thus this algorithm is not applicable to our problem. In [8], the vertex set is partitioned in clusters, and precomputed cluster distances are used to accelerate a Dijkstra search; the authors state that their method can be used in a time-dependent scenario, but they do not provide experimental results for this case. We first discussed the idea of *guarantee regions* in [9]; however, in that case we addressed a different scenario, where the cost function was not time-dependent. In this paper we extend those concepts to the time-dependent case, present a different and more efficient query algorithm, and provide a detailed experimental evaluation.

The rest of this paper is organized as follows: in Sect. 2 we define a guarantee region with an approximation property for the point-to-point shortest path problem on an unclustered graph; then we extend those concepts to a clustered graph, so as to make them useful in practice. In Sect. 3 we describe our query algorithm. In Sect. 4, we discuss some practical issues, such as how to effectively store guarantee regions, and give computational results. In the rest of this section we will give a problem definition and introduce our notation.

Problem definition. We consider the POINT-TO-POINT TIME-DEPENDENT SHORTEST PATH PROBLEM (PPTDSPP): given a directed graph $G = (V, A)$, two distinct vertices $s, t \in V$, a set of time instants T, a starting time $\tau_0 \in T$ and an arc weight function $c : A \times T \to \mathbb{R}_+$, find a path $p = (s = v_1, \ldots, v_k = t)$ in G such that the *time dependent path cost*, defined recursively as follows:

$$\phi(v_1, v_2) = c((v_1, v_2), \tau_0)$$
$$\phi(v_1, \ldots, v_i) = \phi(v_1, \ldots, v_{i-1}) + c((v_{i-1}, v_i), \phi(v_1, \ldots, v_{i-1}))$$

for all $2 \leq i \leq k$, is minimum.

In practice, the time-dependent cost function c is not necessarily fixed: there may be different cost functions for different day types (e.g. weekdays, holidays), or there may be updates to c based on real-time traffic information and traffic forecasting. For the sake of simplicity, throughout this paper we will use a fixed cost function.

We assume some lower and upper bounding functions $\lambda, \mu : A \to \mathbb{R}$ for c are known. This is a reasonable assumption, since the travel time prediction function c is modeled on historical data and is known in advance. For each arc $(u, v) \in A$, let $\lambda(u, v) = \min_{\tau \in T} c((u, v), \tau)$ be the minimum possible travel time on that arc, and let $\mu(u, v) = \max_{\tau \in T} c((u, v), \tau)$ be the maximum travel time; we assume that these values are known for all arcs in A. If the time-dependent cost function c is not fixed, we have to make sure that the lower and upper bounds are still valid. We naturally extend this functions to be defined on a whole path p, i.e. $\lambda(p) = \sum_{(u,v) \in p} \lambda(u, v)$, $\mu(p) = \sum_{(u,v) \in p} \mu(u, v)$. We call $G_\lambda = (V, A, \lambda)$ and $G_\mu = (V, A, \mu)$ the graph G weighted respectively by the lower and upper bounding functions λ, μ.

For $s, t \in V$ we denote the set of all paths $(s, \ldots, t)$ from s to t by $P(s, t)$, the set of all shortest paths from s to t with departure time τ_0 by $P^*_{\tau_0}(s, t)$, and the set of all shortest paths from s to t on the graph weighted by function λ (respectively, μ) as $P^*_\lambda(s, t)$ (respectively, $P^*_\mu(s, t)$). Given $U \subseteq V$ such that $s, t \in U$, let $G[U]$ be the subgraph of G induced by U. We denote the set of all paths between s and t in $G[U]$ by $P[U](s, t)$, the set of all shortest paths between s and t in $G[U]$ with departure time τ_0 by $P^*_{\tau_0}[U](s, t)$ and the set of all shortest path between s and t in $G[U]$ weighted by function λ (respectively, μ) by $P^*_\lambda[U](s, t)$ (respectively, $P^*_\mu[U](s, t)$).

We define $\phi^* : V \times V \times T \to \mathbb{R}^+$, $\phi^*(s, t, \tau_0) = \phi(p^*)$ for $p^* \in P^*_{\tau_0}(s, t)$ as the time-dependent cost of the optimal path from s to t with departure time τ_0.

2   Guarantee Regions

Given two nodes s and t, we can define a subset of V which guarantees an approximation property when computing a time-dependent path between those two nodes. The basic idea is as follows: we consider a path p between s and t; its cost, weighted by the upper bounding function $\mu(p)$, is an upper bound on the cost of the shortest path from s to t for any possible departure time. Then, for each node v, we consider a lower bound on the cost of the shortest $s \to t$ path passing through v: if the lower bound is greater or equal to $\mu(p)/K$, then v does not have to be explored to compute a K-approximated solution between s and t. This is formally stated in Defn. 2.1 and Prop. 2.2. We remark that these are the same concepts described in [9], but here we extend them to the time-dependent case.

Definition 2.1. *For $K > 1$, $s, t \in V$ and any path $p \in P(s, t)$, we define the guarantee region between nodes s and t as:*

$$\gamma_{st}(K, p) = \{v \in V | v \in p \lor \exists q \in P(s, t) \, (v \in q \land \lambda(q) < \frac{1}{K} \mu(p))\}.$$

Proposition 2.2. *For $K > 1$, $s, t \in V$, $p \in P(s,t)$ and $r^* \in P_{\tau_0}^*[\gamma_{st}(K,p)](s,t)$, we have $\phi(r^*) \leq K\phi^*(s,t,\tau_0)$ for any departure time τ_0.*

A natural choice for the "seed" path p is to consider the shortest path between s and t on the G_μ; this allows us to minimize $\mu(p)/K$.

Proposition 2.3. *Let $p^* \in P_\mu^*(s,t)$ be a shortest $s \to t$ path in G_μ, and $p \in P(s,t)$ be another (different) $s \to t$ path. If $p^* \subset \gamma_{st}(K,p)$ then $\gamma_{st}(K,p^*) \subseteq \gamma_{st}(K,p)$.*

Although the result only holds if $p^* \in P_\mu^*(s,t), p^* \subset \gamma_{st}(K,p)$, Prop. 2.3 is useful because it states that choosing the initial path p as the shortest path in G_μ is a good choice, even if it is not necessarily the best one. The trouble with the guarantee regions defined above is that, although only a pre-processing step, building all guarantee regions for all node pairs in a very large graph is not a feasible task with current technology. We deal with this problem by covering V with clusters $V_1, \ldots, V_k$.

Definition 2.4. *A covering $V_1, \ldots, V_k$ of V is valid if for all $i \leq k$ there is a vertex c_i such that for all other vertices $v \in V_i$ there is a path $p \in P(c_i, v)$ entirely contained in V_i. We call c_i the center of cluster i.*

For all $i \leq k$ let $\sigma_i = \max_{v \in V_i, p \in P_\mu^*(v,c_i)} \mu(p)$ and $\delta_i = \max_{v \in V_i, p \in P_\mu^*(c_i,v), p \subset V_i} \mu(p))$ be the cost of the longest shortest path in G_μ from v to c_i over all $v \in V_i$ and the cost of the longest shortest path in G_μ entirely contained in V_i from c_i to v over all $v \in V_i$. These values are finite because we assumed that G is strongly connected, and Defn. 2.4 ensures that a path entirely contained in V_i from the central node c_i to all other vertices in V_i exists.

To define guarantee regions that are valid for any two nodes in the source and destination cluster, we will proceed in the same way as before; in order to compute a valid upper bound on the cost of the shortest path between any node in the source cluster V_i and any node in the destination cluster V_j, we will need to consider not only the cost of a path between the centers of the two clusters, but also the radii σ_i and δ_j.

Definition 2.5. *Given a valid covering $V_1, \ldots, V_k$ of V, for $K > 1$, $i \neq j \leq k$ and any path $p \in P(c_i, c_j)$, we define the guarantee region between V_i and V_j as:*

$$\Gamma_{ij}(K,p) = \{v \in V | v \in p \lor v \in V_j \lor \exists q \in P(c_i, c_j) \ (v \in q \land \lambda(q) < \frac{1}{K}(\mu(p) + \sigma_i + \delta_j))\}.$$

Theorem 2.6. *Given a valid covering $V_1, \ldots, V_k$ of V, for all s, t in V and for $K > 1$ let $i = \arg\min_{n=1,\ldots,k}\{\phi^*(s, c_n, \tau_0)\}$, $j : t \in V_j$; suppose $i \neq j$, and let $p \in P(c_i, c_j)$. Then we have*

$$\min_{v \in \Gamma_{ij}(K,p)} \{\phi(q) + \phi(r) | q \in P_{\tau_0}^*(s,v), r \in P_{\phi(q)}^*[\Gamma_{ij}(K,p)](v,t), \phi(q) \leq \phi^*(s,c_i,\tau_0)\}$$

$$\leq K\phi^*(s,t,\tau_0)$$

for any departure time τ_0.

Thm. 2.6 suggests a query algorithm to compute valid paths between two nodes; the idea is to find, from the source node, the closest cluster center assuming departure time τ_0, and then "hop on" a guarantee region at that center. That is, if i is the index of the cluster whose center is the closest to s assuming departure time τ_0, and j is the index of the cluster which contains t, then after settling c_i we constrain the search to explore only nodes in the guarantee region between V_i and V_j. The query algorithm is described in Sect. 3.

A result similar to Prop. 2.3 holds for $\Gamma_{ij}(K, p^*)$ when $p^* \in P_\mu^*(c_i, c_j)$, and serves as a hint to choose our initial path. Unfortunately, guarantee regions defined this way may fail to be minimal (see [9]).

Proposition 2.7. *Given a valid covering $V_1, \ldots, V_k$ of V, for $i, j \leq k, i \neq j$ let $p^* \in P_\mu^*(c_i, c_j)$ be a shortest $c_i \to c_j$ path in G_μ, and $p \in P(c_i, c_j)$ be another (different) $c_i \to c_j$ path. If $q^* \subset \Gamma_{ij}(K, p)$ then $\Gamma_{ij}(K, p^*) \subseteq \Gamma_{ij}(K, p)$.*

3   Query Algorithm

Given a valid covering $V_1, \ldots, V_k$ for V, Thm. 2.6 points at a way to compute a K-approximated time-dependent path between any pair of nodes $s, t \in V$. Suppose we have already computed $\Gamma_{ij}(K, p^*) \ \forall i \neq j, 1 \leq i, j \leq k$, and for $p^* \in P_\mu^*(c_i, c_j)$; let us define $\Gamma_{ii}(K, p^*) = V \ \forall i, 1 \leq i \leq k$. We will use a slightly modified version of time-dependent Dijkstra's algorithm, where we will call $l[v]$ Dijkstra's algorithm label of a node $v \in V$, and we denote by $p[v]$ the parent node for node v. We assume to set $l[v] := \infty, p[v] := \text{nil} \ \forall v \in V$. Algorithm 1 respects the theorem's conditions; for simplicity, we will assume the departure time to be $\tau_0 = 0$.

Proposition 3.1. *Algorithm 1 computes a path p from s to t such that $\phi(p) \leq K\phi^*(s, t, \tau_0)$.*

In order to provide an upper bound on the computational time of each shortest path computation, we have to provide an upper bound on the number of nodes that are explored. The required upper bound on computational time can then be derived considering the maximum time spent per node (i.e. while settling the node with maximum degree in the graph) and the maximum time for a priority queue operation. It is straightforward to note that, once Algorithm 1 has switched to phase 2, then the number of nodes that can be explored is bounded from above by $|\Gamma_{ij}(K, p^*)| + |V_j|$, where i and j are, respectively, the index of the source and of the destination cluster. We have to provide a bound on the number of nodes explored before switching to phase 2: in order to do so we note that, if we restrict the algorithm in phase 1 to explore only nodes within V_i, where $s \in V_i$, then the approximation guarantee is still valid, although the solution quality may decrease. Thus we require that, if b nodes have already been explored in phase 1, then the algorithm is restricted to explore only nodes in V_i, until it switches to phase 2. It is easy to prove correctness of this approach. An upper bound on the number of explored nodes is then

$$b + |V_i| + |\Gamma_{ij}(K, p^*)| + |V_j|. \tag{1}$$

Algorithm 1. Compute a K-approximation of the time-dependent shortest path from a node s to a node t

```
 1: Let j : t ∈ Vⱼ
 2: Q ← {s}
 3: l[s] ← 0
 4: S ← φ
 5: stop ← false
 6: phase ← 1
 7: i ← 0
 8: while ¬stop do
 9:     extract x ← arg min_{q∈Q}{l[q]}
10:     S ← x
11:     if x = t then
12:         stop ← true
13:     if phase = 1 ∧ ∃n : x = cₙ then
14:         i ← n
15:         phase ← 2
16:     for all arcs (x, y) ∈ A do
17:         if phase = 1 ∨ y ∈ Γᵢⱼ(K, p*) then
18:             if y ∉ S then
19:                 if y ∉ Q then
20:                     l[y] ← l[x] + c((x, y), l[x])
21:                     p[y] ← x
22:                     Q ← Q ∪ {y}
23:                 else if l[x] + c((x, y), l[x]) < l[y] then
24:                     l[y] ← l[x] + c((x, y), l[x])
25:                     p[y] ← x
26: return t, p[t], p[p[t]], ..., s
```

The size of $\Gamma_{ij}(K, p^*)$ can be decreased by increasing K. A polynomial time algorithm that computes guarantee regions is described in [9].

4   Computational Experiments

Once all guarantee regions have been computed, we have to store them efficiently in memory for a fast access. This issue is crucial for performance, since the query algorithm has to test, for each node, whether it belongs to a given guarantee region or not, and thus the algorithm's efficiency depends on how quickly this answer can be given. Assuming that we know each node's position on a plane, a natural way to store node sets would be to define a geometric container for each guarantee region, e.g. an ellipse; however, with this approach the routine which tests if a node belongs to a given guarantee region yields too many false positive answers, which is due to the fact that guarantee regions are an union of paths, and thus their shape is not necessarily easy to model (see Fig. 1). Our approach to solve this problem is to associate, with each node, a bit table, or "bit flags", which are used to determine if a node belongs to the guarantee region between

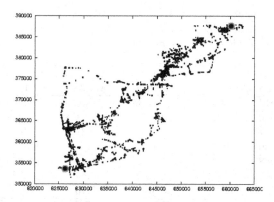

Fig. 1. Graphical representation of a guarantee region on a plane. Gray circled dots represent source and destination node, while small black square dots represent nodes within the guarantee region.

clusters i and j, which we call Γ_{ij}, for given $i \neq j$. Suppose we have covered V with k clusters $V_1, \ldots, V_k$; then we associate a table T of $k \times (k-1)$ bits each with each node v, with the property that the j-th bit of the i-th row of T is 1 if and only if: $v \in \Gamma_{ij}$ if $j < i$, or $v \in \Gamma_{i(j+1)}$ if $j \geq i$. Since $\Gamma_{ii} = V_i$, the corresponding information does not have to be stored, thus each row can have only $k - 1$ elements.

We used a subgraph of the French road network, corresponding to Île-de-France (i.e. Paris and surroundings), to validate our approach with a prototype. This subgraph has ≈ 400000 vertices and ≈ 800000 arcs. Time-dependent costs were modeled as piecewise linear functions of time (expressed in seconds); that is, on each edge we stored 24 breakpoint values, one for each hour over a day, and the arc cost for a given second τ was computed via a linear interpolation of the breakpoints preceding and following τ. For a subset of arcs (8374 arcs in total, all of them corresponding to highways or high importance roads) we used real historical data to compute the breakpoint values for weekdays; for the remaining arcs we generated breakpoint values using the traffic-free speed value for that arc over a day, and then generating two bendings in the speed profile so as to slow the arc down by a factor of 1.5–3 during peak hours, with each drop centered at 8 AM or 6 PM and lasting 3-5 hours. This empirical way to generate time-dependent costs was not meant to be completely realistic, but at least it should provide "reasonable" data. Then, for each query, we randomly generated with a uniform distribution a departure time in seconds between 7 AM and 7 PM, so that the optimal time-dependent solution had a very high probability of being different than the traffic-free static solution.

To validate the clustered approach we generated a k-center clustering over V, with $k = 100$ clusters, using k'-oversampling with $k' = 200$ (see [8]); that is, we picked 200 random nodes, we connected them to a "dummy" central node, and we grew clusters of neighbouring nodes around each of the 200 centers. Then, when all nodes had been assigned to a cluster, we progressively deleted

Table 1. Computational results on clustered graph: average values. A * in the first column indicates that the value for K has been adaptively chosen, and we report the starting value, which is also the maximum one.

MAX K	# SETTLED NODES		SOLUTION COST INCREASE		CPU TIME	IMPROVED
	DIJKSTRA	APPROX	NAIVE	APPROX	SAVINGS	PATHS
3	185514	60045	4.56%	1.10%	53.99%	91.8%
3.5	194640	35561	4.43%	4.91%	74.66%	74.5%
4	190077	15597	4.58%	9.27%	87.69%	53.2%
4.5	193240	9943	4.46%	16.51%	91.51%	38.2%
3.5*	188988	29341	4.38%	1.75%	78.22%	76.4%
4*	184327	17256	4.79%	4.54%	86.28%	67.2%
4.5*	190944	12675	4.40%	5.40%	91.72%	58.2%

the smallest remaining cluster, i.e. the one with the smallest radius, allowing other clusters to grow into the deleted one. We iterated this procedure until 100 clusters were left. We compared the number of explored and settled nodes between a Dijkstra search and Algorithm 1, where source and destination node were chosen at random. We also compared the results with respect to the naive algorithm of computing the shortest path in the static traffic-free graph, i.e. G_λ, and then applying time-dependent costs. Results are reported in Table 1. For each value of K (first column), we indicate the average number of settled nodes in 1000 Dijkstra searches on the full graph, the average number of settled nodes with Algorithm 1 and the same source-destination pairs, the average percentage increase P of the naive solution value with respect to the optimum (that is, if p^* is the optimal solution and p is the naive solution, the average value of $(1 - \phi(p)/\phi(p^*))$), the average percentage increase of the approximated solution value with respect to the optimum, the average CPU time savings of Algorithm 1 in percentage of the CPU times saved with respect to the exact algorithm (0% means as slow as the exact algorithm; a negative value means that there is an increase in CPU time, while a positive value means that CPU time decreased), and the percentage of shortest path computations where the approximated a solution had a cost smaller or equal than the cost of the naive solution. We do not provide exact query times because those are highly dependent on the implementation; what is most interesting, here, is the speed-up with respect to plain Dijkstra's algorithm in terms of number of settled nodes and of relative CPU time. The number of settled nodes for the naive approach is not relevant: many speed-up techniques exist for the static case (see [10]), so we can assume that it is a fast computation.

While for small values of K computational times could increase, due to the overhead for constraining the Dijkstra search within the boundaries of the guarantee region, for large enough values of the approximation constant the savings in CPU time are significant, with a small average decrease of the solution quality with respect to the optimum. We note, however, that the naive solution has a better average behaviour than our approximated solution for values of $K \geq 3.5$. We tried to investigate the reason behind this. We can see that, for $K = 3.5$,

Table 2. Computational results on clustered graph with a maximum number of settled nodes for each computation: average values

MAX NODES	# SETTLED NODES		SOLUTION COST INCREASE		CPU TIME SAVINGS	IMPROVED PATHS
	DIJKSTRA	APPROX	NAIVE	APPROX		
50000	189994	23238	3.77%	3.98%	81.91%	74.8%
65000	190698	27159	3.73%	3.13%	78.54%	77.0%
80000	196529	37771	3.72%	2.71%	71.32%	81.4%

in 74.5% of the shortest path computations the approximated solution is better than the naive one, but in the remaining cases the approximated solution is very far from the optimum, while the naive one isn't. This is due to the fact that, if K is too large, then the guarantee region between two clusters i and j may consist of only the shortest path between c_i and c_j on G_μ. Any approximated solution between those two clusters will pass through that path, which leads to poor performance. For $K = 3$, in 91.8% of the shortest path computations the approximated solution has a cost which is smaller than the cost of the naive solution, so the average behaviour of the approximated solution is satisfying. However, the computation is only 54% faster than a full (unconstrained) Dijkstra search.

To deal with this issue, we adaptively chose the value of the approximation constant K as follows: for each cluster pair, we started with the maximum value for K ($K_{curr} \leftarrow K_{max}$), and if the computed guarantee region included only the path between the cluster centers we decreased K by 10% ($K_{curr} \leftarrow 0.9K_{curr}$). We iterated until the guarantee region for that cluster pair had a cardinality which was greater than the number of nodes on the shortest path on G_μ between the cluster centers. Results for this approach are reported in Table 1, on the rows with a * in the first column. With this modification in the guarantee regions generation process, we see that the solution quality significantly increases, while still yielding a speed-up in computational time. Moreover, we are able to obtain a better average behaviour than the naive approach for larger values of K that allow for an increased speed-up factor, which is a necessary requirement to state that our approach can be useful in practice. We can also see that, if we compare the number of paths where the approximated solution is better than the naive one, there is an improvement with respect to the basic version of the algorithm. Although for $K = 3.5$ the naive solution will be better than the one computed with our algorithm almost 25% of the times, from a practical point of view the approximated solution has much more value with respect to the naive one, because it changes dynamically reflecting traffic changes; on the other hand, the naive solution between two points is always fixed regardless of the time of the day, which is negatively perceived by users.

We also tested the performance of the approach described in Sect. 3 with a maximum number of settled nodes for each point-to-point computation. In order to do this, we initially set $K = 3$, and if necessary we increased its value until all regions comprised a number of nodes smaller than a given threshold. In (1), we set $b = 4000$. Results are reported in Table 2 (same columns as in previous tables). We can easily observe that the algorithm's performance does not decrease, and if we are willing to settle up to 65000 nodes for each shortest path computation then our method finds a path which is on average better than the naive solution, while still yielding a speed-up factor of almost 5 with respect to plain Dijkstra's algorithm.

5 Conclusion

We proposed an algorithm for the point-to-point time-dependent shortest path problem; our approach is based on the definition of a covering of the vertex set, and then for each pair of covering sets we use lower and upper bounding functions for arc costs to define the set of nodes that has to be explored during a Dijkstra search in order to compute a K-approximated path for chosen K. With this approach we are able to give an upper bound on the number of settled nodes during each shortest path computation; this, in turn, can be used to provide an upper bound on the total time of the computation, which is very desirable in some industrial applications. We have proposed and tested an algorithm to implement this method in practice. Computational results show that our approach results in a speed-up with respect to plain Dijkstra's algorithm, with a small average decrease in the solution quality.

References

1. Dijkstra, E.: A note on two problems in connexion with graphs. Numerische Mathematik 1, 269–271 (1959)
2. Dreyfus, S.: An appraisal of some shortest-path algorithms. Operations Research 17(3), 395–412 (1969)
3. Kaufman, D.E., Smith, R.L.: Fastest paths in time-dependent networks for intelligent vehicle-highway systems application. Journal of Intelligent Transportation Systems 1(1), 1–11 (1993)
4. Orda, A., Rom, R.: Shortest-path and minimum delay algorithms in networks with time-dependent edge-length. Journal of the ACM 37(3), 607–625 (1990)
5. Goldberg, A., Kaplan, H., Werneck, R.: Reach for A^*: Efficient point-to-point shortest path algorithms. In: Proceedings of the 8th Workshop on Algorithm Engineering and Experiments (ALENEX 2006). LNCS, pp. 129–143. Springer, Heidelberg (2006)
6. Sanders, P., Schultes, D.: Engineering fast route planning algorithms. In: [11], pp. 23–36
7. Delling, D., Wagner, D.: Landmark-based routing in dynamic graphs. In: [11], pp. 52–65
8. Maue, J., Sanders, P., Matijevic, D.: Goal directed shortest path queries using precomputed cluster distances. In: Alvarez, C., Serna, M.J. (eds.) WEA 2006. LNCS, vol. 4007, pp. 316–327. Springer, Heidelberg (2006)
9. Nannicini, G., Baptiste, P., Krob, D., Liberti, L.: Fast paths in dynamic road networks. In: Quillot, A., Mahey, P. (eds.) Proceedings of ROADEF 2008, Clermont-Ferrand, Université Blaise Pascal (2008)
10. Wagner, D., Willhalm, T.: Speed-up techniques for shortest-path computations. In: Thomas, W., Weil, P. (eds.) STACS 2007. LNCS, vol. 4393, pp. 23–36. Springer, Heidelberg (2007)
11. Demetrescu, C. (ed.): WEA 2007. LNCS, vol. 4525. Springer, Heidelberg (2007)

Ant Colony Optimization Metaheuristic for the Traffic Grooming in WDM Networks

Xiangyong Li, Yash Aneja, and Fazle Baki

Odette School of Business, University of Windsor, Windsor City, Ontario, Canada
N9B 3P4

Abstract. This paper studies the routing strategy in non-bifurcated traffic grooming in WDM networks. It is to optimally route the specified traffic over a given logical topology to minimize the congestion of the WDM network. We first present the node-arc formulation. To overcome the computational complexity by implementing exact algorithms, we present an ant colony optimization (ACO) metaheuristic. The computational results compared to those of exact algorithms demonstrate that ACO is a computationally efficient and suitable approach for obtaining high-quality routing strategy in non-bifurcated traffic grooming problem in WDM networks.

Keywords: WDM networks, traffic grooming, routing strategy, ant colony optimization, metaheuristic.

1 Introduction

Wavelength Division Multiplexing (WDM) is emerging as a dominant and successful technology for use in backbone networks. Generally, many small telecommunications flows are grouped into larger units, which can be processed as single entities. This is called traffic grooming, which is to find the optimum strategy to handle a set of request with deterministic source and destination nodes for data communication. In general, traffic grooming can be divided into two categories, i.e. the bifurcated model of traffic grooming [1] and the non-bifurcated model of traffic grooming. In the first approach, the communication $t(s, d)$ from end-nodes s to d can be split into a number of components, and the different components may be communicated using different logical paths from s to d. The latter approach deals with the case where each communication request can only be communicated by a single lightpath [2]. The bifurcated model routing strategy with splitting traffic flow has been widely discussed in [3,4]. In [2], the non-bifurcated traffic grooming is studied, which has the objective of minimizing the cost of the network by minimizing the number of lightpaths and maximizing the throughput of the network. Hu and Leida [5] studied traffic grooming, routing, and wavelength assignment in optimal WDM mesh networks (GRWA), where the objective is to minimize the number of transponders. The non-bifurcated traffic grooming has three subproblems: (1) find a logical topology; (2) carry out RWA for each lightpath; (3) find an optimal traffic routing strategy [6,2].

B. Yang, D.-Z. Du, and C.A. Wang (Eds.): COCOA 2008, LNCS 5165, pp. 235–245, 2008.

This paper focuses on the routing problem in non-bifurcated traffic grooming with the objective of minimizing the network congestion. The value of network congestion is defined as the maximum traffic load on a logical link [4,6]. For a given logical topology of a WDM network, the MILP of the routing subproblem can be derived based on the node-arc representation. Such formulation can be directly solved by a mathematical programming tool. But this formulation can only work well for smaller networks. To the best of our knowledge, no study has worked on the non-bifurcated traffic grooming minimizing the congestion.

The aim of this paper to deal with the routing strategy in non-bifurcated traffic grooming to minimize the congestion in the WDM network by a metaheuristic. In a WDM network, there exists a communication request between each pairs of end-nodes. The total number of binary variables thus will become very large. This makes the traffic grooming problem computationally challenging. It is not practical to get optimal routing strategy to minimize congestion for large WDM network. A feasible alternative way is to quickly produce good, although not necessarily optimal, routing strategy by implementing the heuristics or metaheuristics. In this paper, we present an ant colony optimization metaheuristic to deal with the non-bifurcated traffic grooming problem with the objective of minimizing congestion. The remainder of this paper is organized as follows. In section 2, we give the arc-node formulation of the traffic grooming problem. In section 3, we present an ACO-based metaheuristic. Some experiments are carried out to investigate the performance of ACO in Section 4. Our conclusions are summarized in Section 5.

2   Problem Formulation

We first outline some notations as follows:

- N: the node set, $N = \{1, 2, 3, \cdots, n\}$, where n is the number of nodes;
- A: the arc set, $A = \{(i, j) | i, j \in N, i \neq j\}$;
- K: the set of traffic demands, $K = \{1, 2, 3, \cdots, |K|\}$;
- $|K|$: number of traffic demands; generally total number of traffic demands is $n \times (n - 1)$ in WDM networks;
- q^k: the quantity of traffic demand k;
- s_k: the source node of traffic demand k;
- d_k: the destination node of traffic demand k;
- x_{ij}^k: a binary decision variable; it equals 1 if the entire quantity of traffic demand k is transferred on arc (i, j), and equals 0 otherwise;
- λ_{max}: the congestion in the network, which represents the maximum traffic load on the arcs in the network.

The node-arc formulation of the grooming problem is defined as follows:

$$\min \lambda_{max} \tag{1}$$

$$s.t. \quad \sum_{k \in K} q^k x_{ij}^k \leq \lambda_{max}, \quad \forall (i, j) \in A \tag{2}$$

$$\sum_{j:(i,j)\in A} x_{ij}^k - \sum_{j:(j,i)\in A} x_{ji}^k = \begin{cases} 1, & \text{if } i = s_k \\ -1, & \text{if } i = d_k \quad \forall i \in N, \forall k \in K \\ 0, & \text{otherwise} \end{cases} \tag{3}$$

$$x_{ij}^k \in \{0,1\}, \quad \forall (i,j) \in A, \forall k \in K \tag{4}$$

where constraint (2) defines the value of congestion in the WDM network.

3 ACO-Based Metaheuristic for Traffic Grooming Problems

We consider ant colony optimization (ACO) as an alternative method to solve traffic grooming in WDM networks, especially for large networks. ACO is a nature-inspired metaheuristic for hard CO problems [7]. The inspiring source underlying ACO is the foraging behavior of real ants. The central mechanism of ACO algorithms is to probabilistically construct solutions for the CO problem by a parameterized probability model indicated by the pheromone trails. For the detailed description of ACO and its variants, the readers can refer to [7].

The ACO algorithm for traffic grooming problem is a $\mathcal{MAX}$-$\mathcal{MIN}$ ant system. It forces the pheromone values in $[\tau_{min}, \tau_{max}]$ that increases the diversification and prevents the algorithm from converging to a solution. To automatically scale the objective functions, we implement the algorithm in the hyper-cube framework [8], where the pheromone trails values are restricted in $[0,1]$. Henceforth, we refer to our proposed ACO algorithm as ACO-TG. ACO-TG acts in a similar way as simple-ACO (S-ACO) [7]. Considering special network properties, each traffic demand k is associated with one artificial ant, which independently searches a path joining its source and destination nodes. At each iteration, one ant starts from its origin and is launched toward the destination, where a greedy stochastic policy is used to choose the next node to move to. Such policy is indicated by the pheromone trails and heuristic information. The ants act independently and communicate in an indirect way through pheromone trails and network's current traffic status. After one ant builds its path, the traffic state of the network is dynamically updated and the traffic information is used for path constructions for other commodities. Algorithm 1 gives the outline of ACO-TG. Before discussing ACO-TG in detail, we first define some symbols used.

- p^k: constructed path of traffic demand k linking its origin and destination;
- P: the set of paths already constructed at each iteration, $P = \{p^k | k \in K\}$;
- λ^*: the minimum congestion of the network since the start of ACO-TG;
- P^*: the set of paths of the network with minimum congestion;
- D: the matrix of current traffic status of the WDM network;
- i_{no}: maximum allowed iterations without solution improvement found;

(1) Definition of pheromone trails and heuristic information. In traffic grooming, we need to find an independent path for each traffic demand. Accordingly, we maintain an independent pheromone matrix $\mathcal{T}_k$ associated with each

Algorithm 1.   ACO for traffic grooming (ACO-TG) in WDM networks

```
input the data of the problem instance
initialize pheromone trails T = {T_k|k ∈ K} and statistical data
set algorithmic parameters
while (termination conditions not met) do
    P ← ∅
    InitialNetworkTraffic(A)
    for k = 1 to |K| do
        ComputeHeuristic(D, A)
        p^k ← PathConstruction(T_k, η, k)    {See Algorithm 2}
        P ← P ∪ p^k
        UpdateTraffic(D, p^k)
    end for
    LocalSearch(P, λ_max)
    λ_max = EvaluateSolution(P)
    if P* = ∅ or λ_max < λ* then
        i_0 = 0, P* ← P and λ* = λ_max
    else if
        i_0 ← i_0 + 1
    end if
    if i_0 ≥ i_no then
        ResetPherromoneTrails(T, A)
    else
        PheromoneTrailsUpdate(T, P*, ρ)
    end if
end while
output P* and λ*
```

traffic demand k, which is defined as $T_k = \{\tau_{ijk}|(i,j) \in A\}$. The entry τ_{ijk} indicates the learned desirability for an ant to move to node j immediately after i in the path of traffic demand k. The pheromone trails represent the search memory and are set to 0.5 at the start of ACO-TG.

The heuristic information η_{ij} is assigned to each arc (i,j). It is a problem-specific information, which represents the attractability of visiting node j as the next node in current partial path when the ant is located at i. j is one of the neighboring nodes, i.e. $j \in \mathcal{N}(p^k)$. We first describe the definition of $\mathcal{N}(p^k)$, which is also used in the path construction (see section 3.4). Let i be the end-node of partial path p^k of traffic demand k. Suppose $\mathcal{N}_1 = \{j|(i,j) \in A, j \notin p^k\}$ and $\mathcal{N}_2 = \{j|(i,j) \in A\}$. $\mathcal{N}(p^k)$ defines the set of all neighboring nodes j, which can be included in partial path p^k of commodity k. It can be defined as follows:

$$\mathcal{N}(p^k) = \begin{cases} \mathcal{N}_1, \text{ if } \mathcal{N}_1 \neq \emptyset \\ \mathcal{N}_2, \text{ if } \mathcal{N}_1 = \emptyset \end{cases} \quad (5)$$

The ant first chooses the next node among all unvisited nodes. If the ant reaches a dead node, then we allow the ant to go back to one node already visited.

We consider two kinds of information to define the heuristic information. The heuristic information is first to reflect current traffic state of the network. Namely, the ants will prefer arcs with less traffic load. Since there is a communication request between each pair of end-nodes, the total number of commodities will become very large. To get a better routing strategy, we should reduce the influence of one path on possible congestion status of the network. With this thought in mind, we naturally make the ants prefer shorter paths. Here the shortest path is the one containing minimum number of arcs, which connect the next node j to move and the destination of current traffic demand. All ants share

Algorithm 2. Path construction in ACO-TG — PathConstruction($\mathcal{T}_k$, η, k)
    $p^k = \{s_k\}$, $current \leftarrow s_k$
    while $(current \neq d_k)$ **do**
        Choose next node $\nu \in \mathcal{N}(p^k)$ according to the policy in Equations (7)-(8)
        $p^k \leftarrow p^k \bigcup \nu$;
    end while
    PathCycleDelete(p^k)
    output p^k

the same heuristic information. It is computed as follows:

$$\eta_{ij} = (\eta_{ij}^1)^\beta (\eta_{ij}^2)^\gamma$$
$$= \left(1 - \frac{D_{ij}}{\sum\limits_{j' \in \mathcal{N}(p^k)} D_{ij'}}\right)^\beta \left(\frac{1}{W_{jd_k}}\right)^\gamma \tag{6}$$

where η_{ij} is the function of η_{ij}^1 and η_{ij}^2. η_{ij}^1 reflects current traffic state of the network. It is a normalized value in $[0, 1]$. η_{ij}^1 contains two special cases. The first is the situation where all the neighboring arcs except one arc (i, j_0) have 0 traffic. With the above equation, $\eta_{ij_0}^1 = 0$. In order to select this arc with a positive probability, we can set it to a small value, e.g. $\eta_{ij_0}^1 = 0.001$. Moreover, at the start of the iteration, we set all the η_{ij}^1 equal to 1. η_{ij}^2 is inversely proportional to the shortest distance W_{jd_k} between j and the destination node d_k of current traffic demand k. When j is the destination node, W_{jd_k} is set to 1. β and γ are two coefficients, which weigh the relative importance of two components.

(2) ComputeHeuristic(D, A). The heuristic information is updated dynamically based on the traffic state of the networks. After the path is built for one demand, its quantity will be placed on the path and current traffic status D will thus be updated. Such traffic status will be used by other ants in path construction. Before constructing a new path, the heuristic information is first computed that is achieved by the function ComputeHeuristic(D, A).

(3) InitialNetworkTraffic(A). It is to initialize the traffic state of the network. In ACO-TG, the network's current traffic state is the result of indirect cooperation of the ants and the decision base for path construction. At the start of each iteration, the network is initialized as a null network.

(4) PathConstruction($\mathcal{T}_k$, k). The most important component of ACO-TG is the greedy stochastic policy applied to construct the path of each traffic demand in the network. Such selection policy is a probability model that is indicated by the pheromone trails and heuristic information. In the path construction, an ant independently and incrementally builds a path for traffic demand k by sequentially adding a node to the partial path p^k until it reaches the destination. The ant starts from the origin of traffic demand k and moves step by step toward its destination. The procedure of path solution is given in Algorithm 2. We take traffic demand k as an example to explain how its path is built probabilistically.

Located at current node i, the ant selects the next node j to move to by a pseudorandom proportional rule as follows:

$$j = \begin{cases} \arg\max_{\nu \in \mathcal{N}(p^k)} \{\tau_{i\nu k} \cdot \eta_{i\nu}\}, & \text{if } rand() \leq p_0 \\ J, & \text{otherwise} \end{cases} \tag{7}$$

where $rand()$ is a random number uniformly distributed in $[0, 1]$. p_0 is parameter controlling the balance between diversification and intensification. When $rand() > p_0$, the next node j is probabilistically determined by some selection strategies. Here the probabilistic choice of the next node is performed analogous to the roulette wheel selection procedure of evolutionary computation. The probability distribution is defined as follows:

$$p_{ij}^k = \begin{cases} \dfrac{\tau_{ijk} \cdot \eta_{ij}}{\sum\limits_{j' \in \mathcal{N}(p^k)} \tau_{ij'k} \cdot \eta_{ij'}}, & \text{if } j \in \mathcal{N}(p^k) \\ 0, & \text{otherwise} \end{cases} \tag{8}$$

where p_{ij}^k denotes the probability of node j being chosen as the target in the partial path of traffic demand k. Since an ant is allowed to return to the already visited nodes, the cycle may occur. This further results in the situation where the path may last for a long time. That is to say a path covers many nodes. We consider a parameter max_l to control the maximum lifetime of each path. While containing more than max_l nodes, the ant deletes the path and the memory about the path, and re-constructs the corresponding path. In our experiment, max_l is set equal to 25. In order to minimize the influence of one path on the traffic state in the network, we delete the cycle in each path. It is achieved by the function of PathCycleDelete(p^k) in Algorithm 2.

After the path is constructed for one traffic demand, the corresponding quantity will be placed on the network. And the local traffic of the network is accordingly updated, which is achieved by the procedure UpdateTraffic(D, p^k). The new traffic information will be used in building the paths for other commodities.

(5) **PheromoneTrailsUpdate($\mathcal{T}, \eta, \rho$).** The aim of update of pheromone trails is that the information in some good solutions should be indicated by the pheromone trails and the nodes in these solutions will be biased by other ants in sequent path constructing. Such update contains two subprocedures, i.e., evaporation and reinforcement. Pheromone trails on all arcs will first be reduced by evaporation ratio. Second, some additional pheromone trails will be deposited on the arcs in some good solutions. Since we define an independent pheromone matrix for each traffic demand, an independent update is associated with each traffic demand that is very different from the implementation in general ACO methods. In the traffic grooming, the only task of the ants is to find a path for each traffic demand. Therefore, we cannot associate an evaluation function with the path of each traffic demand. In fact, all ants cooperate with each other to find a traffic pattern with minimum congestion. As a result, the ants should share the collective experience. As a natural idea, more additional reinforcement should be given to the paths when a better traffic mode (less congestion) is

found. Therefore, we define a best path p_{best}^k for each traffic demand k. It is the path when the network has a best traffic mode. Since ACO-TG is implemented in hyper-cube framework, the following update rule is considered:

$$\tau_{ijk} \leftarrow \tau_{ijk} + \rho(\chi(i, j, p_{best}^k) - \tau_{ijk}), \forall \tau_{ijk} \in \mathcal{T}_k \qquad (9)$$

The function χ is given as

$$\chi(i, j, p_{best}^k) = \begin{cases} 1, \text{ if arc } (i, j) \text{ is included in path } p_{best}^k \text{ of traffic demand } k \\ 0, \text{ Otherwise} \end{cases}$$
$$(10)$$

where ρ is the reinforcement that is set to $\rho = 0.1$ in our implementation. By the above procedure, the pheromone matrix associated with each traffic demand is sequentially renewed. It is important to note that using the best path to update the pheromone trails is an exploitation procedure. By such approach, the search memory can be used and reinforced that is an intensification strategy.

As indicated previously, ACO-TG is a $\mathcal{MAX}$-$\mathcal{MIN}$ ant system. An upper bound τ_{max} and lower bound τ_{min} are added to the pheromone values. In our implementation, we have set the lower bound to 0.001 and upper bound to 0.999. After the pheromone update, we set the pheromone values in $[\tau_{min}, \tau_{max}]$:

$$\tau_{ijk} \leftarrow \min\{\tau_{max}, \max\{\tau_{min}, \tau_{ijk}\}\} \qquad (11)$$

(6) Diversification and intensification strategies. The diversification and intensification strategies are considered to enhance the performance of ACO-TG. We consider the following simple local search, which works as follows.

(1) Find the arc (i^*, j^*) with traffic flow λ_{max};
(2) Find the set K^* of commodities which use arc (i^*, j^*) in their paths;
(3) Sort the commodities in K^* in descending order of the quantities;
(4) For each traffic demand k' in K^*, we consider three different operations to compute the maximum congestion saving $\Delta_{k'}$: (a) Break the arc (i^*, j^*), and insert an unvisited node between in i^* and j^* with maximum congestion saving; (b) Break the path and reconnect the predecessor node of i^* to node j^*; (c) Break the path and reconnect node i^* to the successor node of i^*;
(5) If the maximum congestion saving in K^* is bigger than 0, we perform the operations and reconstruct a path for the corresponding traffic demand.

As an intensification strategy, the best path is used to update the pheromone trails. This may make ACO-TG trapped in local optimum. To lead the algorithm escape from local optimum, ResetPherromoneTrails($\mathcal{T}, A$) sets all the pheromone values back to 0.5 when the number of iterations without improvement found reaches to the value of i_{no}.

4 Experiment and Computational Results

In this section, we carry out a series of experiments to investigate the performance of ACO-TG on some generated instances with practical WDM networks.

ACO-TG is coded in C language and complied on VC++ 6.0. All the experiments are implemented on a PC with Intel CPU 2.2G under Windows XP.

4.1 Benchmark Testing Problems

The testing instances are generated based on two real-world networks: the US National Science Foundation network, the NSF network, and the British Synchronous Digital Hierarchy network, the SDH network. The NSF network is the old US T1 backbone that is composed of 14 nodes, 21 bidirectional links (42 arcs). The SDH network contains 30 nodes, 55 bidirectional links (110 arcs). Based on NSF and SDH networks, we randomly generated two sets of instances, respectively. The first set consists of 20 instances with NSF network. The second set contains 10 instances with SDH network. For all the instances, there exists one communication request between each pair of nodes. The quantity of each traffic demand is an integer uniformly distributed in $[5, 50]$. The arc-node formulation for the NSF network has 7644 binary variables and 2590 constraints. However for the SDH network, there are 95700 binary variables and 26210 constraints. Therefore, it takes a long time to solve the arc-node formulation by exact algorithms. This is also verified in the following experiments.

4.2 Parameters Tuning

The preliminary experiment focuses on the tuning of main parameters. We consider the following candidate set, i.e. $\beta \in \{1, 2, 3, 5, 8, 10\}$, $\gamma \in \{1, 2, 3, 4, 5, 6\}$, $p_0 \in \{0, 0.2, 0.4, 0.6, 0.7\}$. With each configuration, ACO-TG is implemented on 100 tuning instances, which is randomly generated using the method in the previous section. The computational results show that ACO-TG statistically has best performance with the following configuration: $\beta = 2$, $\gamma = 3$, and $p_0 = 0.4$.

4.3 Computational Results

In this section, we investigate the performance of ACO-TG on two sets of instances. The first experiment is carried out to check how well ACO-TG works on the instances of the NSF network. In addition to the determined parameters previously, the stopping criteria of ACO-TG is maximum number of iterations, i.e., 100000 iterations. Two versions of ACO-TG are considered, namely, ACO-TG with local search (ACO-TG), and ACO-TG without local search (ACO-TG$_w$). Each version of ACO-TG is run on all instance of NSF network with 20 trials. To give an indicative comparison, we also use LINGO 9.0 to solve the node-arc formulation on these 20 instances. Table 1 reports the results of ACO-TG on the NSF network. For each instance, we show the best solutions (Best), the worst solution (Worst), average solution (Avg.) and standard deviation (Std.) over 20 trials for ACO-TG. The CPU times (CPUs) taken to find best solutions, are also reported. Since LINGO 9.0 can not find an optimal solution for the node-arc formulation in reasonable times, we ran it for 19 hours for each instance and report

Table 1. Computational results of ACO-TG on the NSF network

No.	LB	Node-arc method		ACO-TG					ACO-TG$_w$				
		Best	CPUs	Best	Worst	Avg.	Std.	CPUs	Best	Worst	Avg.	Std.	CPUs
1	347.5	353	19 h[1]	352	374	362.4	5.98	1200.4 s	355	385	367.15	7.47	236.9 s[2]
2	347.75	351	19 h	351	383	367.05	6.52	265.3 s	355	383	371.1	6.75	211.8 s
3	347	352	19 h	352	379	367.85	6.47	160.3 s	354	376	367.4	6.63	187.3 s
4	354	358[3]	19 h	355	371	363.5	5.46	672.1 s	357	377	365.35	5.83	152.4 s
5	343.25	350	19 h	345	382	359.9	8.50	881.9 s	348	376	365.4	7.60	190.1 s
6	347.25	351	19 h	355	373	366	4.90	382.3 s	356	378	369.6	6.35	282.6 s
7	362	377	19 h	364	383	373.8	6.21	179.2 s	365	388	378.05	5.63	170.3 s
8	336.25	344	19 h	344	372	358.15	6.85	224.8 s	346	371	359.5	7.91	253.5 s
9	345.75	355	19 h	353	388	367	10.24	195.2 s	356	377	368.2	7.18	85.3 s
10	362.5	367	19 h	364	383	373.9	6.24	396.4 s	369	391	380.15	7.51	148.0 s
11	325.5	330	19 h	330	345	339.8	4.69	486.4 s	336	360	346.15	6.07	239.9 s
12	344.5	349	19 h	349	363	361.4	5.85	376.5 s	357	380	368.3	5.97	265.1 s
13	326.6	332	19 h	329	348	335.5	5.68	307.0 s	331	350	340.9	5.43	256.6 s
14	345.75	350	19 h	349	382	363.85	7.23	99.7 s	351	373	362.2	7.50	81.3 s
15	306.5	311	19 h	315	344	328.45	7.01	390.6 s	323	350	333.45	6.29	104.0 s
16	360	362	19 h	361	375	367.85	4.23	297.5 s	363	376	368.6	3.47	191.1 s
17	373.75	380	19 h	376	400	391.6	8.29	819.8 s	387	413	397.6	7.84	112.4 s
18	365.75	369	19 h	366	389	374.5	8.41	699.7 s	368	390	377.8	7.53	189.1 s
19	339.67	345	19 h	347	376	365.3	7.96	174.0 s	351	383	362.3	8.34	143.1 s
20	321.75	327	19 h	329	361	348.05	7.64	162.8 s	331	365	347.2	8.10	119.0 s

[1] "h" denotes hours.
[2] "s" denotes seconds.
[3] Based on the node-arc formulation, LINGO 9.0 can not find an optimal solution and still returns solution 358 after it is implemented for 305 hours.

the best feasible solutions at that time. Additionally, we also solve the linear programming relaxation of the node-arc formulation. Accordingly the solution of linear programming relaxation gives a lower bound (LB), which is also shown in Table 1. Form computational results, we can draw the following conclusions:

(1) In general, ACO-TG works very well and is a computationally efficient approach for traffic grooming; it has found high-quality routing strategy quickly;

(2) The computational time of node-arc method is very expensive. It will take a long time to find good solutions by exact algorithm; in contrast to node-arc method, ACO-TG can find better solutions very quickly;

(3) Compared to the solutions found by exact algorithm in 19 hours, ACO-TG can found better solutions within small CPU times. For example, on instance 4, the best feasible solution, which is returned by LINGO within 305 hours, has the objective value 358. ACO-TG, however, can find a better solution within 335.7 seconds that has smaller objective values, 355. Same conclusion also holds for instances 5, 7, 9, 10, 13, 16, 17, and 18. For instance 18, we found that ACO-TG, in fact, had already found a better solution 368 within 211 seconds;

(4) The computational results also indicate the merit of local search, which has improved the performance of ACO-TG. But on the other hand, the consideration of local search increases the computational time;

(5) It can also be seen that both versions of ACO-TG are an approach with high robustness, which is indicated by the standard deviation column.

The second experiment is to check ACO-TG's performance on the problems of the SDH network. The SDH network is a much bigger network. ACO-TG is implemented to solve all instances of SDH network with 10 runs. The results are reported in Table 2. LINGO 9.0 is also used to get the lower bound (LB) and feasible integer solutions on all instances. Since the SDH network is a bigger network, the computation time is very much more. Therefore, LINGO 9.0

Table 2. Computational results of ACO-TG on the SDH network

No.	Linear relaxation		Node-arc method		ACO-TG				
	LB	CPUs	LINGO-first	CPUs	Best	Worst	Avg.	Std.	CPUs
1	1963.17	19:10:21[1]	2020	19:11:04	2041	2223	2111.4	43.38	2481.2 s[2]
2	1969.33	19:11:11	2019	19:07:01	2082	2205	2131.20	31.58	4017.0 s
3	1953.25	18:58:32	2001	19:03:20	2059	2085	2094.25	28.16	3618.0 s
4	2002.5	19:01:12	2046	19:01:11	2086	2210	2133.10	34.68	3937.7 s
5	2002	19:14:15	2062	18:58:27	2104	2228	2158.00	34.53	3818.3 s
6	2010.33	19:10:09	2060	18:56:52	2074	2151	2120.20	36.07	1524.0 s
7	1956.33	19:22:21	2012	18:53:34	2047	2176	2115.60	33.88	4047.5 s
8	2053	18:59:07	2133	18:52:28	2127	2240	2178.65	28.44	2451.7 s
9	2105	19:06:12	2163	19:40:13	2185	2249	2206.50	28.82	3150.0 s
10	1935.17	19:18:45	1999	19:37:15	1991	2076	2022.45	22.42	3613.2 s

[1] x:xx:xxx denotes x hours, xx minutes, and xxx seconds.
[2] "s" denotes seconds.

is only performed to get the first feasible integer solution (LINGO-first) and corresponding CPU times (CPUs) is reported. In the context of ACO-TG, the best solution (Best), the worst solution (Worst), average solution (Avg.) and standard deviation (Std.) over 10 trials are reported. Additionally, the computational times (CPUs) to find the best solution, is also given. The results in Table 2 indicate that the SDH network is more complex than the NSF network, and therefore more difficult to solve by exact algorithms. In fact, it took about 19 hours for LINGO 9.0 to find the first feasible solution for each instance. However, ACO-TG can return high-quality solutions relatively quickly. Our future work will focus on the improvement of implementation speed on large WDM networks. Based on the computational results in Tables 1 and 2, ACO-TG, in general, can produce high-quality routing strategy in reasonable computation times.

5   Conclusions

This paper is aimed at dealing with non-bifurcated traffic grooming problem in WDM networks, where most of nodes communicate with each other. The objective is to find a communication path for each traffic demand minimizing the congestion. To overcome the computational complexity of exact algorithms, we propose an ACO-based metaheuristic (ACO-TG). The computational results indicate that ACO-TG is an effective approach to find high-quality routing strategy within moderate computational effort. The future work will focus on the performance improvement of ACO-TG on large and practical WDM networks.

References

1. Dutta, R., Rouskas, G.N.: On optimal traffic grooming in wdm rings. IEEE Journal on Selected Areas in Communications 20, 110–121 (2002)
2. Bandyopadhyay, S., Klasing, R.: Dissemination of information in optical networks: from technology to algorithms. Springer, Heidelberg (2007)
3. Krishnaswamy, R., Sivarajan, K.: Design of logical topologies: A linear formulation for wavelength routed optical networks with no wavelength changers. IEEE/ACM Transactions on Networking 9, 186–198 (2001)

4. Aneja, Y., Bandyopadhyay, S., Jaekel, A.: On routing in large wdm networks. Optical Switching and Networking 3, 219–232 (2006)
5. Hu, J., Leida, B.: Traffic grooming, routing, and wavelength assignment in optical wdm mesh networks. In: Proceedings of 2004 IEEE INFOCOM. IEEE Press, Los Alamitos (2004)
6. Dutta, R., Rouskas, G.: A survey of virtual topology design algorithms for wavelength routed optical networks. Optical Networks Magazine 1, 73–89 (2000)
7. Dorigo, M., Stützle, T.: Ant Colony Optimization. The MIT Press, Massachusetts (2004)
8. Blum, C., Dorigo, M.: The Hyper-Cube Framework for Ant Colony Optimization. IEEE Transactions On Systems, Man, And Cybernetics- Part B 34, 1161–1172 (2004)

Elementary Approximation Algorithms for Prize Collecting Steiner Tree Problems

Shai Gutner[*]

School of Computer Science, Tel-Aviv University, Tel-Aviv, 69978, Israel
gutner@tau.ac.il

Abstract. This paper deals with approximation algorithms for the prize collecting generalized Steiner forest problem, defined as follows. The input is an undirected graph $G = (V, E)$, a collection $T = \{T_1, \ldots, T_k\}$, each a subset of V of size at least 2, a weight function $w : E \to \mathbb{R}^+$, and a penalty function $p : T \to \mathbb{R}^+$. The goal is to find a forest F that minimizes the cost of the edges of F plus the penalties paid for subsets T_i whose vertices are not all connected by F.

Our main result is a combinatorial $(3 - \frac{4}{n})$-approximation for the prize collecting generalized Steiner forest problem, where $n \geq 2$ is the number of vertices in the graph. This obviously implies the same approximation for the special case called the prize collecting Steiner forest problem (all subsets T_i are of size 2).

The approximation ratio we achieve is better than that of the best known combinatorial algorithm for this problem, which is the 3-approximation of Sharma, Swamy, and Williamson [13]. Furthermore, our algorithm is obtained using an elegant application of the local ratio method and is much simpler and practical, since unlike the algorithm of Sharma et al., it does not use submodular function minimization.

Our approach gives a $(2 - \frac{1}{n-1})$-approximation for the prize collecting Steiner tree problem (all subsets T_i are of size 2 and there is some root vertex r that belongs to all of them). This latter algorithm is in fact the local ratio version of the primal-dual algorithm of Goemans and Williamson [7]. Another special case of our main algorithm is Bar-Yehuda's local ratio $(2 - \frac{2}{n})$-approximation for the generalized Steiner forest problem (all the penalties are infinity) [3]. Thus, an important contribution of this paper is in providing a natural generalization of the framework presented by Goemans and Williamson, and later by Bar-Yehuda.

Keywords: Approximation algorithms, prize collecting Steiner tree problem, local ratio, primal-dual.

1   Introduction

There is substantial literature dealing with approximation algorithms for prize collecting Steiner tree problems. The purpose of this paper is to present elegant combinatorial algorithms for these problems. The local ratio technique

---

[*] This paper forms part of a Ph.D. thesis written by the author under the supervision of Prof. N. Alon and Prof. Y. Azar in Tel Aviv University.

[2,3,4] that we employ enables us to present simple algorithms together with a straightforward analysis.

The main focus of the paper in on the prize collecting generalized Steiner forest (PCGSF) problem, defined as follows. The input is an undirected graph $G = (V, E)$, a collection $T = \{T_1, \ldots, T_k\}$, each a subset of V of size at least 2, a weight function $w : E \to \mathbb{R}^+$, and a penalty function $p : T \to \mathbb{R}^+$, where $\mathbb{R}^+$ denotes the set of positive real numbers. The objective is to compute a forest F that minimizes the cost of the edges of F and the sum of the penalties of the subsets T_i whose vertices are not all connected by F. Thus, all the vertices of a subset T_i must be in the same connected component of F in order to avoid the penalty. Note that we intentionally define costs and penalties to be positive, as this will turn out to be convenient later. During intermediate stages of the algorithm, zero cost edges are contracted, whereas zero penalties can be ignored.

Previous Results. The special case of the PCGSF problem called the prize collecting Steiner forest problem (all subsets T_i are of size 2) has received considerable attention lately. A modification of the LP rounding algorithm in [6] implies a 3-approximation for this problem. This was improved in [9] to give an LP based 2.54-approximation for the problem as well as a primal-dual combinatorial $(3 - \frac{2}{n})$-approximation using Farkas' Lemma. The authors of [8] give a 3-budget-balanced and group-strategyproof mechanism for the game-theoretic version of the prize collecting Steiner forest problem, which is an extension of the method presented in [12]. Their result also provides a primal-dual 3-approximation algorithm for this problem.

A generalized framework of the prize collecting problems with an arbitrary $0 - 1$ connectivity requirement function and a submodular penalty function is studied by Sharma, Swamy, and Williamson in [13]. Their model captures both the PCGSF problem defined in this paper as well as the problems of [10,9]. The authors give a complicated primal-dual 3-approximation algorithm together with an LP rounding algorithm with a performance ratio of 2.54.

Two classical primal-dual algorithms, relevant to this paper, are due to Goemans and Williamson [7]. They give a $(2 - \frac{1}{n-1})$-approximation for the prize collecting Steiner tree problem (all subsets T_i are of size 2 and there is some root vertex r that belongs to all of them) as well as a $(2 - \frac{2}{n})$-approximation for the generalized Steiner forest problem (all the penalties are infinity) that simulates an algorithm of Agrawal, Klein, and Ravi [1]. For the latter problem, a simple $(2 - \frac{2}{n})$-approximation based on the local ratio technique is presented by Bar-Yehuda in [3].

Our Results. The main result is a local ratio $(3 - \frac{4}{n})$-approximation for the prize collecting generalized Steiner forest problem, where $n \geq 2$ is the number of vertices in the graph. This obviously implies the same approximation for the special case of the prize collecting Steiner forest problem, which was previously studied in [9,8].

The approximation ratio of our algorithm is slightly better than that of the 3-approximation of Sharma et al. [13], which is the best known combinatorial approximation algorithm for the problem. The algorithm we present makes an elegant use of the local ratio method and therefore, unlike the algorithm of Sharma et al., does not use submodular function minimization. This makes our approach much simpler and practical. We also note that the main algorithm presented in this paper is not the local ratio version of the primal-dual 3-approximation algorithm from [13] and cannot be obtained from it using the equivalence between the primal-dual schema and the local ratio technique [5].

There are two interesting special cases of our main algorithm. We present a $(2 - \frac{1}{n-1})$-approximation for the prize collecting Steiner tree problem (all subsets T_i are of size 2 and there is some root vertex r that belongs to all of them). This latter algorithm is in fact the local ratio version of the primal-dual algorithm of Goemans and Williamson [7]. Another special case of our main algorithm is Bar-Yehuda's local ratio $(2 - \frac{2}{n})$-approximation for the generalized Steiner forest problem (all the penalties are infinity) [3]. Thus, an important contribution of this paper is in providing a natural generalization of the framework presented by Goemans and Williamson, and later by Bar-Yehuda.

2   The Prize Collecting Generalized Steiner Forest Problem

In this section we present the algorithm for the PCGSF problem. The following are some definitions that are needed for presenting the algorithm.

Definition 1. *Given an instance* $(G, T = \{T_1, \ldots, T_k\}, w, p)$ *of the PCGSF problem, a vertex is said to be an **active vertex** if it belongs to at least one of the subsets* T_i.

Definition 2. *A solution F to the PCGSF problem is a **minimal solution** if every leaf (a vertex of degree 1) of the forest F is an active vertex.*

Definition 3. *Suppose $G = (V, E)$ is an undirected graph and $T = \{T_1, \ldots, T_k\}$ is a collection of subsets of V of size at least 2. For every active vertex v, let $t(v)$ be some arbitrary subset T_i for which $v \in T_i$. The **degree-weighted instance** corresponding to G,T, and t is the quadruple (G, T, w, p), defined as follows.*

- *For each $e \in E$, $w(e)$ is the number of its endpoints that are active vertices.*
- *For each $T_i \in T$, $p(T_i) = |\{v \in T_i | t(v) = T_i\}|$.*

Note that in the previous definition, the weight of an edge $w(e)$ can take a value of 0, 1, or 2, whereas the penalty of a subset T_i satisfies $0 \leq p(T_i) \leq |T_i|$. The following definition describes the important and natural operation of an edge contraction.

Definition 4. *Let (G, T, w, p) denote an instance of the PCGSF problem. The **contraction** of an edge $\{u, v\}$ into a new vertex x results in a new instance (G', T', w', p'), defined as follows.*

- For a vertex y, if y is adjacent in G to both u and v, then it is adjacent to x in G' and $w'(\{y, x\}) = \min\{w(y, u), w(y, v)\}$. We say that the edge $\{y, x\}$ of the graph G' **corresponds** to the original edge in the graph G for which the minimum is attained.
- For any subset $T_i \in T$, if $T_i \cap \{u, v\} \neq \emptyset$, then define $T'_i = (T_i \cup \{x\}) - \{u, v\}$. In case $T_i \neq \{u, v\}$, it follows that $|T'_i| \geq 2$ and the new subset T'_i is added to T'. During this process, there could be two subsets T_i and T_j for which $T'_i = T'_j$. Obviously, the two subsets can be joined and their new penalty in p' is defined to be $p(T_i) + p(T_j)$.

The following two lemmas establish the important property of minimal solutions to degree-weighted instances.

Lemma 1. *Every solution to a degree-weighted instance of the PCGSF problem has a total cost of at least n, where $n \geq 2$ is the number of active vertices.*

Proof. Consider some specific active vertex v. There is some subset $T_i \in T$ for which $t(v) = T_i$. In case the vertex v is not connected in the solution to all the other vertices of T_i, then we pay a penalty of 1 because of v. Otherwise, the solution must contain at least one edge incident with v. This means that the solution pays a cost of 1 for this edge due to vertex v. Every active vertex incurs a cost or penalty of at least 1, and therefore the total cost of the solution is at least n.   □

Lemma 2. *Every minimal solution to a degree-weighted instance of the PCGSF problem has a total cost of at most $3n - 4$, where $n \geq 2$ is the number of active vertices.*

Proof. Examine some specific connected component of the solution that contains q active vertices. Since this is a minimal solution, all of its leaves are active vertices. We prove by induction on q that the cost of the edges in this connected component is at most $2(q - 1)$. For $q = 1$, the connected component is a vertex with no edges, so this is obviously true. For $q = 2$, the connected component must be a path, whose two endpoints are active vertices, and the claim holds.

Suppose that $q > 2$. Take some leaf v, which by our assumption must be an active vertex. Now examine the path that starts from v and continues until the first time that either a vertex of degree at least 3 or another active vertex is reached. The cost of the edges in this path is at most 2. After removing this path from the solution, we are left with a connected component with $q - 1$ active vertices whose leaves are all active vertices. The result now follows from the induction hypothesis.

For proving the lemma, we distinguish between two cases. If all the active vertices are in one connected component of the solution, then the solution is actually a tree and no penalties are paid. The cost of the edges in the solution is at most $2(n - 1)$. The total cost, including the penalties, is also $2(n - 1) \leq 3n - 4$, since $n \geq 2$. Otherwise, there are at least two connected components in the solution. The cost of the edges in all the components can be at most $2(n-2)$. Note that the sum of all penalties in the instance is exactly n. The solution pays at most n for penalties, so the total cost is at most $2(n-2) + n = 3n - 4$.   □

The last two lemmas determine the approximation ratio of the algorithm. The purpose of the next lemma is to show that the analysis is indeed tight.

Lemma 3. *For every $n \geq 2$, there exists a degree-weighted instance of the PCGSF problem on a graph with n vertices for which the optimal solution has total cost n whereas some minimal solution has a total cost of $3n - 4$.*

Proof. Consider the following instance with vertices $v_1, \ldots, v_n$, which are all active. Between every two vertices there is an edge of cost 2. Define $p(\{v_1, v_2\}) = 2$ and $p(\{v_1, v_i\}) = 1$ for every $3 \leq i \leq n$. All other penalties are zero.

The optimal solution has no edges. The cost of the edges is zero and the sum of penalties paid is n for a total cost of n. A possible minimal solution is the path $v_2, v_3, \ldots, v_n$. The cost of the edges is $2(n-2)$ and the payment of penalties is still n for a total cost of $3n - 4$.   □

We now present the main algorithm of the paper.

Algorithm 1. $PCGSF(G, T, w, p)$

Input: Graph $G = (V, E)$, collection $T = \{T_1, \ldots, T_k\}$, each a subset of V of size
　　　　at least 2, weight function $w : E \to \mathbb{R}^+$, penalty function $p : T \to \mathbb{R}^+$
Output: A forest $F \subseteq E$
if $T = \emptyset$ **then**
　└ **return** $\emptyset$
else
　│　The set of active vertices is defined as $Active \leftarrow \{v \in V | \exists i \ \ v \in T_i\}$
　│　For every edge $e \in E$, let $d(e)$ be the number of its active endpoints
　│　For every active vertex v, let $t(v)$ be an arbitrary subset T_i for which $v \in T_i$
　│　For every $T_i \in T$, define $d(T_i) = |\{v \in T_i | t(v) = T_i\}|$
　│　$\epsilon \leftarrow \min(\{w(e)/d(e) | e \in E, d(e) \neq 0\} \cup \{p(T_i)/d(T_i) | T_i \in T, d(T_i) \neq 0\})$
　│　Define a weight function w' as follows: $w'(e) = w(e) - d(e) \cdot \epsilon$ for all $e \in E$
　│　Define a penalty function p' as follows: $p'(T_i) = p(T_i) - d(T_i) \cdot \epsilon$ for all $T_i \in T$
　│　Let Z be the set of all edges $e \in E$ for which $w'(e) = 0$ and let Z' be a
　│　spanning forest of (V, Z)
　│　Let $(G', \{T_1', \ldots, T_k'\}, w'', p'')$ be obtained from (G, T, w', p') by contracting
　│　the edges in Z'
　│　Let T' be the collection of subsets T_i' of size at least 2 for which $p''(T_i') > 0$
　│　$F' \leftarrow PCGSF(G', T', w'', p'')$
　│　Let F'' be the forest obtained from the edges in G corresponding to those in
　│　F' together with Z'
　│　**while** *there is a leaf in F'' which is not an active vertex* **do**
　│　└ Remove from F'' the edge incident with that leaf
　└ **return** F''

Theorem 1. *There is a local ratio $(3 - \frac{4}{n})$-approximation algorithm for the PCGSF problem, where $n \geq 2$ is the number of vertices in the graph.*

Proof. The pseudocode of algorithm $PCGSF(G, T, w, p)$ that solves this problem appears above. The proof is by induction on $|E| + k$. Given an instance (G, T, w, p) of the problem, the functions w and p are decomposed by the algorithm, so that $w = w' + \delta$ and $p = p' + \gamma$, where the quadruple (G, T, δ, γ) is a constant multiple of a degree-weighted instance. It follows from Lemmas 1 and 2 that the solution F'' computed by the algorithm is a $(3 - \frac{4}{n})$-approximation for the instance (G, T, δ, γ).

Before the next recursive call, the algorithm either contracts an edge or reduced some penalty to zero. By the induction hypothesis, the recursive call to $PCGSF(G', T', w'', p'')$ returns a solution F' which is a $(3 - \frac{4}{n})$-approximation. Adding edges of cost zero to F' does not change the cost of the solution. Removing leaves that are not active vertices can only reduce the cost. Thus, it is easy to verify that the solution F'' computed by the algorithm is also a $(3 - \frac{4}{n})$-approximation for the instance (G, T, w', p'). It follows from the basic local ratio decomposition observation that F'' is a $(3 - \frac{4}{n})$-approximation for the instance $(G, T, w' + \delta, p' + \gamma)$, as needed.

As for the time complexity of the algorithm, the argument above shows that the algorithm performs at most $|E| + k$ recursive calls, and thus runs in polynomial time. □

3   The Prize Collecting Steiner Tree Problem

This section introduces the algorithm for the prize collecting Steiner tree (PCST) problem. An instance of this problem consists of a graph $G = (V, E)$, a root vertex $r \in V$, a subset $U \subseteq V - \{r\}$ of active vertices, a weight function $w : E \to \mathbb{R}^+$, and a penalty function $p : U \to \mathbb{R}^+$. Given an instance (G, r, U, w, p), the goal is to compute a tree rooted at r that minimizes the cost of the edges of the tree plus the penalties paid for vertices not in the tree. An *active vertex* is simply a vertex with positive penalty (the root vertex is not active). The set of active vertices is denoted by U in the problem instance. A *minimal solution* is a tree rooted at r whose leaves are active vertices (and possibly also r). In a *degree-weighted instance* corresponding to a graph $G = (V, E)$, a root r and a subset $U \subseteq V - \{r\}$ of active vertices, the weight function $w(e)$ is equal to the number of active endpoints of the edge e, whereas the penalty function satisfies $p(v) = 1$ for every $v \in U$ and $p(v) = 0$ otherwise. When the edge $\{u, v\}$ is *contracted*, the penalty of the new vertex created is $p(u) + p(v)$. This is except for when an edge $\{r, v\}$ incident with the root is contracted. In this case, the new vertex created is also called r and it still has zero penalty.

Lemma 4. *Every solution to a degree-weighted instance of the PCST problem has a total cost of at least n, where n is the number of active vertices.*

Proof. Let q be the number of active vertices connected to r in the solution. Each such vertex must have some edge incident with it in the solution and therefore a cost of at least 1 is paid for this vertex. For each of the $n - q$ active vertices that are not connected to r, a penalty of 1 is paid.   □

Lemma 5. *Every minimal solution to a degree-weighted instance of the PCST problem has a total cost of at most $2n - 1$, where n is the number of active vertices.*

Proof. Let q be the number of active vertices connected to r in the solution. An argument similar to the one used in the proof of Lemma 2 gives that the cost of the edges in a minimal solution is at most $2q - 1$. The penalty paid is $n - q$ for a total cost of $2q - 1 + n - q = n + q - 1 \leq 2n - 1$, since $q \leq n$.   □

The last two lemmas determine the approximation ratio of the algorithm. The purpose of the next lemma is to show that the analysis is indeed tight.

Lemma 6. *For every $n \geq 1$, there exists a degree-weighted instance of the PCST problem on a graph with n active vertices for which the optimal solution has total cost n whereas some minimal solution has a total cost of $2n - 1$.*

Proof. Consider the following instance with a root vertex r together with the vertices $v_1, \ldots, v_n$. For every $1 \leq i \leq n$, the vertex v_i is active and has a penalty of 1. Between every two vertices there is an edge of cost 2, except for edges between the root r and a vertex v_i that have a cost of 1.

The optimal solution has no edges. The cost of the edges is zero and the sum of penalties paid is n for a total cost of n. A possible minimal solution is the path $r, v_1, v_2, \ldots, v_n$. The cost of the edges is $1 + 2(n - 1)$ and the payment of penalties is zero for a total cost of $2n - 1$.   □

Theorem 2. *There is a local ratio $(2 - \frac{1}{n-1})$-approximation algorithm for the PCST problem, where $n \geq 2$ is the number of vertices in the graph.*

Proof. The pseudocode of algorithm $PCST(G, r, U, w, p)$ that solves this problem appears below. The proof is analogues to that of Theorem 1 using Lemmas 4 and 5 instead of Lemmas 1 and 2. Note that since the root is not an active vertex, the maximum number of active vertices is $n - 1$.   □

4   Concluding Remarks

- The integral solution computed by our algorithm for the prize collecting generalized Steiner forest problem can be compared with the optimal solution of the natural LP for this problem. It can be verified that the algorithm is still a $(3 - \frac{4}{n})$-approximation, even with respect to this optimal fractional solution. This is also implied by the equivalence between the local ratio method and the primal-dual schema.
- The algorithms presented in this paper use the local ratio method which seems like the most natural, simple and time-efficient framework for addressing prize collecting Steiner tree problems. This should be further explored to

---

Algorithm 2. $PCST(G, r, U, w, p)$

Input: Graph $G = (V, E)$, root vertex $r \in V$, subset $U \subseteq V - \{r\}$ of active
    vertices, weight function $w : E \to \mathbb{R}^+$, penalty function $p : U \to \mathbb{R}^+$
Output: A tree $T \subseteq E$ rooted at r
if $U = \emptyset$ **then**
   L **return** $\emptyset$
else
  | For every edge $e \in E$, let $d(e)$ be the number of its active endpoints
  | $\epsilon \leftarrow \min(\{w(e)/d(e) | e \in E, d(e) \neq 0\} \cup \{p(v) | v \in U\})$
  | Define a weight function w' as follows: $w'(e) = w(e) - d(e) \cdot \epsilon$ for every $e \in E$
  | Define a penalty function p' as follows: $p'(v) = p(v) - \epsilon$ for every $v \in U$
  | Let Z be the set of all edges $e \in E$ for which $w'(e) = 0$ and let Z' be a
  | spanning forest of (V, Z)
  | Let (G', r, U', w'', p'') be obtained from (G, r, U, w', p') by contracting the
  | edges in Z'
  | $T' \leftarrow PCST(G', r, U', w'', p'')$
  | Let the tree T'' be the connected component of r in the union of the edges in
  | G corresponding to those in T' together with Z'
  | **while** *there is a leaf in* T'' *which is not an active vertex (or the root r)* **do**
  | L Remove from T'' the edge incident with that leaf
  L **return** T''

---

determine whether this approach has applications for facility location problems and for the multicommodity rent-or-buy (MRoB) problem. The techniques presented in [11] might be helpful in enhancing the time performance of our algorithms.

- An interesting open problem is to decide whether there is a combinatorial algorithm for the prize collecting Steiner forest problem with an approximation factor better than 3. Improving upon the performance guarantee of the LP rounding 2.54-approximation algorithm is another intriguing challenge.

Acknowledgements

I would like to thank Noga Alon and Yossi Azar for helpful discussions.

References

1. Agrawal, A., Klein, P., Ravi, R.: When trees collide: An approximation algorithm for the generalized Steiner problem on networks. SIAM Journal on Computing 24(3), 440–456 (1995)
2. Bar-Noy, A., Bar-Yehuda, R., Freund, A., Naor, J.(S.), Schieber, B.: A unified approach to approximating resource allocation and scheduling. Journal of the ACM 48(5), 1069–1090 (2001)
3. Bar-Yehuda, R.: One for the price of two: a unified approach for approximating covering problems. Algorithmica 27(2), 131–144 (2000)

4. Bar-Yehuda, R., Bendel, K., Freund, A., Rawitz, D.: Local ratio: A unified framework for approximation algorithms: In memoriam: Shimon Even 1935–2004. ACM Computing Surveys 36(4), 422–463 (2004)
5. Bar-Yehuda, R., Rawitz, D.: On the equivalence between the primal-dual schema and the local ratio technique. SIAM Journal on Discrete Mathematics 19(3), 762–797 (2005)
6. Bienstock, D., Goemans, M.X., Simchi-Levi, D., Williamson, D.P.: A note on the prize collecting traveling salesman problem. Math. Program. 59, 413–420 (1993)
7. Goemans, M.X., Williamson, D.P.: A general approximation technique for constrained forest problems. SIAM Journal on Computing 24(2), 296–317 (1995)
8. Gupta, A., Könemann, J., Leonardi, S., Ravi, R., Schäfer, G.: An efficient cost-sharing mechanism for the prize-collecting Steiner forest problem. In: Proceedings of the Eighteenth Annual ACM-SIAM Symposium on Discrete Algorithms, pp. 1153–1162 (2007)
9. Hajiaghayi, M., Jain, K.: The prize-collecting generalized Steiner tree problem via a new approach of primal-dual schema. In: Proceedings of the Seventeenth Annual ACM-SIAM Symposium on Discrete Algorithms, pp. 631–640 (2006)
10. Hayrapetyan, A., Swamy, C., Tardos, É.: Network design for information networks. In: Proceedings of the Sixteenth Annual ACM-SIAM Symposium on Discrete Algorithms, pp. 933–942 (2005)
11. Johnson, D.S., Minkoff, M., Phillips, S.: The prize collecting Steiner tree problem: theory and practice. In: Proceedings of the Eleventh Annual ACM-SIAM Symposium on Discrete Algorithms, pp. 760–769 (2000)
12. Könemann, J., Leonardi, S., Schäfer, G.: A group-strategyproof mechanism for Steiner forests. In: Proceedings of the Sixteenth Annual ACM-SIAM Symposium on Discrete Algorithms, pp. 612–619 (2005)
13. Sharma, Y., Swamy, C., Williamson, D.P.: Approximation algorithms for prize-collecting forest problems with submodular penalty functions. In: Proceedings of the Eighteenth Annual ACM-SIAM Symposium on Discrete Algorithms, pp. 1275–1284 (2007)

Polynomial Time Approximation Scheme for Connected Vertex Cover in Unit Disk Graph

Zhao Zhang[1,*], Xiaofeng Gao[2,**], and Weili Wu[2,**]

[1] College of Mathematics and System Sciences, Xingjiang University,
Urmuqi, Xinjiang, China
zhzhao@xju.edu.cn
[2] Department of Computer Science, University of Texas at Dallas,
Richardson, TX 75083, USA
{xxg05200,weiliwu}@utdallas.edu

Abstract. Connected Vertex Cover Problem (CVC) is an $\mathcal{N}P$-hard problem. The currently best known approximation algorithm for CVC has performance ration 2. This paper gives the first Polynomial Time Approximation Scheme for CVC in Unit Disk Graph.

Keywords: Connected Vertex Cover, Unit Disk Graph.

1 Introduction

Minimum Vertex Cover Problem (MVC) is a classical optimization problem in graph and combinatorial theory. For a undirected graph $G = (V, E)$, a subset $C \subseteq V$ is called a *vertex cover* of G (VC) if for any $(v, w) \in E$, either $v \in C$ or $w \in C$. MVC is to find a vertex cover of G with the minimum number of vertices. This problem has many real-world applications [3], including many in the field of bioinformatics. It can also be used in the construction of phylogenetic trees, in phenotype identification, and in analysis of microarray data. MVC has been studied extensively in the literature [8]. It is known to be $\mathcal{N}\mathcal{P}$-hard [9] for a long time. Papadimitriou et al. [13] proved that VC is APX-complete, and Monien et al. [12,3] gave an approximation algorithm for VC with ratio $1 - \frac{\log \log n}{2 \log n}$ (n is the number of vertices).

If furthermore, a vertex cover C induces a connected subgraph $G[C]$, then C is called a *connected vertex cover* (CVC). The Minimum Connected Vertex Cover Problem (MCVC) is to find a CVC with minimum cardinality. MCVC problem is an enforced version of MVC when certain connectivity constraints are needed in some applications. For example, in routing and wavelength assignment (RWA) problem for optical networks, people select a suitable path and wavelength among the many possible choices with the help of CVC. MCVC is

---

* Support in part by NSFC (60603003) and XJEDU. This work was done while this author visited at University of Texas at Dallas.
** Support in part by National Science Foundation under grants CCF-9208913 and CCF-0728851.

B. Yang, D.-Z. Du, and C.A. Wang (Eds.): COCOA 2008, LNCS 5165, pp. 255–264, 2008.
© Springer-Verlag Berlin Heidelberg 2008

also $\mathcal{NP}$-hard. In fact, Garey and Johnson [7] showed that MCVC is as hard to approximate as MVC. The currently best known approximation algorithms for MCVC have performance ratio 2, which were given by Arkin et al. [2,14].

In this paper, we consider MCVC problem in *Unit Disk Graphs* (UDG). A graph G is an UDG if each vertex of G is associated with the center of a disk with diameter 1 on the plane, and two vertices u, v of G are adjacent if and only if the two disks corresponding to u and v have non-empty intersection. In another word, $(u, v) \in E(G)$ if and only if the Euclidean distance between the centers corresponding to u and v is at most 1. Such a set of unit disks on the plane is called the *geometric representation of G*. When talking about a unit disk graph in this paper, we assume that the geometric representation is given, since it has been proved in [10] that determining whether a graph is a UDG is $\mathcal{NP}$-complete. UDG is widely used in wireless networks, where each vertex represents an idealized multi-hop radio based station, and the corresponding disk is the communication range of the station. For MVC in UDG, there exists Polynomial Time Approximation Scheme (PTAS). That is, for any positive real number ε, there exists a $(1 + \varepsilon)$-approximation. In fact, Erlebach et al. [5,11] presented a PTAS for Minimum Weight Vertex Cover (MWVC) in *Disk Graph* (DG), where DG is a generalization of UDG, in which disks have different radiuses.

In this paper, we present the first PTAS for CVC on UDG, using partition technique and shifting strategy. Such an approach was used for Steiner trees in the plane [15]. A more complicated approach was used for connected dominating set [4]. It should be noted that in [4], the technique is heavily based on the property of 2-dimension. In other words, the technique cannot be applied to higher dimensional space, e.g., unit ball graphs, which is also an important model for wireless sensor networks. The technique presented in this paper can be applied to any dimension. Therefore, it is actually proved in this paper that there exists PTAS in unit n-dimensional ball graphs for any n.

The idea of the algorithm is: First, we take an area containing all vertices of the graph, and partition it into small squares. For each small square, define the inner area and the boundary area, such that the inner area and the boundary area of a same small square has an overlap. For each component of the inner area, compute a minimum CVC. To cover edges not in the inner area, use a constant-approximation algorithm to compute a connected vertex cover C_0 of G, and union those vertices of C_0 which belong to the 'boundary area' of the partition into the above CVC's. The overlap of inner area and boundary area ensures the connectivity of the output. The shifting strategy is used to select a partition such that the number of vertices of C_0 falling into the boundary area of this partition is small enough relative to ε.

The rest of this paper is organized as follows. In Section 2, we introduce some terminologies used to describe the algorithm. In Section 3, the algorithm is presented. In Section 4 we show the correctness of our algorithm, analyze the time complexity, and prove that it is a PTAS. A conclusion is given in Section 5.

2   Preliminaries

In this section we introduce the symbols and definitions used for algorithm description.

For a given UDG $G = (V, E)$, where $|V| = n$, we assume that all the disks are located in a square plane $Q = \{(x, y)|0 \leq x \leq q, 0 \leq y \leq q\}$, where q is related to n. Using partition strategy, we divide Q into squares each with side length $m \times m$. We set $m = \lceil \frac{48\rho}{\varepsilon} \rceil$, where ρ is a constant which is the approximation ratio of an APX for CVC (for example, ρ can be taken as 2 if we use the 2-approximation algorithm in [6]), and ε is an arbitrary positive number. Let $p = \lfloor \frac{q}{m} \rfloor + 1$. Since we shall use shifting policy, we widen Q into a bigger region $\widetilde{Q} = \{(x, y)| - m \leq x \leq pm, -m \leq y \leq pm\}$ (see Fig. 1).

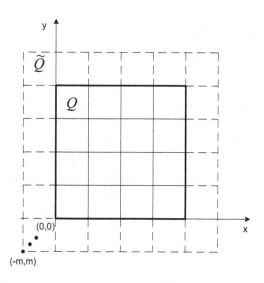

Fig. 1. Partition for Graph G

Name this partition as $P(0)$, and denote by $P(a)$ the partition obtained from $P(0)$ by shifting it such that the left-bottom corner of $P(a)$ is at $(a - m, a - m)$, for $a = 0, 1, \cdots, m - 1$.

For each square e, we define the *Inner* area I_e and *Boundary* area B_e (see Fig. 2). If $e = \{(x, y)|im \leq x \leq (i + 1)m, jm \leq y \leq (j + 1)m\}$,

$$I_e = \{(x, y)|im + 1 \leq x \leq (i + 1)m - 1, jm + 1 \leq y \leq (j + 1)m - 1\},$$

$$B_e = e - \{(x, y)|im + 2 \leq x \leq (i + 1)m - 2, jm + 2 \leq y \leq (j + 1)m - 2\}.$$

Note that there is an overlap of I_e and B_e.

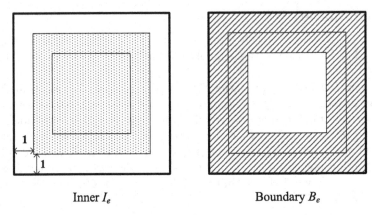

Inner I_e Boundary B_e

Fig. 2. Inner Region and Boundary Region for a Square

3   Algorithm Overview

For a partition $P(a)$, denote by $B(P(a)) = \bigcup_{e \in P(a)} B_e$. The algorithm is executed in two phases.

Phase I. Use a ρ-approximation to compute a CVC C_0 for graph G. Let $C_0(a) = C_0 \cap B(P(a))$ be the set of vertices of C_0 lying in the boundary area of partition $P(a)$. Choose a^* such that $|C_0(a^*)| = \min |C_0(a)|$.

Phase II. For any square $e \in P(a^*)$, denote by G_e the subgraph of G induced by the vertices in I_e, and $Comp(G_e)$ the set of connected components in G_e. For each square e and each component $H \in Comp(G_e)$, use exhaust search to find a minimum CVC C_H of H. Set $C_e = \bigcup_{H \in Comp(G_e)} C_H$.

Final Result Output $C = C_0(a^*) \cup (\bigcup_{e \in P(a^*)} C_e)$.

4   Analysis of the Algorithm

In this section, we firstly prove the correctness of our algorithm, and then discuss the overall time complexity, that is, we prove that our algorithm runs in polynomial time. Finally, we give the performance ratio of the algorithm, which is $(1 + \varepsilon)$.

4.1   Correctness

To prove that the output C of our algorithm is a CVC for graph G, we firstly prove that C is a vertex cover for G, then prove that the induced subgraph $G[C]$ is connected.

Lemma 1. *C is a vertex cover for $G = (V, E)$.*

Proof. For each square e, the inner area I_e and the boundary area B_e have an overlap with width 1. Since for any edge (v, w), the Euclidean distance between v and w is less than or equal to 1, we see that both v and w belong to the inner area I_e for some square e, or belong to the boundary area $B(P(a^*))$. In the former case, the edge (v, w) is in a component H of G_e. By Phase II of the algorithm, either $v \in C_H$ or $w \in C_H$, meaning that (v, w) can be covered by $C_e \subseteq C$. In the second case, by Phase I of the algorithm, C_0 is a CVC of G. Therefore either $v \in C_0(a^*)$ or $w \in C_0(a^*)$, meaning that (v, w) can be covered by $C_0(a^*) \subseteq C$. Thus we have proved that any edge in G is covered by C. So C is a vertex cover of G.

Lemma 2. *The induced subgraph $G[C]$ is connected.*

Proof. We prove this lemma by two steps. In step 1, we show that distinct connected components in $G[C_0(a^*)]$ (if exist) can be connected through vertices in $\bigcup_{e \in P(a^*)} C_e$. In step 2, we show that there is no other components of $G[C]$ left after step 1.

Step 1. Let H_1 and H_2 be two components in $G[C_0(a^*)]$ which are 'closest' in $G[C_0]$ with each other. Then, there is a path $P = (v_1, v_2, \cdots, v_t)$ of $G[C_0]$ connecting H_1 and H_2 through the inner area of 'one' square e. Without loss of generalization, we may assume that $v_1 \in V(H_1)$, $v_t \in V(H_2)$ and $\{v_2, \cdots, v_{t-1}\} \subseteq I_e$. Fig. (see 3 for illustration).

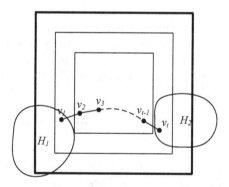

Fig. 3. An Illustration for H_1 and H_2

It is easy to see that v_1 and v_t belong to $I_e \cap B_e$, so P is in a connected component H of G_e. Based on Phase II of our algorithm, P is covered by C_H. It follows that [either $v_1 \in C_H$ or $v_2 \in C_H$], and [either $v_{t-1} \in C_H$ or $v_t \in C_H$]. Since $G[C_H]$ is connected, we see that H_1 and H_2 are connected through $G[C_H]$.

Step 2. Let $\tilde{G}$ be the component of $G[C]$ containing all vertices of $C_0(a^*)$. Such $\tilde{G}$ exists because of step 1. Suppose $\tilde{G} \neq G[C]$, then there exists a square e and a connected component H of G_e such that

(i) $C_H \cap C_0(a^*) = \emptyset$ and

(ii) no vertex of C_H is adjacent with any vertex in $C_0(a^*)$.

Let x be a vertex in C_H. Then either $x \in C_0$ or x is adjacent with a vertex $y \in C_0$. We firstly assume that $x \in C_0$. From (i), we know that $x \notin C_0(a^*)$, so $x \in e \backslash B_e$. Since $G[C_0]$ is connected, there is a path P in $G[C_0]$ connecting x to the other parts of G outside of e. Suppose $P = (v_0, v_1, ..., v_t)$, where $v_0 = x$, $v_t \notin e$, and $\{v_1, ..., v_{t-1}\} \subseteq e$. Let i be the index such that v_i is the first vertex on P with $v_i \in B_e$. Then

(iii) $v_i \in C_0(a^*)$;

(iv) $v_i \in I_e$ and thus v_i and x belong to a same component of G_e, which is H;

(v) both v_{i-1} and v_i are in I_e, and hence the edge (v_{i-1}, v_i) are in H (note that $i \geq 1$ since $v_0 = x \notin B_e$).

By (v) and Phase II of the algorithm, either $v_i \in C_H$ or $v_{i-1} \in C_H$. But this contradicts (i) (ii) and (iii).

The case that $x \notin C_0$ but is adjacent with a vertex $y \in C_0$ can be proved similarly.

Therefore, we have proved that $\widetilde{G} = G[C]$. $\qquad\square$

Based on the conclusions from Lemma 1 and Lemma 2, we obtain the following theorem showing the correctness of our algorithm.

Theorem 1. *The output C of our algorithm is a connected vertex cover for G.*

4.2   Time Complexity

In this subsection we consider the time complexity of our algorithm. Phase I of the algorithm uses a polynomial time ρ-approximation to compute C_0. Phase II uses exhaust search which is the most time consuming part. We shall prove that this part can also be executed in polynomial time, by implementing the relation between vertex cover and independent set.

Lemma 3. *The number of independent unit disks in an $m \times m$ square is at most $\lfloor \frac{(m+2)^2}{\pi} \rfloor$.*

Proof. Enlarge the $m \times m$ square to an $(m + 2) \times (m + 2)$ square by adding a boundary with width one. Then all the disks whose centers are in the $m \times m$ square lie completely in the $(m + 2) \times (m + 2)$ square. Since each unit disk occupies area π, the result follows from the independence assumption.

With the help of Lemma 3, we have the following theorem.

Theorem 2. *The running time of our algorithm is $n^{O(1/\varepsilon^2)}$, where n is the number of vertices in the graph.*

Proof. It is well known that a vertex set S is a vertex cover of a graph if and only if its complement is an independent set. Thus by Lemma 3, each $V(H) \backslash C_H$ contains at most $\lfloor \frac{(m+2)^2}{\pi} \rfloor$ independent vertices, and therefore the exhaust search for C_H (which can be done by considering the complement of each independent set in H) takes time at most $\sum_{k=0}^{\lfloor \frac{(m+2)^2}{\pi} \rfloor} \binom{n_H}{k} = n_H^{O(m^2)}$, where n_H is the number of vertices in H, and the total running time for phase II is at most $\sum_{e,H} n_H^{O(m^2)} = \left(\sum_{e,H} n_H \right)^{O(m^2)} = n^{O(m^2)} = n^{O(1/\varepsilon^2)}$.

4.3   Performance

Here we prove that our algorithm is a $(1 + \varepsilon)$-approximation.

Definition 1. For two subgraphs G_1, G_2 of G, the *distance between G_1 and G_2* is the length of a shortest path of G connecting G_1 and G_2 (where 'length' means the number of edges on the path), denoted by $dist(G_1, G_2)$.

In another word, if $dist(G_1, G_2) = k$, then G_1 and G_2 can be connected through $k - 1$ vertices. If a vertex cover of a connected graph is not a connected vertex cover, the distance between connected components of the subgraph induced by the vertex cover is not far, as can be seen from the following lemma.

Lemma 4. *Suppose H is a connected graph, and C is a vertex cover of H. If $H[C]$ is not connected, then there exist two components R_1, R_2 of $H[C]$ such that $dist(R_1, R_2) = 2$.*

Proof. Let R_1, R_2 be two 'closest' connected components of $G[C]$, and $P = (v_0, v_1, ..., v_t)$ be a shortest path of H connecting R_1 and R_2, $v_0 \in V(R_1)$ and $v_t \in V(R_2)$. If $t \geq 3$, consider the edge (v_1, v_2). Since C covers H, we have either $v_1 \in C$ or $v_2 \in C$. Suppose, without loss of generality, that $v_1 \in C$. Let R_3 be the component of $G[C]$ containing v_1. Then $R_3 \neq R_1$ and R_2, and $dist(R_3, R_2) < dist(R_1, R_2)$, contradicting our choice of R_1 and R_2.

The following lemma is well known in unit disk graph.

Lemma 5. *Let G be a unit disk graph and u be a vertex in $V(G)$. The there are at most 5 independent vertices in $N(u)$, where $N(u)$ is the set of vertices adjacent with u in G.*

The following theorem shows that our algorithm is a PTAS.

Theorem 3. *Let C^* be an optimal CVC for G, and C be the output of our algorithm. Then $|C| \leq (1 + \varepsilon)|C^*|$.*

Proof. Firstly, we prove that

$$|C_0(a^*)| \leq \frac{\varepsilon}{6}|C^*|. \tag{1}$$

When the partition shifts, a vertex of C_0 belongs to at most 8 boundary areas of $B(P(a))$'s. Therefore, we have,

$$|C_0(0)| + |C_0(1)| + \cdots + |C_0(m-1)| \leq 8|C_0|,$$

and thus

$$|C_0(a^*)| \leq \frac{8\rho|C^*|}{m} \leq \frac{\varepsilon}{6}|C^*|.$$

Next, we shall add some vertices to C^* such that the resulting set $\widetilde{C}$ satisfies: for each square e and

$$\text{for each component } H \in Comp(G_e), \ \widetilde{C} \cap V(H) \text{ is a CVC of } H. \tag{2}$$

For a square e, let $\widetilde{C}_e = C^* \cap I_e$. It is easy to see that for each component $H \in Comp(G_e)$, $\widetilde{C}_e \cap V(H)$ covers H. Suppose there exists a component $H \in Cpomp(G_e)$ such that requirement (2) is not satisfied. By Lemma 4, there are two components R_1, R_2 of $G[\widetilde{C}_e \cap V(H)]$ such that R_1 and R_2 can be connected through one vertex in $V(H) \backslash \widetilde{C}_e$. Add this vertex to $\widetilde{C}_e$. If the new $\widetilde{C}_e$ still does not satisfy requirement (2), continue as above to add vertices to merge components. Suppose this is done k times before $\widetilde{C}_e$ satisfies (2), then

$$|\widetilde{C}_e| \leq |C^* \cap e| + k. \tag{3}$$

On the other hand, we can show that

$$|C_0(a^*) \cap e| \geq \frac{k}{5}. \tag{4}$$

For this purpose, we suppose that the components merged are in the order that R_1 with R_2, R_3 with R_4, $\cdots$, R_{2k-1} with R_{2k}. For simplicity of presenting the idea, we firstly assume that all the above R_j's are distinct components of $G[C^* \cap I_e]$. For each $i = 1, 2, \cdots, k$, let x_i be a vertex in $V(R_{2i-1}) \cap B_e \cap I_e$, such that x_i is adjacent with a vertex $y_i \in B_e \backslash I_e$. Such x_i exists since R_{2i-1} is connected to the outer parts of e through C^*. Then either $x_i \in C_0$ or $y_i \in C_0$. Set $z_i = x_i$ if $x_i \in C_0$ and $z_i = y_i$ otherwise. Note that both $x_i, y_i \in B_e$. Hence $z_i \in C_0(a^*) \cap e$. A vertex may serve more than once as z_i's. For example, it is possible that there are two indices $i \neq j$ such that the vertex of C_0 covering edges (x_i, y_i) and (x_j, y_j) is the same $y_i = y_j \in C_0$. In this case, we see that x_i and x_j are independent since they belong to different components of $G[C^* \cap I_e]$. Then by Lemma 5, such vertex serves at most 5 times as z_i's, and inequality (4) follows. Next, suppose the R_j's are not all distinct. For example, suppose R_3 is the component obtained by merging R_1 and R_2. Then x_3 can be chosen such that $x_3 \in V(H_2) \cap B_e \cap I_e$, which is independent with x_1. In general we can find k independent vertices $x_1, x_3, ..., x_{2k-1}$ and thus (4) also holds in this case.

Combining inequalities (3) and (4), we have

$$|\widetilde{C}_e| \leq |C^* \cap e| + 5|C_0(a^*) \cap e|. \tag{5}$$

Since in Phase II of the algorithm, C_e is a 'minimum' vertex set satisfying requirement (2) for each square e, we have $|C_e| \leq |\widetilde{C}_e|$. Combining this with (1) and (5), we have

$$\left| \bigcup_{e \in P(a^*)} C_e \right| = \bigcup_{e \in P(a^*)} |C_e| \leq \sum_{e \in P(a^*)} |\widetilde{C}_e|$$

$$\leq \sum_{e \in P(a^*)} \left(|C^* \cap e| + 5|C_0(a^*) \cap e| \right)$$

$$= |C^*| + 5|C_0(a^*)| \leq (1 + \frac{5\varepsilon}{6})|C^*|.$$

Hence

$$|C| \leq |C_0(a^*)| + \left| \bigcup_{e \in P(a^*)} C_e \right| \leq (1 + \varepsilon)|C^*|.$$

5 Conclusion

In this paper, we presented the first polynomial time approximation scheme to compute a connected vertex cover of a graph. The method used in this paper can be applied to CVC problems in n-dimensional ball graphs. In a *unit ball graph*, each vertex corresponds to the center of a unit ball in the n-dimensional space, and two vertices are adjacent if and only if the Euclidean distance between them is at most 1.

References

1. Abu-Khzam, F.N., et al.: Kernelization Algorithms for the Vertex Cover Problem: Theory and Experiments. In: Proc.6th Workshop on Algorithm Engineering and Experiments & 1st Workshop on Analytic Algorithmics and Combinatorics, pp. 62–69 (2004)
2. Arkin, E.M., Halldorsson, M.M., Hassin, R.: Approximating the Tree and Tour Covers of a Graph. Inform. Process. Lett. 47, 275–282 (1993)
3. Bar-Yehuda, R., Even, S.: A local-ration Theorem for Approximating the Weighted Vertex Cover Problem, Analysis and Design of Algorithms for Combinatorial Problems. Annals of Discrete Mathematics 25, 27–46 (1985)
4. Cheng, X., Huang, X., Li, D., Wu, W., Du, D.: A polynomial-time approximation scheme for minimum connected dominating set in ad hoc wireless networks. Networks 42, 202–208 (2003)
5. Erlebach, T., Jansen, K., Seidel, E.: Polynomial-Time Approximation Schemes for Geometric Intersection Graphs. SIAM J. Comput. 34(6), 1302–1323 (2005)
6. Fujito, T., Doi, T.: A 2-Approximation NC Algorithm for Connected Vertex Cover and Tree Cover. Inform. Process. Lett. 90, 59–63 (2004)
7. Garey, M.R., Johnson, D.S.: The Rectilinear Steiner-Tree Problem is $\mathcal{NP}$-Complete. SIAM J. Appl. Math. 32, 826–834 (1977)
8. Hochbaum, D.H.: Approximation Algorithm for $\mathcal{NP}$-hard Problems, PWS, Boston, MA (1996)

9. Karp, R.M.: Reducibility among Combinatorial Problems. In: Miller, R.E., Thatcher, J.W. (eds.) Complexity of Computer Computations, pp. 85–103. Plenum Press, New York (1972)

10. Kratochvil, J.: Intersection graphs of noncrossing arc-connected sets in the plane. In: North, S.C. (ed.) GD 1996. LNCS, vol. 1190, pp. 257–270. Springer, Heidelberg (1997)

11. Li, X.Y., Wang, Y.: Simple Approximation Algorithms and PTASs for Various Problems in Wireless Ad-Hoc Networks. Journal of Parallel and Distributed Computing 66, 515–530 (2006)

12. Monien, B., Speckenmeyer, E.: Ramsey Numbers and an Approximation Algorithm for the Vertex Cover Problem. Acta Informatica 22, 115–123 (1985)

13. Papadimitriou, C.H., Yannakakis, M.: Optimization, Approximation, and Complexity Classes. J. Computer and System Sciences 43, 425–440 (1991)

14. Savage, C.: Depth-First Search and the Vertex Cover Problem. Inform. Process. Lett. 14, 233–235 (1982)

15. Wang, L.S., Jiang, T.: An approximation scheme for some Steiner tree problems in the plane. Networks 28(4), 187–193 (1996)

Improved Primal-Dual Approximation Algorithm
for the Connected Facility Location Problem[*]

Hyunwoo Jung, Mohammad Khairul Hasan, and Kyung-Yong Chwa

Division of Computer Science, Korea Advanced Institute of Science and Technology,
Daejeon, Republic of Korea
{solarity,shaon,kychwa}@tclab.kaist.ac.kr

Abstract. In the Connected Facility Location(ConFL) problem, we are given a graph $G = (V, E)$ with nonnegative edge cost c_e on the edges, a set of facilities $\mathcal{F} \subset V$, a set of demands, i.e., clients $\mathcal{D} \subset V$, and a parameter $M \geq 1$. Each facility i has a nonnegative opening cost f_i and each client j has d_j units of demand. Our objective is to open some facilities, say $F \subset \mathcal{F}$, assign each demand j to some open facility $i(j) \in F$ and connect all open facilities using a Steiner tree T such that the total cost, which is $\sum_{i \in F} f_i + \sum_{j \in \mathcal{D}} d_j c_{i(j)j} + M \sum_{e \in T} c_e$, is minimized.

We give an improved primal-dual 6.55-approximation algorithm for the ConFL problem which improves the Swamy and Kumar's primal-dual 8.55-approximation algorithm [1].

Keywords: Approximation algorithms, Primal-Dual algorithms, Facility location problem.

1   Introduction

For the past ten years, researches related to facility location problems have attracted new interests of many researchers. After the first constant factor approximation algorithm of Shmoys, Tardos and Aardal [2] for the uncapacitated facility location problem(UFLP), various approximation algorithms by using technique of LP-rounding [2,3,4], primal-dual [5] and local-search [6] have been developed. Dual-fitting combined with greedy augmentation gives 1.52-approximation algorithm [7]. Byrka [8] get an approximation algorithm that touches $(r_f, 1 + 2e^{-r_f})$ approximability limiting curve found by Jain et al. [9] for $r_f \geq 1.6774$ for UFLP by extending Chudak and Shmoys' algorithm [10]. He gave a bi-factor $(1.6774, 1.3738)$-approximation algorithm. By mixing this result with $(1.11, 1.7764)$-approximation algorithm of Jain et al. [9], he gave the current best 1.5-factor approximation algorithm for UFLP. This is very close to current lower bound 1.463 of UFLP by Guha and Khuller [4] unless $NP \subseteq DTIME(n^{\log \log n})$. Under the development of approximation algorithms of UFLP, variants of UFLP like hard capacity facility [11], k-median [5] and universal facility [12] have been considered.

---

[*] This research is supported by KOSEF Grant R01-2007-000-11905-0.

B. Yang, D.-Z. Du, and C.A. Wang (Eds.): COCOA 2008, LNCS 5165, pp. 265–277, 2008.

Connected Facility Location(ConFL) problem is a variant of UFLP. In an instance of the ConFL problem, we are given a graph $G = (V, E)$ with nonnegative edge cost c_e on each edge $e \in E$, a set of facilities $\mathcal{F} \subset V$, a set of demands, i.e., clients $\mathcal{D} \subset V$, and a parameter $M \geq 1$. Each facility $i \in \mathcal{F}$ has a nonnegative opening cost f_i and each client j has d_j units of demand. A solution of this problem opens a set of facilities $F \subset \mathcal{F}$ and assigns each demand j to some open facility $i(j) \in F$. In addition, the solution connects all open facilities using a Steiner tree T. The assignment cost of demand j is $d_j c_{i(j)j}$, where c_{kl} is the distance between vertices k and l. We assume that the distance follows triangle inequality. The facility connection cost is M times the cost of Steiner tree T which connects all open facilities. An optimal solution of this problem is a solution where total cost, $\sum_{i \in F} f_i + \sum_{j \in \mathcal{D}} d_j c_{i(j)j} + M \sum_{e \in T} c_e$ is minimized.

1.1 Related Works

The first constant factor approximation algorithm for ConFL problem was given by Karger and Minkoff [13]. Gupta et al. [14] improved this result by giving a 10.66-approximation algorithm based on LP rounding. The first primal-dual algorithm was given by Swamy and Kumar [1]. Their algorithm is combinatorial having approximation factor 8.55. Hasan, Jung and Chwa [15] gave an LP rounding 8.29-approximation algorithm for this problem. The authors in the same paper gave a LP rounding based 7-approximation algorithm for a special case of the ConFL problem where all facilities have equal opening costs.

Very recently Eisenbrand et al. [16] gave a 4-factor Randomized Approximation Algorithm (RAA) for the ConFL problem. This algorithm can be derandomized using the technique of Williams and Zuylen [17] that uses conditional expectation by solving a exponential sized linear program. This means that the algorithm of [16] is not so practical. Our algorithm is more efficient in time complexity than the most recent algorithm which needs to solve an exponential size linear program.

In section 2, we give our result and high level idea of our algorithm. In section 3, well-known linear programming formulation is given. In section 4, we describe the improved primal-dual 6.55-approximation algorithm. Finally in section 5, we give an analysis for the approximation factor of the algorithm.

2   High Level Idea of the Algorithm

In this paper we propose a primal-dual 6.55-approximation algorithm for ConFL problem which outperforms the best known primal-dual algorithm [1] with approximation factor 8.55. As we described before, the only deterministic approach [16] which can give approximation factor better than our result performs worse in time complexity.

The high level idea of Swamy and Kumar's algorithm [1] is like following.

1. Each demand pays for gathering at least M demands.
2. Each demand pays for being connected to a facility and temporarily opening a facility.

3. Select independent locations and open facilities
4. Construct Steiner tree for open facilities

Our algorithm proceeds like following.

1. Each demand pays for being connected to a facility and temporarily opening a facility.
2. Each demand pays for gathering at least M demands.
3. Construct Steiner tree for open facilities

The main difference between our algorithm and Swmay and Kumar's [1] lies in that our algorithm first pays for opening facilities, then pays for gathering demands. Also in our algorithm, we open facilities incrementally to be independent. So, in our algorithm there is no extra step to pick independent set of locations. These things together gives 2-factor decrease in the approximation factor.

3   Linear Programming Formulation

Given an instance of ConFL problem, we assume that a particular facility, say v, has zero opening cost and it belongs to an optimal solution. This assumption does not affect the approximation factor of the algorithm. Also we assume that for each demand $j \in \mathcal{D}$, $d_j = 1$. With these assumptions, the LP relaxation of ConFL problem and its dual can be written as:

$$\min \sum_{i \in \mathcal{F}, i \neq v} f_i y_i + \sum_{j \in \mathcal{D}} \sum_{i \in \mathcal{F}} c_{ij} x_{ij} + M \sum_{e \in E} c_e z_e \tag{P1}$$

$$\text{s.t} \sum_{i \in \mathcal{F}} x_{ij} \geq 1 \qquad \text{for all } j \in \mathcal{D}$$

$$x_{ij} \leq y_i \qquad \text{for all } i \in \mathcal{F}, i \neq v, j \in \mathcal{D}$$

$$x_{vj} \leq 1 \qquad \text{for all } j \in \mathcal{D}$$

$$\sum_{i \in S} x_{ij} \leq \sum_{e \in \delta(S)} z_e \qquad \text{for all } S \subseteq V, v \notin S, j \in \mathcal{D}$$

$$x_{ij}, y_i, z_e \geq 0$$

$$\max \sum_{j \in \mathcal{D}} \alpha_j - \sum_{j \in \mathcal{D}} \beta_{vj} \tag{D1}$$

$$\text{s.t} \quad \alpha_j \leq c_{ij} + \beta_{ij} + \sum_{s \subseteq V : i \in S, v \notin S} \theta_{Sj} \qquad \text{for all } i \neq v, j \in \mathcal{D} \tag{1}$$

$$\alpha_j \leq c_{vj} + \beta_{vj} \qquad \text{for all} j \in \mathcal{D} \tag{2}$$

$$\sum_{j \in \mathcal{D}} \beta_{ij} \leq f_i \qquad \text{for all } i \in \mathcal{F}, i \neq v \tag{3}$$

$$\sum_{j \in \mathcal{D}} \sum_{S \subseteq V : e \in \delta(S), v \notin S} \theta_{Sj} \leq M c_e \qquad \text{for all } e \in E \tag{4}$$

$$\alpha_j, \beta_{ij}, \theta_{Sj} \geq 0$$

Here x_{ij} indicates whether a client j is connected to facility i. y_i indicates whether facility i is open. And z_e indicates whether edge e is included in the Steiner tree.

In the dual program, α_j is the payment of client j for opening a facility, for being connected to a facility, and for constructing the Steiner tree connecting all open facilities. β_{ij} is the payment of client j for opening facility i. And $\theta_{S,j}$ is the payment of client j for constructing Steiner tree.

4   Algorithm Description

We use the term *location* to refer to a vertex in V or an internal point on an edge. A facility, on the other hand, refers only to a vertex in $\mathcal{F}$. The concept of location has been described in [1]. For convenience we are repeating the properties of location. For an edge $e = (u, w)$, we assume that $c_e = c_{uw} =$ the shortest path distance between u and w. That is, c_e satisfies metric property. For a point p on e, c_{up} increases continuously from 0 to c_e as p moves from u to w. For a vertex $q \in V - \{u, w\}$ and p on (u, w) we assume that $c_{pq} = \min(c_{qu} + c_{up}, c_{qw} + c_{wp})$. For any two points p and $\hat{p}$ on $e = (u, w)$ and $\hat{e}$ respectively, we assume that $c_{p\hat{p}} = \min(c_{up} + c_{u\hat{p}}, c_{wp} + c_{w\hat{p}})$.

Our algorithm executes in two phases. In phase 1, the algorithm mainly determines the set of facilities to open and the assignment of each client to some open facility. Phase 1 also determines part of the Steiner tree which connects all open facilities. Our main contribution is phase 1. In phase 1 of Swamy and Kumar's algorithm, there is a demand that needs to pay 7-times of its dual cost to be connected to an open facility. But in our algorithm, in phase 1, demands pays for at most 5-times of its dual cost for being connected to an open facility. So in fact, in phase 1, there appears 2-factor decrease in the approximation factor. Phase 2, which is similar to that of the algorithm described in [1], basically determines the remaining part of the Steiner tree used to connect all open facilities.

4.1   Phase 1

A demand is in one of two states: (a) frozen and (b) unfrozen. Initially all demands are unfrozen. Each facility is in one of three states (a) active (b) inactive, and (c) admin. Initially all facilities are active except v. v is always inactive. For each inactive facility except v, we will assign an *admin* facility which is called the *master* facility of that inactive facility. Throughout the algorithm we update dual variables α, β, θ and an extra variable γ, where γ_{ij} is defined for each facility demand pair $(i, j) \in (\mathcal{F} \times \mathcal{D})$. Initially $\gamma_{i,j} = -1$ for each (i, j) pair and all other variables are zero. At a particular time, we say that a demand j is *tight* with a location l iff $\alpha_j \geq c_{lj}$, a facility i is *paid for* iff $\sum_{j \in \mathcal{D}} \beta_{ij} = f_i$, a demand j *spans* location l through facility i iff $\gamma_{ij} \geq c_{il}$, and we say that demand j spans location l iff $\exists_{i \in \mathcal{F}}[j$ spans l through $i]$.

We always raise the dual variable α_j at the same speed for each unfrozen demand j. A demand becomes frozen when it gets connected to some facility.

A demand can be connected to some facility in two ways (i) directly and (ii) indirectly. It should be noticed that a demand can be connected to different facilities at different times and in different ways (directly or indirectly). In that case the connection established at last will be the final connection. For example, if demand j is indirectly connected to facility i_1 and later the same demand j is directly connected to facility i_2 then after the execution of phase 1, demand j is said to be directly connected to facility i_2.

We will use a notion of time t. At $t = 0$ we start raising dual variable α_j in unit rate for every demand j until j gets frozen. As we raise α_j, demand j may become tight with some facility i. Starting from this time we also raise β_{ij} in unit rate until j becomes frozen or facility i becomes paid for. Let t'_i be the earliest time when a facility i is paid for and t'_{ij} be the earliest time when demand j is tight with facility i. Suppose, $t_{ij} = \max\{t'_i, t'_{ij}\}$ which is the earliest time that a facility i is paid for and a demand j is tight with a facility i. For each facility-demand pair $(i, j) \in \mathcal{F} \times \mathcal{D}$, at t_{ij}, we set $\gamma_{ij} = 0$ (previously $\gamma_{ij} = -1$). Starting from time at t_{ij}, we raise γ_{ij} in unit rate until j becomes frozen. This implies that if demand j spans a location l through facility i then $c_{il} \leq \gamma_{ij} \leq \alpha_j$. Throughout phase 1 we raise θ_{Sj} in unit rate for each non-empty S and unfrozen j pair, where $S = \{i \in \mathcal{F} - \{v\} : j \text{ spans location } i\}$.

We continue raising all the variables until one of the following events occurs (if several events occur simultaneously then ties are broken according to the order of events given below):

1. A demand j becomes tight with some admin facility i: Indirectly connect j to i. Make j frozen.
2. A demand j becomes tight with some inactive facility k: if $k = v$ then indirectly connect j to v. Otherwise, indirectly connect j to i, where i is the master facility of k. Make j frozen.
3. At least M distinct demands (frozen or unfrozen) span a location l: A facility-demand pair (i, j) is chosen like this: If there is a pair (i', j') such that j' spans l through i' and i' is an admin facility then $i = i'$ and $j = j'$. Or, if there is a pair (i'', j'') such that j'' spans l through i'' and i'' is an inactive facility then $i = i''$ and $j = j''$. Otherwise, j is a demand that spans l with minimum α_j value and i is a facility such that j spans l through i. There are three cases:

   (a) i is an admin facility: Indirectly connect each unfrozen demand that spans l to i and make it frozen.
   (b) i is an inactive facility: Let $\hat{i}$ be the master of i. Indirectly connect each unfrozen demand that spans l to $\hat{i}$ and make it frozen.
   (c) i is an active facility: Make facility i admin. Declare this demand set (frozen or unfrozen) spanning l as a *village* D_i for an admin facility i. Additionally any demand $\hat{j}$ with $\beta_{i\hat{j}} > 0$ is also added to D_i. Note that $\hat{j}$ may not span l because otherwise it has already been added. We say that location l is the *admin location* of D_i. Directly connect all the demands of village D_i to admin facility i. Let I be the set of active facilities such that each of them are tight with some demand of this village ($i \notin I$ since

i is an admin facility now). Make each facility of I inactive and make i the master facility of each facility in I. Also facility i is the master facility of itself. Each demand j' which is tight with some facility of I but $j' \notin D_i$ is indirectly connected to i. Make all demands which are directly or indirectly connected to i frozen.

If a demand j gets frozen, we stop raising α_j, β_{ij}, γ_{ij} and θ_{Sj} for all $i \in \mathcal{F}$ and $S \subset V$. We continue this process until all the demands get frozen.

4.2   Phase 2

Let $\hat{G} = (V, \hat{E})$ be a graph augmented from G by including edges adjacent to admin location of each village D_i since an admin location may not be a vertex in G. Let $T' = (V, E')$ be a subgraph of $\hat{G}$ such that E' contains only the shortest distance path edges between the admin facility and admin location of each village. We use ρ approximation algorithm ($\rho \leq 2$) for Steiner tree problem to form the Steiner tree T'' on v and the component of each admin facility in T'. We take the union of edges of T' and T'' to form $\hat{T}$. Finally a Steiner tree T on v and the vertices corresponding to the admin facilities is found by removing each edge (l_i, i) whenever l_i is a leaf in $\hat{T}$.

After the execution of phase 1 and phase 2, we open v and all admin facilities. Demands indirectly connected to v are assigned to v and for each admin facility, all demands which are directly or indirectly connected to it is assigned to this facility. Finally the edges of T are selected to connect all open facilities.

To give a intuition why the approximation factor reduces in our algorithm, we give an example for phase 1. See the figure 1. If we run the Swamy and Kumar's algorithm, after phase 1 of their algorithm, one of facilities among i_1, i_2 and i_3 will be open after the clients being gathered at i_4 location. We suppose that i_2 is open, then the client j_1 needs $3 \cdot \alpha_j$ to be connected. But if we run our algorithm, after paying for facilities(although it is 0), the clients are gathered at the facility i_4 after being spanned from each facilities. After that we can choose i_2 as open facility. So clients j_2 needs at most $2 \cdot \alpha_j$ to be connected.

So far we have assumed that for each demand j, $d_j = 1$. It is not very difficult to handle arbitrary nonnegative demand by controlling the rate at which we raise the variables α_j, β_{ij}, γ_{ij} and θ_{Sj}. We omit the detail here.

5   Analysis

Lemma 1. (α, β, θ) *is a dual feasible solution.*

Proof. The idea of the proof of this lemma is similar to the proof of Lemma 3.1 in [1]. For a facility-demand pair $(i, j) \in \mathcal{F} \times \mathcal{D}$ and $i \neq v$, it is very easy to see that (1) is satisfied up to t_{ij} where t_{ij} is the earliest time such that at t_{ij}, i is paid for and j is tight with i. But at t_{ij}, γ_{ij} is set to zero and one can observe that after that time j spans location i through facility i. Since after t_{ij} we raise θ_{Sj} where $i \in S$, (1) is always satisfied. It can be checked very easily that both (2) and (3) are also satisfied.

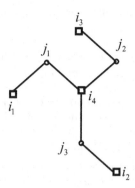

Fig. 1. This is a local information of a whole graph. We assume that all other facilities including v and other clients are placed far enough. The edge cost seen is all 1 and cost of all facilities except i_4 is 0. The facility cost of i_4 is ∞.

To check the satisfiability of (4), let us consider an edge $e = (u, w)$ and a demand j such that $\sum_{S \subseteq V : e \in \delta(S), v \notin S} \theta_{Sj} > 0$. Also assume that j spans u before w. Let p be a point on (u, w) such that $c_{up} = x$. Let us define $f(j, x)$ as 1 if j spans p and j was not frozen at the earliest time when j spanned p , otherwise $f(j, x)$ is 0. Then, $\sum_{S \subseteq V : e \in \delta(S), v \notin S} \theta_{S,j} \leq \int_0^{c_e} f(j, x) dx$. Now,

$$\sum_j \sum_{S \subseteq V : e \in \delta(S), v \notin S} \theta_{S,j} \leq \sum_j \int_0^{c_e} f(j, x) dx = \int_0^{c_e} \sum_j f(j, x) dx \leq M c_e$$

The last inequality holds because $\sum_j f(j, x)$ is at most M. Otherwise, we have more than M demands that span a location but none of them are frozen which is a contradiction. □

Lemma 2. *Let j be a demand and $I_j = \{i \in \mathcal{F} : \beta_{ij} > 0\}$. Then, at most one facility in I_j is open in our solution.*

Proof. Since only v and the admin facilities created in phase 1 are opened and $v \notin I_j$ because $\beta_{vj} = 0$, it is enough to prove that at most one facility in I_j is selected as an admin facility in phase 1.

For a contradiction, let us assume that i_1 and i_2 are two admin facilities such that $i_1, i_2 \in I_j$. Without loss of generality let us assume that i_1 has been selected as an admin facility earlier than i_2 and i_1 has been made admin at t_1. Clearly, $\beta_{i_1 j} > 0$ at t_1. And this implies that $\beta_{i_2 j} > 0$ at t_1 because j is frozen after t_1. Then, facility i_2 is made inactive at t_1. This is a contradiction because once a facility is inactive it can never become an admin facility. □

As we have discussed before, a demand can be connected (directly or indirectly) to several facilities. In that case the connection established at the latest time is the final connection. A demand gets frozen only when it is connected to some facility. Since the algorithm continues until all the demands become frozen, it is easy to see that after the execution of phase 1, the facility to which demand j is connected (directly or indirectly) is well defined. The following lemma proves that once a demand is directly connected to some facility it will never be connected to any other facility.

Lemma 3. *Once a demand j is directly connected to a facility i it will never be connected (directly or indirectly) to some other facility.*

Proof. It is obvious that a frozen demand can not be indirectly connected to some other facility by phase 1. Since a demand becomes frozen after being connected for the first time, once j is directly connected to i, it will never be indirectly connected to some other facility. It remains to prove that j will not be directly connected to some other facility also.

For deriving a contradiction, let us assume that the algorithm directly connects j twice first time to i and then to i', where $i \neq i'$. Then both i and i' are two admin facilities. Let l_i and $l_{i'}$ be the corresponding admin locations. Let us assume that t_i and $t_{i'}$ are the times when j is directly connected to i and i' respectively, where $t_i \leq t_{i'}$. Now, since j is directly connected to i', either $\beta_{i'j} > 0$ at $t_{i'}$ or j spans i' at $t_{i'}$. If $\beta_{i'j} > 0$ at $t_{i'}$ then $\beta_{i'j} > 0$ at t_i because j is frozen after t_i. But it is not possible because in that case j is tight with i' at t_i and i' is made inactive. So, we can assume that j spans $l_{i'}$ at $t_{i'}$ which implies that j spans l_{i} at t_i since j is frozen after t_i. Then, at $t_{i'}$ there is a facility say k, such that j spans $l_{i'}$ through k and k is inactive because at t_i all the facilities tight with j are made inactive. But the algorithm has chosen facility i', where some client spans $l_{i'}$ through k and k is inactive. This is a contradiction because the algorithm can not chose i' when k is present. $\square$

Lemma 4. *At the end of phase 1 for each village D_i, $|\{j \in D_i : j \text{ spans } l_i \}| \geq M$ where i and l_i are the admin facility and the admin location of village D_i, respectively.*

Proof. When a village is created, there are at least M demands spanning the admin location l_i. These demands are added to D_i and are directly connected to the admin facility i at the time when the village has been created. According to the previous lemma these demands remain directly connected at the end of phase 1. $\square$

Lemma 5. *If a client j is directly connected to facility i then $c_{ij} \leq 2\alpha_j$.*

Proof. Since j is directly connected to i, i is the admin facility of a village D_i such that $j \in D_i$. This implies, either j spans an admin location l_i or $\beta_{ij} > 0$ when a village D_i has been created. If $\beta_{ij} > 0$ at that time then $c_{ij} < \alpha_j$ by the constraint(1). So let us assume that $\beta_{ij} = 0$.

Now, when D_i has been created, the algorithm has chosen (i, j') as the facility-demand pair such that j' spans l_i through facility i and $\alpha_{j'} \leq \alpha_j$. Since j also spans l_i, $c_{ij} \leq c_{il_i} + c_{l_ij} \leq \alpha_{j'} + \alpha_j \leq 2\alpha_j$ $\square$

Lemma 6. *If client j is indirectly connected to a facility i then $c_{ij} \leq 5\alpha_j$.*

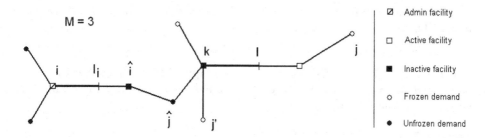

Fig. 2. j is indirectly connected to i with $c_{ij} \leq 5\alpha_j$: $\hat{j}$ spans l_i through $\hat{i}$, j' spans l through k, $\hat{j}$ is tight with k ($\hat{j}$ may not span l) and j spans l

Proof. There are several cases depending on how j has been indirectly connected to i.

Case 1 (j becomes tight with an admin facility i): In this case $c_{ij} \leq \alpha_j$.

Case 2 (j becomes tight with v): In this case $i = v$ and $c_{ij} \leq \alpha_j$

Case 3 (j becomes tight with an inactive facility k such that i is the master of k): In this case, when k was made inactive by i there is a demand $j' \in D_i$ such that j' was tight with k. Then, $c_{ij'} \leq 2\alpha_{j'}$ (by lemma 5) and $c_{kj'} \leq \alpha_{j'}$. Since j becomes tight with k, $c_{kj} \leq \alpha_j$. Clearly $\alpha_{j'} \leq \alpha_j$ because j is unfrozen. So, $c_{ij} \leq c_{ij'} + c_{kj'} + c_{kj} \leq 4\alpha_j$.

Case 4 (At least M demands span a location l and one of these demands spans l through i which is already an admin facility): In this case the algorithm chooses (i, j') pair where j' spans l through i. Since only unfrozen demands of these M spanning demands are indirectly connected to i, j is unfrozen. This implies, $\alpha_{j'} \leq \alpha_j$. Now, since j' spans l through i and j spans l, $c_{il} \leq \alpha_{j'}$ and $c_{lj} \leq \alpha_j$. So, $c_{ij} \leq c_{lj} + c_{il} \leq 2\alpha_j$

Case 5 (At least M demands span a location l and one of these demands spans l through inactive facility k such that i is the master of k): In this case the algorithm chooses (k, j') pair where j' spans l through k. Since only unfrozen demands of these M spanning demands are indirectly connected to the master of k, j is unfrozen. Since i is the master of k, there is a client $\hat{j} \in D_i$ such that $\hat{j}$ is tight with k. Now, $\alpha_{j'} \leq \alpha_j$, $\alpha_{\hat{j}} \leq \alpha_j$ since j is unfrozen and $c_{i\hat{j}} \leq 2\alpha_{\hat{j}}$ (by lemma 5). Because j' spans l through k and j spans l, $c_{kl} \leq \alpha_{j'}$ and $c_{lj} \leq \alpha_j$. Thus, $c_{ij} \leq c_{i\hat{j}} + c_{k\hat{j}} + c_{kl} + c_{lj} \leq 2\alpha_{\hat{j}} + \alpha_{\hat{j}} + \alpha_{j'} + \alpha_j \leq 5\alpha_j$ (see Figure 2). □

Let F' be the set of admin facilities selected by the algorithm at the end of phase 1. For a facility $i \in F'$, assume that $L_i = \{j \in D_i : j \text{ spans } l_i \}$ which is the set of demands that span l_i, where l_i is the admin location of village D_i. Also let us assume that $C_i = \{j \in \mathcal{D} : \beta_{ij} > 0\}$ which is the set of demands that pays for facility cost of i. Lemma 2 shows that $C_i \subseteq D_i$ for each $i \in F'$. Let $D' = \cup_{i \in F'} D_i$ which is the set of demands each of which is directly connected. $C' = \cup_{i \in F'} C_i$ which is the set of demands each of which pays for the facility costs of some admin facility. Then, $C' \subseteq D'$. We start raising β_{ij} when j becomes tight with i and stop raising β_{ij} when either i becomes paid for or j becomes frozen. Thus,

$$j \in C_i \Rightarrow c_{ij} + \beta_{ij} \le \alpha_j. \tag{5}$$

For a graph $T_0 = (V, E_0)$, which is a subgraph of G, let us define the cost of T_0, $cost(T_0) = M \sum_{e \in E_0} c_e$. Next two lemmas bound the cost of T' and T'' generated in phase 2, respectively.

Lemma 7. $cost(T') \le \sum_{j \in D'} \alpha_j$

Proof. Let us consider a village D_i. Then, there is a demand $\hat{j} \in D_i$ such that $\hat{j}$ spans l_i through i and $\alpha_{\hat{j}} = \min_{j \in L_i} \alpha_j$, where l_i is the admin location of village D_i. So, $Mc_{il_i} \le M\alpha_{\hat{j}} \le \sum_{j \in L_i} \alpha_j \le \sum_{j \in D_i} \alpha_j$, where the second last inequality holds because of Lemma 4 and the fact that $\alpha_{\hat{j}} = \min_{j \in L_i} \alpha_j$. Now, $cost(T') = M \sum_{i \in F'} c_{il_i} \le \sum_{i \in F'} \sum_{j \in D_i} \alpha_j = \sum_{j \in D'} \alpha_j$. $\quad\square$

Lemma 8. *Let C^* and S^* be the assignment cost and Steiner tree cost of an optimal solution. Then, $cost(T'') \le \rho(S^* + C^* + \tau)$, where $\rho \le 2$ is the approximation factor of the Steiner tree algorithm used in phase 2 and $\tau = \sum_{i \in F'} \sum_{j \in L_i} c_{l_i j}$.*

Proof. The idea of this proof is similar to that of Swamy and Kumar [1]. We can obtain a Steiner tree $\hat{T}'' = G(V, \hat{E}'')$ on v and the component of each admin facility in T' like this: for each village with admin facility i we find the shortest $l_i - j - i^*(j)$ path for $j \in L_i$, where $i^*(j)$ is the facility to which demand j is assigned in the optimal solution. We add these shortest paths found for all villages to T'. Then we add the edges of T^* to this graph where T^* is the Stener tree connecting all open facilities in the same optimal solution. Finally cycles are broken by removing arbitrary edges so that the resultant graph becomes tree $\hat{T}''$. It is easy to see that the $cost(T'') \le \rho \cdot cost(\hat{T}'')$. Now,

$$
\begin{aligned}
cost(\hat{T}'') &= S^* + M \sum_{i \in F'} (\text{shortest } l_i - j - i^*(j) \text{ path for } j \in L_i) \\
&\le S^* + \sum_{i \in F'} \sum_{j \in L_i} (c_{l_i j} + c_{i^*(j)j}) \tag{6} \\
&= S^* + \sum_{i \in F'} \sum_{j \in L_i} c_{i^*(j)j} + \sum_{i \in F'} \sum_{j \in L_i} c_{l_i j} \\
&\le S^* + C^* + \tau
\end{aligned}
$$

where (6) follows from Lemma 4. Thus, $cost(T'') \le \rho(S^* + C^* + \tau)$. $\quad\square$

Lemma 9. *Let a demand j be assigned to $i(j)$ in our solution, $F = \sum_{i \in F'} f_i$, $C = \sum_{j \in D} c_{i(j)j}$, and $\tau = \sum_{i \in F'} \sum_{j \in L_i} c_{l_i j}$. Then, $F + C + cost(T') + 2\tau \le 5 \sum_{j \in D} \alpha_j$, where l_i and i are the admin location and admin facility of village D_i respectively.*

Proof.

$$F + C + cost(T') + 2\tau = \sum_{i \in F'} f_i + \sum_{j \in \mathcal{D}} c_{i(j)j} + cost(T') + 2 \sum_{i \in F'} \sum_{j \in L_i} c_{l_ij}$$

$$= \sum_{i \in F'} f_i + \sum_{j \in D'} c_{i(j)j} + \sum_{j \in \mathcal{D} \setminus D'} c_{i(j)j} + cost(T') + 2 \sum_{i \in F'} \sum_{j \in L_i} c_{l_ij}$$

$$= \sum_{i \in F'} \sum_{j \in C_i} \beta_{ij} + \sum_{i \in F'} \sum_{j \in D_i} c_{ij} + \sum_{j \in \mathcal{D} \setminus D'} c_{i(j)j} + cost(T') + 2 \sum_{i \in F'} \sum_{j \in L_i} c_{l_ij} \tag{7}$$

$$= \sum_{i \in F'} \left(\sum_{j \in C_i} \beta_{ij} + \sum_{j \in D_i} c_{ij} + 2 \sum_{j \in L_i} c_{l_ij} \right) + \sum_{j \in \mathcal{D} \setminus D'} c_{i(j)j} + cost(T')$$

$$= \sum_{i \in F'} \left(\sum_{j \in C_i} (\beta_{ij} + c_{ij}) + \sum_{j \in D_i \setminus C_i} c_{ij} + 2 \sum_{j \in L_i} c_{l_ij} \right) + \sum_{j \in \mathcal{D} \setminus D'} c_{i(j)j} + cost(T')$$

$$\leq \sum_{i \in F'} \left(\sum_{j \in C_i} \alpha_j + \sum_{j \in D_i \setminus C_i} 2\alpha_j + 2 \sum_{j \in L_i} \alpha_j \right) + \sum_{j \in \mathcal{D} \setminus D'} 5\alpha_j + \sum_{j \in D'} \alpha_j \tag{8}$$

$$\leq \sum_{i \in F'} \left(\sum_{j \in C_i} 2\alpha_j + \sum_{j \in D_i \setminus C_i} 2\alpha_j + \sum_{j \in D_i} 2\alpha_j \right) + \sum_{j \in \mathcal{D} \setminus D'} 5\alpha_j + \sum_{j \in D'} \alpha_j \tag{9}$$

$$= \sum_{j \in D'} 4\alpha_j + \sum_{j \in \mathcal{D} \setminus D'} 5\alpha_j + \sum_{j \in D'} \alpha_j$$

$$= 5 \sum_{j \in \mathcal{D}} \alpha_j$$

where (7) follows from Lemma 2, (8) follows from Lemma 5, Lemma 6, Lemma 7 and Equation 5, finally (9) follows from the fact that $L_i \subseteq D_i$. $\square$

Theorem 1. *The algorithm gives a solution of cost at most $6.55 \cdot OPT$, where OPT is the cost of optimal solution.*

Proof. Let F, C and S be the facility opening cost, demand assignment cost, and Steiner tree cost of the solution obtained by the algorithm, respectively. For $\rho \leq 2$ we get:

$$F + C + S \leq F + C + (cost(T') + cost(T''))$$

$$\leq F + C + cost(T') + 2\tau + \rho(S^* + C^*) \tag{10}$$

$$\leq 5 \sum_{j \in \mathcal{D}} \alpha_j + \rho(S^* + C^*) \tag{11}$$

$$\leq 5 \cdot OPT + \rho \cdot OPT$$

$$= (5 + \rho) \cdot OPT$$

(10) follows from Lemma 8 and (11) follows from Lemma 9. Plugging the value $\rho = 1.55$ [18] which is the current best factor for approximation algorithm of Steiner tree problem gives us a solution of cost at most $6.55 \cdot OPT$.   □

6   Concluding Remarks

In this paper we deal with the Connected Facility Location problem and give a 6.55 factor primal-dual approximation algorithm for this problem. This problem has important applications specially in the field of computer networks and mobile communication. Our algorithm is simple and combinatorial in structure.

The Capacitated Facility Location problem is a generalization of the Facility Location problem. LP-based algorithm for this problem with uniform facility cost is given by Levi et al [19]. There is no known primal-dual technique for capacitated facility location problem. Prominent future works are to find primal-dual algorithms for the Capacitated Facility Location and the Connected Capacitated Facility Location problems by extending the algorithm of this paper.

References

1. Swamy, C., Kumar, A.: Primal-dual algorithms for connected facility location problems. Algorithmica 40, 245–269 (2004)
2. Shmoys, D.B., Tardos, É., Aardal, K.: Approximation algorithms for facility location problems (extended abstract). In: STOC 1997: Proceedings of the twenty-ninth annual ACM symposium on Theory on Computing, pp. 265–274 (1997)
3. Goemans, M.X., Williamson, D.P.: A general approximation technique for constrained forest problems. SIAM Journal on Computing 24, 296–317 (1995)
4. Guha, S., Khuller, S.: Greedy strikes back: improved facility location algorithms. Journal of Algorithms 31, 228–248 (1999)
5. Jain, K., Vazirani, V.V.: Approximation algorithms for the metric facility location and k-median problems using the primal-dual schema and lagrangian relaxation. Journal of the ACM 48, 274–296 (2001)
6. Korupolu, M.R., Plaxton, C.G., Rajaraman, R.: Analysis of a local search heuristic for facility location problems. In: Proceedings of the 9th Annual ACM-SIAM Symposium on Discrete Algorithms, pp. 1–10 (1998)
7. Mahdian, M., Ye, Y., Zhang, J.: Improved approximation algorithms for metric facility location problems. In: Proceedings of the 5th International Workshop on Approximation Algorithms for Combinatorial Optimization, pp. 229–242 (2002)
8. Byrka, J.: An optimal bifactor approximation algorithm for the metric uncapacitated facility location problem. In: APPROX-RANDOM, pp. 29–43 (2007)
9. Jain, K., Mahdian, M., Saberi, A.: A new greedy approach for facility location problems. In: STOC 2002: Proceedings of the thiry-fourth annual ACM symposium on Theory of computing, pp. 731–740 (2002)
10. Chudak, F.A., Shmoys, D.: Improved approximation algorithms for the uncapacitated facility location problem. SIAM Journal on Computing 33(1), 1–25 (2003)
11. Pal, M., Tardos, E., Wexler, T.: Facility location with nonuniform hard capacities. In: Proceedings of The 42nd Annual IEEE Symposium on Foundations of Computer Science, pp. 329–338 (2001)

12. Mahdian, M., Pal, M.: Universal facility location. In: Proceedings of 11the European Symposim on Algorithms, pp. 409–422 (2003)
13. Karger, D.R., Minkoff, M.: Building steiner trees with incomplete global knowledge. In: FOCS 2000: Proceedings of 41st Annual Symposium on Foundations of Computer Science, pp. 613–623 (2000)
14. Gupta, A., Kleinberg, J., Kumar, A., Rastogi, R., Yener, B.: Provisioning a virtual private network: a network design problem for multicommodity flow. In: Proceedings of the 33rd Annual ACM Symposium on Theory of Computing, pp. 389–398 (2001)
15. Hasan, M.K., Jung, H., Chwa, K.-Y.: Improved approximation algorithm for connected facility location problems. In: Proceedings of The First International Conference on Combinatorial Optimization and Applications, pp. 311–322 (2007)
16. Eisenbrand, F., Grandoni, F., Rothvoß, T., Schäfer, G.: Approximating connected facility location problems via random facility sampling and core detouring. In: SODA 2008: Proceedings of the nineteenth annual ACM-SIAM symposium on Discrete algorithms, pp. 1174–1183 (2008)
17. Williamson, D.P., van Zuylen, A.: A simpler and better derandomization of an approximation algorithm for single source rent-or-buy. Operations Research Letters 35(6), 707–712 (2007)
18. Robins, G., Zelikovsky, A.: Improved Steiner tree approximation in graphs. In: Proceedings of The 11th Annual ACM-SIAM Symposium on Discrete Algorithms, pp. 329–338 (2000)
19. Levi, R., Shmoys, D.B., Swamy, C.: LP-based spproximation slgorithms for capacitated facility location. In: Proceedings of IPCO, pp. 206–218 (2004)

Two Constant Approximation Algorithms for Node-Weighted Steiner Tree in Unit Disk Graphs

Feng Zou[1], Xianyue Li[2], Donghyun Kim[1], and Weili Wu[1,⋆]

[1] Department of Computer Science, University of Texas at Dallas,
Richardson, TX, 75080
{phenix.zou,donghyunkim}@student.utdallas.edu, weiliwu@utdallas.edu
[2] School of Mathematics and Statistics, Lanzhou University,
Lanzhou, Gansu, P.R. China, 730000
lixianyue@lzu.edu.cn

Abstract. Given a graph $G = (V, E)$ with node weight $w : V \to R^+$ and a subset $S \subseteq V$, find a minimum total weight tree interconnecting all nodes in S. This is the node-weighted Steiner tree problem which will be studied in this paper. In general, this problem is NP-hard and cannot be approximated by a polynomial time algorithm with performance ratio $a \ln n$ for any $0 < a < 1$ unless $NP \subseteq DTIME(n^{O(\log n)})$, where n is the number of nodes in s. In this paper, we show that for unit disk graph, the problem is still NP-hard, however it has polynomial time constant approximation. We will present a 4-approximation and a 2.5ρ-approximation where ρ is the best known performance ratio for polynomial time approximation of classical Steiner minimum tree problem in graphs. As a corollary, we obtain that there is polynomial time (9.875+ε)-approximation algorithm for minimum weight connected dominating set in unit disk graphs.

1 Introduction

Given a graph $G = (V, E)$ with weight function w on E and a subset S, *Steiner tree problem* (STP) is to find a minimum subgraph of G interconnecting all nodes in S. We call the set S as *terminal set*. For any Steiner tree T for S and node $u \in V(T)$, we call u as a *terminal node* if $u \in S$, otherwise, we call it as a *Steiner node*. The Steiner tree problem is a classical problem in networks, which is very interesting nowadays. The problem is known to be NP-hard, as well as in most metrics[6]. Lots of effort have been devoted to study the approximation algorithms for this problem[3,8,11,14,17] and have successfully achieved constant ratios for this problem.

Node-Weighted Steiner Tree problem (NWST) is a variation of the classical STP. Given a graph $G = (V, E)$ with node weight $w : V \to R^+$ and a subset

---

⋆ Support in part by National Science Foundation under grant CCF-0514796 and CCF-0750992.

B. Yang, D.-Z. Du, and C.A. Wang (Eds.): COCOA 2008, LNCS 5165, pp. 278–285, 2008.

S of V, the node-weighted Steiner tree problem is to find a Steiner tree for the set S such that its total weighted is minimum. Since all of nodes in terminal set will be contained in any Steiner tree constructed, for convenience, we usually set $w(u) = 0$ for all nodes $u \in S$.

In this paper, We study NWST problem in a special type of graphs called unit disk graph, which has a wide application in networks. A *unit disk graph* is associated with a set of unit disks in the Euclidean plane. Each node is the center of a unit disk. An edge exists between two nodes u and v if and only if $|uv| \le 1$, where $|uv|$ is the Euclidean distance between u and v. For this special type of graphs, we propose two constant approximation algorithms for NWST problem from different perspectives with approximation ratio 4 and 2.5ρ respectively, where $\rho = 1 + \frac{\ln 3}{2} \approx 1.55$ is the best known approximation for the classical Steiner tree problem[14]. As an application on Minimum Weighted Connected Dominating Set (MSCDS) problem, we obtain a $(9.875+\varepsilon)$-approximation algorithm for this problem, which improves the best previous approximation ratio $10+\varepsilon$[9].

The rest of this paper is organized as follows. In section 2, we introduce the related work for NWST problem. In section 3, we first present the main idea of our algorithms. Then, we give some useful definitions and denotations. In subsection 3.2 and 3.3, we introduce the algorithms and prove their approximation ratios, respectively. As a corollary, we obtain a $(9.875+\varepsilon)$-approximation algorithm for MSCDS problem in unit disk graphs. Final, we conclude our results.

2 Related Work

Quite a few work has been done for NWST problem starting from early 80s. Most of the early work[1,2,13,15,16] focus on solving the problem and its variations using linear programming. For instance, Aneja [1] used a specialized integer programming (set covering) formulation to represent the STP and used the row generation scheme to solve the exponential increase of the number of constraints in the formulation related to the size of the problem. Segev [15] proposed an integer programming solution for an extension of the standard Steiner Tree problem using lagarangian relaxation and subgradient optimization. Though these work presented solutions for Steiner tree problem from the perspective of linear programming, they did not provide solid theoretical proof for the approximation ratios of their algorithms at all.

In 1991, Berman (see ref. in [10]) proved that NWST problem can not be approximated within a factor of $o(\ln k)$ by giving an approximation-preserving reduction from Set Cover[5] to NWST problem, where k is the size of the terminal set. Later, Klein and Ravi [10] presented the first asymptotically optimal solution of approximation ratio $2 \ln k$, by constructing the Steiner tree with spiders, a tree with at most one node of degree greater than two. Later, this ratio is improved to be $1.35 \ln k$ by Guha and Khuller [7] by introducing a new concept called branch-spider. This is the best known ratio up till now.

As a special case, when the weights of all nodes are same, NWST problem is equivalent to Steiner Tree with Minimum Number of Steiner Nodes (ST-MSN) problem. Given a set of n terminals S in the Euclidean plane and a positive constant c, the ST-MSN problem is to find a Steiner tree for S with minimum number of Steiner nodes such that each edge in the tree has a length no more than c. In [4], they showed that there is a 3-approximation algorithm for ST-MSN. In 2006, M.Min et al.[12] presented this problem in unit disk graph and gave a 3-approximation algorithm for a special terminal set.

3   Node-Weighted Steiner Tree Problem (NWST)

In this section, we study the NWST problem in unit disk graphs and propose two approximation algorithms for solving this problem. The main idea of the first one is to construct the Minimum Spanning Tree (MST) to solve the NWST. In this algorithm, we first construct an edge-weighted complete graph G' on terminal set such that the weight of every edge in G' is equal to the length of the shortest path between its endpoints in the original graph G. Then, compute a MST of the new graph G'. By replacing each edge in the MST with the corresponding shortest path in the original graph, we obtain a connected subgraph of G containing all nodes in the terminal set. We prove that this yields a valid NWST which has a total weight of no more than 4 times the weight of the optimal Node-Weight Steiner Tree of G.

The second algorithm has a better approximation ratio than the first one, while rooted from a totally different methodology. The main idea is based on the classical Steiner Tree problem. Firstly, we construct an edge-weighted graph G'' with the same node-set and edge-set as the original graph G. Then, we define the weight of every edge in G'' as the half of the sum of its endpoints' weights in G. Finally, we use ρ-approximation algorithm to obtain a Steiner tree of G''. From the Steiner tree of G'', we can get a Steiner Tree of G with approximation ratio $2.5\rho \approx 3.875$. As a corollary, we obtain a polynomial time algorithm for minimum weight connected dominating set in unit disk graphs with approximation ratio $9.875+\varepsilon$.

3.1   Preliminaries

In this section, we give some useful definitions and denotations. For a node-weighted (or edge-weighted) graph G with weight function w and a subgraph H of G, denote $w(H)$ be the the weight sum of all nodes (or edges) in H. For any two nodes u and v of G, denote $dist_G(u, v)$ as the weight of the shortest path between u and v, which is calculated as $min \sum_{v_k \in p} w(v_k)$ among all the possible paths between u and v. Here v_k is the internal node on every possible path p.

Given a edge-weighted graph G and a node subset S, we denote the ρ-approximation algorithm for Steiner Minimum Tree(SMT) as **SMT(G,S)** and the algorithm for finding Minimum Spanning Tree(MST) as **MST(G)**.

3.2   4-Approximation Algorithm

In this algorithm, we first construct a complete graph G' on terminal set S with a weight function c on its edges such that $c(u, v) = dist_G(u, v)$ for any two nodes u and v in S. Denote the optimal Node-Weighted Steiner Tree of the graph G on S as T_{OPT_NWS} and the minimum spanning tree of G' as T_{MST}. We can obtain the following theorem:

Theorem 1. *Suppose that for any terminal set S in G, there always exists a minimum spanning tree T_{MST} for the newly constructed complete graph G' such that the maximum degree of it is at most Δ. Then, the weight of T_{MST} is no more than $(\Delta - 1)$-times of the weight of T_{OPT_NWS}.*

Proof. Suppose that T_{OPT_NWS} contains k Steiner points $\{s_1, s_2, \ldots, s_k\}$ in the ordering of the breadth-first search starting from a vertex of S. Meanwhile, we define $N(S)$ as the weight of the minimum spanning tree of the new complete graph for the terminal set S. We claim that

$$N(S \cup \{s_1, s_2, \ldots, s_i\}) \leq N(S \cup \{s_1, s_2, \ldots, s_{i+1}\}) + (\Delta - 1) * w(s_{i+1}).$$

To show this claim, let G_i be the newly constructed complete graph for the terminal set $S \cup \{s_1, s_2, \ldots, s_i\}$. Consider a minimum spanning tree T for G_{i+1} with degree at most Δ. Suppose s_{i+1} has adjacent vertices $v_1, v_2, \ldots, v_d (d \leq \Delta)$. Then at least one of the edges $(s_{i+1}, v_j)(j = 1, \ldots, d)$ has weight 0. Otherwise, by BFS, there exists a vertex $u \in S \cup \{s_1, s_2, \ldots, s_i\}$ such that (s_{i+1}, u) is an edge of $E(G)$. This implies that the weight of this edge in the graph G_{i+1} is 0. If all weight of edges $(s_{i+1}, v_j)(j = 1, \ldots, d)$ are more than 0, adding edge (s_{i+1}, u) and deleting one of them, we can get a new tree with weight less than the current one, which is a contradiction. So without loss of generality, assume that the weight of (s_{i+1}, v_1) is 0. Delete all edges $(s_{i+1}, v_j)(j = 2, \ldots, d)$, and add $(d - 1)$ edges $(v_1, v_j)(j = 2, \ldots, d)$. We obtain a spanning tree T' for G_i. Note that according to the triangle inequality of the distance between two vertices, $c_{G_i}(v_1, v_j) \leq c_{G_{i+1}}(s_{i+1}, v_j) + w(s_{i+1})$. Hence, the claim holds from $N(S \cup \{s_1, s_2, \ldots, s_i\}) \leq c(T') \leq N(S \cup \{s_1, s_2, \ldots, s_{i+1}\}) + (\Delta - 1) * w(s_{i+1})$. Since $N(S \cup \{s_1, s_2, \cdots, s_k\}) = 0$ and by the claim and recurrence, we have

$$c(T_{MST}) = N(S) \leq N(S \cup \{s_1\}) + (\Delta - 1) * w(s_1)$$
$$\leq N(S \cup \{s_1, s_2\}) + (\Delta - 1)(w(s_1) + w(s_2))$$
$$\vdots$$
$$\leq N(S \cup \{s_1, s_2, \ldots, s_k\}) + (\Delta - 1)(\sum_{i=1}^{k} w(s_i))$$
$$= (\Delta - 1)w(T_{OPT\_NWS}). \qquad \square$$

For every edge of T_{MST}, replacing it with the corresponding shortest path between its endpoints in G, we obtain a connected subgraph H of G containing

---

Algorithm 1. NWST1 (G=(V,E,w,S))

1: Initialize a complete graph $G^{'} = (V^{'}, E^{'}, c)$ of G by setting $V^{'} = S$
2: **for** each pair of vertex (v_i, v_j) in graph $G^{'}$ **do**
3:     Set the $c(v_i, v_j) = dis_G(v_i, v_j)$, the shortest path between v_i and v_j
4: **end for**
5: $H = MST(G^{'})$
6: **for** each edge (v_i, v_j) in graph T **do**
7:     Replace it with the corresponding shortest path in graph G
8: **end for**
9: Output H

Algorithm 2. NWST2 $(G = (V, E, w, S))$

1: Initialize an edge-weighted graph $G^{'} = (V^{'}, E^{'}, w^{'}, S^{'})$ by setting $V^{'} = V, S^{'} = S$
   and $E^{'} = E$
2: **for** each edge (v_i, v_j) in graph $G^{'}$ **do**
3:     Assign the weight of this edge $w^{'}(v_i, v_j) = (w(v_i) + w(v_j))/2.$
4: **end for**
5: $T = \mathbf{SMT}(G^{'}, S)$
6: Output T

---

all nodes of S. Obviously, the weight sum of all nodes in H is no more than the weight of T_{MST}.

The detailed algorithm is presented in algorithm 1. Since in any unit disk graph, there always exists a spanning tree with maximum degree no more than 5[12], we can obtain the following corollary:

Corollary 1. *The connected subgraph we obtained at the end of the algorithm $NWST1$ is 4-approximation for T_{OPT_NWS} in unit disk graph.*

3.3   2.5ρ-Approximation Algorithm

The idea of this algorithm is to convert the node-weighted Steriner tree problem to the classical Steiner tree problem. Firstly, we construct an edge-weighted graph $G^{''}$ from G as follows by initializing $G^{''}$ with the same node-set and edge-set as G and an edge weight function $w^{'}$. For every edge $e = (u, v)$ in $G^{''}$, let the edge weight $w^{'}(u, v) = \frac{1}{2}(w(u) + w(v))$. The second step of this algorithm is to compute a Steiner tree T of $G^{''}$ on S through the ρ-approximation algorithm. Final, view T as the node-weighted Steiner tree of G on S and output it. The pseudo-code of this algorithm is presented in algorithm 2.

To better illustrate this algorithm, we give an example. Given a node-weighted graph G and a terminal set S as in figure1. a, firstly, we transform it into graph G' 1. b according to the weight assignment. The numbers besides each vertex in G are the weights associated with them and we set the weight of all vertices in the terminal set as 0. Secondly, we calculate the SMT for graph G', which is the subgraph in figure 1. c. Figure 1. d presents the NWST for original graph G.

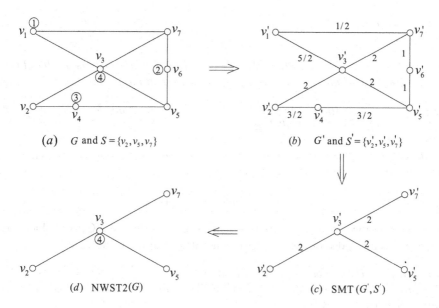

(a) G and $S = \{v_2, v_5, v_7\}$

(b) G' and $S' = \{v_2', v_5', v_7'\}$

(d)   NWST2(G)

(c)   SMT(G', S')

Fig. 1. An example of 2.5ρ algorithm

Since G is a unit disk graph, for any terminal set S, there always exists a Steiner minimum tree with maximum degree no more than 5[12]. The following lemma proves the relationship between the weight of the optimal SMT in graph G'' and the weight of the optimal NWST we want.

Lemma 1. *Denote T_{OPT_SM} as the optimal Steiner Minimum Tree for G'' on the set S and T_{OPT_NWS} as the optimal Node-weighted Steiner Tree of G on the same terminal set S, respectively. Then $w'(T_{OPT\_SM}) \leq 2.5w(T_{OPT\_NWS})$ when G is a unit disk graph.*

Proof. Consider T_{OPT_NWS} as a Steiner tree on S of G''. For convenience, denote T as T_{OPT_NWS}, V as $V(T_{OPT\_NWS})$, E as $E(T_{OPT\_NWS})$ and $d_T(u)$ as the degree of node u in tree T, we have

$$w'(T_{OPT\_SM}) \leq w'(T_{OPT\_NWS})$$
$$= \sum_{e=uv \in E} (\frac{1}{2}(w(u) + w(v)))$$
$$= \sum_{u \in V} \frac{d_T(u)}{2} w(u)$$
$$\leq \frac{5}{2} \sum_{u \in V} (w(u)) = 2.5w(T).$$

Hence, the lemma holds. $\square$

If we consider the Steiner tree T of G'' as a subgraph of G, we can get the following lemma:

Lemma 2. *For any Steiner tree T of G'' on terminal set S, if we view it as a subgraph G, then T is also a Steiner tree of G on set S and $w(T) \leq w'(T)$.*

Proof. Since for any Steiner tree, the degree of each Steiner node is not less than 2 and all weight of nodes in terminal set is 0, the lemma holds.   □

Theorem 2. *Algorithm NWST2 is a 2.5ρ-approximation for node-weighted Steiner tree problem in unit disk graph.*

Proof. Let T be the output of NWST2. Denote T_{OPT_SM} and T_{OPT_NWS} as Lemma 1. By Lemmas 1 and 2, we have

$$w(T) \leq w'(T) \leq \rho w'(T_{OPT\_SM}) \leq 2.5\rho w(T_{OPT\_NWS}).$$ □

Since the node-weighted Steriner tree can be used in the MWCDS problem to interconnect all nodes of the CDS, we can obtain the following corollary.

Corollary 2. *Using node-weighted Steriner tree to interconnect all nodes of the CDS produces $(9,875+\varepsilon)$-approximation for minimum weighted connected dominating set in unit disk graph.*

Proof. For any node-weighted graph G and a terminal set S, denote T_{OPT} and T_{OPT_CDS} be the optimal Steiner tree of G on S and optimal CDS of G, respectively. Since the induced graph $G[S \cup T_{OPT\_CDS}]$ is connected, this graph contain a Steiner tree of G on S. Thus, $w(T_{OPT}) \leq w(T_{OPT\_CDS})$. By[9], we can get a dominating set C of G with $w(C) \leq (6+\varepsilon)w(T_{OPT\_CDS})$. Then, using algorithm NWST2 for C, we can obtain a Steiner tree T with $w(T) \leq 2.5\rho w(T_{OPT})$. Clearly, $V(T)$ is a connected dominating set of G and

$$\begin{aligned}
w(V(T)) &= w(C) + w(T) \\
&\leq (6+\varepsilon)w(T_{OPT\_CDS}) + 2.5\rho w(T_{OPT}) \\
&\leq (9.875+\varepsilon)w(T_{OPT\_CDS}).
\end{aligned}$$ □

4   Conclusion

In this paper, we propose two constant approximation algorithms for NWST problem in unit disk graph . The approximation ratios of these two algorithms are 4 and 2.5ρ, respectively. As a corollary, we improve the approximation of MWCDS in unit disk graphs from $10+\varepsilon$ to $9.875+\varepsilon$.

References

1. Aneja, Y.P.: An integer linear programming approach to the Steiner problem in graphs. Networks 10, 167–178 (1980)
2. Beasley, J.E.: An algorithm for the Steiner problem in graphs. Networks 14, 147–159 (1984)

3. Berman, P., Ramaiyer, V.: Improved approximations for the Steiner tree problem. Journal of Algorithms 17, 381–408 (1994)

4. Chen, D., Du, D.Z., Hu, X.D., Lin, G.H., Wang, L., Xue, G.: Approximation for Steiner tree with minimum number of Steiner points. Theoretical Computer Science 262, 83–99 (2001)

5. Feige, U.: A threshold of lnn for approximating set cover. J. ACM 45, 634–652 (1998)

6. Garey, M.R., Johnson, D.S.: Computers and Intractability:A Guide to the Theory of NP-Completeness. Freeman, San Fransico (1978)

7. Guha, S., Khuller, S.: Improved Methods for Approximating Node Weighted Steiner Trees and Connected Dominating Sets. Information and Computation 150, 57–74 (1999)

8. Hougardy, S., Prömel, H.J.: A 1.598 Approximation Algorithm for the Steiner Problem in Graphs. SODA, 448–453 (1998)

9. Huang, Y., Gao, X., Zhang, Z., Wu, W.: A Better Constant-Factor Approximation for Weighted Dominating Set in Unit Disk Graph (preprint)

10. Klein, P., Ravi, R.: A nearly best-possible approximation algorithm for node-weighted steiner trees. Journal of Algorithms 19, 104–115 (1995)

11. Kou, L.T., Markowsky, G., Berman, L.: A Fast Algorithm for Steiner Trees, pp. 141–145 (1981)

12. Min, M., Du, H., Jia, X., C.X.H, Huang, S.C.H., Wu, W.: Improving Construction for Connected Dominating Set with Steiner Tree in Wireless Sensor Networks. Journal of Global Optimizatio 35, 111–119 (2006)

13. Moss, A., Rabani, Y.: Approximation Algorithms for Constrained Node Weighted Steiner Tree Problems. In: STOC (2001)

14. Robins, G., Zelikovski, A.: Improved Steiner Tree Approximation in Graphs. In: Proc. of 11th. ACM-SIAM Symposium on Discrete. Algorithms, pp. 770–779 (2000)

15. Segev, A.: The node-weighted steiner tree problem. Networks 17, 1–17 (1987)

16. Shore, M.L., Foulds, L.R., Gibbons, R.B.: An algorithm for the Steiner problem in graphs. Networks 12, 323–333 (1982)

17. Zelikovsky, A.: An 11/6 approximation algorithm for the network Steiner problem. Algorithmica 9, 463–470 (1993)

An Improved Approximation Algorithm for the Capacitated Multicast Tree Routing Problem

Zhipeng Cai[1,*], Zhi-Zhong Chen[2,**], Guohui Lin[1,***], and Lusheng Wang[3,†]

[1] Department of Computing Science, University of Alberta
Edmonton, Alberta T6G 2E8, Canada
[2] Department of Mathematical Sciences, Tokyo Denki University
Hatoyama, Saitama 350-0394, Japan
[3] Department of Computer Science, City University of Hong Kong
Tat Chee Avenue, Kowloon, Hong Kong

Abstract. The Capacitated Multicast Tree Routing Problem is considered, in which only a limited number of destination nodes are allowed to receive data in one routing tree and multiple routing trees are needed to send data from the source node to all destination nodes. The goal is to minimize the total cost of these routing trees. An improved approximation algorithm is presented, which has a worst case performance ratio of $\frac{8}{5} + \frac{5}{4}\rho$. Here ρ denotes the best approximation ratio for the Steiner Minimum Tree problem, and it is about 1.55 at the writing of the paper. This improves upon the previous best having a performance ratio of $2 + \rho$.

Keywords: Capacitated Multicast Tree Routing, Approximation Algorithm, Steiner Minimum Tree, Tree Partitioning.

1   Introduction

Multicast consists of concurrently sending the same data from a single source node to multiple destination nodes. Such a service plays an important role in computer and communication networks supporting multimedia applications [8,10,14]. It is well known that multicast can be easily implemented on local area networks (LANs) since nodes connected to a LAN usually communicate over a broadcast network, yet quite challenging to implement in wide area networks (WANs) as nodes connected to a WAN communicate via a switched/routed network [5,15].

* Supported by NSERC and the 2007 Queen Elizabeth II Doctoral Award. Email: zhipeng@cs.ualberta.ca.

** Supported in part by the Grant-in-Aid for Scientific Research of the Ministry of Education, Science, Sports and Culture of Japan, under Grant No. 20500021. chen@r.dendai.ac.jp.

*** To whom correspondence should be addressed. Tel: (780) 492 3737; Fax: (780) 492 1071. Supported by CFI and NSERC. Email: ghlin@cs.ualberta.ca.

† Fully supported by a grant from the RGC of the Hong Kong Special Administrative Region, China [Project No. CityU 121207]. cswangl@cityu.edu.hk.

B. Yang, D.-Z. Du, and C.A. Wang (Eds.): COCOA 2008, LNCS 5165, pp. 286–295, 2008.
© Springer-Verlag Berlin Heidelberg 2008

In order to perform multicast communication in WANs, the source node and all the destination nodes must be interconnected. The problem of multicast routing in WANs is thus equivalent to finding a multicast tree in a network that spans the source and all the destination nodes, with its goal to minimize the *cost* of the multicast tree which is the total weight of edges in the tree.

In this paper, the *Capacitated Multicast Routing Problem* is studied in which only a limited number of destination nodes can be assigned to receive the packets sent from the source node during each transmission. The switches or routers in the underlying network are assumed to have the broadcasting ability. For simplicity, such a routing model is called the *multi-tree model* [7,6]. Multi-tree model has its origin in WDM optical networks with limited light-splitting capabilities. Under this model, we are interested in finding a set of trees such that each tree spans the source node and a limited number of destination nodes that are assigned to receive data and every destination node must be designated to receive data in one of the trees. Compared with the traditional multicast routing model without the capacity constraint (called the *Steiner Minimum Tree* problem which allows any number of receivers in the routing tree), this simpler model makes multicast easier and more efficient to be implemented, at the expense of increasing the cost of the routing tree. Specifically, when the number of destination nodes in a tree is limited to at most k, we call it the *Multicast k-Tree Routing* (kMTR) problem, which is formally defined in the following.

For a graph G, we denote its node set by $V(G)$. We model the underlying communication network as a triple (G, s, D), where G is a simple, undirected, and edge-weighted complete graph, $s \in V(G)$ is the *source* node, and $D \subseteq V(G) - \{s\}$ is the set of *destination* nodes. The weight of each edge e in G, denoted by $w(e)$, is nonnegative and represents the routing cost of e. The additive edge weight function $w(\cdot)$ generalizes to subgraphs of G in a natural way. That is, if T is a subgraph of G, then the *weight* (or *cost*) of T, denoted by $w(T)$, is the total weight of edges in T. A subgraph T of G is said to be a *D-marked Steiner tree* if T is a tree, at least one node in T is marked, and each marked node in T is contained in D. For each D-marked Steiner tree T, we use $D \cap T$ to denote the set of marked nodes in T. Note that some nodes in both D and T may not be marked. The *size* of T is the number of marked nodes in T. A set $\mathcal{T}$ of D-marked Steiner trees are *disjointly-D-marked* if $(D \cap T_1) \cap (D \cap T_2) = \emptyset$ for every two trees T_1 and T_2 in $\mathcal{T}$. Let k be a given positive integer. A *k-tree routing* in network (G, s, D) is a set $\{T_1, \ldots, T_\ell\}$ of disjointly-D-marked Steiner trees such that each T_i ($1 \leq i \leq \ell$) contains s and is of size at most k and $D = \bigcup_{i=1}^{\ell}(D \cap T_i)$. The *weight* (or *cost*) of a k-tree routing is the total weight of trees in the routing. Given a network (G, s, D), the multicast k-tree routing (kMTR) problem asks for a k-tree routing in (G, s, D) whose weight is minimized over all k-tree routings in (G, s, D).

For the kMTR problem, the cases where $k = 1, 2$ can be solved efficiently [6]. The general case of kMTR, where k is not fixed, is NP-hard [5]. In [1,11], kMTR is proven to be NP-hard when k is a fixed integer greater than 2. The best known approximation algorithm for kMTR ($k \geq 3$) has a worst case performance ratio of $(2 + \rho)$ [1,3,9], where ρ is the approximation ratio for the Steiner Minimum Tree problem, and it is about 1.55 [4,13] at the writing of this paper. Recently, Morsy and Nagamochi presented an approximation algorithm for kMTR ($k \geq 3$) having a worst case performance ratio of $(\frac{3}{2} + \frac{4}{3}\rho)$ [12], which constitutes an improvement only when $\rho < 1.5$.

In this paper, we take advantage of the weight averaging technique introduced in [1,3] to facilitate the design and analysis of a better approximation algorithm for kMTR. We extend another technique for partitioning routing trees in [1,3] to guarantee better quality subtrees. Combining them, we achieve an $(\frac{8}{5} + \frac{5}{4}\rho)$-approximation algorithm. This improves upon the previous best approximation ratio of $(2 + \rho)$ [1,3,9]. It is also an improvement over $(\frac{3}{2} + \frac{4}{3}\rho)$ [12] as long as $\rho \geq 1.2$. In the next section, we present the tree partitioning process in details, the complete algorithm, and its performance analysis.

Due to the space constraint, proofs of the technical lemmas are left out. The interested readers may refer to [2] for the details.

2   An $(\frac{8}{5} + \frac{5}{4}\rho)$-Approximation Algorithm for kMTR

Throughout this section, fix a positive integer k and an instance (G, s, D) of the kMTR problem. For ease of explanation, we assume that k is a multiple of 12. Recall that G is a simple, undirected, and edge-weighted complete graph, $s \in V(G)$ is the source node, and $D \subseteq V(G) - \{s\}$ is the set of destination nodes. The nodes in $V(G) - (D \cup \{s\})$ can be used as intermediate nodes in a routing to save the routing cost.

For each pair (u, v) of nodes in G, we use $w(u, v)$ to denote the weight of the edge between u and v. If $\{u, v\}$ is an edge in G such that $w(u, v)$ is larger than the weight of the shortest path between u and v in G, then $\{u, v\}$ is useless in any routing and hence can be ignored. So, we can assume that for each pair (u, v) of nodes in G, $w(u, v)$ equals the weight of the shortest path between u to v in G. Then, the edge weight function of G satisfies the triangle inequality.

Let $\mathcal{T}^*$ be an optimal k-tree routing in network (G, s, D). Let $R^* = \sum_{T \in \mathcal{T}^*} w(T)$. Note that R^* is the weight of the k-tree routing $\mathcal{T}^*$. Moreover, if d is a marked node in a tree $T \in \mathcal{T}^*$, then clearly $w(s, d) \leq w(T)$. Thus, we have

$$\sum_{d \in D} w(s, d) \leq \sum_{T \in \mathcal{T}^*} \sum_{d \in D \cap T} w(s, d) \leq k \times \sum_{T \in \mathcal{T}^*} w(T) \leq k \times R^*. \qquad (1)$$

In the following $(\frac{8}{5} + \frac{5}{4}\rho)$-approximation algorithm, we first apply the currently best approximation algorithm for the Steiner Minimum Tree problem (which has a worst-case performance ratio of ρ) to obtain a Steiner tree T^0 on $\{s\} \cup D$ in network (G, s, D). Recall that T^0 is a subgraph of G that is D-marked Steiner tree with $D \cap T^0 = D$. Since the weight of an optimal Steiner tree is a lower bound on R^*, the weight of tree T^0 is upper bounded by ρR^*, that is, $w(T^0) \leq \rho R^*$. We now root tree T^0 at source s. Note that tree T^0 does not necessarily correspond to a k-tree routing, because the subtree rooted at some child of s in T^0 may contain more than k marked nodes.

In the following, for a D-marked Steiner tree T in G and a node v in T, we use T_v to denote the subtree of T rooted at v. For a child u of an internal node v in T, the subtree T_u together with edge (v, u) is called the *branch rooted at v and containing u*. Recall that $D \cap T$ denotes the set of marked nodes in T and the size of T is $|D \cap T|$. If $|D \cap T| \le k$, then T can be used in a k-tree routing to route those nodes in $D \cap T$. If source s is not in T, then we can add s and the edge $\{s, u\}$ to T, where u is a node in T such that $w(s, u) = \min_{v \in V(T)} w(s, v)$. Let $c(T)$ denote $\min_{v \in V(T)} w(s, v)$. Note that $c(T) = 0$ if $s \in V(T)$. We call $c(T)$ the *connection cost* of T and define the *routing cost* of T to be $w(T) + c(T)$. Moreover, since $c(T) \le \min_{d \in D \cap T} w(s, d)$, we have

$$c(T) \le \frac{1}{|D \cap T|} \sum_{d \in D \cap T} w(s, d). \tag{2}$$

Although tree T^0 does not necessarily correspond to a k-tree routing, it serves as a good starting point because $w(T^0) \le \rho R^*$. Our idea is to transform T^0 into a k-tree routing without increasing its weight significantly. Basically, the transformation is done by case analysis. Each case corresponds to a lemma in Section 2.1. With these lemmas, we will define several types of operations in Section 2.2 that can be applied to T^0 (to turn it into a k-tree routing). An outline of the whole algorithm is given in Section 2.3.

2.1   Several Lemmas

This section proves several lemmas that will help us transform T^0 into a k-tree routing.

Lemma 1. [1,3] *Given a D-marked Steiner tree T such that*

- $k < |D \cap T| \le \frac{3}{2}k$,

we can compute two disjointly-D-marked Steiner trees X_1 and X_2 from T in polynomial time such that both X_1 and X_2 are of size at most k, $D \cap T = (D \cap X_1) \cup (D \cap X_2)$, and the total routing cost of X_1 and X_2 is at most $w(T) + 2 \times \frac{1}{k} \sum_{d \in D \cap T} w(s, d)$.

Lemma 2. *If T is a D-marked Steiner tree such that*

- $\frac{2}{3}k \le |D \cap T| \le k$,

then the routing cost of T is at most $w(T) + \frac{3}{2} \times \frac{1}{k} \sum_{d \in D \cap T} w(s, d)$.

Proof. This is trivial since the size of T is at least $\frac{2}{3}k$, following Equation 2.

Lemma 3. *Suppose that T is a D-marked Steiner tree satisfying the following conditions:*

- $\frac{3}{2}k \le |D \cap T| \le 2k$.
- The root r of T has exactly three children v_1, v_2, and v_3.
- $|D \cap T_{v_1}| < \frac{2}{3}k$, $|D \cap T_{v_2}| < \frac{2}{3}k$, and $|D \cap T_{v_1}| + |D \cap T_{v_2}| > k$.

Given T, we can compute disjointly-D-marked Steiner trees $X_1, \ldots, X_p$ with $2 \le p \le 3$ in polynomial time such that each X_i $(1 \le i \le p)$ is of size at most k, $D \cap T = \bigcup_{i=1}^{p}(D \cap X_i)$, and the total routing cost of X_1 through X_p is at most $\frac{5}{4}w(T) + \frac{3}{2} \times \frac{1}{k}\sum_{d \in D \cap T} w(s,d)$.

Proof. See [2].

Lemma 4. *Suppose that T is a D-marked Steiner tree satisfying the following conditions:*

- $\frac{5}{2}k \le |D \cap T| \le 3k$.
- The root r of T has exactly two children v_1 and v_2.
- $k < |D \cap T_{v_1}| \le \frac{3}{2}k$ and $k < |D \cap T_{v_2}| \le \frac{3}{2}k$.
- For $i \in \{1,2\}$, there is a node u_i in T_{v_i} (possibly $u_i = v_i$) such that u_i has exactly two children $x_{i,1}$ and $x_{i,2}$ in T_{v_i}, $|D \cap T_{x_{i,1}}| < \frac{2}{3}k$, $|D \cap T_{x_{i,2}}| < \frac{2}{3}k$, and $|D \cap T_{x_{i,1}}| + |D \cap T_{x_{i,2}}| > k$.

Given T, we can compute disjointly-D-marked Steiner trees $X_1, \ldots, X_p$ with $3 \le p \le 4$ in polynomial time such that each X_i $(1 \le i \le p)$ is of size at most k, $D \cap T = \bigcup_{i=1}^{p}(D \cap X_i)$, and the total routing cost of X_1 through X_p is at most $\frac{5}{4}w(T) + \frac{8}{5} \times \frac{1}{k}\sum_{d \in D \cap T} w(s,d)$.

Proof. See [2].

Lemma 5. *Suppose that T is a D-marked Steiner tree satisfying the following conditions:*

- $2k < |D \cap T| \le \frac{5}{2}k$.
- The root r of T has exactly two children v_1 and v_2.
- $k < |D \cap T_{v_1}| < \frac{4}{3}k$ and $k < |D \cap T_{v_2}| < \frac{4}{3}k$.
- For each $i \in \{1,2\}$, there is a node u_i in T_{v_i} (possibly $u_i = v_i$) such that u_i has exactly two children $x_{i,1}$ and $x_{i,2}$, $|D \cap T_{x_{i,1}}| < \frac{2}{3}k$, $|D \cap T_{x_{i,2}}| < \frac{2}{3}k$, and $|D \cap T_{x_{i,1}}| + |D \cap T_{x_{i,2}}| > k$.

Given T, we can compute disjointly-D-marked Steiner trees X_1, X_2, and X_3 in polynomial time such that each X_i $(1 \le i \le 3)$ is of size at most k, $D \cap T = \bigcup_{i=1}^{3}(D \cap X_i)$, and the total routing cost of X_1, X_2, and X_3 is at most $\frac{5}{4}w(T) + \frac{3}{2} \times \frac{1}{k}\sum_{d \in D \cap T} w(s,d)$.

Proof. See [2].

Lemma 6. *Suppose that T is a D-marked Steiner tree satisfying the following conditions:*

- $\frac{4}{3}k \leq |D \cap T| \leq \frac{3}{2}k$.
- *The root r of T has exactly three child nodes v_1, v_2, and v_3.*
- $|D \cap T_{v_1}| < \frac{2}{3}k$, $|D \cap T_{v_2}| < \frac{2}{3}k$, *and* $|D \cap T_{v_1}| + |D \cap T_{v_2}| > k$.

Given T, we can compute disjointly-D-marked Steiner trees X_1 and X_2 in polynomial time such that both X_1 and X_2 are of size at most k, $D \cap T = (D \cap X_1) \cup (D \cap X_2)$, and the total routing cost of X_1 and X_2 is at most $\frac{5}{4}w(T) + \frac{8}{5} \times \frac{1}{k} \sum_{d \in D \cap T} w(s, d)$.

Proof. See [2]. □

2.2   Operations to Be Applied to T^0

We are now ready to describe how to transform the initial Steiner tree T^0 (rooted at the source node s) into a k-tree routing. The transformation will be done by performing eight types of operations (namely, type-i operations with $i \in \{0, \ldots, 7\}$) on T^0 until T^0 becomes empty. When performing theses operations on T^0, we will maintain the following invariants:

(I1) A type-i operation is applied to T^0 only when no type-j operations with $j < i$ can be applied.

(I2) The source node s always remains in T^0.

We define a *big* node in T^0 to be an internal node v in T^0 with $|D \cap T_v^0| > k$, and define a *huge* node in T^0 to be an internal node v in T^0 with $|D \cap T_v^0| > 2k$. Note that a big node in T^0 may be a huge node or not. A big node in T^0 is *extreme* if all its children in T^0 are not big. Similarly, a huge node in T^0 is *extreme* if all its children in T^0 are not huge.

We next proceed to the definition of the operations on T^0. A type-0 operation can be applied on T^0 if $|D \cap T^0| \leq k$ or every branch rooted at s and containing a child of s is of size at most k. In the former case, a *type-0 operation* on T^0 includes T^0 in the output k-tree routing and then deletes the whole tree. In the latter case, a *type-0 operation* on T^0 includes each branch rooted at the root of T^0 (and containing a child of the root) in the output k-tree routing and then deletes the whole tree. In either case, the total routing cost equals $w(T^0)$ (i.e., no connection cost is needed when a type-0 operation is applied) because $s \in V(T^0)$ by Invariant (I2). Note that if no type-0 operations can be applied to T^0, then s is a big node in T^0 but is not an extreme big node in T^0 for $s \notin D$, implying that extreme big nodes always exist in T^0 and they are different from s.

If T^0 has an internal node v that has at least three children and has two children x_1 and x_2 with $|D \cap T_{x_1}^0| + |D \cap T_{x_2}^0| \leq k$, then a *type-1 operation* modifies T^0 as follows:

1. Make a copy v_c of v (without marking v_c even if v is marked in T^0).
2. Delete the edges (v, x_1) and (v, x_2).
3. Add three edges (v, v_c), (v_c, x_1), and (v_c, x_2) so that v_c becomes a new child of v while x_1 and x_2 become the children of v_c. (*Comment:* (v, v_c) is a dummy edge of weight 0.)

If T^0 has an internal node v with $\frac{2}{3}k \le |D \cap T_v^0| \le k$, then a *type-2 operation* modifies T^0 as follows:

1. Include T_v^0 in the output k-tree routing (cf. Lemma 2).
2. Remove v and all its descendants from T^0.

Note that if no type-2 operations can be applied to T^0, then every extreme big node in T^0 has at least two children because $k > k - 1 \ge \frac{2}{3}k$.

If T^0 has an extreme big node u with at least three children, then a *type-3 operation* modifies T^0 as follows:

1. Pick three arbitrary children v_1, v_2, and v_3 of u in T^0. (*Comment*: Since u is an extreme big node in T^0 and no type-2 operations can be applied to T^0, $|D \cap T_{v_j}^0| < \frac{2}{3}k$ for each $j \in \{1, 2, 3\}$. Moreover, since no type-1 operations can be applied to T^0, $|D \cap T_{v_i}^0| + |D \cap T_{v_j}^0| > k$ for every pair (i, j) with $1 \le i < j \le 3$.)
2. Let T be the union of the three branches rooted at u and containing v_1, v_2, or v_3.
3. Use T to obtain a set of D-marked Steiner trees as described in Lemma 3, and include them in the output k-tree routing.
4. Remove v_1, v_2, v_3, and their descendants from T^0.
5. If u is marked in T^0, then unmark it in T^0.

Note that if neither type-2 nor type-3 operations can be applied to T^0, then every extreme big node v in T^0 has exactly two children and hence satisfies that $k < |D \cap T_v^0| < \frac{4}{3}k$. Moreover, we can claim that every huge node in T^0 has a descendant that is a big but not huge node, if neither type-2 nor type-3 operations can be applied to T^0. For a contradiction, assume that the claim does not hold. Then, there is an extreme huge node v in T^0 whose children are not big nodes. So, v is an extreme big node in T^0. Thus, $k < |D \cap T_v^0| < \frac{4}{3}k$, contradicting the assumption that v is huge.

If T^0 has an extreme big vertex v such that the path from s to v contains a node u with $\frac{4}{3}k \le |D \cap T_u^0| \le \frac{3}{2}k$, then a *type-4 operation* modifies T^0 as follows:

1. Construct a D-marked Steiner tree T by initializing it as T_u^0 and re-rooting it at v.
2. Use T to obtain two D-marked Steiner trees as described in Lemma 6, and include them in the output k-tree routing.
3. Remove u and its descendants from T^0.

If T^0 has an extreme big node v such that the path from s to v contains a node u with $\frac{3}{2}k \le |D \cap T_u^0| \le 2k$, then a *type-5 operation* modifies T^0 in the same way as a type-4 operation does except that Lemma 3 is used instead of Lemma 6.

If T^0 has a huge node, then a *type-6 operation* modifies T^0 as follows:

1. Select an (arbitrary) extreme huge node u in T^0.
2. Find an extreme big node v_1 that is a descendant of u in T^0 (*Comment*: As claimed before, v_1 is big but not huge, implying that $v_1 \ne u$.)

3. Let u_1 be the child of u in T^0 that is v_1 itself or an ancestor of v_1 in T^0. (*Comment:* $|D \cap T^0_{u_1}| < \frac{4}{3}k$ because u_1 is not huge and neither type-4 nor type-5 operations can be applied to T^0. Consequently, u has at least two children in T^0.)

4. If every child u_2 of u in T^0 with $u_2 \neq u_1$ satisfies that $|D \cap T^0_{u_2}| \leq \frac{2}{3}k$, then modify T^0 as follows:

   (a) Construct a D-marked Steiner tree T by initializing it as T^0_u and then repeatedly deleting a child $u_2 \neq u_1$ and the descendants of u_2 until $|D \cap T| \leq 2k$. (*Comment:* $|D \cap T| \geq \frac{4}{3}k$ because $|D \cap T^0_{u_2}| < \frac{2}{3}k$ for each child u_2 of u in T^0 with $u_2 \neq u_1$.)

   (b) Re-root T at v_1.

   (c) If $|D \cap T| > \frac{3}{2}k$, then use T to obtain two or three D-marked Steiner trees as described in Lemma 3 and include them in the output k-tree routing. Otherwise, use T to obtain two D-marked Steiner trees as described in Lemma 6 and include them in the output k-tree routing.

   (d) Remove the nodes in $V(T) - \{u\}$ from T^0.

   (e) If u is marked in T^0, then unmark it in T^0.

5. If some child u_2 of u in T^0 with $u_2 \neq u_1$ satisfies that $|D \cap T^0_{u_2}| > \frac{2}{3}k$, then modify T^0 as follows:

   (a) Find an extreme big node v_2 in $T^0_{u_2}$. (*Comment:* Since u is an extreme huge node in T^0, $|D \cap T^0_{u_2}| \leq 2k$. Consequently, u_2 must be a big node in T^0 because $|D \cap T^0_{u_2}| > \frac{2}{3}k$ no type-2 operations can be applied to T^0. Moreover, $|D \cap T^0_{u_2}| < \frac{4}{3}k$ because neither type-4 nor type-5 operations can be applied to T^0. Possibly, $v_2 = u_2$.)

   (b) Construct a D-marked Steiner tree T by setting it to be the union of the two branches rooted at u and containing u_1 or u_2. (*Comment:* Clearly, $2k < |D \cap T| < \frac{8}{3}k$.)

   (c) If $|D \cap T| \leq \frac{5}{2}k$, then use T to obtain three D-marked Steiner trees as described in Lemma 5 and include them in the output k-tree routing. Otherwise, use T to obtain three or four D-marked Steiner trees as described in Lemma 4 and include them in the output k-tree routing.

   (d) Remove the nodes in $V(T) - \{u\}$ from T^0.

   (e) If u is marked in T^0, then unmark it in T^0.

Suppose that no type-i operations with $0 \leq i \leq 6$ can be applied to T^0. Then, $k < |D \cap T^0| < \frac{4}{3}k$. Consequently, there is only one extreme big node u in T^0. As mentioned before, s is a big but not extreme big node in T^0. So, $u \neq s$. Let v_1 and v_2 be the children of u in T^0, and let v_3 be the parent of u in T^0 (possibly, $v_3 = s$). Now, a *type-7 operation* modifies T^0 as follows:

1. Re-root T^0 at u (so that v_3 becomes a child of u, too). (*Comment:* $|D \cap T^0_{v_3}| < \frac{1}{3}k$ because $k < |D \cap T^0| < \frac{4}{3}k$ and $|D \cap T^0_{v_1}| + |D \cap T^0_{v_1}| > k$.)

2. Among the nodes in $(D \cap T^0_{v_1}) \cup (D \cap T^0_{v_2})$, find the closest node d' to s. (*Comment:* $w(s, d') < \frac{1}{k} \sum_{d \in (D \cap T^0_{v_1}) \cup (D \cap T^0_{v_2})} w(s, d)$.)

3. Let $i \in \{1, 2\}$ be the integer with $d' \in T^0_{v_i}$.

4. Include $T^0_{v_i}$ as a D-marked Steiner tree in the output k-tree routing. (*Comment:* $c(T^0_{v_i}) \leq w(s, d') < \frac{1}{k} \sum_{d \in D \cap T} w(s, d)$.)

5. Obtain a tree T by deleting v_i and its descendants from T^0. (*Comment:* $|D \cap T| < k$ because $|D \cap T^0_{v_3}| < \frac{1}{3}k$ and $|D \cap T^0_{v_j}| < \frac{2}{3}k$, where j is the integer in $\{1, 2\} - \{i\}$.)

6. Include T as a D-marked Steiner tree in the output k-tree routing. (*Comment:* Since s remains in T^0 after Step 5, the connection cost of T^0 is 0. Thus, the total routing cost of $T^0_{v_i}$ and T is at most $w(T^0) + \frac{1}{k}\sum_{d \in D \cap T} w(s, d)$.)

7. Remove the whole tree T^0.

2.3   Summary of the Algorithm

A high-level description of the complete approximation algorithm is depicted in Figure 1.

INPUT: A network (G, s, D).
OUTPUT: a k-tree routing in (G, s, D).

1. Compute a Steiner tree T^0 on $\{s\} \cup D$, using the currently best approximation algorithm;
2. Root T^0 at s;
3. While (T^0 is not empty) do:
3.1. Let i be the smallest j such that a type-j operation, $0 \le j \le 7$, can be applied to T^0;
3.2. Perform a type-i operation on T^0;
4. Output the k-tree routing.

Fig. 1. A high-level description of the approximation algorithm for kMTR

Theorem 1. *kMTR ($k \ge 3$) admits an $(\frac{8}{5} + \frac{5}{4}\rho)$-approximation algorithm, where ρ is the currently best performance ratio for approximating the* Steiner Minimum Tree *problem.*

Proof. Notice that whenever we cut a subtree T out of the base Steiner tree T^0 by performing a type-i operation with $i \in \{0, \ldots, 7\}$, we maintain the following invariants:

- We construct a set $\mathcal{T}$ of disjointly-D-marked Steiner trees from T and include them in the output k-tree routing, where the total routing cost of the trees in $\mathcal{T}$ is at most $\frac{5}{4}w(T) + \frac{8}{5} \times \frac{1}{k}\sum_{d \in D \cap T} w(s, d)$.
- After cutting T out of T^0, T^0 may share a node with T but does not share an edge with T, and no node of $D \cap T$ is marked in T^0.

By the above invariants, the total routing cost of the trees in the output k-tree routing is $R \le \frac{5}{4}w(T^0) + \frac{8}{5} \times \frac{1}{k}\sum_{d \in D} w(s, d) \le \frac{5}{4}w(T^0) + \frac{8}{5}R^*$, where T^0 is the initial Steiner tree obtained in Step 1 of the algorithm and the last inequality follows from Equation 1. Since $w(T^0) \le \rho R^*$, we have $R \le (\frac{5}{4}\rho + \frac{8}{5})R^*$.

3 Conclusions

We have presented an $(\frac{8}{5} + \frac{5}{4}\rho)$-approximation algorithm for kMTR. This improves the previous best $(2+\rho)$-approximation algorithm. It would be interesting to know whether the algorithm can be further improved along this approach via finer analysis.

References

1. Cai, Z.: Improved algorithms for multicast routing and binary fingerprint vector clustering. Master's thesis, Department of Computing Science, University of Alberta (June 16, 2004)
2. Cai, Z., Chen, Z.-Z., Lin, G.-H., Wang, L.: An improved approximation algorithm for the capacitated multicast tree routing problem. Technical Report TR08-06, Department of Computing Science, University of Alberta (May 2008)
3. Cai, Z., Lin, G.-H., Xue, G.L.: Improved approximation algorithms for the capacitated multicast routing problem. In: Wang, L. (ed.) COCOON 2005. LNCS, vol. 3595, pp. 136–145. Springer, Heidelberg (2005)
4. Gröpl, C., Hougardy, S., Nierhoff, T., Prömel, H.J.: Approximation algorithms for the Steiner tree problem in graphs. In: Du, D.-Z., Cheng, X. (eds.) Steiner Trees in Industries, pp. 235–279. Kluwer Academic Publishers, Dordrecht (2001)
5. Gu, J., Hu, X.D., Jia, X., Zhang, M.-H.: Routing algorithm for multicast under multi-tree model in optical networks. Theoretical Computer Science 314, 293–301 (2004)
6. Gu, J., Hu, X.D., Zhang, M.-H.: Algorithms for multicast connection under multi-path routing model. Information Processing Letters 84, 31–39 (2002)
7. Hadas, R.L.: Efficient collective communication in WDM networks. In: Proceedings of IEEE ICCCN 2000, pp. 612–616 (2000)
8. Huitema, C.: Routing in the Internet. Prentice Hall PTR, Englewood Cliffs (2000)
9. Jothi, R., Raghavachari, B.: Approximation algorithms for the capacitated minimum spanning tree problem and its variants in network design. ACM Transactions on Algorithms 1, 265–282 (2005)
10. Kuo, F., Effelsberg, W., Garcia-Luna-Aceves, J.J.: Multimedia Communications: Protocols and Applications. Prentice Hall, Inc., Englewood Cliffs (1998)
11. Lin, G.-H.: An improved approximation algorithm for multicast k-tree routing. Journal of Combinatorial Optimization 9, 349–356 (2005)
12. Morsy, E., Nagamochi, H.: An improved approximation algorithm for capacitated multicast routings in networks. Theoretical Computer Science 390, 81–91 (2008)
13. Robins, G., Zelikovsky, A.Z.: Improved Steiner tree approximation in graphs. In: Proceedings of the 11th Annual ACM-SIAM Symposium on Discrete Algorithms (SODA 2000), pp. 770–779 (2000)
14. Wang, Z., Crowcroft, J.: Quality-of-service routing for supporting multimedia applications. IEEE Journal on Selected Areas in Communications 14, 1228–1234 (1996)
15. Zhang, X., Wei, J., Qiao, C.: Constrained multicast routing in WDM networks with sparse light splitting. In: Proceedings of IEEE INFOCOM 2000, March 26–30, pp. 1781–1790 (2000)

Covering Arrays Avoiding Forbidden Edges

Peter Danziger[1], Eric Mendelsohn[2], Lucia Moura[3], and Brett Stevens[4]

[1] Ryerson University
[2] University of Toronto
[3] University of Ottawa
[4] Carleton University

Abstract. Covering arrays (CAs) can be used to detect the existence of faulty pairwise interactions between parameters or components in a software system. The generalization considered here applies to the situation in which some input combinations are invalid. In this paper, we study covering arrays avoiding forbidden edges (CAFEs), where certain pairwise interactions are forbidden while all others must be covered. We study the complexity of the problem and give an algorithm for the case of binary alphabets.

1 Introduction

This paper addresses the application problem of testing a complex system whose behavior depends on the values of k parameters or *factors*. Suppose each of the k factors may take any of g values. To exhaustively test the system, we would need g^k tests, which is too costly in practice, even for moderate number of factors k. Instead, covering arrays have been extensively used in this context since they offer effectiveness at a substantially lower cost, having applications ranging from software and hardware testing [9,15,16,18,24] to genomics [25] and material sciences [4]. In realistic situations, constraints involving the factors will restrict certain configurations of the parameters. In this paper, we investigate test plans given by covering arrays that avoid some fixed set of configurations of the parameters.

Covering arrays provide an alternative to exhaustive testing, by offering a much smaller test suite that guarantees coverage of all possible interactions from any t factors. In this paper, we focus on $t = 2$, where pairwise interactions of factors are tested. A *covering array* (CA) is a matrix with symbols from a g-ary alphabet, with n rows and k columns. Each of its columns represents a parameter and each of its rows gives a test to be performed. The number of rows, n, is called the *size* of the array. The array must offer a pairwise ($t = 2$) coverage of factor values, that is, for any pair of the k factors, the corresponding columns exhaustively cover all possible g^2 combinations of values. In practice, small interaction coverage ($t = 2, 3, 4$) has been shown very powerful for software failure detection, and in particular $t = 2$ offers an excellent compromise between failure detection and size. Empirical results show that testing all pairwise interactions in a software system finds most of its faults [9,15,16]; some authors also link pairwise coverage to good "code coverage" [3,10]. The minimal size of a covering

B. Yang, D.-Z. Du, and C.A. Wang (Eds.): COCOA 2008, LNCS 5165, pp. 296–308, 2008.
© Springer-Verlag Berlin Heidelberg 2008

Product line options (factors):	1) display	2) email viewer	3) camera	4) video camera	5) video ringtones
possible values:	A=16 Million colours B=8 Million colours C=Black and White	D=graphical E=text F=none	G=2 Megapixels H=1 Megapixel I=none	J=Yes K=No	L=Yes M=No

Constraints on valid configurations: Forbidden edges:
(C1) Graphical email viewer **requires** a colour display $\{C, D\}$
(C2) 2 Megapixel camera **requires** a color display $\{C, G\}$
(C3) Graphical email viewer **not supported** with the 2 Megapixel camera $\{D, G\}$
(C4) 8 Million colour display **does not support** a 2 Megapixel camera $\{B, G\}$
(C5) Video camera **requires** a camera and a colour display $\{I, J\}, \{C, J\}$
(C6) Video ringtones **cannot occur** with No video camera $\{L, K\}$

Graph G for given constraints: Graph $\hat{G}$ for given and implied constraints:

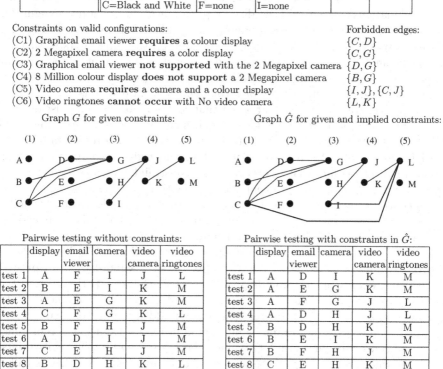

Pairwise testing without constraints:

	display	email viewer	camera	video camera	video ringtones
test 1	A	F	I	J	L
test 2	B	E	I	K	M
test 3	A	E	G	K	M
test 4	C	F	G	K	L
test 5	B	F	H	J	M
test 6	A	D	I	J	M
test 7	C	E	H	J	M
test 8	B	D	H	K	L
test 9	C	D	G	J	L

(a)

Pairwise testing with constraints in $\hat{G}$:

	display	email viewer	camera	video camera	video ringtones
test 1	A	D	I	K	M
test 2	A	E	G	K	M
test 3	A	F	G	J	L
test 4	A	D	H	J	L
test 5	B	D	H	K	M
test 6	B	E	I	K	M
test 7	B	F	H	J	M
test 8	C	E	H	K	M
test 9	C	F	I	K	M

(b)

Fig. 1. Mobile phone product line

array with k factors and g possible values per factor is denoted by $CAN(k, g)$. For fixed g, this number is known to asymptotically grow as $(g \log k)/2$ [12], offering a practical solution over exhaustively going through g^k tests.

There is a vast literature on covering arrays and their constructions (see Colbourn's surveys [6,7]), including generalizations. There are situations in which considering t-way interactions for $t > 2$ gives additional testing benefits [15] over $t = 2$, so we can look at covering arrays of strength t [6]. When each factor i has g_i possible values, *mixed covering arrays* (MCAs) [22] have been considered. Another generalization of CAs and MCAs targets situations where some factor combinations need not be tested [21,20], the so called *covering arrays on graphs*.

Figure 1 shows an example of a system with $k = 5$ factors. Let us first consider the case without constraints on valid configurations. Testing exhaustively such a system would require $72 = 3 \times 3 \times 2 \times 2 \times 2$ tests, but an MCA of size 9 allows us to test all pairwise interactions (see MCA in Figure 1a). If the system fails due to a single factor value or due to a combination of two factor values on the system, at least one of the tests given by this MCA will fail.

In reliability testing, we often encounter extra constraints on feasible parameter combinations. Cohen et.al. [5] discuss the application in highly configurable systems, where these types of constraints are common. The example in Figure 1 with constraints (C1) to (C6) is extracted from [5]; the constraints in this hypothetical mobile phone example can be modeled as pairs that cannot appear as part of a valid test. Other applications frequently present similar constraints such as adverse drug interactions or forbidden chemical reactions [4].

The problem of constructing a test set that covers all allowed pairwise interactions while avoiding a set of forbidden pairs has not been satisfactorily addressed in the literature nor by the combinatorial interaction testing tools available [5]. With the exception of the paper by Cohen et al., which gives a comparison of various heuristic algorithms that incorporate such constraints, we have not found any other works addressing this problem nor seen any theoretical results. The present paper concentrates on establishing initial theoretical results, algorithms for building covering arrays with forbidden configurations as well as bounds on their sizes.

In general, the forbidden interactions may be of any size; for instance, in the example of Figure 1, we could add the following constraint: "(C7) The combination of 16 Million colours, Text and 2 Megapixel camera will **not be supported**" [5], which is equivalent to forbidding the interaction $\{A, E, G\}$. However, in the present paper we concentrate on the case of forbidden **pairwise** interactions like in constraints (C1) to (C6). In this case, the forbidden interactions can be represented by the edges of a graph as illustrated in Figure 1. Graph G represents the given constraints and graph $\hat{G}$ includes both given and implied constraints. An array whose rows cover all the non-edges of $\hat{G}$ and avoids its edges is called a *Covering Array avoiding Forbidden Edges* (CAFE), and is exemplified for this case in Figure 1b.

Next, we summarize our contributions and the organization of the paper. In Section 2, we give the main definitions and basic bounds. In Section 3, we study the complexity of the problem. In particular, we show that determining whether G admits a CAFE can be decided in polynomial time for $g = 2$, while it is NP-complete for $g \geq 5$. On the other hand, it remains open whether determining its minimum size is NP-hard even for g=2. In Section 4, we give a general recursive construction for CAFEs based on CAFEs for the connected components of the graph. In Section 5, we give an algorithm for $g = 2$ which constructs a CAFE whose size is bounded by $k + 1$ plus the size of an edge clique covering number of a related, smaller graph. In particular, when G is bipartite, our algorithm produces a covering array of size at most $k + 2$.

2   Definitions and Preliminaries

Let us define the testing problem more formally. Consider a system with k *factors* (parameters or components) $1, \ldots, k$. Each factor i can take one of g_i possible *values*, which we consider w.l.o.g. to be in the set $\{0, \ldots, g_i - 1\}$, denoted by $[0, g_i - 1]$. A *test* is an assignment of values to factors, i.e., a k-tuple in $[0, g_1 - 1] \times \cdots \times [0, g_k - 1]$. The execution of a test can have two outcomes: *pass* or *fail*; we call it a *passing* or a *failing* test, respectively. An *interaction* is a set of values assigned to distinct factors: $I = \{(f_1, a_1), \ldots, (f_t, a_t)\}$, $f_i \neq f_j$ for $i \neq j$, and $a_i \in [0, g_{f_i} - 1]$, $1 \leq i \leq t$. An interaction I is a *t-way interaction* if $|I| = t$. We say that a test (or a k-tuple) $T = (T_1, \ldots, T_k)$ *covers* interaction $I = \{(f_1, a_1), \ldots, (f_t, a_t)\}$, if $T_{f_i} = a_i$ for $1 \leq i \leq t$. Thus, a test covers exactly $\binom{k}{t}$ t-way interactions, $1 \leq t \leq k$. We assume that failures are caused by faulty interactions, that is, the execution of a test fails if and only if it covers one or more faulty interactions. Covering arrays are combinatorial designs that correspond to test suites that cover all t-way interactions of factor values, and consequently all s-way interactions with $1 \leq s \leq t$.

Definition 1. *A* mixed covering array, A, *denoted by* $MCA(n; t, (g_1, \ldots, g_k))$, *is an $n \times k$ array, such that each column i (corresponding to a factor) has values from the alphabet $[0, g_i - 1]$, and every possible t-way interaction is covered by some row, or in other words, for every t-set of factors $\{f_1, \ldots, f_t\}$ and every t-tuple of values $(a_1, \ldots, a_t) \in [0, g_{f_1} - 1] \times \cdots \times [0, g_{f_t} - 1]$, there exists at least one row r (corresponding to a test) such that $A[r, f_j] = a_j$ for all $j \in [1, t]$. Given t and $g_1, \ldots, g_k$, the* mixed covering array number, *denoted by* $MCAN(t, (g_1, \ldots, g_k))$, *is the smallest n for which an $MCA(n; t, (g_1, \ldots, g_k))$ exists. When $g_i = g$ for all $1 \leq i \leq k$, we call the objects simply* covering arrays *and simplify the notation to* $CA(n; t, k, g)$, *and* $CAN(t, k, g)$.

The test suite in Figure 1a gives a $MCA(9; 2, (3, 3, 2, 2, 2))$ which is optimal since $g_1 g_2 = 9$.

Consider a graph whose edges represent the forbidden pairwise interactions in a system like in Figure 1. Let $G = G_{(g_1, \ldots, g_k)}$ denote a graph with k parts of sizes $g_1, \ldots, g_k$ that is k-partite except for the possible existence of loops (forbidden single settings). The vertices of G are v_{i,a_i}, where $i \in [1, k]$ and $a_i \in [0, g_i - 1]$. If $g_1 = \cdots = g_k = g$, then we simplify the notation to $G = G_{k,g}$. We define a graph $G^|$ on the same vertex set as G and including the edges from $E(G)$ but also containing all the edges $\{v_{i,a}, v_{i,b}\}$ for $a \neq b \in [0, g_i - 1]$. A graph G is said to be *factor connected* if $G^|$ is connected; *factor-connected components* of G correspond to components of $G^|$.

A k-tuple $T = (T_1, \ldots, T_k) \in [0, g_1 - 1] \times \cdots \times [0, g_k - 1]$ is said to *avoid* $G = G_{(g_1, \ldots, g_k)}$ if for all $i, j \in [1, k]$, we have $\{v_{i,T_i}, v_{j,T_j}\} \notin E(G)$. We say that an interaction $\{(i, a), (j, b)\}, i \neq j$, such that $\{v_{i,a}, v_{j,b}\} \notin E(G)$ is *consistent* with G if there exists a k-tuple T with $T_i = a$ and $T_j = b$ that avoids G. A graph is *consistent* if all pairwise interactions $\{(i, a), (j, b)\}, i \neq j$, with $\{v_{i,a}, v_{j,b}\} \notin E(G)$ are consistent. This definition is equivalent to saying that all forbidden interactions implied by the edges of the graph are also edges of the graph.

Definition 2. *(CAFE) A* covering array with forbidden edges *for a graph $G = G_{g_1, \ldots, g_k}$ is an $n \times k$ array A with each column i having symbols from the alphabet $[0, g_i - 1]$, and denoted by $CAFE(n, G)$, such that*

1. *each row of A forms a k-tuple avoiding G;*
2. *for all $v_{i,a}, v_{j,b} \in V(G)$ with $i \neq j$, if $\{v_{i,a}, v_{j,b}\} \notin E(G)$, then there exists a row r such that $A_{r,i} = a$ and $A_{r,j} = b$.*

We note that this definition does not have to specifically mention loops because of the requirement that each row avoids G. We denote by $CAFEN(G)$ the minimum n for which there exists a $CAFE(n, G)$, if such an object exists, or $+\infty$ otherwise.

Similarly, $CAFE^1(n, G)$ and $CAFEN^1(G)$ are defined by substituting the second condition above by a pointwise coverage requirement: for all $v_{i,a} \in V(G)$ such that $\{v_{i,a}, v_{i,a}\} \notin E(G)$, there exists a row r such that $A_{r,i} = a$.

It is easy to see that there exists a CAFE for G if and only if G is consistent. The minimal supergraph of G that is consistent is called the *avoidance closure* of G and denoted by $\hat{G}$. So, a graph is consistent if and only if $G = \hat{G}$. Figure 1 gives an example of G and $\hat{G}$. Let $E^{i,j}(G)$ denote the set of edges with an end in factor i and the other in factor j. When a CAFE exists it is also easy to calculate a lower and upper bound on its size,

$$\max_{1 \leq i < j \leq k} (g_i g_j - |E^{i,j}(G)|) \leq CAFEN(G) \leq \sum_{1 \leq i < j \leq k} (g_i g_j - |E^{i,j}(G)|), \quad (1)$$

since each nonedge between two factors must be covered and any row whose pairs are all covered elsewhere can be discarded.

As we add edges to $G = G_{k,g}$, $CAFEN(G)$ may decrease or increase. For fixed g, when G is empty we know $CAFEN(G) = CAN(k, g) \rightarrow \frac{g}{2} \log k$, as $k \rightarrow \infty$. The upper bound in Eq. 1 gives $CAFEN(G) \leq g^2 \binom{k}{2} - |E(G)|$. A graph consisting of two cliques induced by $\{v_{0,i} \in V(G) : 1 \leq i \leq \lfloor k/2 \rfloor\}$ and $\{v_{0,i} \in V(G) : \lfloor k/2 \rfloor + 1 \leq i \leq k\}$, respectively, requires $k^2/4$ rows just to cover the non-edges involving one vertex in each clique. This example shows that quadratic growth in k is unavoidable.

CAFEs are closely related to edge clique coverings of graphs.

Definition 3. *Let G be a graph. An* edge clique cover *of G is a set of cliques $\{K_i\}_{i \in I}$ of G such that for every $e \in E(G)$ there exists an $i \in I$ such that $e \in K_i$. The* edge clique covering number *of G, $\theta'(G)$ is defined to be the size of the minimum edge clique cover.*

In Section 5 we show that $CAFEN(G) = \theta'(\overline{G})$, when $G = G_{k,2}$ (Theorem 6). Therefore, we direct the reader to some of the many results known about this graph parameter [1,2,11,13,17].

3   Computational Complexity Results

In this section, we establish main computational complexity results for the problems under study (Theorem 4). Consider the following languages, associated to decision problems of interest to us:

- AVOID= $\{< G = G_{(g_1,...,g_k)} >:$ *there exists a* $k-$*tuple avoiding* $G\}$;
- ONE-COVER&AVOID= $\{< G = G_{(g_1,...,g_k)} >:$ *for some* n *there exists a* $CAFE^1(n, G)\}$;
- COVER&AVOID= $\{< G = G_{(g_1,...,g_k)} >:$ *for some* n *there exists a* $CAFE(n, G)\}$;
- CAFEN= $\{< G = G_{(g_1,...,g_k)}, N >:$ there exists a $CAFE(N, G)\}$.

For each language L defined above, we define g-$L = \{< G = G_{(g_1,...,g_k)} >\in L : g_1 = \cdots = g_k = g\}$.

Theorem 4. *Consider the decision problems defined above. Then,*

1. *2-AVOID is in P.*
2. *g-AVOID is NP-complete for $g \geq 3$, and so AVOID is NP-complete.*
3. *2-ONE-COVER&AVOID and 2-COVER&AVOID are in P.*
4. *$(g-1)$-ONE-COVER&AVOID and g-COVER&AVOID are NP-complete for $g \geq 5$.*
5. *g-CAFEN is NP-complete, for $g \geq 5$, and so CAFEN is NP-complete. (Note that the NP-hardness of 2-CAFEN is still open!)*

Proof. 1. Solve 2-AVOID by treating each forbidden edge as a disjunction of literals.

2. To prove that 3-AVOID is NP-complete, we reduce from 3-SAT. Let $\phi = (l_{1,0} \vee l_{1,1} \vee l_{1,2}) \wedge \cdots \wedge (l_{k,0} \vee l_{k,1} \vee l_{k,2})$ be a formula with k clauses with 3 literals each. Build G, a k-partite graph with 3 vertices per part, corresponding to the literals in each of the clauses, and such that $\{v_{i,a}, v_{j,b}\} \in E(G)$ if and only if $i \neq j$ and $l_{i,a} = \neg l_{j,b}$. It is easy to see that ϕ is satisfiable if and only if there exists a k-tuple avoiding G.

3. 2-COVER&AVOID can be solved by one call for each non-edge, $\{v_{i,a}, v_{j,b}\}$, to a 2-AVOID oracle on the graph $G^{\{v_{i,a}, v_{j,b}\}}$ obtained from G by replacing the vertices in parts i and j with g_i copies of $v_{i,a}$ and g_j copies of $v_{j,b}$, respectively, as well as copy of their associated edges. A similar argument works for 2-ONE-COVER&AVOID.

4. We first transform an instance of g-AVOID into an instance of $(g+1)$-ONE-COVER&AVOID. Given G, an instance of g-AVOID, build an instance G' for $(g+1)$-ONE-COVER&AVOID in the following way. Append to G one new part indexed by $k+1$ with a vertex $v_{k+1,0}$ plus $g-1$ isolated vertices. Add a new vertex, v_i^A, per part i, $1 \leq i \leq k+1$. Add edges between $v_{k+1,0}$ and each v_i^A, $1 \leq i \leq k$. The second transformation is identical.

5. We can reduce from g-COVER&AVOID, by just using N large enough (e.g. the upper bound in Eq. 1) when deciding whether $< G, N >\in g$-CAFEN, and the result follows from 4. □

4   A Recursive Construction for CAFEs

We give a construction of CAFEs for a graph based on CAFEs for the factor connected components of the graph. The proof is omitted due to space considerations.

Theorem 5. *Let $G = G_{(g_1,\ldots,g_k)}$ be a graph with non-trivial factor-connected components $G_1,\ldots,G_s$ with associated factors $F_1,\ldots,F_s$, respectively, and with $\ell \geq 0$ trivial factor-connected components, corresponding to single factors $1,\ldots,\ell$. (So $\{1,\ldots,k\} = \{1,\ldots,\ell\}\cup F_1\cup\ldots\cup F_s$). Let $k_i = |F_i|$, $1 \leq i \leq s$. If the following designs exist:*

1. *P_i: a $CAFE^1(p_i, G_i)$, $1 \leq i \leq s$, that is, a covering array of strength 1 avoiding graph G_i with k_i columns and p_i rows;*
2. *A_i: an $a_i \times k_i$ array, such that the array C_i obtained by appending the rows of P_i and A_i is a $CAFE(p_i + a_i, G_i)$;*
3. *P: an $MCA(p, 1, (g_1,\ldots,g_\ell))$ on factors $1,\ldots,\ell$; (possibly empty ($p = 0$) if $\ell = 0$)*
4. *Q: a $q \times \ell$ array, such that the array C obtained by appending the rows of P and Q is an $MCA(p + q; 2, (g_1,\ldots,g_\ell))$;*
5. *M: an $MCA(m; 2, (p_1,\ldots,p_s,p))$, that is, a mixed covering with m rows, $s + 1$ (or s columns if $p = 0$) columns with alphabets sizes $p_1,\ldots,p_s,p$, respectively.*

Then, there exists a $CAFE(n, G)$ with $n = m + \max\{a_1,\ldots,a_s,q\}$.   □

5   An Algorithm for Constructing Binary CAFEs

The following proposition characterizes consistent graphs for $g = 2$, that is, graphs with $g = 2$ for which $G = \hat{G}$, or equivalently graphs that are in 2-COVER&AVOID. Its proof is essentially derived from transitively closing a related directed graph, $\overrightarrow{G}$: for every undirected edge $\{v_{i,a}, v_{j,b}\} \in E(G)$, put $(v_{i,a}, v_{j,1-b}), (v_{j,b}, v_{i,1-a}) \in A(\overrightarrow{G})$. The statement is equivalent to saying that $G = \hat{G}$ if and only if G does not contain the subgraphs in Figure 2 as induced subgraphs. The necessity is clear, for the dashed lines (non-edges) in Figure 2 would be non-consistent.

Proposition 1. *Let $G = G_{k,2}$ be a graph with vertex set $V(G) = \{v_{i,a}|1 \leq i \leq k, a \in \{0,1\}\}$. Then, $G = \hat{G}$ if and only if $\{v_{i,a}, v_{j,b}\} \in E(G)$ whenever*

- *there exist vertices in the same part, $v_{l,c}$ and $v_{l,1-c}$ such that edges $\{v_{i,a}, v_{l,c}\}$ and $\{v_{j,b}, v_{l,1-c}\}$ exist; or*
- *there exist vertices in the same part, $v_{l,0}$ and $v_{l,1}$ such that edges $\{v_{i,a}, v_{l,0}\}$ and $\{v_{i,a}, v_{l,1}\}$ exist; or*
- *there is a loop $\{v_{i,a}, v_{i,a}\} \in E(G)$.*

and $\{v_{i,a}, v_{i,a}\} \in E(G)$ whenever there exist vertices in the same part, $v_{l,0}$ and $v_{l,1}$ such that edges $\{v_{i,a}, v_{l,0}\}$ and $\{v_{i,a}, v_{l,1}\}$ exist.

The avoidance closure of G, $\hat{G}$, can be efficiently computed in $O(k \times |E(G)|)$, the same complexity as calculating the transitive closure of $\overrightarrow{G}$ [23].

Our next result is that for $g = 2$ a CAFE for a graph corresponds to an edge cover by cliques of its complement.

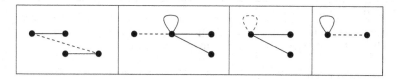

Fig. 2. Forbidden induced subgraphs for binary CAFEs

Theorem 6. *Let $G = G_{k,2}$ such that $G = \hat{G}$. Then, $CAFEN(G) = \theta'(\overline{G^|})$.*

Proof. Since every row of a $CAFE(n, G)$ must be an independent set of size k in $G^|$, the maximum size of an independent set is k and every non-edge of $G^|$ must be covered by some row of the $CAFE(n, G)$, it is easy to see that $\theta'(\overline{G^|}) \leq CAFEN(G)$. Given any clique in $\overline{G^|}$ (independent set in $G^|$), we show that it can be extended to a clique of size k in $\overline{G^|}$. Suppose that the vertices $\{v_{i_1,a_1}, v_{i_2,a_2}, \ldots, v_{i_\ell,a_\ell}\}$ induce a clique in $\overline{G^|}$. Suppose that neither vertex for part j can be added and still induce a clique. This means that some edge $\{v_{i_\ell,a_\ell}, v_{j,0}\}$ and $\{v_{i_m,a_m}, v_{j,1}\}$ must both be missing from $\overline{G^|}$, or in other words that these two edges are present in $G^|$. But in this case since $G = \hat{G}$ would force edge $\{v_{i_\ell,a_\ell}, v_{i_m,a_m}\}$ to appear in G which contradicts the fact that this was an edge in a clique in $\overline{G^|}$. Hence, every clique can be extended by one point and induction shows that all maximal cliques are of size k. This establishes that $CAFEN(G) \leq \theta'(\overline{G^|})$. □

Calculating $\theta'(\overline{G^|})$ may be an NP-hard problem so Theorem 6 does not necessarily give a good method to determine either $CAFEN(G)$ or produce a $CAFE(n, G)$. However, since the extension of cliques to k-cliques can be done greedily (see proof above), any heuristics or approximation algorithms for the edge clique covering problem adapt directly to produce $CAFE(n, G)$ for $g = 2$.

In Algorithm 1, we give a method that builds a CAFE for a graph $G = G_{k,2} = \hat{G}$. This algorithm reduces the problem to finding an edge covering by cliques of a subgraph of $\overline{G^|}$. At a first glance, this might not seem much better than what Theorem 6 gives us. However, in certain situations the subgraph obtained may be much simpler, as in the case of bipartite G (see Corollary 2).

Theorem 7. *Let $G = G_{k,2}$ such that $G = \hat{G}$. Let G_1 be the graph calculated in BUILDCAFE, which is obtained by removing all edges incident to the vertices corresponding to a k-tuple avoiding G. Then, Algorithm 1 is correct, and BUILDCAFE returns a $CAFE(n, G)$ for $n = k + 1 + m$, where m is the size of an edge clique cover of $\overline{G_1}$.*

Proof. If $G = \hat{G} = K_{k,2}$ then the required $CAFE$ is empty. So from now on we assume that $G = \hat{G} \neq K_{k,2}$. Since $G = \hat{G} \neq K_{k,2}$, there exists a feasible row via the reduction from 2-AVOID to 2-SAT which yields a k-tuple T (test) that avoids G. Thus, we relabel the values of 0 and 1 in each factor so that

---

Algorithm 1. Build $CAFE(n, G)$ for $g = 2$.

Require: $G = G_{k,2}$, $G^|$ connected and G avoidance closed.
  procedure BUILDCAFE(G)   **** (main procedure) ****
      $CAFE \leftarrow \emptyset$
      if $G \neq K_{k,2}$ **then**
          Using $2 - SAT$ formulation find an independent set of size k, $I \in G^|$
          Swapping values in each factor if necessary, let $I = \{v_{i,0} | 1 \leq i \leq k\}$
          Remove factors that contain a loop
          Contract pairs of factors with parallel edges
          Order the factors so all 0-1 edges go down and to the right ▷ Topologically
             sort $\vec{G}$
          $V(G_B) \leftarrow V(G); \; E(G_B) \leftarrow \{\{v_{i,0}, v_{j,1}\} \in E(G) : i, j \in [1, k]\}$
          $V(G_1) \leftarrow \{v_{i,1} \in V(G) : i \in [1, k]\}; \; E(G_1) \leftarrow \{\{v_{i,1}, v_{j,1}\} \in E(G) : i, j \in$
$[1, k]\}$
          $B \leftarrow$ BUILDBIPARTITE(G_B)
          $C \leftarrow$ BUILDONES(G_1)
          $CAFE \leftarrow \{0^k\} \cup B \cup C$
          for all i, a factor contracted to j **do**
             add new column i in $CAFE$
             set values in column i as forced by values in column j
          for all i, a factor with a loop **do**
             add new column i in $CAFE$
             set values in column i as forced by loop.
      return $CAFE$
  procedure BUILDBIPARTITE(G)
      for all $1 \leq i \leq k$ **do**
          for all $k \geq j > i$ **do**
             $b_j \leftarrow 0$
          $b_i \leftarrow 1$
          for all $i > j \geq 1$ **do**
             if $\{v_{j,0}, v_{i,1}\} \in E(G)$ **then**
                $b_j \leftarrow 1$
             else
                $b_j \leftarrow 0$
          $B \leftarrow B \cup \{b\}$
      return B
  procedure BUILDONES(G)
      edge-cover $\overline{G}$ by cliques $(K^1, K^2, \ldots, K^m)$
      for all $1 \leq i \leq m$ **do**
          for all $j \in V(K^i)$ **do**
             $c_j \leftarrow 1$
          for all $j \notin V(K^i)$ **do**
             if $\{v_{j,0}, v_{l,1}\} \in E(G)$ for some $l \in V(K^i)$ **then**
                $c_j \leftarrow 1$
             else
                $c_j \leftarrow 0$
          $C \leftarrow C \cup \{c\}$
      return C

$T = (0, \ldots, 0)$. After relabeling, all edges of G are between a vertex valued 0 (i.e. $v_{i,0}$ for some i) and a vertex valued 1 or are between two vertices valued 1.

Next, we remove any factors which have their values forced. This happens to factor i if and only if there is a loop at $v_{i,1}$. Note that after we are done building a $CAFE$ for the graph without these factors, we simply put these factors back into the array as new columns and set their value to 0 in every row. Next, we iteratively contract any pair of factors which contain two *parallel edges* between them, that is two edges of the form $\{v_{i,0}, v_{j,1}\}$ and $\{v_{i,1}, v_{j,0}\}$. These edges imply that the value in factor i must be equal to the value in factor j. Note that after we are done building a $CAFE$ for the graph with contracted factors, we simply put the extra factors back into the array as new columns and set to their forced value in every row.

We can now assume that our graph is closed, has no loops and no pair of parallel edges and thus at most one edge between any two factors. It has no 0-0 edges and thus only 0-1 edges and 1-1 edges. We reorder the factors so that all 0-1 edges slope down and to the right, that is, if $\{v_{i,0}, v_{j,1}\} \in E(G)$ then $j > i$. This is always possible and can be verified using the fact that there is at most one edge between any two factors and that $G = \hat{G}$ (alternatively, it is equivalent to the fact that the corresponding directed graph is acyclic and can be topologically sorted).

We edge-decompose the graph into the subgraphs containing exclusively each of these two kinds of edges: the *bipartite subgraph* (G_B) and the *1-1 subgraph* (G_1). We now construct three partial covering arrays each of which avoids the whole graph. They will cover the non-forbidden 0-0 pairs, 0-1 pairs and 1-1 pairs, respectively. The first covering array, A, is a single row containing only 0's.

The second partial covering array, B, will have exactly k rows with the values $B_{i,i} = 1$ and $B_{i,j} = 0$ for all $i < j$ already set:

$$
\begin{array}{cccccc}
1 & 0 & 0 & \cdots & 0 & 0 \\
  & 1 & 0 & \cdots & 0 & 0 \\
  &   & \ddots & & & \vdots \\
  &   &   & & 1 & 0 \\
  &   &   & & & 1.
\end{array}
$$

Then for $i > j$, we set $B_{i,j} = 1$ if $\{v_{j,0}, v_{i,1}\} \in E(G_B)$ and 0 otherwise.

We must show that this covers all required 0-1 pairs and avoids all forbidden 0-1 and 1-1 edges. For every pair of factors $i < j$, we must always cover the pair $\{v_{i,1}, v_{j,0}\}$ and this is covered in row i. If we also need to cover the pair $\{v_{i,0}, v_{j,1}\}$ this is done in row j; $B_{j,i} = 0$ since this pair is not a forbidden edge.

We now check that we do not cover any forbidden edges. Suppose $B_{\ell,i} = 0$ and $B_{\ell,j} = 1$. By construction the pair $\{v_{i,0}, v_{\ell,1}\}$ is not an edge of G, $\{v_{j,0}, v_{\ell,1}\} \in E(G)$ and by closure, $\{v_{i,0}, v_{j,1}\}$ is not an edge of G. Similarly, suppose $B_{\ell,i} = 1$ and $B_{\ell,j} = 1$. Again by construction, both $\{v_{i,0}, v_{\ell,1}\}, \{v_{j,0}, v_{\ell,1}\} \in E(G)$ and thus by closure, $\{v_{i,1}, v_{j,1}\}$ is not a forbidden edge.

The third partial covering array, C, will have one row for each clique in an edge clique cover of the complement of G_1 (only on the vertices at the 1-level).

These cliques in the complement are independent sets in G_1 and thus avoid the 1-1 edges and every 1-1 pair that is not an edge will be covered. Theses are also cliques in the graph $\overline{G^l}$ and thus by the proof of Theorem 6 they can be extended to maximal cliques of size k, which is equivalent to completing this row of C while avoiding G.   □

Next, we want to apply the algorithm to bipartite graphs. We need the following lemma which we state without proof.

Lemma 8. Let $G = \hat{G}$. Then G is bipartite if and only if G^l is bipartite.   □

The two colour classes are equivalent to two disjoint rows of a $CAFE$, thus we get the next result.

Corollary 1. G is bipartite if and only if it admits a $CAFE(n, G)$ which contains two disjoint rows.

Corollary 2. Let $G = G_{k,2}$, $G = \hat{G}$ and suppose that G is bipartite. Then $CAFEN(G) \leq k + 2$.

PROOF. Since G is bipartite, so is G^l by Lemma 8. Thus, we can choose the k-tuple avoiding G in Theorem 7, as one of the two parts of G^l, which yields G_1 with no edges and $m = 1$. Thus the CAFE produced by Algorithm 1 has size $k + 2$.   □

We know that the graph G which has edges $\{v_{i,0}, v_{j,1}\}$ for all $1 \leq i < j \leq k$ has $CAFEN(G) = k + 1$ so the upper bound given by this algorithm is close to best possible in this case.

If every factor connected component of G is bipartite then using Theorem 5 we get:

Corollary 3. Let $G = G_{k,2}$, $G = \hat{G}$ and suppose the factor connected components of G are $\{G_1, G_2 \ldots G_s\}$ with k_i factors in G_i. Assume also that each G_i is bipartite. Then

$$CAFEN(G) \leq CAN(s, 2) + \max_{1 \leq i \leq s}\{k_i\} - 1 \in O(\log s + \max_{1 \leq i \leq s}\{k_i\}).$$

PROOF. Algorithm 1 produces P_i which all have size 2 and can be used in Theorem 5.   □

For the case of non-bipartite graphs, Algorithm 1 requires us to build an edge clique cover of $\overline{G_1}$. In this case, we can use one of the constructive results on edge clique covers [1,2,11,13,17]. We hope that an edge clique cover of $\overline{G_1}$ will be much smaller than one of $\hat{G^l}$, as it was the case of a bipartite G. A direction of further research is to extend the results obtained for bipartite graphs to other interesting classes of graphs.

References

1. Brigham, R.C., Dutton, R.D.: Upper bounds on the edge clique cover number of a graph. Discrete Math. 52, 31–37 (1984)
2. Brigham, R.C., Dutton, R.D.: A compilation of relations between graph invariants. Supplement I. Networks 21, 421–455 (1991)
3. Burr, K., Young, W.: Combinatorial test techniques: Table-based automation, test generation, and code coverage. In: Proc. Intl. Conf. on Soft. Test. Anal. and Rev., October 1998, pp. 503–513. ACM, New York (1998)
4. Cawse, J.N.: Experimental design for combinatorial and high throughput materials development. GE Global Research Technical Report 29, 769–781 (2002)
5. Cohen, M.B., Dwyer, M.B., Shi, J.: Interaction testing of highly-configurable systems in the presence of constraints. In: International Symposium on Software Testing and Analysis (ISSTA), London, July 2007, pp. 129–139 (2007)
6. Colbourn, C.J.: Combinatorial aspects of covering arrays. Le Matematiche(Catania) 58, 121–167 (2004)
7. Colbourn, C.J.: Covering arrays. In: Colbourn and Dinitz [8], pp. 361–364
8. Colbourn, C.J., Dinitz, J.H. (eds.): Handbook of combinatorial designs, 2nd edn. Discrete Mathematics and its Applications. Chapman & Hall/CRC, Boca Raton (2007)
9. Dalal, S.R., Karunanithi, A.J.N., Leaton, J.M.L., Patton, G.C.P., Horowitz, B.M.: Model-based testing in practice. In: Proc. Intl. Conf. on Software Engineering (ICSE 1999), pp. 285–294 (1999)
10. Dunietz, S., Ehrlich, W.K., Szablak, B.D., Mallows, C.L., Iannino, A.: Applying design of experiments to software testing. In: Proc. Intl. Conf. on Software Engineering (ICSE 1997), October 1997, pp. 205–215. IEEE, Los Alamitos (1997)
11. Erdős, P., Goodman, A.W., Pósa, L.: The representation of a graph by set intersections. Canad. J. Math. 18, 106–112 (1966)
12. Gargano, L., Körner, J., Vaccaro, U.: Sperner capacities. Graphs Combin. 9, 31–46 (1993)
13. Gyárfás, A.: A simple lower bound on edge coverings by cliques. Discrete Math. 85, 103–104 (1990)
14. Kou, L.T., Stockmeyer, L.J., Wong, C.K.: Covering edges by cliques with regard to keyword conflicts and intersection graphs. Comm. ACM 21, 135–139 (1978)
15. Kuhn, D.R., Reilly, M.: An investigation of the applicability of design of experiments to software testing. In: Proc. 27th Annual NASA Goddard/IEEE Software Engineering Workshop, October 2002, pp. 91–95. IEEE, Los Alamitos (2002)
16. Kuhn, D.R., Wallace, D.R., Gallo, A.M.: Software fault interactions and implications for software testing. IEEE Trans. Soft. Eng. 30, 418–421 (2004)
17. Lovász, L.: On covering of graphs. In: Theory of Graphs (Proc. Colloq., Tihany, 1966), pp. 231–236. Academic Press, New York (1968)
18. Mandl, R.: Orthogonal latin squares: An application of experiment design to compiler testing. Communic. of the ACM 28, 1054–1058 (1985)
19. Martinez, C., Moura, L., Panario, D., Stevens, B.: Algorithms to locate errors using covering arrays. In: Proc. LATIN 2008 - 8th Latin American Theoretical INformatics conference (April 2008)
20. Meagher, K., Moura, L., Zekaoui, L.: Mixed covering arrays on graphs. Journal of Combinatorial Designs 15, 393–404 (2007)
21. Meagher, K., Stevens, B.: Covering arrays on graphs. J. Combin. Theory. Ser. B 95, 134–151 (2005)

22. Moura, L., Stardom, J., Stevens, B., Williams, A.W.: Covering arrays with mixed alphabet sizes. J. Combin. Des. 11, 413–432 (2003)
23. Nuutila, E.: Efficient transitive closure computation in large digraphs. Acta Polytechnica Scandinavica, Mathematics and Computing in Engineering Series, Finnish Academy of Technology, vol. 74 (1995)
24. Seroussi, G., Bshouty, N.H.: Vector sets for exhaustive testing of logic circuits. IEEE Transactions on Information Theory 34, 513–522 (1988)
25. Shasha, D.E., Kouranov, A.Y., Lejay, L.V., Chou, M.F., Coruzzi, G.M.: Using combinatorial design to study regulation by multiple input signals: A tool for parsimony in the post-genomics era. Plant Physiology 127, 1590–2594 (2001)
26. Tang, D.T., Chen, C.L.: Iterative exhaustive pattern generation for logic testing. IBM Journal Research and Development 28, 212–219 (1984)
27. Williams, A.W., Probert, R.L.: A measure for component interaction test coverage. In: Proc. ACS/IEEE Intl. Conf. Comput. Syst. & Applic., pp. 301–311 (2001)

The Robot Cleans Up

Margaret-Ellen Messinger* and Richard J. Nowakowski*

Department of Mathematics and Statistics,
Dalhousie University, Halifax NS, Canada
{messnger,rjn}@mathstat.dal.ca

Abstract. Imagine a large building with many corridors. A robot cleans these corridors in a greedy fashion, the next corridor cleaned is always the dirtiest to which it is adjacent. We determine bounds on the minimum $s(G)$ and maximum $S(G)$ number of time steps (over all edge weightings) before every edge of a graph G has been cleaned. We show that Eulerian graphs have a self-stabilizing property that holds for any initial edge weighting: after the initial cleaning of all edges, all subsequent cleanings require $s(G)$ time steps. Finally, we show the only self-stabilizing trees are a subset of the superstars.

Keywords: cleaning process, searching, greedy algorithm, edge traversing.

1 Introduction

The *robot cleaning process* is a variant of the cleaning and searching processes (see [3,4]) in which there is one cleaning agent, the *robot*, available to clean the edges of the graph. In this model, the contaminant is immobile, so recontamination of edges is not an issue. For example, imagine a set of water pipes with algae, or, as in the Abstract, a robot vacuum randomly cleaning a set of rooms. A robot would have some, possibly limited, processing power. In the model considered here, we allow the robot to make a local, greedy choice. One formulation of this model is to assign a weight to each edge of the graph, which corresponds to the level of contamination. At each time step, the robot cleans (i.e. traverses) the incident edge with the highest level of contamination and re-sets that contamination level to zero, while the level of contamination for all other edges is increased by one.

In Section 2, the problem is formulated in an equivalent way: each edge has a time associated with it, the time at which it was last cleaned. If an edge has not been cleaned, then it has a negative number assigned. We assume that it takes one time step to clean an edge. At time t, the robot cleans (traverses) the incident edge of minimum weight, sets the weight on that edge to t while all other edge weights remain unchanged.

We first show, Theorem 1, that for any finite connected graph, there is some finite time step by which every edge of the graph has been cleaned. The robot doesn't know that it has cleaned the graph, indeed, it is supposed to *continually*

---

* Research partially supported by NSERC.

B. Yang, D.-Z. Du, and C.A. Wang (Eds.): COCOA 2008, LNCS 5165, pp. 309–318, 2008.
© Springer-Verlag Berlin Heidelberg 2008

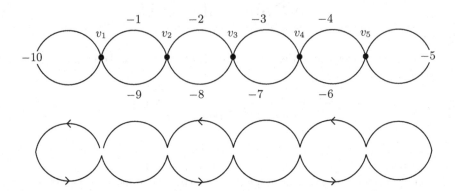

Fig. 1. An initial weighting of G for which the robot has cleaned every edge after $|E(G)|$ steps

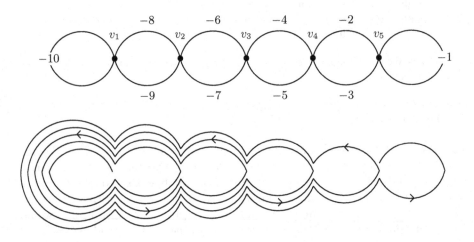

Fig. 2. An initial weighting of G for which the robot has cleaned every edge after $(1/4)|E(G)|^2 + (1/2)|E(G)|$ steps

clean the graph. The focus is then on $s(G)$ and $S(G)$, the minimum and maximum number of steps required to clean the graph on the first and on subsequent cleanings.

Given an initial set of edge weights and initial location of the robot, the robot acts very much like a cellular automaton, see the *automatic ant* [2]. The robot problem is also similar to some edge-traversal (also known as arc-routing) problems, which generally require a least-cost traversal of a subset of edges of a network. In Section 3, an algorithm that solves one such edge-traversal problem (the Chinese Postman Problem) is exploited to find a lower bound for $s(G)$.

Consider the two weightings of G, given in Figures 1 and 2, and suppose the robot is initially placed at v_1. In Figure 1, every edge of G has been cleaned after $|E(G)|$ steps (and the path of the robot is traced in the lower graph). In Figure 2, every edge of G has been cleaned after $(1/4)|E(G)|^2 + (1/2)|E(G)|$ steps (and the path of the robot is traced in the lower graph). If G is a cycle, then $s(G) = S(G) = |E(G)|$, but Figures 1 and 2 illustrate that the difference between $S(G)$ and $s(G)$ can be at least $(1/4)|E(G)|^2 - (1/2)|E(G)|$. Theorem 5 in Section 4 provides an upper bound for $S(G)$.

What happens if the robot traverses the edges of G a second, third, or fourth time? Eventually the process must settle into some reoccurring sequence of initial weightings as there are only a finite number of edge sequences. Theorem 6 of Section 5 shows that for Eulerian graphs, no matter how many steps are required before every edge of the graph G has been cleaned at least once, every subsequent cleaning requires only $|E(G)|$ steps. For a tree T, Theorem 7 shows the only trees with this self-stabilizing property are a subset of the superstars.

Throughout this paper, we will assume that *every graph is finite and connected*.

2 Definitions and Preliminaries

For a graph $G = (V, E)$, an *initial weighting* of G is a bijection $\mathcal{C} : E \to \{-1, -2, \ldots, -|E|\}$. Initially, the robot is placed at some vertex $v \in V$. Let $\mathcal{C}_t(e)$ denote the weight of edge e at step t. Thus, if edge e' is traversed at step t, then set $\mathcal{C}_{t+1}(e') := t+1$ and set $\mathcal{C}_{t+1}(e) := \mathcal{C}_t(e)$ for all $e \in E \backslash \{e'\}$. As the number of steps before every edge of a graph is cleaned is the main concern of this paper, set $\mathcal{C}_0 = E$ and let $\mathcal{C}_t$ denote the set of edges which have *not* been traversed by step t. Consequently, the process will terminate when $\mathcal{C}_t = \emptyset$ for some step t (once every edge has been traversed) and return the time step $\sigma(G, \mathcal{C}, v) = t$.

Definition 1. *For a graph $G = (V, E)$ with an initial weighting $\mathcal{C}$ and the robot initially located at some vertex $v \in V$, the* **one-pass robot** $R_1(G, \mathcal{C}, v)$ *is a process defined as follows:*

(0) *Set $t := 0$, $\mathcal{C}_0 := E$ and $\mathcal{C}(e) := \mathcal{C}_0(e)$ for every $e \in E$.*
(1) *If $\mathcal{C}_t = \emptyset$ then STOP, return $\sigma(G, \mathcal{C}_0, v) = t$.*
(2) *Let $e' = \{u, v\}$ such that $\mathcal{C}_t(e') = \min\{\mathcal{C}_t(e) : e = \{u, v\}$ for all $w \in N(v)\}$.*
(3) *The robot traverses edge e' from v to u, setting $\mathcal{C}_{t+1} := \mathcal{C}_t \backslash \{e'\}$, $\mathcal{C}_{t+1}(e') := t + 1$ and $\mathcal{C}_{t+1}(e) := \mathcal{C}_t(e)$ for all $e \in E \backslash \{e'\}$.*
(4) *Relabel vertex u (where the robot is located) as v, set $t := t + 1$, and return to (1).*

Given the initial weightings of Figures 1 and 2, suppose the robot is initially located at v_1. Then the process of Figure 1 terminates after 10 steps and the process of Figure 2 terminates after 30 steps. Although the termination was apparent in these examples, Theorem 1 shows the process always terminates in a finite number of steps.

Given an arbitrary initial weighting of graph G and an arbitrary initial location of the robot, an *arbitrary application of R_1 to G* is an application of the one-pass robot process to G.

Lemma 1. *During an arbitrary application of R_1 to graph G, suppose the robot first visits v after traversing the edge $\{u, v\}$. The robot will not traverse $\{u, v\}$ from v to u until all other edges incident with v have been traversed.*

Suppose the robot first visits v after traversing $\{u, v\}$. By Definition 1, at each subsequent visit to v, the robot may not traverse $\{u, v\}$ unless every other incident edge has been traversed.

Theorem 1. *An arbitrary application of R_1 to (a finite connected) graph G will terminate after a finite number of steps.*

Proof. For a finite connected graph $G = (V, E)$, let $\mathcal{C}$ be an arbitrarily initial weighting and let $v \in V$ be an arbitrary initial location of the robot. Assume that R_1 does not terminate after a finite number of steps. As the process terminates once every edge has been traversed, there is some edge $e = \{u, v\}$ which is not traversed, but is adjacent to an edge that is. There must be a maximum cardinality subset S of vertices, each of which is visited by the robot an infinite number of times. Suppose $u \in S$. By Lemma 1, all vertices adjacent to u must be visited by the robot an infinite number of times. Since S is a maximum connected component thus $G = S$. Thus, R_1 must terminate after a finite number of steps.

The fact that every edge in a graph will be traversed raises several questions. Specifically, what is the minimum or maximum number of time steps before every edge has been traversed, taken over all initial weightings and initial locations of the robot?

Definition 2. *Let C be the set of all bijections $\mathcal{C} : E \to \{-1, -2, \ldots, -|E|\}$ for a graph $G = (V, E)$. Then*

$$s(G) = \min_{v \in V, \mathcal{C} \in C} \sigma(G, \mathcal{C}, v) \quad and \quad S(G) = \max_{v \in V, \mathcal{C} \in C} \sigma(G, \mathcal{C}, v).$$

Let $G = (V, E)$ be a graph and suppose no edge is traversed twice in the first k steps. Label the first k edges traversed as $e_1, e_2, \ldots, e_k$ such that e_i was traversed at step i. We may now assume that the edges were weighted in that order originally, thus we do not have to give the weighting in advance. In essence, for $s(G)$, we try to use untraversed edges whereas for $S(G)$, we try to reuse edges as often as possible.

3   Results for $s(G)$

Clearly, $s(G) \geq |E(G)|$ for every graph G. If a graph $G = (V, E)$ is Eulerian, then Lemma 2 shows the edges can be traversed in exactly $|E|$ steps; however, many initial weightings will return a larger number of steps to traverse the edges as shown in Figure 2. Let $d_o(G)$ denote the number of vertices of odd degree in a graph G. Then graph G is *semi-Eulerian* if $d_o(G) = 2$.

Lemma 2. *If $G = (V, E)$ is an Eulerian graph, then $s(G) = |E|$ and the initial and final locations of the robot are the same. If G is a semi-Eulerian graph, then $s(G) = |E|$ and the initial and final locations are the two vertices of odd degree.*

Proof. Let $G = (V, E)$ be an Eulerian graph and v be an arbitrary vertex at which the robot is initially located. There must exist an initial weighting that permits the robot to follow an Eulerian circuit through G. Note the robot must return to v after traversing the $|E|^{th}$ edge.

Suppose $G = (V, E)$ is semi-Eulerian with odd vertices u, v. Let $G' = (V', E')$ be the supergraph of G formed by the addition of the edge $e = \{u, v\}$. Assign a weight of $-|E'|$ to e and start the robot at u. The result follows immediately from the previous argument.

If G is semi-Eulerian, then both the initial and final locations of the robot must be the two vertices of odd degree. However, for every vertex of odd degree of a graph that is neither the initial or final vertex for the robot, at least one incident edge must be traversed at least twice.

Theorem 2. *For a graph $G = (V, E)$ with $d_o(G) > 0$, $s(G) \geq |E| + (d_o(G) - 2)/2$.*

Proof. Let $G = (V, E)$ be a graph with an arbitrary application of R_1 to G. If $d_o(G) = 2$, then the result is known from Lemma 2; assume $d_o(G) > 2$. Assume that the initial and final locations of the robot are both vertices of odd degree (if either or both are vertices of even degree the lower bound will be increased). Each of the remaining $d_o(G) - 2$ vertices of odd degree must be incident to an edge which is traversed twice. But such an edge could be incident with two vertices of odd degree, so at least $(d_o(G) - 2)/2$ edges must be traversed twice.

Corollary 1 shows that the bound of Theorem 2 can be tight.

Corollary 1. *Let $K_n = (V, E)$ be a complete graph on n vertices. Then*

$$s(K_n) = \begin{cases} |E| & \text{if } n \text{ is odd} \\ |E| + (n-2)/2 & \text{if } n \text{ is even.} \end{cases}$$

Proof. If n is odd, the result follows from Lemma 2. Suppose n is even. Label the vertices of K_n as $v_1, v_2, \ldots, v_n$ and initially place the robot at v_1. The robot can first traverse an $n - 1$-cycle such that $C_i(v_i, v_{i+1}) = i$ for $i = 1, 2, \ldots, n - 2$ and $C_{n-1}(v_{n-1}, v_1) = n - 1$. The robot can now traverse the remaining edges of the subgraph induced by the vertices $\{v_1, v_2, \ldots, v_{n-1}\}$. As this subgraph is Eulerian, the robot can traverse each edge once before returning to v_1 at step $t = |E(K_{n-1})|$. Note that at this step, the untraversed edges of K_n are precisely those edges incident with v_n.

At step $t = |E(K_{n-1})| + 1$, the robot traverses edge $\{v_1, v_n\}$. At step $t = |E(K_{n-1})| + 2$, it traverses edge $\{v_n, v_{n-1}\}$. At step $t = |E(K_{n-1})| + 3$, the robot is located at vertex v_{n-1} and the minimum weighted edge incident with v_{n-1} is $\{v_{n-2}, v_{n-1}\}$ (as it was last traversed at step $t = n - 2$). The robot traverses $\{v_{n-2}, v_{n-1}\}$, followed by edge $\{v_{n-2}, v_n\}$. This argument can be repeated to traverse the remaining edges incident with v_n.

Finally, every edge of K_n has been traversed in $(n-1) + |E(K_{n-1})| + (n-2)/2 = |E(K_n)| + (n-2)/2$ steps. By Theorem 2, this is the minimum.

Let $W(G)$ denote the length of the shortest walk which traverses every edge of a graph $G = (V, E)$ at least once. It is clear that $s(G) \geq W(G) \geq |E|$ with equality if and only if G is Eulerian or semi-Eulerian. The Chinese Postman Problem is to find the shortest *closed* walk of a graph such that each edge is traversed at least once. For a graph G, the solution of the Chinese Postman Problem may be obtained by creating a new graph G': minimally adding new edges to G, such that G' is Eulerian [1]. We use this construction for $W(G)$: let $M(G)$ be the minimum number of edges added to G to ensure it is semi-Eulerian (which will require less edges than ensuring it is Eulerian). That is, $M(G)$ is a least-cost augmentation of G into G' where G' is semi-Eulerian. Although $W(G) = |E| + M(G)$ provides a lower bound for $s(G)$, the difference between $s(G)$ and $W(G)$ can be large as shown in Figure 3 (the order in which edges are cleaned is shown for each).

Theorem 3. *For any graph G, $s(G) \geq W(G) = |E| + M(G)$.*

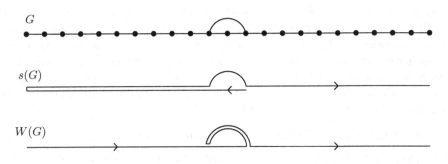

Fig. 3. An example for which $s(G) = W(G) + (\mathrm{diam}(G) - 3)/2$

To determine $s(T)$ for a tree T, we must alter the termination point of the one-pass robot. The *vertex-recurrent* one-pass robot process is simply the one-pass robot process extended such that it terminates once every edge has been traversed *and* the robot has returned to its initial location. Thus, for an initial weighting $\mathcal{C}$ and initial location v of the robot, the vertex-recurrent process will be denoted by $VR_1(G, \mathcal{C}, v)$. Lemma 3 shows that for a tree $T = (V, E)$, the vertex-recurrent process terminates after $2|E|$ steps.

Lemma 3. *Let $T = (V, E)$ be a tree with an arbitrary initial weighting $\mathcal{C}$ and the robot initially placed at an arbitrary vertex $v \in V$. Then $VR_1(T, \mathcal{C}, v)$ terminates after exactly $2|E|$ steps.*

Proof. Let $T = (V, E)$ be a tree and initially place the robot at an arbitrary vertex $v \in V$. If v is not a leaf, then $\deg(v) = k \geq 2$. Let $T_1, T_2, \ldots, T_k$ be the maximal subtrees of T that have v as a leaf. By Lemma 1, when the robot leaves v and traverses an edge to T_i, it will only return to v after having traversed every edge of T_i. By induction, the number of steps until every edge has been traversed and the robot has returned to v is $2 \sum_{i=1}^{k} |E(T_i)| = 2|E|$.

If v is a leaf with stem u, the robot begins at v and first traverses $\{u, v\}$ to u. By induction, the rest of the tree is traversed in $2(|E| - 1)$ steps with the robot returning to u. Finally, the robot traverses $\{u, v\}$ back to v. Then $VR_1(T, \mathcal{C}, v)$ will terminate after $2|E|$ steps.

Theorem 4. *For any tree* $T = (V, E)$ *with diameter* $diam(T)$,

$$s(T) = 2|E| - diam(T).$$

Proof. Let $T = (V, E)$ be a tree with diameter $diam(T)$. As T is a tree, the maximum number of edges that can be traversed exactly once, is $diam(T)$. As the remainder of the edges must be traversed at least twice, $s(T) \geq diam(T) + 2(|E| - diam(T)) = 2|E| - diam(T)$.

Let $P = \{p_1, p_2, \ldots, p_{d+1}\}$ be a path of length $d = diam(T)$. In the subgraph obtained by the removal of the edges of P, let T_{p_i} be the tree which includes vertex p_i. In T, initially place the robot at p_1; it sequentially traverses the edges of P, such that when it arrives at a vertex p_i, it traverses the subtree T_{p_i} before traversing the edge $\{p_i, p_{i+1}\}$. By Lemma 3, it traverses T_{p_i} and returns to p_i in exactly $2|E(T_{p_i})|$ steps. In this manner, the robot will traverse all edges of the tree in $d + \sum_{i=1}^{d+1} 2|E(T_{p_i})| = 2|E| - diam(T)$ steps.

Corollary 2. *Let* $G = (V, E)$ *be a unicyclic graph with cycle* C *and let* P *be the longest path in* G *that contains only one vertex of* C. *Then*

$$s(G) = 2|E| - |E(C)| - |E(P)|.$$

4   Results for $S(G)$

Theorem 1 showed that for any initial weighting and any initial location of the robot, the process will terminate after a finite number of steps. This motivates the question, what is the maximum number of steps before the process terminates (over all possible initial weightings and initial robot locations)?

Corollary 3 follows directly from Lemma 3. Clearly $S(T) \leq 2|E| - 1$ for a tree $T = (V, E)$ as the final edge traversed is only traversed once. Suppose v is a leaf on T and the robot is initially located at some vertex u (where u is adjacent to v). Then it can traverse the edges of $T \backslash \{v\}$ in exactly $2|E| - 2$ steps, followed by edge $\{u, v\}$.

Corollary 3. *For any tree* $T = (V, E)$, $S(T) = 2|E| - 1$.

Let H_n be the graph on n vertices that consists of a path $P = \{v_1, v_2, \ldots, v_n\}$; an additional edge between v_i and v_{i+1} for $i = 1, 2, \ldots, n - 1$; and a loop at both v_1 and v_n. The graph H_5 is illustrated in Figure 2. If the robot is initially located at v_1, it can traverse the edges of H_5 in the order given in Figure 2. Clearly the robot could traverse the edges of H_n in a similar manner. For H_n, as $|E| = 2n - 2$, the number of steps before the process terminates is $2 \sum_{i=1}^{n-1} i = n^2 - n = (1/4)|E|^2 + (1/2)|E|$. Consequently, a general upper bound on $S(G)$ must be at least $O(|E|^2)$.

Theorem 5. *Let $G = (V, E)$ be a graph and let $u \in V$ with $\deg(u) = \Delta(G)$. Then*

$$S(G) \leq 1 + 2(\deg(u) - 1)\Big(\prod_{v \in V:\ v \neq u} \deg(v)\Big).$$

Further, if G contains no leaves,

$$S(G) \leq 1 + (\deg(u) - 1)\Big(\prod_{v \in V:\ v \neq u} \deg(v)\Big).$$

Proof. Let $G = (V, E)$ be a graph with an arbitrary application R_1 to G. Label the vertices of G as $v_1, v_2, \ldots, v_n$ and let u be a vertex with $\deg(u) = \Delta(G)$.

For each time step $t > 0$ we associate with each vertex, the incident edge with minimum weight to form a list $\mathcal{L}_t$ (with n entries) such that the ith entry in the list is the edge incident with v_i of minimum weight. If at step t, the edge incident with a leaf is traversed for the first time, then $\mathcal{L}_t = \mathcal{L}_{t+1}$. Otherwise, $\mathcal{L}_t$ and $\mathcal{L}_{t+1}$ must be distinct. In order to get an upper bound, we allow the process to continue well past the time step t_1 (when R_1 normally terminates) to when all edges have been traversed. By some time step $t' > t_1$, the event $(\star)$ *must* have occurred:

$(\star)$ For every $v \in V$, each of the $\deg(v)$ edges has appeared in a list as the edge with minimum weight incident to v.

Although there are a total of $\prod_{v \in V} \deg(v)$ possible lists, if G contains no leaves, then $(\star)$ must occur by at *most* step $t' = (\deg(u) - 1)(\prod_{v \in V:\ v \neq u} \deg(v)) + 1$. This implies that every edge of the graph has been traversed.

Recall that for every step t in which the edge incident to a leaf is first traversed, $\mathcal{L}_t = \mathcal{L}_{t+1}$. So if G contains leaves, the result is $1 + 2(\deg(u) - 1)(\prod_{v \in V:\ v \neq u} \deg(v))$ time steps.

5   Self-stabilizing Graphs

What happens if the process is applied a second time, using the final weighting (and final robot location) of the first application as the initial weighting (and initial robot location) of the second application. What if the process is applied i times? Eventually the process must settle into some reoccurring sequence of initial weightings as there are only a finite number of edge sequences; however, we first formalize the idea of consecutive applications of the robot process.

Given an initial weighting $\mathcal{C}$ and an initial location of the robot at some vertex $v(t_0)$, the one-pass robot process terminates after $\sigma(G, \mathcal{C}, v) = t_1$ steps at which time the robot is located at vertex $v(t_1)$. Then the ith application of the process $R_1(G, \mathcal{C}_{t_{i-1}}, v(t_{i-1}))$ terminates at step $t_i = \sigma(G, \mathcal{C}_{t_{i-1}}, v(t_{i-1}))$. For $i > 1$, the ith application of the process will use the final weighting of the i-1th process as its initial weighting.

Definition 3. *For every initial weighting and every initial robot location, if $t_i = s(G)$ for every $i > 1$, then graph G is self-stabilizing.*

If a graph G is self-stabilizing, then no matter how many steps pass before every edge of G has been traversed in the first application, the process will require only $s(G)$ steps in subsequent applications. For example, given the initial weighting of graph G in Figure 2 where the robot is initially placed at v_1, the process requires $(1/2)|E(G)|^2 + (1/2)|E(G)|$ steps before every edge has been traversed. One may easily verify that for the graph in Figure 2, for each subsequent application, the process only requires $|E(G)|$ steps.

Theorem 6. *Every Eulerian graph is self-stabilizing.*

Proof. Let $G = (V, E)$ be an Eulerian graph with an arbitrary initial weighting C^1 and an arbitrary initial location v_1 for the robot. Let t be the first time step in the process when the robot is located at a vertex and every edge incident with that vertex has been traversed. By Lemma 1 and the fact that all vertex degrees are even, the robot must be located at v_1 at step t. There are two cases to consider. First, if all the edges have been traversed then the robot will follow the same path in subsequent applications of the process.

Second, if not all edges have been traversed then the robot has traversed a circuit A_1, starting and ending at v_1 (i.e. vertices, but not edges, can be repeated on A_1). There must be some vertex on A_1 that has untraversed incident edges. Let v_2 be the first such vertex on A_1. Let $\{e_1, e_2, \ldots, e_j\} \in A_1$ be the first j edges of G the robot traverses such that the robot traverses e_j to visit v_2 for the first time. Consider a new weighting: let $C^2(e_i) = C^1_t(e_i)$ for $i = 1, 2, \ldots, j$ and $C^2(e) = C^1(e)$ for all other edges e. Now consider a new process with initial weighting C^2 and the robot initially located at v_2. By changing the weighting in this way, the robot must follow the edges of A_1 from v_2 to v_1 and then back to v_2. So all the edges that were traversed starting at v_1 with weighting C^1 are also traversed starting at v_2 with weighting C^2. As v_2 is incident to at least two untraversed edges, the robot will traverse these and create a larger circuit. If this does not include all the edges then apply this procedure again. Each time a larger circuit is produced so (eventually) there must be some vertex v_i with circuit A_i such that there are no untraversed edges incident with A_i.

Given k paths, the *superstar* is the graph obtained by merging k leaves, one from each path. The reader may easily verify that both paths and superstars, where each ray is of equal length, are self-stabilizing trees. Theorem 7 shows these are the only self-stabilizing trees.

Theorem 7. *A tree T is self-stabilizing if and only if it is either a path or a superstar where each ray is of equal length.*

Proof. Since a path is self-stabilizing, consider $T = (V, E)$, a self-stabilizing tree with more than 2 leaves. We show that for the given initial weighting and location of the robot, T must be a superstar (where each ray is of equal length). Let x be the initial location of the robot and y be the final location of the robot after an application of the process R_1 where x and y are leaves. Let $P = \{p_0, p_1, p_2, \ldots, p_{a-1}, p_a\}$ be the path from $x = p_0$ to $y = p_a$. Let z be a leaf of T such that $d = d(y,z) \leq d(y,\ell)$ for all other leaves $\ell \in V$. In the subgraph obtained by the removal of the edges of P from T, let T_{p_i} be the subtree containing vertex p_i, for each $i \in \{1,2,\ldots,a-1\}$. Let T_{p_k} be the subtree containing vertex z.

The robot now traverses the edges of T such that when it arrives at vertex p_i, it traverses the subtree T_{p_i} before traversing the edge $\{p_i, p_{i+1}\}$ (this may be done by Lemma 3). Once the robot arrives at p_k, it traverses the edges of T_{p_k} such that the first leaf in T_{p_k} the robot reaches is z. (Again, this may be done by Lemma 3). Finally, the robot traverses the remaining edges of P from p_k to $y = p_a$ and every edge has been traversed.

Apply the process a second time to T with the robot initially located at $y = p_a$, using the final weighting of the first application as the initial weighting in the second. As the weight on edge $\{p_{i-1}, p_i\}$ is less than the weight on all other edges incident with p_i, the robot first traverses the edges of P. After the robot has traversed P, it is located at $x = p_0$ and must subsequently traverse the edges of P again; however when it arrives at vertex p_i, it now traverses the subtree T_{p_i} before traversing edge $\{p_i, p_{i+1}\}$ for each $i \in \{1,2,\ldots,k-1\}$. When the robot arrives at p_k, it traverses the edges of T_{p_k}; however, the final edge traversed will be the edge incident with z.

By Lemma 3, the second process requires exactly $2|E| - d(y,z)$ steps and as T is self-stabilizing, Theorem 4 implies $d(y,z) = \mathrm{diam}(T)$. Then $d(y,\ell) = \mathrm{diam}(T)$ for all leaves $\ell \in V \setminus \{y\}$ and further, $d(u,v) = \mathrm{diam}(T)$ for any leaves $u,v \in V$. Thus, T must be a superstar with each ray of equal length.

References

1. Edmonds, J., Johnson, E.L.: Matching, Euler Tours and the Chinese Postman. Mathematical Programming 5, 88–124 (1973)
2. Gale, D.: Tracking the automatic ant and other mathematical explorations. Springer, New York (1998)
3. Messinger, M.E., Nowakowski, R.J., Prałat, P.: Cleaning a Network with Brushes. Theoretical Computer Science 399, 191–205 (2008)
4. Messinger, M.E., Nowakowski, R.J., Prałat, P., Wormald, N.C.: Cleaning random d-regular graphs with brushes using a degree–greedy algorithm. In: Janssen, J., Prałat, P. (eds.) CAAN 2007. LNCS, vol. 4852, pp. 13–26. Springer, Heidelberg (2007)

On Recovering Syntenic Blocks
from Comparative Maps

Zhixiang Chen[1], Bin Fu[1], Minghui Jiang[2,*], and Binhai Zhu[3]

[1] Department of Computer Science, University of Texas - Pan American, Edinburg, TX
78539-2999, USA
{chen,binfu}@cs.panam.edu
[2] Department of Computer Science, Utah State University, Logan, UT 84322-4205, USA
mjiang@cc.usu.edu
[3] Department of Computer Science, Montana State University, Bozeman, MT 59717, USA
bhz@cs.montana.edu

Abstract. A genomic map is represented by a sequence of gene markers, and
a gene marker can appear in several different genomic maps, in either positive
or negative form. A *strip* (syntenic block) is a sequence of distinct markers that
appears as subsequences in two or more maps, either directly or in reversed and
negated form. Given two genomic maps G and H, the problem *Maximal Strip
Recovery* (MSR) is to find two subsequences G' and H' of G and H, respectively,
such that the total length of disjoint strips in G' and H' is maximized. Previously
only a heuristic was provided for this problem, which does not guarantee finding
the optimal solution, and it was unknown whether the problem is NP-complete
or polynomially solvable. In this paper, we develop a factor-4 polynomial-time
approximation algorithm for the problem, and show that several close variants of
the problem are intractable.

1 Introduction

In comparative genomics, a starting point is to decompose two given genomes into
syntenic blocks—segments of chromosomes which are deemed to be homologous in the
two input genomes. Various methods have been proposed, but they are very vulnerable
to ambiguities and errors. Recently, a heuristic method was proposed to eliminate noise
and ambiguities in genomic maps, through handling a problem called Maximal Strip
Recovery (see below for the formal definition) [6,16]. But it was unknown whether the
problem can be solved in polynomial time or is NP-complete. In this paper, we design
a factor-4 polynomial-time approximation algorithm for the problem, and show that
several close variants of the problem are intractable.

A genomic map is represented by a sequence of gene markers, and a gene marker can
appear in several different genomic maps, in either positive or negative form. A *strip*
(syntenic block) is a sequence of distinct markers that appears as subsequences in two or
more maps, either directly or in reversed and negated form. Given two genomic maps G
and H, the problem *Maximal Strip Recovery* (MSR) [6,16] is to find two subsequences

---

* Partially supported by NSF grant DBI-0743670.

B. Yang, D.-Z. Du, and C.A. Wang (Eds.): COCOA 2008, LNCS 5165, pp. 319–327, 2008.
© Springer-Verlag Berlin Heidelberg 2008

G' and H' of G and H, respectively, such that the total length of disjoint strips in G' and H' is maximized. Intuitively, those gene markers not included in G' and H' are noise and ambiguities.

We give a precise formulation of the generalized problem MSR-d: Given d signed permutations (genomic maps) G_i of $\langle 1, \ldots, n \rangle$, $1 \leq i \leq d$, find k sequences (strips) S_j of length at least two, and find d signed permutations π_i of $\langle 1, \ldots, k \rangle$, such that each sequence $G'_i = S_{\pi_i(1)} \ldots S_{\pi_i(k)}$ (here S_{-j} denotes the reversed and negated sequence of S_j) is a subsequence of G_i, and the total length of the strips S_j is maximized. Note that the problem Maximal Strip Recovery (MSR) [6,16] is exactly the problem MSR-2 in our new formulation. We refer to Fig. 1 for an example.

$$G_1 = \langle 1, 2, 3, 4, 5, 6, 7, 8, 9, 10, 11, 12 \rangle$$
$$G_2 = \langle -8, -5, -7, -6, 4, 1, 3, 2, -12, -11, -10, 9 \rangle$$
$$S_1 = \langle 1, 3 \rangle$$
$$S_2 = \langle 6, 7, 8 \rangle$$
$$S_3 = \langle 10, 11, 12 \rangle$$
$$\pi_1 = \langle 1, 2, 3 \rangle$$
$$\pi_2 = \langle -2, 1, -3 \rangle$$
$$G'_1 = \langle 1, 3, 6, 7, 8, 10, 11, 12 \rangle$$
$$G'_2 = \langle -8, -7, -6, 1, 3, -12, -11, -10 \rangle$$

Fig. 1. An example for the problem MSR

A heuristic based on Maximum Clique (and its complement Maximum Independent Set) was previously given for the problem MSR (MSR-2) [6,16], which does not guarantee finding the optimal solution. It was unknown whether MSR is NP-complete or polynomially solvable. In this paper, we show that the previous heuristic [6,16] can be modified to achieve a factor-4 approximation for MSR; we also show that the problem becomes intractable when the number of genomic maps is increased from two to three. In Section 2 and Section 3, we prove the following two theorems:

Theorem 1. *There is a factor-4 polynomial-time approximation algorithm for MSR.*

Theorem 2. *MSR-3 is NP-complete.*

1.1 Weight Constraint on Markers

When building genomic maps, a priori information about the gene markers can be derived from comparative analysis. For example, certain genes that are responsible for important genetic functions in several close species can often be identified. It is reasonable to give the corresponding gene markers larger weights. Denote by *MSR-WT* the problem MSR with the additional weight constraint WT:

WT: The total weight of markers in the strips is between two positive integers w_1 and w_2.

In Section 4, we prove the following theorem:

Theorem 3. *MSR-WT is NP-complete.*

1.2 Number of Non-breaking Points as Score Function

A careful reader will notice that our definition of the problem MSR-2 is slightly different from the original definition of the problem Maximal Strip Recovery (MSR) [16, Page 517]: we require a minimum length of two for each strip, i.e., each strip must contain at least two distinct markers. We believe this requirement is indeed necessary (but overlooked) in the original definition. Indeed, the problem would become trivial otherwise: if a strip can have length one, then we simply need to count the number of common markers, which can be done in $O(n \log n)$ time by sorting.

$$G_1 = \langle 1, 2, 3, 4, 5, 6, 7, 8, 9, 10, 11, 12 \rangle$$
$$G_2 = \langle 1, 2, -4, 9, -3, 5, 10, 6, -8, 11, -7, 12 \rangle$$
$$G_1' = \langle 1, 2, 3, 4, 5, 6, 7, 8 \rangle$$
$$G_2' = \langle 1, 2, -4, -3, 5, 6, -8, -7 \rangle$$
$$G_1'' = \langle 1, 2, 9, 10, 11, 12 \rangle$$
$$G_2'' = \langle 1, 2, 9, 10, 11, 12 \rangle$$

Fig. 2. An example for the two score functions

An alternative way to avoid such short strips is to define the score of each strip in a different way: instead of using the strip length as the score, we can use the number of non-breaking points (or adjacencies) in a strip [3,15]. Then a strip of length l will have a score of $l - 1$ when $l \geq 2$, and a zero score when $l = 1$. We believe that, in practice, the two score functions can lead to different levels of effectiveness of the strip recovery algorithm. We refer to Fig. 2 for an example. For the two subsequences G_1' and G_2', the total strip length (there are four strips) is 8 and the number of non-breaking points is 4. For the two subsequences G_1'' and G_2'', the total strip length (there is only one strip) is 6 and the number of non-breaking points is 5. In the biological context, both solutions may be desirable.

Denote by *MSR-NB* the problem MSR with the alternative score function NB:

NB: The score function uses the number of non-breaking points instead of the strip length.

Also define two problems MSR-NB-3 and MSR-NB-WT, analogous to the two problems MSR-3 and MSR-WT. The difference between the two score functions, the strip length and the number of non-breaking points, has led to contrasting computational complexities (NP-hard versus polynomially solvable) of another biological problem [12,10]. For our three problems MSR-NB, MSR-NB-3, and MSR-NB-WT, however, we can obtain results similar to those for MSR, MSR-3, and MSR-WT. We have the following three theorems (the proofs are omitted to avoid repeating trivial technical details):

Theorem 4. *There is a factor-4 polynomial-time approximation algorithm for MSR-NB.*

Theorem 5. *MSR-NB-3 is NP-complete.*

Theorem 6. *MSR-NB-WT is NP-complete.*

1.3 Duplicate Markers

In our definition of MSR-d, each marker appears exactly once in each genomic map and hence at most once in the strips. In the biological context, however, duplicate markers are rare but still possible, as so-called *paralogy set*: "Two or more strips may contain exactly the same markers but differ in where they appear, by virtue of the paralogy sets their markers belong to." [16, Page 516].

Define the following variation DU of the problem MSR:

DU: Duplicate markers are allowed in the genomic maps and the strips.

Note that although duplicate markers may appear in the strips, each strip in itself does not contain duplicate markers since it is a sequence of *distinct* markers. That is, duplicate markers may only appear in different strips.

Denote by *MSR-NB-WT-DU* the problem MSR with the score function NB, the weight constraint WT, and the variation DU. It is not surprising that recovering syntenic blocks becomes harder when duplicate markers are allowed. Recognizing the similarity between the problem MSR-NB-WT-DU and the recently studied problem Exemplar Non-breaking Similarity [3], we prove the following theorem:

Theorem 7. *MSR-NB-WT-DU is W[1]-complete. Moreover, unless P equals NP, MSR-NB-WT-DU does not admit any approximation of factor $n^{1-\epsilon}$ for some ϵ, $0 < \epsilon < 1$.*

Several more variants of the problem MSR can be defined analogously, in particular, MSR-DU and MSR-NB-DU. Our factor-4 approximations for MSR and MSR-NB, with slight modifications, also hold for MSR-DU and MSR-NB-DU; however, the complexities of MSR and its several variants remain unknown. In Section 6, we conclude with two open problems.

2 A Factor-4 Approximation for MSR

We prove Theorem 1 in this section. A heuristic for the problem Maximal Strip Recovery (MSR) was previously proposed [6,16]. This simple heuristic works as follows:

1. Extract a set of pre-strips from the two sequences;
2. Compute an independent set of strips from the pre-strips.

This approach is inefficient because the number of pre-strips could be exponential in the sequence length, and furthermore the problem Maximum Weight Independent Set (MWIS) is NP-hard.

Our factor-4 approximation algorithm for MSR is slightly modified from the previous heuristic [6,16]. For a sequence S and two indices i and j, denote by $S[i,j]$ the substring of S starting at i and ending at j. The algorithm works as follows:

1. Compose $O(n^4)$ 2-intervals, one for each pair of substrings of the two genomic maps G_1 and G_2. For each 2-interval with indices i_1 and j_1 in G_1 and indices i_2 and j_2 in G_2, assign it a weight equal to the maximum length of a common subsequence (may be reversed and negated) of the two substrings $G_1[i_1, j_1]$ and $G_2[i_2, j_2]$.

2. Compute a 4-approximation for MWIS in the intersection graph of the 2-intervals using the fractional local-ratio algorithm for split interval graphs [1].

This completes the proof of Theorem 1.

We note that the problem MWIS in 2-interval graphs is also known as the problem *2-Interval Pattern* [14], which has been extensively studied [1,2,5,7,9,10,11,14] because of its application to RNA secondary structure prediction.

3   MSR-3 Is NP-Complete

We prove Theorem 2 in this section. It is clear that MSR-3 is in NP. We show that MSR-3 is NP-hard by a reduction from the NP-hard problem Separated 2-Interval Pattern [1]. Let L_1 and L_2 be two parallel lines. Denoted by $D = (I, J)$ a *separated 2-interval* that is the union of two closed intervals $I \subset L_1$ and $J \subset L_2$. Given a set of n separated 2-intervals, the problem *Separated 2-Interval Pattern* is to find a maximum independent set in the corresponding intersection graph. By a standard technique in interval graphs, we can assume without loss of generality that the $4n$ endpoints of the $2n$ intervals of the n separated 2-intervals are distinct.

Our construction uses $2n^2 + 2n$ distinct markers: $2n^2$ *peg markers* $p_{i,j}$ for $1 \leq i \leq n$ and $1 \leq j \leq 2n$, and $2n$ *interval markers* u_i and v_i for $1 \leq i \leq n$.

Use the $2n^2$ peg markers to construct n pairs of *peg sequences* P_i and Q_i of equal length $2n$, for $1 \leq i \leq n$:

$$P_i = p_{i,1} \ldots p_{i,2n},$$
$$Q_i = -p_{i,2n} \ldots - p_{i,1}.$$

Let S be the sequence $u_1 v_1 \ldots u_n v_n$ of $2n$ interval markers. For each separated 2-interval $D_i = (I_i, J_i)$, label the two left endpoints of I_i and J_i with the marker u_i, and label the two right endpoints of I_i and J_i with the marker v_i. Then the $2n$ markers for the $2n$ endpoints of the n intervals I_i, ordered along the line L_1, is a permutation S_1 of S; similarly, the $2n$ markers for the $2n$ endpoints of the n intervals J_i, ordered along the line L_2, is another permutation S_2 of S.

Construct three genomic maps:

$$G_0 = u_1 v_1 P_1 \ldots u_n v_n P_n,$$
$$G_1 = S_1 Q_1 \ldots Q_n,$$
$$G_2 = S_2 Q_1 \ldots Q_n.$$

Note that each genomic map is a signed permutation of the $2n^2 + 2n$ distinct markers. We will show that the set of n separated 2-intervals has an independent set of size at least

k if and only if G_1, G_2, and G_3 have three subsequences G_1', G_2', and G_3', respectively, with a total strip length of at least $2n^2 + 2k$.

We note the following important property of our construction:

Proposition 1. *If a strip of G_1', G_2', and G_3' contains a peg marker $p_{i,j}$, then it does not contain any interval marker or any peg marker $p_{i',j'}$ such that $i' \neq i$.*

We first prove the "only if" direction. Suppose that the set of n separated 2-intervals has an independent set of size k, that is, there are k disjoint separated 2-intervals $D_{i_1}, \ldots, D_{i_k}$. Let G_1', G_2', and G_3', respectively, be the subsequences of G_1, G_2, and G_3 that contain the $2n^2$ peg markers and the $2k$ interval markers $u_{i_1}, v_{i_1}, \ldots, u_{i_k}, v_{i_k}$. Then G_1', G_2', and G_3' have $n + k$ strips: the n strips $P_1, \ldots, P_n$, each of length $2n$, and the k strips $u_{i_1} v_{i_1}, \ldots, u_{i_k} v_{i_k}$, each of length 2. The total strip length is $2n^2 + 2k$.

We next prove the "if" direction. Suppose that G_1, G_2, and G_3 have three subsequences G_1', G_2', and G_3', respectively, with a total strip length of at least $2n^2 + 2k$, $k > 0$. We say that a sequence S *contributes* to a strip R if R contains a marker in S. If a strip contains two interval markers for two different separated 2-intervals, then the two interval markers would enclose a peg sequence P_i in the genomic map G_0. Hence, by Proposition 1, the peg sequence P_i would not contribute to any strip. Note that the total length of the strips is at most the length of each genomic map, which is $2n^2 + 2n$. Also note that the length of each peg sequence is $2n$. If the peg sequence P_i does not contribute to any strip, then the total length of the strips would be at most $2n^2$, which is less than $2n^2 + 2k$, a contradiction. Therefore, if a strip contains an interval marker u_i or v_i of a separated 2-interval D_i, then the strip must contain only the two interval markers u_i and v_i. The total length of strips of peg markers is at most the total length of the n peg sequences, which is $2n^2$. The remaining strip length of at least $2k$ must come from at least k strips of interval markers, which correspond to an independent set of k separated 2-intervals.

The reduction time is clearly polynomial in the size of the Separated 2-Interval Pattern instance. This completes the proof of Theorem 2.

4   MSR-WT Is NP-Complete

We prove Theorem 3 in this section. It is easy to see that MSR-WT is in NP. We show that MSR-WT is NP-hard by a reduction from the NP-hard problem One-In-Three 3SAT [13]. Let $\phi = f_1 \wedge f_2 \wedge \ldots \wedge f_m$ be a boolean formula with m clauses in conjunctive normal form, with n variables $x_1, x_2, \ldots, x_n$. Each clause f_i is the disjunction of exactly three literals, like $(x_2 \vee x_5 \vee \bar{x}_7)$. We will construct two genomic maps G and H and show that ϕ is one-in-three satisfiable (i.e., each clause has exactly one true literal) if and only if G and H have two subsequences G' and H', respectively, such that

1. The total length of the strips in G' and H' is at least some integer ℓ;
2. The total weight of the markers in G' and H' is equal to some integer w (note that we set $w_1 = w_2 = w$ in this proof).

Our construction uses $3m + n$ distinct markers: $3m$ clause markers

$$f_{1,1}, f_{1,2}, f_{1,3}, \ldots, f_{m,1}, f_{m,2}, f_{m,3},$$

and n peg markers

$$g_1, \ldots, g_n.$$

For each clause f_i, label its three literals with the three clause markers $f_{i,1}, f_{i,2}, f_{i,3}$. For each variable x_i, let F_i and $\bar{F}_i$, respectively, be the two sequences of markers for the literals x_i and $\bar{x}_i$:

$$F_i = f_{i_1,j_1} \ldots f_{i_u,j_u},$$
$$\bar{F}_i = f_{i'_1,j'_1} \ldots f_{i'_v,j'_v}.$$

Put

$$S_i = F_i \bar{F}_i,$$
$$T_i = \bar{F}_i F_i.$$

Then construct two genomic maps

$$G = S_1 g_1 \ldots S_n g_n,$$
$$H = T_1 g_1 \ldots T_n g_n.$$

Assign each peg marker g_i a weight that is a decimal number with a one followed by $n - i$ zeros. Assign each clause marker for f_i a weight that is a decimal number with a one followed by $m + n - i$ zeros. Then set the threshold weight w to a decimal number with $m + n$ consecutive ones, and set the threshold strip length ℓ to $m + n$. Note that each genomic map is a permutation of the $3m + n$ markers. Also note that w is exactly the total weight of m clause markers, one for each clause, and the n peg markers.

We first prove the "only if" direction. Let $x_1 = b_1, \ldots, x_n = b_n$ be a one-in-three truth assignment that satisfies ϕ. For each i, obtain two subsequences S'_i and T'_i, respectively, from the two sequences S_i and T_i:

$$S'_i = T'_i = \begin{cases} F_i & \text{if } b_i = \text{true}, \\ \bar{F}_i & \text{if } b_i = \text{false}. \end{cases}$$

Then let

$$G' = S'_1 g_1 \ldots S'_n g_n,$$
$$H' = T'_1 g_1 \ldots T'_n g_n.$$

There is only one strip in G' and H' since $G' = H'$. By the definition of one-in-three truth assignment, each clause f_i contains exactly one true literal, hence exactly one of the three clause markers $f_{i,1}, f_{i,2}, f_{i,3}$ appears in G' (H'). Therefore the total strip length is exactly ℓ, and the total weight of markers is exactly w.

We next prove the "if" direction. Let G' and H' be two subsequences of G and H, respectively, with total strip length at least ℓ and total marker weight exactly w. The weight condition ensures that exactly one of the three markers for each clause appears in G' and H'. The strip length condition then implies that there is only one strip of length exactly ℓ. Indeed $G' = H'$. The distribution of the clause markers among the "buckets" S'_i and T'_i corresponds to a one-in-three truth assignment for ϕ.

The reduction time is clearly polynomial in the length of ϕ. This completes the proof of Theorem 3.

5   Complexity of MSR-NB-WT-DU

We prove Theorem 7 in this section. We achieve this by showing that the problem MSR-NB-WT-DU contains another difficult problem as a special case. Given a genome $\mathcal{G}$, which is a sequence of genes possibly with duplicates, a genome $\mathcal{G}'$ is *exemplar* of $\mathcal{G}$ if $\mathcal{G}'$ contains the same set of genes as $\mathcal{G}$ does but has no duplicates. Given two genomes $\mathcal{G}$ and $\mathcal{H}$, the problem *Exemplar Non-Breaking Similarity* (ENBS) [3] is to compute two exemplar genomes $\mathcal{G}'$ and $\mathcal{H}'$ such that the number of non-breaking points between $\mathcal{G}'$ and $\mathcal{H}'$ is maximized.

Let $g_1, \ldots, g_n$ be the n distinct genes in the two genomes $\mathcal{G}$ and $\mathcal{H}$. Use n distinct markers, one marker g'_i for each gene g_i, $1 \leq i \leq n$, and construct two genomic maps G and H of markers corresponding to the two genomes $\mathcal{G}$ and $\mathcal{H}$ of genes. The optimization goal of maximizing non-breaking similarity in ENBS corresponds to exactly the score function NB in MSR-NB-WT-DU. To ensure that each gene g_i appears exactly once in $\mathcal{G}'$ and in $\mathcal{H}'$, assign each gene marker g'_i a special weight that is a decimal number with a one followed with $n - i$ zeros, then set the threshold weights w_1 and w_2 to an integer that is a decimal number with n ones.

ENBS is a very difficult problem [3]: (i) ENBS is W[1]-complete; (ii) even if each of the n genes appears exactly once in $\mathcal{G}$ and at most twice in $\mathcal{H}$, ENBS still cannot be approximated within a factor of $n^{1-\epsilon}$ for some ϵ, $0 < \epsilon < 1$, unless P equals NP. Since ENBS is a special case of MSR-NB-WT-DU, these lower bounds automatically apply to MSR-NB-WT-DU. This completes the proof of Theorem 7.

We note that the W[1]-completeness [8] of MSR-NB-WT-DU has the following implication [4]: Let p be the optimal solution value for MSR-NB-WT-DU. Then, unless an unlikely collapse occurs in the parameterized complexity theory, MSR-NB-WT-DU is not solvable in time $f(p)n^{o(p)}$ for any function f.

6   Open Problems

We conclude the paper with two open problems:

1. Are the four problems MSR (MSR-2), MSR-NB, MSR-DU, and MSR-NB-DU NP-complete? (We conjecture that at least MSR-DU and MSR-NB-DU are NP-complete.)
2. Are the two problems MSR-WT and MSR-NB-WT-DU still intractable with only one-sided weight constraint, i.e., the total weight of the strips is at least w_1? (Note that our proofs of Theorem 3 and Theorem 7 use the property that the weight constraint is from both directions.)

References

1. Bar-Yehuda, R., Halldórsson, M.M., Naor, J.(S.), Shachnai, H., Shapira, I.: Scheduling split intervals. SIAM Journal on Computing 36, 1–15 (2006)
2. Blin, G., Fertin, G., Vialette, S.: Extracting constrained 2-interval subsets in 2-interval sets. Theoretical Computer Science 385, 241–263 (2007)
3. Chen, Z., Fu, B., Yang, B., Xu, J., Zhao, Z., Zhu, B.: Non-breaking similarity of genomes with gene repetitions. In: Ma, B., Zhang, K. (eds.) CPM 2007. LNCS, vol. 4580, pp. 119–130. Springer, Heidelberg (2007)
4. Chen, J., Huang, X., Kanj, I., Xia, G.: Linear FPT reductions and computational lower bounds. In: Proceedings of the 36th ACM Symposium on Theory of Computing (STOC 2004), pp. 212–221 (2004)
5. Chen, E., Yang, L., Yuan, H.: Improved algorithms for largest cardinality 2-interval pattern problem. Journal of Combinatorial Optimization 13, 263–275 (2007)
6. Choi, V., Zheng, C., Zhu, Q., Sankoff, D.: Algorithms for the extraction of synteny blocks from comparative maps. In: Proceedings of the 7th International Workshop on Algorithms in Bioinformatics (WABI 2007), pp. 277–288 (2007)
7. Crochemore, M., Hermelin, D., Landau, G.M., Rawitz, D., Vialette, S.: Approximating the 2-interval pattern problem. In: Brodal, G.S., Leonardi, S. (eds.) ESA 2005. LNCS, vol. 3669, pp. 426–437. Springer, Heidelberg (2005)
8. Downey, R., Fellows, M.: Parameterized Complexity. Springer, Heidelberg (1999)
9. Jiang, M.: A 2-approximation for the preceding-and-crossing structured 2-interval pattern problem. Journal of Combinatorial Optimization 13, 217–221 (2007)
10. Jiang, M.: Improved approximation algorithms for predicting RNA secondary structures with arbitrary pseudoknots. In: Kao, M.-Y., Li, X.-Y. (eds.) AAIM 2007. LNCS, vol. 4508, pp. 399–410. Springer, Heidelberg (2007)
11. Jiang, M.: A PTAS for the weighted 2-interval pattern problem over the preceding-and-crossing model. In: Dress, A.W.M., Xu, Y., Zhu, B. (eds.) COCOA. LNCS, vol. 4616, pp. 378–387. Springer, Heidelberg (2007)
12. Lyngsø, R.B.: Complexity of pseudoknot prediction in simple models. In: Díaz, J., Karhumäki, J., Lepistö, A., Sannella, D. (eds.) ICALP 2004. LNCS, vol. 3142, pp. 919–931. Springer, Heidelberg (2004)
13. Schaefer, T.: The complexity of satisfiability problem. In: Proceedings of the 10th ACM Symposium on Theory of Computing (STOC 1978), pp. 216–226 (1978)
14. Vialette, S.: On the computational complexity of 2-interval pattern matching problems. Theoretical Computer Science 312, 223–249 (2004)
15. Watterson, G., Ewens, W., Hall, T., Morgan, A.: The chromosome inversion problem. Journal of Theoretical Biology 99, 1–7 (1982)
16. Zheng, C., Zhu, Q., Sankoff, D.: Removing noise and ambiguities from comparative maps in rearrangement analysis. IEEE/ACM Transactions on Computational Biology and Bioinformatics 4, 515–522 (2007)

Automatic Generation of Symmetry-Breaking Constraints

Leo Liberti

LIX, École Polytechnique, F-91128 Palaiseau, France
`liberti@lix.polytechnique.fr`

Abstract. Solution symmetries in integer linear programs often yield long Branch-and-Bound based solution processes. We propose a method for finding elements of the permutation group of solution symmetries, and two different types of symmetry-breaking constraints to eliminate these symmetries at the modelling level. We discuss some preliminary computational results.

1 Introduction

We consider a Mixed Integer Linear Program (MILP) in the following form:

$$\left.\begin{array}{r} \min c^\top x \\ Ax \leq b \\ x \in [x^L, x^U] \\ \forall i \in Z \quad x_i \in \mathbb{Z}. \end{array}\right\} \tag{1}$$

where $c, x, x^L, x^U \in \mathbb{R}^n$, $b \in \mathbb{R}^m$, A is a real $m \times n$ matrix and $Z \subseteq \{1, \ldots, n\}$. Throughout the paper, elements of groups are represented by means of permutations of either the column or the row space; permutations on the row space are denoted by left multiplication, and permutations on the column space by right multiplication. Because a solution x of (1) has as many elements as the columns of A, a permutation π on x is likened to a permutation of the column space, and hence denoted by right multiplication $x\pi$.

Problems (1) having many symmetries are known to be very difficult to solve to global optimality with Branch-and-Bound (BB) techniques. These converge slowly in presence of symmetries because many leaf nodes in the BB tree may contain (symmetric) global optima: hence, no node in the paths leading from the root to these leaf nodes can ever be pruned. Despite the practical difficulties given by solution symmetries, there are relatively few group theory based methods in mathematical programming. These may be classified in three broad categories: (a) the abelian group approach proposed by Gomory to writing integer feasibility conditions for x; (b) symmetry-breaking techniques for specific problems, whose symmetry group can be computed in advance; (c) general-purpose symmetry group computations and symmetry-breaking techniques implemented via branching strategies and local cuts in a typical BB solution algorithm.

Category (a) was established by R. Gomory [7]: given a basis B of the constraint matrix A, it considers the (abelian) group $\mathscr{G} = \mathbb{Z}^n / \langle \text{col}(B) \rangle$, where $\mathbb{Z}^n$ is

B. Yang, D.-Z. Du, and C.A. Wang (Eds.): COCOA 2008, LNCS 5165, pp. 328–338, 2008.

the additive group of integer n-sequences and $\langle \mathrm{col}(B) \rangle$ is the additive group generated by the columns of the (nonsingular) matrix B. We consider the natural group homomorphism $\varphi : \mathbb{Z}^n \to \mathcal{G}$ with $\ker \varphi = \langle \mathrm{col}(B) \rangle$: letting (x_B, x_N) be a basic/nonbasic partition of the decision variables, we apply φ to the standard form constraints $Bx_B + Nx_N = b$ to obtain $\varphi(Bx_B) + \varphi(Nx_N) = \varphi(b)$. Since $\varphi(Bx_B) = 0$ if and only if $x_B \in \mathbb{Z}^n$, setting $\varphi(Nx_N) = \varphi(b)$ is a necessary and sufficient condition for x_B to be integer feasible. Gomory's seminal paper gave rise to further research, among which [22,1]

Category (b) is possibly the richest in terms of number of published papers. Many types of combinatorial problems exhibit a certain amount of symmetry. Symmetries are usually broken by means of specific branching techniques (e.g. [16]), appropriate global cuts (e.g. [21]) or special formulations [11,2] based on the problem structure. The main limitation of the methods in this category is that they are difficult to generalize and/or to be rendered automatic.

Category (c) contains two main research streams. The first was established by F. Margot in the early 2000s [14,15], and is applicable to problems in general form (1) where $x^L = 0, x^U = 1$, i.e. Binary Linear Programs (BLPs). Margot defines the *symmetry group* of a BLP as:

$$\{\pi \in S_n \mid c^\top \pi = c^\top \wedge \exists \sigma \in S_n \, (\sigma b = b \wedge \sigma A \pi = A)\}, \qquad (2)$$

or, in other words, all relabellings of problem variables for which the objective function and constraints are the same. The symmetry group (2) is used to derive effective BB pruning strategies by means of isomorphism pruning and isomorphism cuts local to some selected BB tree nodes (Margot extended his work to general integer variables in [17]). Stronger results of the same type can be obtained for covering and packing problems [20], for these have an objective function vector $c = (1, \ldots, 1)$ and a RHS vector $b = (1, \ldots, 1)$ fixed by all elements of S_n and S_m respectively, and their constraint matrix is 0-1.

The second was established by V. Kaibel and M. Pfetsch in 2007 [9]. Symmetries in the column space (i.e. permutations of decision variables) of binary ILPs having 0-1 constraint matrices are shown to affect the quality of the linear programming bound. Limited only to permutations in cyclic and symmetric group, complete descriptions of *orbitopes* are provided by means of linear inequalities. Let x' be a point in $\{0,1\}^n$ (the solution space), with $n = pq$, so that we can arrange the components of x' in a matrix C. Given a group G and $\pi \in G$, for all 0-1 $p \times q$ matrices C let $C\pi$ be the matrix obtained by permuting the columns of C according to π. Let $G \cdot C$ be the orbit of C under the action of all $\pi \in G$, $\overline{G \cdot C}$ be the lexicographically maximal matrix in $G \cdot C$ (ordering matrices by rows first) and $\mathcal{M}_{pq}^{\max}(G)$ be the set of all $\overline{G \cdot C}$. Then the *full orbitope* associated with G is $\mathrm{conv}(\mathcal{M}_{pq}^{\max}(G))$. Inspired by the work on orbitopes, E. Friedman very recently proposed a similar but extended approach leading to *fundamental domains* [5]: given a feasible polytope $X \subseteq [0,1]^n$ with integral extreme points and a group G acting as an affine transformation on X (i.e. for all $\pi \in G$ there is a matrix $A \in GL(n)$ and an n-vector d such that $\pi x = Ax + d$ for all $x \in X$), a fundamental domain is a subset $F \subset X$ such that $GF = X$.

The Constraint Programming (CP) community is also concerned with symmetries and some of the results in the CP literature can be extended to mathematical programming (see [3] for a good introduction).

The present work belongs to category (c): it proposes general-purpose methods for identifying (some) solution symmetries and restrict the feasible region so that it does not contain all representatives per equivalence class. This paper contributes two ideas: (i) breaking the same type of symmetries described in [16,14,15,17] at the modelling instead of the algorithmic level (whereas Margot proposes symmetry breaking methods local to each node of the BB tree, we discuss global symmetry breaking constraints which can be added to the original formulation); and (ii) computing symmetries automatically instead of assuming them as given. In Section 2 we propose symmetry breaking constraints derived from cycles; Section 3 describes a mathematical program whose solution encodes a permutation of the symmetry group; in Section 4 we discuss some practical strategies for exploiting the proposed symmetry breaking constraints and some preliminary computational results.

1.1   Notation

For a group $G \leq S_n$ and a set X of row vectors, $XG = \{xg \mid x \in X \wedge g \in G\}$; if Y is a set of column vectors, $GY = \{gy \mid y \in Y \wedge g \in G\}$. If $X = \{x\}$, we denote XG by xG (and similarly for Y). For a mathematical program P we let $\mathcal{F}(P)$ be the feasible region of P and $\mathcal{G}(P)$ be the set of (global) optima of P. For $x \in \mathbb{R}^n$, we let $\text{ran}(x) = \{a \in \mathbb{R} \mid \exists j \leq n \ (x_j = a)\}$ be the range of x. All groups considered in this paper are finite.

2   Theoretical Results

2.1   Efficiency of Symmetry Breaking Constraints

We let S_n be the symmetric group of order $n \in \mathbb{N}$. For a set $X \subseteq \mathbb{R}^n$, a group $G \leq S_n$ and $x, y \in X$, we define an equivalence relation $x \sim_G y \Leftrightarrow \exists \pi \in G \ (x\pi = y)$. The relation $\sim_G$ partitions X into a set $\mathcal{E}(G, X)$ of equivalence classes (each of finite cardinality) such that $X = \bigcup_{Y \in \mathcal{E}(G,X)} Y$ (the cardinality of $\mathcal{E}(G, X)$ itself need not be finite or even countable).

Definition 2.1

A linear constraint $dx \leq d_0$ with $(d, d_0) \in \mathbb{R}^{n+1}$ is symmetry breaking with respect to G and X if for all $Y \in \mathcal{E}(G, X)$ there are $\bar{x} \neq \bar{y} \in Y$ s.t. $d\bar{x} \leq d_0$ and $d\bar{y} > d_0$. The constraint is symmetry breaking of order ℓ if there is at least one equivalence class $Y \in \mathcal{E}(G, X)$ in which there are exactly $\ell - 1$ points y s.t. $d\bar{y} > d_0$. The constraint is maximally symmetry breaking if there is at least one equivalence class $Y \in \mathcal{E}(G, X)$ for which it is symmetry breaking of order $|Y|$.

Defn. 2.1 can easily be extended to systems of constraints. Supposing $X \subseteq \mathbb{Z}^n$, symmetry breaking constraints are not in general valid cuts for any linear polyhedron containing X, because they may also cut off some integral points. However, they guarantee feasibility of at least one integral point per equivalence class. Adding appropriate symmetry breaking constraints to P results in a reformulation of the *narrowing* type [13], i.e. a reformulation Q of P with a map $\psi : \mathcal{F}(Q) \rightarrow \mathcal{F}(P)$ such that $\psi(\mathcal{G}(Q)) \subseteq \mathcal{G}(P)$ [12]. Notice that symmetry breaking constraints of order $\ell \leq 1$ do not break any symmetry at all, as they do not separate any point in any equivalence class of $\mathcal{E}(G, X)$.

2.2   Symmetry Groups Associated to a MILP

We consider symmetries that leave various properties of P invariant.

Definition 2.2
The set

$$G^* = \{\pi \in S_n \mid \forall x \in \mathcal{G}(P) \, (x\pi \in \mathcal{G}(P))\} \tag{3}$$

of automorphisms of $\mathcal{G}(P)$ is called the solution symmetry group of P. The set

$$\tilde{G} = \{\pi \in S_n \mid \forall x \in \mathcal{F}(P) \, (x\pi \in \mathcal{F}(P))\} \tag{4}$$

of automorphisms of $\mathcal{F}(P)$ is called the feasible symmetry group of P

It is easy to show that $\tilde{G}, G^*$ are both subgroups of S_n, and that $G^* \leq \tilde{G}$. Next, we extend (2) to formulation (1).

Definition 2.3
The set

$$G_P = \{\pi \in S_n \mid \forall a \in \{c, x^L, x^U\}$$
$$(a^\top \pi = a^\top) \wedge Z\pi = Z \wedge \exists \sigma \in S_n \, (\sigma b = b \wedge \sigma A\pi = A)\} \tag{5}$$

of permutations that fix the problem formulation is called the problem symmetry group of P.

It is equally easy to show that G_P is a subgroup of S_n. The following useful result states that problem symmetries are solution symmetries.

Proposition 2.4
$G_P \leq G^*$.

2.3   Symmetry Breaking Constraints from Disjoint Cycles

Let R be the relaxation of P obtained by removing the constraints $Ax \leq b$, $\bar{X} = \mathcal{F}(R)$ and $X^* = \mathcal{G}(R)$. Let $\sigma = (\sigma_1, \dots, \sigma_k)$ be a cycle of length $1 < k \leq n$ in S_n. For any $x \in \mathbb{R}^n$, let $x[\sigma] = (x_{\sigma_1}, \dots, x_{\sigma_k})$, and assume that $x[\sigma]$ are constrained to be integer. Let $\bar{G} \leq S_n$ be the group of all permutation automorphisms of $\bar{X}$. If $\sigma \in \bar{G}$, for all $x \in \bar{X}$ and $j \leq k$ we have $x_{\sigma_1}^L \leq x[\sigma]_j \leq x_{\sigma_1}^U$, i.e. there is a unique number of values χ that all variables in $x[\sigma]$ can take. We let $\bar{\chi}$ be the row vector whose j-th component is χ^{k-j} for all $j \leq k$. Consider the following constraints, often cited in the literature for symmetry breaking purposes [21,10]:

$$\forall 1 \leq h \leq k-1 \quad \bar{\chi}(x[\sigma] - x^L[\sigma]) \leq \bar{\chi}(x[\sigma] - x^L[\sigma])\sigma^h \tag{6}$$

$$\Rightarrow \sum_{j=1}^{k} \chi^{k-j}(x_{\sigma_j} - x_{\sigma_j}^L) \leq \sum_{j=1}^{k} \chi^{k-j}(x_{\sigma^h(\sigma_j)} - x_{\sigma^h(\sigma_j)}^L)).$$

Proposition 2.5
Let $\sigma = (\sigma_1, \ldots, \sigma_k)$ be a cycle of length $k \leq n$ in $\bar{G}$. Then constraints (6) are maximally symmetry breaking w.r.t. $\langle \sigma \rangle$, $\bar{X}$.

The practical trouble with (6) is their well-known poor scaling, as the values of the coefficients are of different orders of magnitude.

Next, we consider some well-scaled (though less effective) symmetry breaking constraints.

Proposition 2.6
Let $\sigma = (\sigma_1, \ldots, \sigma_k)$ be a cycle of length $k \leq n$ in $\bar{G}$. For all $x \in \bar{X}$ with $|\mathrm{ran}(x[\sigma])| = \ell$,

$$\forall 2 \leq j \leq k \quad x_{\sigma_1} \leq x_{\sigma_j} \tag{7}$$

are symmetry breaking constraints of order ℓ.

Notice that if $\ell = k$, then by Prop. 2.6 (7) are symmetry-breaking constraints of order k; furthermore, if $\ell = k$ the vector $x \in \mathbb{R}^n$ having distinct components $x[\sigma]$ gives rise to an equivalence class $x\langle\sigma\rangle$ of cardinality k, so (7) are maximally symmetry breaking. If $\ell = 1$ then (7) do not break any symmetry (remark after Defn. 2.1) but then again if $\ell = 1$ it means that $x[\sigma] = (a, \ldots, a)$ for some $a \in \mathbb{R}$, so $|x\langle\sigma\rangle| = 1$, which means there are no symmetric solutions complicating the solution process. The most likely case is that $x[\sigma] \in \{0,1\}^k$ and $\ell = 2$: this is unfortunate as this situation provides the weakest case of Prop. 2.6. We stress, however, that symmetry breaking constraint of order ℓ will cut away *at least* (not *exactly*) $\ell - 1$ symmetric solutions.

Example 2.7
Let $x = (0,1,1,1)$, and $\sigma = (1,2,3,4)$. Then $x\langle\sigma\rangle = \{(0,1,1,1), (1,0,1,1), (1,1,0,1), (1,1,1,0)\}$. Since $|\mathrm{ran}(x)| = |\{0,1\}| = 2$, constraints (7) are symmetry breaking of order 2. However, exactly 3 elements of $x\langle\sigma\rangle$ are cut off by (7) (i.e. all elements of $x\langle\sigma\rangle \setminus \{x\}$). Taking $x = (0,0,0,1)$, on the other hand, results in (7) only cutting off $x\sigma = (1,0,0,0)$ according to Prop. 2.6.

The main insight given by Example 2.7 is that if we make the assumption that optimal solutions of binary problems will contain on average as many 0s as 1s on components indexed by σ, we can expect (7) to cut away $\lfloor k/2 \rfloor$ symmetric solutions even though $\ell = 2$. Another insight is that if we suspect optimal solutions to have a large number of components attaining small values of the range, we might want to change the $\leq$ relation in (7) to a $\geq$ relation to increase the number of cut-off symmetric solutions (modifying the inequality relation in (7) to $x_{\sigma_1} \geq x_{\sigma_j}$ only requires a trivial change to the proof).

3   Finding Symmetries

Let

$$\hat{G}_P = \{(\sigma, \pi) \in S_m \times S_n \mid \forall a \in \{c, x^L, x^U\}$$
$$(a^\top \pi = a^\top) \wedge Z\pi = Z \wedge (\sigma b = b \wedge \sigma A\pi = A)\}. \tag{8}$$

It is easy to see that the projection of $\hat{G}_P$ on the second component S_n is equal to G_P. For $q \in \mathbb{N}$, Let $\vartheta : S_q \to GL(q)$ be the regular (faithful) permutation matrix representation of elements of S_q, i.e. for $\pi \in S_q$, $\vartheta(\pi)$ is a doubly stochastic invertible matrix with entries in $\{0, 1\}$, such that for any row vector $v \in \mathbb{R}^n$, $v\pi = v\vartheta(\pi)$. We can then write the condition of (8) in terms of products of vectors and matrices.

We consider decision variables: σ_{ih}, the (i, h)-th element of the matrix $\vartheta(\sigma)$ for all $i, h \leq m, \sigma \in S_m$; and π_{jk}, the (j, k)-th element of the matrix $\vartheta(\pi)$ for all $j, k \leq n, \pi \in S_n$. Let $z \in \{0, 1\}^n$ be the indicator vector of Z, such that $z_j = 1 \Leftrightarrow j \in Z$, and let $\Gamma(P)$ be the set of binary values of σ, π defined by the following constraints:

$$\forall j \leq n \quad \sum_{k \leq n} x_k^L \pi_{jk} = x_j^L \quad \wedge \quad \sum_{k \leq n} x_k^U \pi_{jk} = x_j^U \tag{9}$$

$$\forall j \leq n \quad \sum_{k \leq n} c_k \pi_{jk} = c_j \quad \wedge \quad \sum_{k \leq n} z_k \pi_{jk} = z_k \tag{10}$$

$$\forall i \leq m, j \leq n \quad \sum_{h \leq m} \sigma_{ih} A_{hj} \quad = \quad \sum_{k \leq n} A_{ik} \pi_{kj} \tag{11}$$

$$\forall i \leq m \quad \sum_{h \leq m} \sigma_{ih} b_h \quad = \quad b_i \tag{12}$$

$$\forall j \leq n \quad \sum_{k \leq n} \pi_{kj} = 1 \quad \wedge \quad \sum_{k \leq n} \pi_{jk} = 1 \tag{13}$$

$$\forall i \leq m \quad \sum_{h \leq m} \sigma_{ih} = 1 \quad \wedge \quad \sum_{h \leq m} \sigma_{hi} = 1 \tag{14}$$

$$\forall j, k \leq n \quad \pi_{kj} \quad \in \quad \{0, 1\} \tag{15}$$

$$\forall i, h \leq m \quad \sigma_{ih} \quad \in \quad \{0, 1\}. \tag{16}$$

It is easy to show that $\Gamma(P) = \{(\vartheta(\sigma), \vartheta(\pi)) \mid (\sigma, \pi) \in \hat{G}_P\}$. In order to exclude the identity from $\Gamma(P)$, we also add the constraint:

$$\sum_{j \leq n} \pi_{jj} \leq n - 1 \tag{17}$$

Since by Sect. 4 we look for long cycles, we arbitrarily choose an index $j' \leq n$ which is likely to belong to a long cycle in some permutation of G_P (this choice should be based on the block structure of A) and minimize the following objective function, which ensures that we select a permutation moving j':

$$\min \pi_{j'j'}. \tag{18}$$

We call the problem of minimizing (18) subject to (9)-(17) the FEASIBLE PERMUTATION PROGRAM associated to P w.r.t. j', denoted by FPP(P, j').

Proposition 3.1
If FPP(P, j') is infeasible, then $G_P = \{e\}$.

Although solving the full FPP(P, j') may be more CPU-intensive than solving the original problem, various improvements based on the block structure of the constraints in $\Gamma(P)$ are possible. A promising one consists in solving relaxations of the FPP where (11) only fix certain rows or blocks, and verifying later than the solution is valid in the general problem. Computational experience shows that although the linear relaxation of the FPP may be fractional, solutions to the FPP are mostly found at the root node of the CPLEX [8] BB tree after cuts addition.

4   Practical Solution Strategies

Our strategy for solving (1) consists in seeking permutations of G_P having long cycles in their disjoint cycle representation, and add symmetry breaking constraints (6) or (7). In this section we restrict the discussion to the well-scaled constraints (7) but the same ideas can be (and were) applied to (6) too.

Ideally, we would like to be able to add symmetry-breaking constraints (7) for all disjoint cycles in all generators of G_P. This, however, may lead to infeasibility, as Example 4.1 shows.

Example 4.1
Suppose $\mathcal{G}(P) = \{(0, 1, 1, 0), (1, 0, 0, 1)\}$ and $G_P = \{e, (1, 2)(3, 4)\}$. Then both (6) and (7) would imply $x_1 \leq x_2$ and $x_3 \leq x_4$, which are satisfied by no point in $\mathcal{G}(P)$.

The trouble arises because for a cycle σ, constraints (7) arbitrarily decide that x_{σ_1} is the component of $x[\sigma]$ having minimum value. At the modelling level, this is similar to the main drawback of the algorithm proposed in [14]: "the branching variable cannot be chosen freely, but always has to be the non-fixed variable with smallest index" ([15], p. 3-4). However, since at the modelling level there is no knowledge of what variables are fixed, (7) can only be imposed for one single cycle. A promising strategy is that of selecting the longest cycle σ from the set of all disjoint cycles in all permutations of G_P. In general, we can change the arbitrary choice of minimum component for any $i \leq |\sigma|$ (the cycle length), and we denote BREAKSYMM2(P, i) the reformulated problem P with (7) adapted to σ and i as added constraints. It is easy to show that BREAKSYMM2(P, i) is a valid narrowing for all cycles σ and $i \leq |\sigma|$, albeit one that is still subject to an arbitrary choice. We circumvent this by introducing continuous variables $y_i^\sigma \geq 0$ whose value is exactly 0 only if x_{σ_i} is the minimum element of σ, and reformulating (7) as follows:

$$\forall i, j \leq |\sigma|, j \neq i \quad x_{\sigma_i} - x_{\sigma_j} \leq y_i^\sigma \tag{19}$$

$$\sum_{i \leq |\sigma|} y_i^\sigma \leq |\sigma| - 1. \tag{20}$$

Constraints (19) express the fact that there may be indices i for which x_{σ_i} is minimum in $x[\sigma]$, and (20) say that there is at least one such i, thus yielding a narrowing BREAKSYMM2(P) that is independent of of the choice of i. We remark that an aggregated version of (19), when combined with (20), produces a narrowing Q' such that $\mathcal{F}(Q') = \mathcal{F}(\text{BREAKSYMM2}(P))$:

$$\forall i \leq k \quad (|\sigma| - 1)x_{\sigma_i} - \sum_{j \neq i} x_{\sigma_j} \leq (|\sigma| - 1)y_i^{\sigma}. \tag{21}$$

Although in general aggregated constraints tend to produce slacker linear relaxations [19], we mention (21) here because in the tested instances they usually improve CPU times.

Since (19)-(20) together simply express the fact that there is a minimum component in each $x[\sigma]$, and this sentence is true for each disjoint cycle and each permutation in G_P, it follows that (19)-(20) can be added to P for each cycle σ appearing in the set of disjoint cycles over all permutations of G_P, yielding a valid narrowing denoted by BREAKSYMM2ALL(P). This, however, adds several variables and constraints to P, which implies that the size of the solution set increases, and each linear relaxation costs more in terms of CPU time; moreover, although BREAKSYMM2ALL(P) is a valid narrowing of P, the relaxation of BREAKSYMM2ALL(P) need not be strictly tighter than that of P.

4.1 Computational Experiments

The results we report should be treated as preliminary experiments rather than full computational results. Although the whole software structure is in place and the process of finding symmetries and then solutions of MILPs has been made fully automatic (by means of several different software packages, some developed on purpose and some off-the-shelf such as AMPL [4], CPLEX [8] and GAP [6]), finding elements of G_P automatically is still too costly in terms of CPU time. Moreover, because of the small size of problems for which it was possible to automatically compute symmetries, the addition of constraints and variables and consequent higher cost at each BB node offset the advantages of the narrowing as regards CPU time.

We employed the test set as described in Table 1. *Instance* is the instance name, *Source* lists the instance library or citation where the instance appears, *Integers* and *Constraints* report the number of integer variables and constraints respectively, *Size* lists the size of the instance in bytes, and *Infeasible* is 1 if the instance is infeasible. Table 2 reports the results. Column *Group* contains the subgroup of G_P found automatically by repeatedly solving a version of the FPP with random coefficients on the objective function (we call this the "randomized FPP procedure"); the group descriptions are non-unique, as there are many possible semi-direct product types ⋉, see e.g. sts27 and stein27. *Longest* contains the size of the longest cycle in the group generators, which is used to formulate SymmBreak2(P). N is the number of nodes in the BB tree of the original problem P, and N' is the number of nodes in the BB tree of BREAKSYMM2(P). The

Table 1. Test set description

Instance	Source	Integers	Constraints	Size	Infeasible
enigma	[18]	100	21	5616	0
jgt18	[11], p. 413	132	105	9399	1
oa66234	[17], Table 1	64	42	5176	0
oa67233	[17], Table 1	128	64	12153	0
oa76234	[17], Table 1	64	42	5176	0
ofsub9	[11], p. 413	203	92	13157	1
stein27	[18]	27	118	5789	0
sts27	[15], p. 17	27	117	5612	0

Table 2. Results of the computational experiments. The last column refers to SYMMBREAK2(P). The group descriptions were computed by GAP [6].

Instance	Group	Longest	N	N'
enigma	C_2	2	3321	**269**
jgt18	$C_2 \times S_4$	6	**573**	1300
oa66234	S_3	2	0	0
oa67233	$C_2 \times S_4$	6	6	**0**
oa76234	S_3	2	0	0
ofsub9	$C_3 \times S_7$	21	1111044	**980485**
stein27	$((C_3 \times C_3 \times C_3) \ltimes PSL(3,3)) \ltimes C_2$	24	**1084**	1843
sts27	$((C_3 \times C_3 \times C_3) \ltimes PSL(3,3)) \ltimes C_2$	26	1317	**968**

instances were solved by CPLEX 10.1 [8] on one core of a 32 bit Intel Core Duo 1.2GHz with 1.5GB RAM running Linux.

Remarks on the experiments.

- Although the instance set is definitely still too small to draw significant conclusions (work is ongoing to enlarge it), the encouraging result is that there was an improvement on the only really difficult instance (ofsub9): even more so as, it being infeasible, BB performance is poor because there are many fewer prunings than with feasible ones (no upper bounding objective function value is ever available).
- The extensions of Sect. 4 applied to (6) did not yield good results due to the increased constraint matrix density and bad scaling.
- Using the straight version of (6) and (7) (with the arbitrary choice on the chosen longest cycle orbit representative) sometimes decreases N' so that even the CPU times are improved, depending on the arbitrary choice; in such cases, (6) were better than (7) as expected.
- Instance stein27 is like sts27 but with an added cardinality constraint (sum of all variables ≥ 13, a constraint which is inactive on all optimal solutions); this (small) difference in problem formulation caused the randomized FPP procedure to find different permutations (although leading to the same group description, the exact group structure is different) and hence to different performance.

– Sometimes long "easy" cycles are overlooked, such as in the case of oa66234 and oa76234: the group is S_3, yet the cycle only has length 2. This happens because we select disjoint cycles from the group generators instead of the group elements themselves to avoid listing all permutations of a group. At the moment we let GAP select the generators automatically, but an improved implementation should take care of selecting generators having long cycles.

5 Conclusion and Future Work

We proposed an automatic way to compute permutations of the symmetry group of a MILP in general form (1) and derived two types of global symmetry breaking constraints designed to reduce the number of symmetric solutions. We exhibited a few preliminary experimental results indicating a positive trend. Future work will concentrate on: (i) reducing the computational effort taken to find permutations by means of exploitation of the block structure of the MILP constraint matrix; (ii) finding a method to reduce the influence of the arbitrary choice of orbit representative in (6), (7) *not* based on adding variables to the problem; (iii) extend the computational results to a more significant instance test set.

References

1. Bell, D.: Constructive group relaxations for integer programs. SIAM Journal on Applied Mathematics 30(4), 708–719 (1976)
2. Boulle, M.: Compact mathematical formulation for graph partitioning. Optimization and Engineering 5, 315–333 (2004)
3. Cohen, D., Jeavons, P., Jefferson, C., Petrie, K., Smith, B.: Symmetry definitions for constraint satisfaction problems. In: van Beek, P. (ed.) CP 2005. LNCS, vol. 3709, pp. 17–31. Springer, Heidelberg (2005)
4. Fourer, R., Gay, D.: The AMPL Book. Duxbury Press, Pacific Grove (2002)
5. Friedman, E.J.: Fundamental domains for integer programs with symmetries. In: Dress, A.W.M., Xu, Y., Zhu, B. (eds.) COCOA. LNCS, vol. 4616, pp. 146–153. Springer, Heidelberg (2007)
6. The GAP Group. GAP – Groups, Algorithms, and Programming, Version 4.4.10 (2007)
7. Gomory, R.: Some polyhedra related to combinatorial problems. Linear Algebra and Its Applications 2(4), 451–558 (1969)
8. ILOG. ILOG CPLEX 10.1 User's Manual. ILOG S.A., Gentilly, France (2006)
9. Kaibel, V., Pfetsch, M.: Packing and partitioning orbitopes. Mathematical Programming (to appear)
10. Lee, J.: All-different polytopes. Journal of Combinatorial Optimization 6, 335–352 (2002)
11. Lee, J., Margot, F.: On a binary-encoded ILP coloring formulation. INFORMS Journal on Computing 19(3), 406–415 (2007)
12. Liberti, L.: Reformulation techniques in mathematical programming, Thèse d'Habilitation à Diriger des Recherches (November 2007)

13. Liberti, L.: Reformulations in mathematical programming: Definitions. In: Ar- inghieri, R., Cordone, R., Righini, G. (eds.) Proceedings of the 7th Cologne-Twente Workshop on Graphs and Combinatorial Optimization, Crema, Università Statale di Milano (2008)
14. Margot, F.: Pruning by isomorphism in branch-and-cut. Mathematical Program- ming 94, 71–90 (2002)
15. Margot, F.: Exploiting orbits in symmetric ILP. Mathematical Programming B 98, 3–21 (2003)
16. Margot, F.: Small covering designs by branch-and-cut. Mathematical Programming B 94, 207–220 (2003)
17. Margot, F.: Symmetric ILP: coloring and small integers. Discrete Optimization 4, 40–62 (2007)
18. Martin, A., Achterberg, T., Koch, T.: MIPLIB 2003. A library of pure and mixed- Integer programs (2003)
19. Nemhauser, G.L., Wolsey, L.A.: Integer and Combinatorial Optimization. Wiley, New York (1988)
20. Ostrowski, J., Linderoth, J., Rossi, F., Smriglio, S.: Orbital branching. In: Fischetti, M., Williamson, D.P. (eds.) IPCO 2007. LNCS, vol. 4513, pp. 104–118. Springer, Heidelberg (2007)
21. Sherali, H., Smith, C.: Improving discrete model representations via symmetry considerations. Management Science 47(10), 1396–1407 (2001)
22. Wolsey, L.: Group representation theory in integer programming. Technical Report Op. Res. Center 41, MIT (1969)

On the Stable Set Polytope of Claw-Free Graphs

Anna Galluccio, Claudio Gentile, and Paolo Ventura

Istituto di Analisi dei Sistemi ed Informatica, CNR, Viale Manzoni 30, 00185 Roma, Italy
{galluccio,gentile,ventura}@iasi.cnr.it

Abstract. We define the class of *geared (fuzzy) line graphs* as the class of graphs obtained by repeated applications of the extended gear composition to a (fuzzy) line graph H. Using the decomposition theorem for claw-free graphs of Chudnovsky and Seymour [2], we show that this class represents a large subclass of claw-free graphs having stability number at least 4.

We provide a complete linear description of the stable set polytope of geared (fuzzy) line graphs. This result gives a first substantial answer to the longstanding open question of finding a defining linear system for the stable set polytope of claw-free graphs [10].

1 Introduction

Let $P \subseteq \mathbb{R}^n$ be a polyhedron; a linear system $Ax \leq b$ is said to be *defining* for P if $P = \{x \in \mathbb{R}_+^V : Ax \leq b\}$. The *facet defining inequalities* (*facets*, for short) for P are those inequalities that constitute the unique (up to positive multiplications) nonredundant defining linear system of P. Given $c \in \mathbb{R}^n$, the *optimization problem* over (P, c) consists in finding the maximum value of $c^T x$ for $x \in P$. So, finding the defining linear system for P is equivalent to transform the original optimization problem into the linear program $\max\{c^T x : Ax \leq b\}$. Given $x^* \in \mathbb{R}^n$, the *separation problem* over (P, x^*) is to find an inequality valid for all points of P and violated by x^*, or prove that $x^* \in P$. A well-known result of Grötschel, Lovász and Schrijver [10] states that the existence of a polynomial time algorithm to optimize over (P, c) for any $c \in \mathbb{R}^n$ is equivalent to the existence of a polynomial time separation algorithm for (P, x^*) for any $x^* \in \mathbb{R}^n$. In practice, a consequence of this result is that a defining linear system for P may be dynamically determined by solving the optimization problem with respect to different objective functions. Therefore, a largely accepted conjecture in the Mixed-Integer Programming community is that if there exists a polynomial time algorithm for optimize over a polyhedron P, then an explicit description of the defining linear system of P can also be found. Only for very few known problems [12] this conjecture is still open and one of them is the stable set problem for claw-free graphs.

Given a graph $G = (V, E)$ and a vector $w \in \mathbb{Q}_+^V$ of node weights, the *stable set problem* is the problem of finding a set of pairwise nonadjacent nodes *(stable set)* of maximum weight. Let $\alpha(G, w)$ denote the maximum weight of a stable set of G; we refer to $\alpha(G) = \alpha(G, \mathbb{1})$ ($\mathbb{1}$ being the vector of all ones) as the *stability number* of G. The *stable set polytope*, denoted by $STAB(G)$, is the convex hull of the incidence vectors of the stable sets of G. Since the stable set problem is NP-hard, it is unlikely to find a defining linear system of $STAB(G)$ for general graphs. Nevertheless there are classes of graphs for which such systems are known, as bipartite graphs, line graphs [3],

B. Yang, D.-Z. Du, and C.A. Wang (Eds.): COCOA 2008, LNCS 5165, pp. 339–350, 2008.
© Springer-Verlag Berlin Heidelberg 2008

series-parallel graphs [13], perfect graphs. For all these classes of graphs, the weighted stable set problem is polynomial time solvable and an explicit linear description of $STAB(G)$ is known.

Claw-free graphs are those graphs such that the neighborhood of each node has no stable set of size three. There exist polynomial time algorithms for solving the maximum weight stable set problem on a claw-free graph [14,5] but, despite many research efforts [8,9,11,16] and many disproved conjectures [9,6], still a linear description of the stable set polytope of a claw free graph is unknown. Finding such a defining linear system may be the first step towards a new algorithm for claw-free graphs, possibly computationally more effective than the existing ones (the algorithm by [14] is $O(|V|^7)$ and the one by [5] is $O(|V|^6)$).

The recent work of Chudnovsky and Seymour [2] on the structure of claw-free graphs set new directions to investigate the problem of finding a defining linear system for $STAB(G)$ when G is claw-free. We consider graphs with stability number at least 4; we denote by $\mathcal{C}$ the set of claw-free graphs with stability number at least 4 and by $\mathcal{Q}$ the set of *quasi-line* graphs with stability number at least 4 (a graph is quasi-line if the neighborhood of each node can be partitioned into two cliques). Clearly, $\mathcal{Q} \subseteq \mathcal{C}$. In [1] Chudnovsky and Seymour proved that the set $\mathcal{Q}$ is partitioned into two sets: $\mathcal{Q}^\ell$ (*fuzzy line graphs*) and $\mathcal{Q}^c$ (*fuzzy circular interval graphs*). Then they showed that any graph in $\mathcal{C}$ either belongs to $\mathcal{Q}^c$ or it can be obtained by composing three types of graphs, called *strips*: linear interval strips, XX-strips and *antihat strips*.

A defining linear system for $STAB(G)$ was given by Chudnovsky and Seymour [1] when $G \in \mathcal{Q}^\ell$ and by Eisenbrand et al. [4] when $G \in \mathcal{Q}^c$. This leaves open the problem of finding a linear description for $STAB(G)$ when $G \in \mathcal{C} \setminus \mathcal{Q}^c$. In this paper we consider the class $\mathcal{XX}$ of graphs obtained by composing only two types of strips: linear interval strips and XX-strips. Clearly $\mathcal{XX} \subseteq \mathcal{C} \setminus \mathcal{Q}^c$.

We provide a linear description of $STAB(G)$ when $G \in \mathcal{XX}$. To this aim we use a graph composition named *gear composition* introduced in [6] that builds a new graph G, called *geared graph*, starting from a given graph H and substituting an edge of H with the fixed graph called *gear*.

The gear composition produces new facets for the stable set polytope [6], called *geared inequalities*, which play an important role in solving the problem of finding a linear description for $STAB(G)$ when $G \in \mathcal{XX}$. In fact, we prove that any graph in $\mathcal{XX}$ can be built from a graph in $\mathcal{Q}^\ell$ via the gear composition and then we show that a defining linear system for $STAB(G)$ when $G \in \mathcal{XX}$ consists of: *rank inequalities, (lifted) 5-wheel inequalities and geared inequalities*.

We now introduce some notations. A linear inequality $\pi^T x \leq \pi_0$ is *valid* for $STAB(G)$ if it holds for all $x \in STAB(G)$ and it will be denoted as (π, π_0). An inequality (π, π_0) is said to be a *rank inequality* if $\pi_i = 1$ for $i \in S \subseteq V_G$, $\pi_i = 0$ for $i \in V_G \setminus S$ and $\pi_0 = \alpha(G[S])$ where $G[S]$ is the subgraph of G induced by S.

We denote by $\delta(v)$ the set of edges of G having v as endnode and by $N(v)$ the set of nodes of V_G adjacent to v. We also denote by $G \setminus A$ the subgraph of G induced by $V_G \setminus A$ where $A \subseteq V_G$ and by $G \setminus e$ ($G + e$) the subgraph of G obtained by removing (adding) the edge e.

A k-hole $C_k = (v_1, v_2, \ldots, v_k)$ is a chordless cycle of length k. A k-antiwheel $W = (h : \overline{C}_k)$ is a graph consisting of a k-antihole $\overline{C}_k$ and a node h (hub of W) adjacent to every node of $\overline{C}_k$. If $k = 5$, then $\overline{C}_5$ is isomorphic to C_5 and we refer to W as a 5-wheel. The inequality $\sum_{i=1}^{5} x_{v_i} + 2x_h \leq 2$ is facet defining for $STAB(W)$ and it is called 5-wheel inequality. A claw is a 3-antiwheel. A gear B is a graph of eight nodes $\{a, b_1, b_2, c, d_1, d_2, h_1, h_2\}$ such that $W_1 = (h_1 : a, d_1, b_1, c, h_2)$ and $W_2 = (h_2 : a, d_2, b_2, c, h_1)$ are 5-wheels; moreover, the edges of these wheels are the only edges of B.

In Section 2, we recall the definition of gear composition and some of its polyhedral properties. In Section 3, we show the graphs in $\mathcal{XX}$ can be built by iteratively applying the gear composition to a graph of $\mathcal{Q}^\ell$. Finally in Section 4, we provide a defining linear system for the stable set polytope of graphs in $\mathcal{C} \setminus \mathcal{Q}^c$.

2   Gear Composition

An edge $v_1 v_2$ of a graph H is said to be simplicial if $K_1 = N(v_1) \setminus \{v_2\}$ and $K_2 = N(v_2) \setminus \{v_1\}$ are cliques of H and both $K_1 \setminus K_2$ and $K_2 \setminus K_1$ are nonempty. Notice that K_1 and K_2 might have nonempty intersection.

Definition 1. Let $H = (V_H, E_H)$ be a graph with a simplicial edge $v_1 v_2$ and let $B = (V_B, E_B)$ be a gear. The gear composition of H and B produces a new graph $G = (H, B, v_1 v_2)$, called geared graph, such that:

$$V_G = V_H \setminus \{v_1, v_2\} \cup V_B,$$
$$E_G = E_H \setminus (\delta(v_1) \cup \delta(v_2)) \cup E_B \cup F_1 \cup F_2, with F_i = \{d_i u, b_i u | u \in K_i\} (i = 1, 2).$$

Definition 2. Let $H = (V_H, E_H)$ be a graph containing the simplicial edge $v_1 v_2$ and let (π, π_0) be a valid inequality for $STAB(H)$ such that $\pi_{v_1} = \pi_{v_2} = \lambda > 0$. Let $B = (V_B, E_B)$ be a gear and $G = (H, B, v_1 v_2)$ a geared graph. Then the inequalities

$$\diamond \quad \sum_{i \in V_H \setminus \{v_1, v_2\}} \pi_i x_i + \lambda \sum_{i \in V_B \setminus \{h_1, h_2\}} x_i + 2\lambda(x_{h_1} + x_{h_2}) \leq \pi_0 + 2\lambda \quad (1)$$

$$\diamond \quad \sum_{i \in V_H \setminus \{v_1, v_2\}} \pi_i x_i + \lambda \sum_{i \in V_B \setminus A} x_i \leq \pi_0 + \lambda \quad (2)$$
$$\text{where} \quad A \in \{\{b_1, c\}, \{b_2, c\}, \{d_1, a\}, \{d_2, a\}, \{a, c\}\}$$

are called geared inequalities associated with (π, π_0). The unique geared inequality that is full support on V_B is (1) and it will be called proper geared inequality.

Definition 3. Let H^e be a graph obtained from $H = (V_H, E_H)$ by subdividing the simplicial edge $e = v_1 v_2$ with a node t. An inequality (π, π_0) which is valid for $STAB(H^e)$ is said to be g-liftable (with respect to $v_1 v_2$) if $\pi_{v_1} = \pi_{v_2} = \pi_t = \lambda > 0$.

Definition 4. *Let $G = (H, B, v_1v_2)$ be a geared graph and (π, π_0) a g-liftable inequality. Then the inequalities*

$$\diamond \qquad \sum_{i \in V_H \setminus \{v_1, v_2\}} \pi_i x_i + \lambda \sum_{i \in V_B} x_i \leq \pi_0 + \lambda, \qquad (3)$$

$$\diamond \qquad \sum_{i \in V_H \setminus \{v_1, v_2\}} \pi_i x_i + \lambda \sum_{i \in V_B \setminus A} x_i \leq \pi_0 \qquad (4)$$

$$where \quad A \in \{\{b_1, c, b_2, h_1, h_2\}, \{d_1, a, d_2, h_1, h_2\}\}$$

are called g-lifted inequalities *associated with (π, π_0). The unique g-lifted inequality that is full support on V_B is (3) and it will be called* proper g-lifted inequality.

In [7], we showed that the linear description of $STAB(G)$ is completely determined by the linear description of $STAB(H)$ and $STAB(H^e)$. In fact, we proved that:

Theorem 1. *[7] Let $G = (H, B, e)$ be a geared graph. Then the stable set polytope $STAB(G)$ is described by the following linear inequalities:*

- *clique-inequalities,*
- *(lifted) 5-wheel inequalities,*
- *geared inequalities associated with facet defining inequalities of $STAB(H)$ having nonzero coefficient on the endnodes of e,*
- *g-lifted inequalities associated with facet defining inequalities of $STAB(H^e)$ having nonzero coefficient on the endnodes of e,*
- *facet of $STAB(H)$ having zero coefficient on the endnodes of e.*

This result implies that, if H and H^e are "well behaved" with respect to the stable set problem (meaning that there exist a defining linear system for their stable set polytopes) then the geared graphs obtained from H do the same. In the following we will extend some of the polyhedral properties of the gear composition in order to prove that a large subclass of claw-free graphs behave well with respect to the stable set problem.

3 Geared (Fuzzy) Line Graphs

In 2006, Chudnovsky and Seymour [2] proved that claw-free graphs with stability number at least 4 that are not quasi-line are obtained by composing only three kinds of graphs, called *strips*. One kind of strips, the *linear interval strips*, is used to generate quasi-line graphs while the other two kind of strips, the *XX-strips* and the *antihat-strips*, contain 5-wheels and so, they are necessary to generate claw-free graphs that are not quasi-line. This structure suggests the idea that claw-free graphs which are not quasi-line and have stability number at least 4 might not be so distant from line graphs in terms of polyhedral description of their stable set polytope. In the following sections we give an evidence of this fact by showing that a defining linear system for the stable set polytope of a large subclass of claw-free graphs with stability number at least 4 is built starting from the defining linear system of the stable set polytope of line graphs and using the gear composition. We actually conjectured that this holds for all claw-free graphs with large stability number [6].

Before stating the decomposition theorem of Chudnovsky and Seymour we recall some of their definitions:

Definition 5. *A strip* (G, a, b) *consists of a claw-free graph G together with two designated simplicial vertices a, b called the* ends *of the strip. Two strips can be composed as follows: let A and B be the nodes of $G \setminus \{a, b\}$ adjacent in G to a and b respectively, and define A' and B' similarly. Take the disjoint union of $G \setminus \{a, b\}$ and $G' \setminus \{a', b'\}$; and let H be the graph obtained from this by adding all possible edges between A and A' and between B and B'.*

Definition 6. *A homogeneous pair of cliques in G is a pair (A, B) such that:*

- *A and B are cliques in G and $A \cap B = \emptyset$,*
- *$|A| \geq 2$ or $|B| \geq 2$,*
- *no vertex of $G \setminus (A \cup B)$ has both a neighbour and a non-neighbour in A, and the same in B.*

Definition 7. *Let T be a graph with vertex set $\{u_1, \ldots, u_{13}\}$ and with adjacency as follows. $(u_1, \ldots, u_6)$ is a hole of G of length 6. Next, u_7 is adjacent to u_1, u_2; u_8 is adjacent to u_4, u_5; u_9 is adjacent to u_6, u_1, u_2, u_3; u_{10} is adjacent to u_3, u_4, u_5, u_6, u_9; u_{11} is adjacent to $u_3, u_4, u_6, u_1, u_9, u_{10}$; u_{12} is adjacent to $u_2, u_3, u_5, u_6, u_9, u_{10}$; u_{13} is adjacent to $u_1, u_2, u_4, u_5, u_7, u_8$. Let $X \subseteq \{u_{11}, u_{12}, u_{13}\}$; then the strip $(T \setminus X, u_7, u_8)$ is called an XX-strip.*

We can now state the decomposition theorem of Chudnovsky and Seymour; the decomposition involves the antihat strips, but we omit their definition since they will never be used in the following.

Theorem 2. *[2] For every claw-free graph G with $\alpha(G) \geq 4$, if G does not admit a 1-join and there is no homogeneous pair of cliques in G, then either G is a circular interval graph, or G is a composition of linear interval strips, XX-strips, and antihat strips.*

Since graphs containing homogeneous pairs cannot be represented with the above strips, Chudnovsky and Seymour were forced to introduce the concept of *fuzziness* [2] and to give a "fuzzy" version of the above theorem were all the strips are fuzzy strips. Since the fuzziness is a very technical concept we do not go into the detail of its definition (we refer the interested reader to [2]). To our purpose it suffices to observe that (fuzzy) linear interval strips are quasi-line graphs and to refer to a quasi-line graph that is a composition of (fuzzy) linear interval strips as a *(fuzzy) line graph*. The class of (fuzzy) line graphs is denoted by $\mathcal{Q}^\ell$ and the class of (fuzzy) circular interval graphs is denoted by $\mathcal{Q}^c$.

It is worth noticing that the fuzziness does not have much relevance from the polyhedral point of view. This was already noticed by Chudnovsky and Seymour who proved that:

Theorem 3. *[1] If G is a (fuzzy) line graph, then $STAB(G)$ is described by the Edmonds' inequalities.*

And it was further confirmed by the work of Eisenbrand et al. on (fuzzy) circular interval graphs.

Lemma 1. *[4] Let F be a facet of $STAB(G)$ where G is a fuzzy circular interval graph. Then F is also a facet of $STAB(G')$, where G' is a circular interval graph obtained from G by removing some edges.*

By Theorem 2 and its "fuzzy" version, finding a linear description of $STAB(G)$ for claw-free graphs with $\alpha(G) \geq 4$ is equivalent to finding a linear description of $STAB(G)$ for (fuzzy) circular interval graphs, namely the graphs in $\mathcal{Q}^c$, and for graphs that are composition of (fuzzy) linear interval strips, XX-strips and antihat strips. Since the first case has been solved in [4] and the antihat strips do not seem to produce "interesting" facet defining inequalities for $STAB(G)$, we focus our attention on the graphs that are a

composition of XX-strips and (fuzzy) linear interval strips.

We call these graphs XX-*graphs* and their family will be denoted as $\mathcal{XX}$.

In the following we show that any XX-graph can be obtained by iteratively applying the gear composition defined in Section 2 to a (fuzzy) line graph, i.e., a graph in $\mathcal{Q}^\ell$. We start by showing that the gear is a subgraph of an XX-strip.

Lemma 2. *The graph obtained by composing a strip (G, v_1, v_2) with the XX-strip $(T \setminus \{u_{11}, u_{12}, u_{13}\}, u_7, u_8)$ is a geared graph.*

Proof. Rename the nodes $\{a, b_1, b_2, c, d_1, d_2, h_1, h_2\}$ of a gear B as $\{u_6, u_2, u_4, u_3, u_1, u_5, u_9, u_{10}\}$. Thus, the strip composition of (G, v_1, v_2) and the XX-strip $(T \setminus \{u_{11}, u_{12}, u_{13}\}, u_7, u_8)$), as defined in Definition 5, corresponds to the gear composition of $G' = (V_G, E_G \cup \{v_1 v_2\})$ and the gear B. In fact, being the nodes v_1 and v_2 simplicial, we have that the edge $v_1 v_2$ of G' is simplicial. The graph obtained by applying the above strip composition is precisely the geared graph $(G', B, v_1 v_2)$.

As a consequence of the above lemma and Definition 7 we have that each XX-strip composition produces a geared graph $G = (H, B, e)$ plus an extra set Y of nodes which are properly adjacent to B. This, together with Theorem 2, implies that a large number of claw-free graphs can be seen as geared graphs. We now prove that we can restrict ourselves to consider only XX-strips not containing node u_{13} since this node can be added using an appropriate linear interval strip.

Lemma 3. *The class of XX-graphs coincides with the subclass of claw-free graphs obtained by composition of XX-strips of type $(T \setminus \{u_{13}\}, u_7, u_8)$ and (fuzzy) linear interval strips.*

Proof. Let G be an XX-graph obtained by composing a strip (L, v_1, v_2) and an XX-strip $(T \setminus X, u_7, u_8)$ such that $u_{13} \notin X$. Consider the linear interval strip (L', a_0, b_0) such that $V_{L'} = \{v_1', v_2', u_{13}, a_0, b_0\}$ and the triples $\{v_1', u_{13}, a_0\}$ and $\{v_2', u_{13}, b_0\}$ induce two triangles. It is trivial to see that the graph L'' obtained by composing the strip (L, v_1, v_2) with the linear interval strip (L', a_0, b_0) has v_1' and v_2' as simplicial nodes. Thus (L'', v_1', v_2') is a strip and its composition with the XX-strip $(T \setminus (X \cup \{u_{13}\}), u_7, u_8)$ yields the graph G, as claimed.

We now show that the XX-graphs admit a decomposition different from the strip decomposition: they can be obtained by repeated applications of an "extended" gear composition to a (fuzzy) line graph.

Definition 8. *Let* $B = (V_B, E_B)$ *be a gear and let* u_{11} *and* u_{12} *be two new nodes. Let* $\delta(u_{11}) = \{u_{11}d_1, u_{11}a, u_{11}h_1, u_{11}h_2, u_{11}c, u_{11}b_2\}$ *and* $\delta(u_{12}) = \{u_{12}d_2, u_{12}a, u_{12}h_1, u_{12}h_2, u_{12}c, u_{12}b_1\}$. *Let* $Y \subseteq \{u_{11}, u_{12}\}$ *and* $\delta(Y) = \cup_{i \in Y} \delta(u_i)$.

An **extended gear** B_Y *is a graph with node set* $V_B \cup Y$ *and edge set* $E_B \cup \delta(Y)$ *(see Fig. 1.).*

An extended gear composition *is a gear composition where the gear* B *is replaced by* B_Y *for some* $Y \subseteq \{u_{11}, u_{12}\}$. *The graph resulting from an extended gear composition will still be called geared graph and denoted as* (H, B_Y, e).

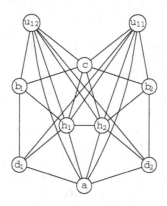

Fig. 1. The extended gear B_Y with $Y = \{u_{11}, u_{12}\}$

In order to show that the gear composition can be used to build XX-graphs, we need to show that the removal of a simplicial edge preserves the property of the graph of being (fuzzy) line.

Proposition 1. *H is a (fuzzy) line graph with a simplicial edge* $e = v_1 v_2$ *if and only if* $(H \setminus e, v_1, v_2)$ *is a strip with* $H \setminus e$ *(fuzzy) line graph.*

Proof. First we prove the "if" direction. It suffices to observe that H is obtained by composing the strip $(H \setminus e, v_1, v_2)$ with a strip (P, a_0, b_0) consisting of a path $P = (a_0, u_1, u_2, b_0)$ and then rename the nodes u_i as v_i, $i = 1, 2$. To prove the other direction observe that a (fuzzy) line graph is a graph in $\mathcal{Q}^\ell$. Hence to prove that $H \setminus e$ is a (fuzzy) line graph we must first show that it is quasi-line, i.e., it contains neither a claw nor an odd-antiwheel. Suppose by contradiction that $H \setminus e$ contains a claw C. Since the only edge which was removed from H is $e = v_1 v_2$, we have that C contains both v_1 and v_2. So, $C = (y : v_1, v_2, w)$ with $y \in K_1 \cap K_2$ and $w \in V_H \setminus (K_1 \cup K_2 \cup \{v_1, v_2\})$. Since e is simplicial, there exists a node $z_1 \in K_1 \setminus K_2$ and a node $z_2 \in K_2 \setminus K_1$. Thus,

$wz_i \in E_H$, $i = 1, 2$, since otherwise $(y : v_1, z_2, w)$ or $(y : v_2, z_1, w)$ would be claws in H, contradicting the hypothesis that H is quasi-line. Hence, the edge e belongs to the 5-wheel $(y : v_1, v_2, z_2, w, z_1)$ contained in H, contradicting the hypothesis that H is quasi-line. Similarly, it can be proved that $H \setminus e$ does not contain odd-antiwheels and so, it is a quasi-line graph. Thus, by the decomposition of quasi-line graphs in [1], $H \setminus e$ either belongs to $\mathcal{Q}^\ell$ or it belongs to $\mathcal{Q}^c$. Since it is possible to prove that if $H \setminus e \in \mathcal{Q}^c$, then $H \in \mathcal{Q}^c$, we have that $H \setminus e$ is a (fuzzy) line graph as claimed.

Definition 9. *Let H be a (fuzzy) line graph which is not a clique and let E_H^* be the set of its simplicial edges. A g-operation on $e \in E_H^*$ is either an extended gear composition or an edge subdivision of e. A graph $G \in \mathcal{G}_H^*$ if and only if*

either $G = H$,
or $G = (L, B_Y, e)$, where $L \in \mathcal{G}_H^*$, B_Y is an extended gear, and $e \in E_H^* \cap E_L$
(i.e., e is a simplicial edge of H on which no g-operation has been performed),
or $G = L^e$, where $L \in \mathcal{G}_H^*$ and $e \in E_H^* \cap E_L$.

The graphs in $\bigcup_{H \in \mathcal{Q}^\ell} \mathcal{G}_H^*$ will be called geared (fuzzy) line graphs.

In the following lemma we show that:

Lemma 4. *The geared (fuzzy) line graphs are the XX-graphs.*

Proof. By Proposition 1 and Lemmas 2 and 3, it is trivial to see that XX-graphs are geared (fuzzy) line graphs. To prove the opposite, suppose by contradiction that there exists a graph G in $\mathcal{G}_H^*$ for some $H \in \mathcal{Q}^\ell$ which is not an XX-graph. In particular, assume G be obtained by performing the smallest number of g-operations. If $G = H$ then, by definition, H is an XX-graph. Hence, either $G = (L, B_Y, e)$ or $G = L^e$, where $e = v_1 v_2$ is a simplicial edge of H.

Suppose first that $G = L^e$. By the minimality of G we know that L is an XX-graph. Since e is simplicial it does not belong to any XX-strip of L. So, we can build a new graph $\tilde{L}$ from L by replacing each XX-strip $(T \setminus X, v_1, v_2)$ with the simple path (a_0, v_1, v_2, b_0). It follows that $\tilde{L}$ is a (fuzzy) line graph and, by Proposition 1, $\tilde{L} \setminus e$ is also a (fuzzy) line graph. Now we reconstruct $L \setminus e$ from $\tilde{L} \setminus e$ by replacing the simple paths previously introduced with the corresponding XX-strips. Thus $L \setminus e$ is obtained as a composition of XX-strips and (fuzzy) linear interval strips, and so, it is an XX-graph. Since G is obtained by composing the strip $(L \setminus e, v_1, v_2)$ and (a, v_1', t, v_2', b) and renaming v_i' as v_i, $i = 1, 2$, we have that G is an XX-graph, as claimed.

Consider now the case $G = (L, B_Y, e)$. As above, we can prove that $L \setminus e$ is an XX-graph. If we add to B_Y two nodes v_1' and v_2' adjacent to b_1, d_1 and b_2, d_2, respectively, we have that G is obtained by composing the strips $(L \setminus e, v_1, v_2)$ and the XX-strip $(B_Y \cup \{v_1', v_2'\}, v_1', v_2')$. Thus the thesis follows.

From the above results it follows that the XX-graphs can be built in two different ways: either using the strip composition defined by Chudnovsky and Seymour in [2] or using the extended gear composition. This result allows us to exploit the polyhedral properties of the gear composition to find a linear description for the stable set polytope of XX-graphs. This will be discussed in the next section.

4   Stable Set Polytope of XX-Graphs

In this section we consider the geared graph $G = (H, B_Y, e)$ obtained as an extended gear composition of the graph H and the extended gear B_Y, with $Y \subseteq \{u_{11}, u_{12}\}$.

Let (γ, γ_0) be a non trivial facet defining inequality of $STAB(G)$ which is not a clique or a (lifted) 5-wheel inequality. If $Y = \emptyset$ then, by Theorem 1, (γ, γ_0) is one of the following inequalities:

- an inequality of type (1) or (2) associated with a facet of $STAB(H)$;
- an inequality of type (3) or (4) associated with a facet of $STAB(H^e)$;
- a facet of $STAB(H)$.

We now show that the properties of the extended gear composition do not change substantially when $Y \neq \emptyset$. In particular we prove that the above inequalitities can be "lifted" to the higher dimensional space containing u_{11} and u_{12} using the *sequential lifting* procedure [15]. We start finding the lifting coefficient of u_{11} and u_{12} for inequalities (1), (2), (3), and (4).

Lemma 5. *Let $G = (H, B_Y, e)$ with $Y = \{u_{11}\}$ and (β, β_0) be a geared inequality that is facet defining for $STAB(G')$, where $G' = (H, B, e)$.*
If (β, β_0) is of type (1) then the node u_{11} is lifted with coefficient $\beta_{u_{11}} = \lambda$.
If (β, β_0) is of type (2) then the node u_{11} is lifted with coefficient $\beta_{u_{11}} = \lambda$ if $A = \{b_1, c\}$ or $A = \{d_2, a\}$, and $\beta_{u_{11}} = 0$ otherwise.

Similarly, we prove that:

Lemma 6. *Let $G = (H, B_Y, e)$ with $Y = \{u_{11}\}$ and (β, β_0) be a g-lifted inequality that is facet defining for $STAB(G')$, where $G' = (H, B, e)$.*
If (β, β_0) is of type (3) then the node u_{11} is lifted with coefficient $\beta_{u_{11}} = \lambda$.
If (β, β_0) is of type (4) then the node u_{11} is lifted with coefficient $\beta_{u_{11}} = 0$.

The next lemma shows the lifting coefficient $\beta_{u_{12}}$ for geared and g-lifted inequalities once the node u_{11} has been already lifted.

Lemma 7. *Let $G = (H, B, e)$, $G' = (H, B_{\{u_{11}\}}, e)$, and $G'' = (H, B_{\{u_{11}, u_{12}\}}, e)$.*
Moreover, let (β', β_0) be a facet defining inequality for $STAB(G')$, obtained by lifting the inequality (β, β_0) of type (1) $\div$ (4) on node u_{11}.
If (β, β_0) is a proper gear inequality (1) or a proper g-lifted inequality (3), then the node u_{12} is lifted with coefficient $\beta_{u_{12}} = \lambda$.
In all other cases, the node u_{12} is lifted with coefficient $\beta_{u_{12}} = 0$.

By symmetry, the results of Lemmas 5÷7 hold if we interchange the role of u_{11} and u_{12}. However the extension of the gear B with the nodes u_{11} and u_{12} does not generate only inequalities that are (sequential liftings of inequalities) defined in definitions 2 and 4. Indeed, new facet defining inequalities are generated. More precisely,

Theorem 4. Let $G = (H, B, e)$, $G' = (H, B_{\{u_{11}\}}, e)$, and $G'' = (H, B_{\{u_{11}, u_{12}\}}, e)$. Let (π, π_0) be a g-liftable facet defining inequality for $STAB(H^e)$, then

$$\sum_{i \in V_H \setminus \{v_1, v_2\}} \pi_i x_i + \lambda(x_{d_1} + x_{u_{11}} + x_{b_2}) \leq \pi_0 \tag{5}$$

is facet defining for both $STAB(G')$ and $STAB(G'')$; moreover,

$$\sum_{i \in V_H \setminus \{v_1, v_2\}} \pi_i x_i + \lambda(x_{b_1} + x_{u_{12}} + x_{d_2}) \leq \pi_0 \tag{6}$$

is facet defining for $STAB(G'')$.

Observe that the above inequalities (5) and (6) have a structure very similar to the g-lifted inequalities (4). In fact, the nodes of the gear associated with the nonzero coefficients of each inequality of type (4) induce the simple paths (d_1, a, d_2) and (b_1, c, b_2). The same holds for inequalities (5) and (6), where the simple paths are (d_1, u_{11}, b_2) and (b_1, u_{12}, d_2), respectively.

We are now ready to prove the main result of the paper.

Definition 10. A facet defining inequality $(\gamma, \gamma_0) \in \mathcal{G}$ if and only if it is (the sequential lifting of)

either a rank inequality,

or a 5-wheel inequality,

or a geared or a g-lifted inequality associated with an inequality in $\mathcal{G}$.

From the definitions 2 and 4, it is not difficult to see that geared inequalities and g-lifted inequalities either contain at least a pair of coefficients equal to 2λ corresponding to the hubs of a gear or all their coefficients equal λ. For the sake of simplicity, from now on, we call geared inequalities the inequalities in $\mathcal{G}$ containing at least a pair of hubs of an extended gear with coefficients 2λ. Thus, all the inequalities of $\mathcal{G}$ that are not geared are either rank-inequalities or (lifted) 5-wheel inequalities. Consider now the polyhedron

$$\mathcal{G}STAB(G) = \{x \in \mathbb{R}_+^V | \ x \ satisfies \ \mathcal{G}\}. \tag{7}$$

By Theorem 1 and the results of this section, it follows that $STAB(G) \subseteq \mathcal{G}STAB(G)$; moreover, a graph G is said to be $\mathcal{G}$-perfect if the equality holds. By Theorem 1, we have that:

Theorem 5. Let $G = (H, B_Y, e)$ with $Y = \emptyset$. If H and H^e are $\mathcal{G}$-perfect, then G is $\mathcal{G}$-perfect.

The above result can be extended to geared graphs obtained by performing an extended gear composition as follows:

Theorem 6. Let $G = (H, B_Y, e)$ with $Y \subseteq \{u_{11}, u_{12}\}$. If H and H^e are $\mathcal{G}$-perfect, then G is $\mathcal{G}$-perfect.

Finally, to prove that geared (fuzzy) line graphs, i.e., graphs in $\cup_{H \in \mathcal{Q}^\ell} \mathcal{G}_H^*$, are $\mathcal{G}$-perfect we need the following:

Theorem 7. *Let H be a graph and E_H^* the set of its simplicial edges. Let H^F be the graph obtained from H by subdividing all the edges in $F \subseteq E_H^*$. If H and H^F are $\mathcal{G}$-perfect for any $F \subseteq E_H^*$, then every graph $G \in \cup_{H \in \mathcal{Q}^\ell} \mathcal{G}_H^*$ is $\mathcal{G}$-perfect.*

We now prove that the claw-free graphs in $\mathcal{X}\mathcal{X}$ are $\mathcal{G}$-perfect. More precisely,

Theorem 8. *If G is an XX-graph then $STAB(G)$ is defined by the (sequential lifting of the) following inequalities:*

- *rank inequalities,*
- *5-wheel inequalities,*
- *geared inequalities.*

Proof. By Lemma 4, the graph G is a geared (fuzzy) line graph, namely it belongs to the family $\mathcal{G}_H^*$ for some $H \in \mathcal{Q}^\ell$. By Proposition 1, the graph $H \setminus e$ is a (fuzzy) line graph. To obtain H^e one just needs to compose $(H \setminus e, v_1, v_2)$ with the strip (P, a, b) where P is the path (a, u_1, t, u_2, b) and then rename u_i as v_i, $i = 1, 2$. Hence, H^e is a (fuzzy) line graph and the same holds for H^F for any subset $F \subseteq E_H^*$ of simplicial edges of H. By Theorem 3, we know that $STAB(H)$ and $STAB(H^F)$ are described only by rank inequalities, i.e., the Edmonds' inequalities [3], and so, H and H^F are $\mathcal{G}$-perfect. Hence, by Theorem 7, every graph $G \in \cup_{H \in \mathcal{Q}^\ell} \mathcal{G}_H^*$ is $\mathcal{G}$-perfect, i.e., $STAB(G)$ is completely described by inequalities in $\mathcal{G}$ and the theorem follows.

References

1. Chudnovsky, M., Seymour, P.: Claw-free graphs VI: Quasi-line graphs (manuscript) (2004)
2. Chudnovsky, M., Seymour, P.: Claw-free graphs IV: Decomposition theorem. J. Comb. Th. B (to appear, 2007)
3. Edmonds, J.: Maximum matching and a polyhedron with 0, 1 vertices. J. Res. of Nat. Bureau of Stand. B 69B, 125–130 (1965)
4. Eisenbrand, F., Oriolo, G., Stauffer, G., Ventura, P.: The stable set polytope of quasi-line graphs. Combinatorica (to appear)
5. Oriolo, G., Pietropaoli, U., Stauffer, G.: A new algorithm for the maximum weighted stable set problem in claw-free graphs. In: Lodi, A., Panconesi, A., Rinaldi, G. (eds.) IPCO 2008. LNCS, vol. 5035, pp. 77–96. Springer, Heidelberg (2008)
6. Galluccio, A., Gentile, C., Ventura, P.: Gear composition and the stable set polytope. Operations Research Letters (to appear, 2008)
7. Galluccio, A., Gentile, C., Ventura, P.: Gear composition of stable set polytopes and $\mathcal{G}$-perfection. Mathematics of Operations Research (submitted, 2006)
8. Galluccio, A., Sassano, A.: The rank facets of the stable set polytope for claw-free graphs. J. Comb. Th. B 69, 1–38 (1997)
9. Giles, R., Trotter, L.E.: On stable set polyhedra for $K_{1,3}$-free graphs. J. Comb. Th. B 31, 313–326 (1981)
10. Grötschel, M., Lovász, L., Schrijver, A.: Geometric algorithms and combinatorial optimization. Springer, Berlin (1988)

11. Liebling, T.M., Oriolo, G., Spille, B., Stauffer, G.: On non-rank facets of the stable set poly-tope of claw-free graphs and circulant graphs. Math. Methods of Oper. Research 59, 25–35 (2004)

12. Rinaldi, G., Conforti, M., Wolsey, L.: On the cut polyhedron. Discrete Mathematics 277, 279–285 (2004)

13. Mahjoub, A.R.: On the stable set polytope of a series-parallel graph. Mathematical Program-ming 40, 53–57 (1988)

14. Minty, G.J.: On maximal independent sets of vertices in claw-free graphs. J. Comb. Th. B 28, 284–304 (1980)

15. Padberg, M.W.: On the facial structure of vertex packing polytope. Mathematical Program-ming 5, 199–215 (1973)

16. Pulleyblank, W.R., Shepherd, B.: Formulations of the stable set polytope. In: Rinaldi, G., Wolsey, L. (eds.) Proceedings Third IPCO Conference, pp. 267–279 (1993)

A Combinatorial Algorithm to Optimally Colour the Edges of the Graphs That Are Join of Regular Graphs

Caterina De Simone and Anna Galluccio

Istituto di Analisi dei Sistemi ed Informatica, CNR, Viale Manzoni 30,
00185 Rome, Italy
{desimone,galluccio}@iasi.cnr.it

Abstract. We prove that the edges of every even graph $G = G_1 + G_2$ that is the join of two regular graphs $G_i = (V_i, E_i)$ can be coloured with $\Delta(G)$ colours, whenever $\Delta(G) = \Delta(G_2) + |V(G_1)|$. The proof of this result yields a combinatorial algorithm to optimally colour the edges of this type of graphs.

1 Introduction

The graphs in this paper are simple, that is they have no loops or multiple edges. Let $G = (V, E)$ be a graph; the degree of a vertex v, denoted by $d_G(v)$, is the number of edges incident to v; the maximum degree of G, denoted by $\Delta(G)$, is the maximum vertex degree in G; G is regular if the degree of every vertex is the same.

An *edge-colouring* of G is an assignment of colours to its edges so that no two edges incident to the same vertex receive the same colour. A t-edge-colouring of G is then a partition of the edge set E into t disjoint matchings.

The *chromatic index* of G, denoted by $\chi'(G)$, is the least t for which G has a t-edge-colouring. Since all the edges incident to a vertex must be in different matchings, we know that $\chi'(G)$ is at least $\Delta(G)$. In fact, Vizing [16] proved that $\chi'(G)$ is at most $\Delta(G) + 1$. Graphs with $\chi'(G) = \Delta(G)$ are said to be *Class 1*, while the others are said to be *Class 2*. Despite the fact that an edge-colouring of any graph G with $\Delta(G) + 1$ colours can be found in polynomial time, the problem of deciding whether G is Class 1 is NP-complete even if $\Delta(G) = 3$ [10]. On the other hand, graphs with "large" maximum degree seem to behave better from the computational point of view [4,14]. Indeed, the study of the edge-colouring problem for this type of graphs is motivated by a famous and still unsolved conjecture of Hilton [3] known as "The Overfull Conjecture": if true, there would exist a polynomial time algorithm to decide whether a graph G (with n vertices and $\Delta(G) > n/3$) is Class 1. This motivates us to study the edge-colouring problem for the class of join graphs.

A graph $G = (V, E)$ is the *join* of two graphs $G_1 = (V_1, E_1)$ and $G_2 = (V_2, E_2)$ (with $V_1 \cap V_2 = \emptyset$), if $V = V_1 \cup V_2$ and $E = E_1 \cup E_2 \cup \{uv : u \in V_1, v \in V_2\}$. When G is the join of G_1 and G_2 we shall write $G = G_1 + G_2$. (Note that $\Delta(G) \geq |V|/2$.)

B. Yang, D.-Z. Du, and C.A. Wang (Eds.): COCOA 2008, LNCS 5165, pp. 351–360, 2008.

The join operation (also known in literature as *graph substitution* or *sum*) plays an important role in solving hard combinatorial problems and in designing efficient algorithms for many optimization problems [2,12]. The class of join graphs has one more characteristics: it properly contains the class of P_4-*free* graphs, i.e., graphs not containing a P_4 as an induced subgraph. These graphs, also known as *cographs*, are very interesting because for them many NP-hard problems can be solved in polynomial time [5,6]. Nevertheless, it is not known whether it is NP-complete to decide if a cograph is Class 1.

Contrary to the intuition suggested by the Overfull Conjecture, it is very convenient to deal with graphs that not only have large maximum degree but are also "dense". Thus, in this paper, we study the edge-colouring problem for the class of graphs that are the join of two regular graphs: these are the join graphs with the highest number of edges. Observe that any polynomial time algorithm that colours the edges of a graph G with $\Delta(G)$ colours yields also a $\Delta(H)$-edge-colouring for any subgraph H (induced or not) of G such that $\Delta(H) = \Delta(G)$.

Besides the classical application to the timetabling problem [13], the edge-colouring problem has several applications in computer network problems, such as the design of wavelength division multiplexing (WDM) networks [1].

Let $G = G_1 + G_2$ be a join graph with an even number n of vertices such that $G_i = (V_i, E_i)$ is regular ($i = 1, 2$). Write $n_i = |V(G_i)|$ and $\Delta(G_i) = \Delta_i$. Clearly, $n = n_1 + n_2$ and $\Delta(G) = max\{n_1 + \Delta_2, n_2 + \Delta_1\}$. Without loss of generality, we shall assume that $n_1 \leq n_2$. If $\Delta_1 = 0$ and $\Delta(G) = n_1 + \Delta_2$ then G is Class 1 [11]; if $\Delta_1 = \Delta_2$ and $n_1 = n_2$ then G is Class 1 [9]; if G is regular then G is Class 1 [8].

The aim of this paper is to prove that, if $\Delta(G) = n_1 + \Delta_2$ then G is Class 1, thus generalizing the previous results.

In Section 2 we give some properties of the equitable edge-colourings of graphs. In Section 3 we show that every join graph $G = G_1 + G_2$ with an even number n of vertices and maximum degree equal to $n_1 + \Delta_2$ is Class 1, whenever both G_1 and G_2 are regular. The proof of this result is combinatorial and provides an $O(n^4)$ time algorithm for finding an optimal edge-colouring of these graphs.

2   Equitable Edge-Colourings of Graphs

Let $\mathcal{C} = \{c_1, \ldots, c_t\}$ be an edge-colouring of a graph G with m edges. The colouring $\mathcal{C}$ is said to be *equitable* if each c_i has size equal to either $\lfloor m/t \rfloor$ or $\lceil m/t \rceil$. Throughout the paper we shall refer to each c_i as both a colour and a matching of G. Equitable edge-colourings always exist and are easy to construct:

Proposition 1 ([7],[13][15]). *Every graph G with $\chi'(G) \leq t$ has an equitable t-edge-colouring which can be found in $O(|V(G)|^4)$ time.*

Before showing the relations between the equitable edge-colourings of two graphs we need two easy observations.

Observation 1. *Let G be a graph with m edges such that $\chi'(G) \leq t$; let $C = \{c_1, \ldots, c_t\}$ be an equitable edge-colouring of G. If m/t is not integral, then the number of colours c_i of size $\lfloor m/t \rfloor$ is equal to $t\lceil m/t \rceil - m$.*

To see the validity of the observation, note that if p denotes the number of colours c_i of size $\lfloor m/t \rfloor$ (and so $t - p$ is the number of colours c_i of size $\lceil m/t \rceil$), then $m = p\lfloor m/t \rfloor + (t - p)\lceil m/t \rceil$.

The next observation is a simple application of the previous one.

Observation 2. *For $i = 1, 2$, let G_i be a graph with n_i vertices, m_i edges, and maximum degree Δ_i; let C_i be an equitable $(\Delta_i + 1)$-edge-colouring of G_i. If $m_i/(\Delta_i + 1)$ is not integral $(i = 1, 2)$ and if*

$$\left\lceil \frac{m_2}{\Delta_2 + 1} \right\rceil - \left\lceil \frac{m_1}{\Delta_1 + 1} \right\rceil = \frac{n_2 - n_1}{2}$$

then $q - p$ is equal to

$$(\Delta_1 + 1)\left(\frac{n_2 - n_1}{2} - \frac{m_2}{\Delta_2 + 1} + \frac{m_1}{\Delta_1 + 1} \right) + (\Delta_2 - \Delta_1)\left(\left\lceil \frac{m_2}{\Delta_2 + 1} \right\rceil - \frac{m_2}{\Delta_2 + 1} \right)$$

where p denotes the number of colours of C_1 that have size equal to $\lfloor m_1/(\Delta_1 + 1) \rfloor$ and q denotes the number of colours of C_2 that have size equal to $\lfloor m_2/(\Delta_2 + 1) \rfloor$.

For every colour c_i, we shall denote by $X(c_i)$ the subset of vertices of $G = (V, E)$ that are missed by colour c_i; in other words, $X(c_i)$ is the set of all vertices that are exposed with respect to the matching c_i. Clearly, $|X(c_i)| = |V| - 2|c_i|$.

Theorem 1. *For $i = 1, 2$, let G_i be a graph with n_i vertices, m_i edges, and maximum degree Δ_i; let $C_1 = \{f_1, \ldots, f_{\Delta_1+1}\}$ and $C_2 = \{h_1, \ldots, h_{\Delta_2+1}\}$ be equitable edge-colourings of G_1 and G_2, respectively. Assume that $n_1 \leq n_2$, $n_1 + n_2$ even, and $\Delta_1 \leq \Delta_2$.*

If $m_2/(\Delta_2 + 1) - m_1/(\Delta_1 + 1) \geq (n_2 - n_1)/2$ then there exists an ordering of the elements of C_2 such that

$$|X(h_i)| \leq |X(f_i)| \quad for \quad every \quad i = 1, \ldots, \Delta_1 + 1.$$

Otherwise, then there exists an ordering of the elements of C_2 such that

$$|X(h_i)| \geq |X(f_i)| \quad for \quad every \quad i = 1, \ldots, \Delta_1 + 1.$$

Proof. Set

$$x = \frac{m_1}{\Delta_1 + 1}, \qquad y = \frac{m_2}{\Delta_2 + 1}.$$

Since C_1 is equitable, each matching f_i has size equal to $\lfloor x \rfloor$ or $\lceil x \rceil$ $(i = 1, \ldots, \Delta_1 + 1)$; since C_2 is equitable, each matching h_i has size equal to $\lfloor y \rfloor$ or $\lceil y \rceil$ $(i = 1, \ldots, \Delta_2 + 1)$. Observe that, for every $i = 1, \ldots, \Delta_2 + 1$,

$$|X(h_i)| = n_2 - 2|h_i| = n_1 - 2\left(|h_i| - \frac{n_2 - n_1}{2} \right).$$

Since $|X(f_i)| = n_1 - 2|f_i|$ $(i = 1, \ldots, \Delta_1 + 1)$, proving the theorem amounts to show that there exists an ordering of the elements of $\mathcal{C}_2$ such that

$$|h_i| \geq |f_i| + \frac{n_2 - n_1}{2} \qquad i = 1, \ldots, \Delta_1 + 1, \tag{1}$$

whenever $y - x \geq (n_2 - n_1)/2$, and such that

$$|h_i| \leq |f_i| + \frac{n_2 - n_1}{2} \qquad i = 1, \ldots, \Delta_1 + 1, \tag{2}$$

whenever $y - x < (n_2 - n_1)/2$.

First assume that

$$y - x \geq \frac{n_2 - n_1}{2}. \tag{3}$$

Since $\lceil y \rceil - \lceil x \rceil \geq y - \lceil x \rceil > y - (x + 1)$, it follows that $\lceil y \rceil - \lceil x \rceil > (n_2 - n_1)/2 - 1$. But then the integrality of both r.h.s and l.h.s in the previous inequality implies that

$$\lceil y \rceil - \lceil x \rceil \geq \frac{n_2 - n_1}{2}. \tag{4}$$

Now if $y = \lceil y \rceil$ then $|h_i| = y$ for every i. Since $|f_i| \leq \lceil x \rceil$ for every i, it follows that $|h_i| - |f_i| \geq \lceil y \rceil - \lceil x \rceil$ for every $i = 1, \ldots, \Delta_1 + 1$, and so (1) holds.

Hence we can assume that $\lfloor y \rfloor < y < \lceil y \rceil$. If $x = \lceil x \rceil$ then $\lceil y \rceil - \lceil x \rceil = \lceil y \rceil - x > y - x$, and so

$$\lceil y \rceil - \lceil x \rceil > \frac{n_2 - n_1}{2}. \tag{5}$$

But then the identity $\lceil y \rceil = \lfloor y \rfloor + 1$ and the integrality of both the r.h.s. and the l.h.s. in (5) imply that $\lfloor y \rfloor - \lfloor x \rfloor \geq (n_2 - n_1)/2$. Since $|h_i| \geq \lfloor y \rfloor$ for every i and since $|f_i| = \lfloor x \rfloor$ for every i, it follows that $|h_i| - |f_i| \geq \lfloor y \rfloor - \lfloor x \rfloor$ for every $i = 1, \ldots, \Delta_1 + 1$. But then again (1) holds.

Hence we can assume that $\lfloor x \rfloor < x < \lceil x \rceil$ and $\lfloor y \rfloor < y < \lceil y \rceil$. In this case $\lceil x \rceil = \lfloor x \rfloor + 1$ and $\lceil y \rceil = \lfloor y \rfloor + 1$, and so (4) implies that

$$\lfloor y \rfloor - \lfloor x \rfloor \geq \frac{n_2 - n_1}{2}. \tag{6}$$

Note that

$$\lceil y \rceil - \lfloor x \rfloor > \frac{n_2 - n_1}{2}. \tag{7}$$

Now, let p denote the number of matchings f_i of size $\lfloor x \rfloor$ and let q denote the number of matchings h_i of size $\lfloor y \rfloor$. If $p \geq q$ then (4), (6), and (7) imply that any q matchings h_i of size $\lfloor y \rfloor$ and any $\Delta_1 + 1 - q$ matchings h_i of size $\lceil y \rceil$ satisfy (1), and we are done.

Hence we may assume that $p < q$. If $\Delta_2 - \Delta_1 \geq q - p$, then $\Delta_2 + 1 - q \geq \Delta_1 + 1 - p$, and so (4) and (6) imply that any p matchings h_i of size $\lfloor y \rfloor$ and any $\Delta_1 + 1 - p$ matchings h_i of size $\lceil y \rceil$ satisfy (1), and again we are done.

Hence we can assume that $p < q$ and $\Delta_2 - \Delta_1 < q - p$. If $\lfloor y \rfloor - \lceil x \rceil \geq (n_2 - n_1)/2$ then the previous inequality along with (4) and (6) imply that any $q + \Delta_1 - \Delta_2$

matchings h_i of size $\lfloor y \rfloor$ and any $\Delta_2 + 1 - q$ matchings h_i of size $\lceil y \rceil$ satisfy (1) and again we are done.

Thus we are left with the case $p < q$, $\Delta_2 - \Delta_1 < q - p$, and $\lfloor y \rfloor - \lceil x \rceil < (n_2 - n_1)/2$. In this case, (4) and (6) imply that $\lfloor y \rfloor - \lfloor x \rfloor = \lceil y \rceil - \lceil x \rceil = (n_2 - n_1)/2$, and so by Observation 2,

$$q - p = (\Delta_1 + 1)\left(\frac{n_2 - n_1}{2} - y + x\right) + (\Delta_2 - \Delta_1)(\lceil y \rceil - y).$$

Since (3) holds and since $\lceil y \rceil - y < 1$, it follows that $q - p < \Delta_2 - \Delta_1$, a contradiction.

Next assume that

$$y - x < \frac{n_2 - n_1}{2}. \tag{8}$$

Since $\lfloor y \rfloor - \lfloor x \rfloor < y - (x - 1)$, it follows that $\lfloor y \rfloor - \lfloor x \rfloor < (n_2 - n_1)/2 + 1$. But then the integrality of both r.h.s and l.h.s in the previous inequality implies that

$$\lfloor y \rfloor - \lfloor x \rfloor \leq \frac{n_2 - n_1}{2}. \tag{9}$$

Now if $y = \lfloor y \rfloor$ then $|h_i| = y$ for every i. Since $|f_i| \geq \lfloor x \rfloor$ for every i, it follows that $|h_i| - |f_i| \leq \lfloor y \rfloor - \lfloor x \rfloor$ for every $i = 1, \ldots, \Delta_1 + 1$, and so (2) holds.

Hence we can assume that $\lfloor y \rfloor < y < \lceil y \rceil$. If $x = \lfloor x \rfloor$ then $\lfloor y \rfloor - \lfloor x \rfloor = \lfloor y \rfloor - x < y - x$, and so $\lfloor y \rfloor - \lfloor x \rfloor < (n_2 - n_1)/2$. But then the identity $\lceil y \rceil = \lfloor y \rfloor + 1$ and integrality of both the r.h.s. and the l.h.s. in the previous inequality imply that

$$\lceil y \rceil - \lfloor x \rfloor \leq \frac{n_2 - n_1}{2}. \tag{10}$$

Since $|h_i| \leq \lceil y \rceil$ for every i and since $|f_i| = \lfloor x \rfloor$ for every i, it follows that $|h_i| - |f_i| \leq \lceil y \rceil - \lfloor x \rfloor$ for every $i = 1, \ldots, \Delta_1 + 1$, and so (2) holds.

Hence we can assume that $\lfloor x \rfloor < x < \lceil x \rceil$ and $\lfloor y \rfloor < y < \lceil y \rceil$. In this case $\lceil x \rceil = \lfloor x \rfloor + 1$ and $\lceil y \rceil = \lfloor y \rfloor + 1$, and so (9) implies that

$$\lceil y \rceil - \lceil x \rceil \leq \frac{n_2 - n_1}{2}. \tag{11}$$

Note that

$$\lfloor y \rfloor - \lceil x \rceil < \frac{n_2 - n_1}{2}. \tag{12}$$

Now, let p denote the number of matchings f_i of size $\lfloor x \rfloor$ and let q denote the number of matchings h_i of size $\lfloor y \rfloor$. If $p \leq q$ and $\Delta_1 + 1 - p \leq \Delta_2 + 1 - q$, then (9) and (11) imply that any p matchings h_i of size $\lfloor y \rfloor$ and any $\Delta_1 + 1 - p$ matchings h_i of size $\lceil y \rceil$ satisfy (2), and we are done.

If $p \leq q$ and $\Delta_1 + 1 - p > \Delta_2 + 1 - q$, then (9), (11) and (12) imply that any $q - \Delta_2 + \Delta_1$ matchings h_i of size $\lfloor y \rfloor$ and any $\Delta_2 + 1 - q$ matchings h_i of size $\lceil y \rceil$ satisfy (2), and again we are done.

Hence, we can assume that $p > q$. If $\lceil y \rceil - \lfloor x \rfloor \leq (n_2 - n_1)/2$ then any q matchings h_i of size $\lfloor y \rfloor$ and any $\Delta_1 + 1 - q$ matchings h_i of size $\lceil y \rceil$ satisfy (2) and again we are done.

Thus we are left with the case $p > q$ and $\lceil y \rceil - \lfloor x \rfloor > (n_2 - n_1)/2$. In this case, (9) and (11) imply that $\lfloor y \rfloor - \lfloor x \rfloor = \lceil y \rceil - \lceil x \rceil = (n_2 - n_1)/2$, and so by Observation 2,

$$q - p = (\Delta_1 + 1)\left(\frac{n_2 - n_1}{2} - y + x\right) + (\Delta_2 - \Delta_1)(\lceil y \rceil - y).$$

Since (8) holds and since $\lceil y \rceil > y$, it follows that $q - p > 0$, a contradiction. Thus the theorem follows.

3   Even Join Graphs

In this section we shall show how Theorem 1 becomes a very useful tool when we want to optimally colour the edges of a graph that has an even number of vertices and it is the join of two regular graphs.

Observation 3. *Let G be a k-regular graph and let $C = \{c_1, \ldots, c_{k+1}\}$ be an equitable edge-colouring of G. Then, each vertex is missed by exactly one colour, and so the vertex set of G can be partitioned into the $k + 1$ subsets $X(c_1), \ldots, X(c_{k+1})$.*

Let $G = G_1 + G_2$ be a join graph with an even number n of vertices such that G_i is k_i-regular with n_i vertices. Without loss of generality we shall assume that $n_1 \leq n_2$. In [9] it was shown that if $k_1 > k_2$, or if $k_1 < k_2$ and $n_1 = n_2$ then G is Class 1. Hence we shall assume that $k_1 \leq k_2$, and that if $n_1 = n_2$ then $k_1 = k_2$. Moreover, we can always assume that $k_i < n_i - 1$ $(i = 1, 2)$, for otherwise $\Delta(G) = n - 1$, and so G would be a subgraph of the complete graph with n vertices which is Class 1 (because n is even).

In [8] it was shown that every regular join graph with an even number of vertices is Class 1. To prove this result, it was crucial to prove that every regular join graph $G = G_1 + G_2$ contains a $(k_1 + 1)$-regular spanning subgraph H such that H is $(k_1 + 1)$-edge-colourable and H contains G_1 as induced subgraph. Note that, when G is regular, $\Delta(G) = k_2 + n_1 = k_1 + n_2$. Here we want to generalize this result by just assuming that $\Delta(G) = k_2 + n_1$ (G is not necessarily regular). To this purpose, we first need a technical lemma that generalizes Theorem 1 in [8] and can be proved in a similar way.

Lemma 1. *Let $G = G_1 + G_2$ be a join graph with an even number of vertices such that G_i has n_i vertices with $n_1 < n_2$, and it is k_i-regular with $k_1 \leq k_2$ and $k_i < n_i - 1$. Let $C_1 = \{f_1, \ldots, f_{k_1+1}\}$ and $C_2 = \{h_1, \ldots, h_{k_2+1}\}$ be equitable edge-colourings of G_1 and G_2, respectively. If there exists an ordering of the elements of C_2 such that $|X(h_i)| \leq |X(f_i)|$ for every $i = 1, \ldots, k_1 + 1$, then G contains a spanning subgraph H that is $(k_1 + 1)$-regular and has the following two properties:*

(a) $H \supset G_1$,
(b) $\chi'(H) = k_1 + 1$.

Proof. (*Sketch*) Let H_2 be the spanning subgraph of G_2 induced by the matchings h_i ($i = 1, \ldots, k_1 + 1$). Note that each vertex of H_2 has degree equal to either k_1 or $k_1 + 1$ (by Observation 3); let A denote the set of vertices having degree k_1 and let B denote the set of vertices having degree $k_1 + 1$. By construction, $A = \cup_{i=1}^{k_1+1} X(h_i)$ and $|A| = \sum_{i=1}^{k_1+1} |X(h_i)|$. Set $\alpha_i = |X(f_i)| - |X(h_i)|$ ($i = 1, \ldots, k_1 + 1$). Since $|X(h_i)| = n_2 - 2|h_i|$ and $|X(f_i)| = n_1 - 2|f_i|$, it follows that each α_i is a nonnegative even integer (because $n_2 - n_1$ is even).
To build the required graph H we shall consider two cases.

1. $\alpha_i = 0$ for every $i = 1, \ldots, k_1 + 1$.

With every vertex u_j of V_1 associate a vertex v_{u_j} of H_2 so that if $u_j \in X(f_i)$ for some i, then $v_{u_j} \in X(h_i)$ (this can be done because $|X(f_i)| = |X(h_i)|$ for every i). Let $e_j = u_j v_{u_j}$, $j = 1, \ldots, n_1$. Then H is the spanning subgraph of G that is formed by G_1, H_2, and the n_1 edges e_j. Clearly H is $(k_1 + 1)$-regular and it satisfies property (a). Now, identify each colour h_i with colour f_i, and colour edge $e_j = u_j v_{u_j}$ with the colour f_i missing both u_j and v_{u_j}. Then $\chi'(H) = k_1 + 1$ and the lemma holds.

2. $\alpha_i > 0$ for some i.

From every matching h_i (with $\alpha_i > 0$) remove precisely $\alpha_i/2$ "special" edges in order to obtain a new matching h_i' such that $|X(h_i')| = |X(h_i)| + \alpha_i = |X(f_i)|$ (the special edges are edges having at least one endpoint of degree $k_1 + 1$ in H_2). Set $h_i := h_i'$ and $\alpha_i = 0$. Proceed as in Case 1.

Corollary 1. *Let $G = G_1 + G_2$ be a join graph with an even number of vertices such that G_i is a k_i-regular graph having n_i vertices, with $n_1 < n_2$, and $k_i < n_i - 1$. If $\Delta(G) = k_2 + n_1$ then G contains a spanning subgraph H that is $(k_1 + 1)$-regular, it properly contains G_1 and it is Class 1.*

Proof. Let $\mathcal{C}_1 = \{f_1, \ldots, f_{k_1+1}\}$ and $\mathcal{C}_2 = \{h_1, \ldots, h_{k_2+1}\}$ be equitable edge-colourings of G_1 and G_2, respectively. Let m_i denote the number of edges of G_i, $i = 1, 2$. By assumption, $\Delta(G) = k_2 + n_1$ and $n_1 < n_2$, and so $k_1 < k_2$. By Lemma 1, it is sufficient to show that there exists an ordering of the elements of $\mathcal{C}_2$ such that $|X(h_i)| \leq |X(f_i)|$ for every $i = 1, \ldots, k_1 + 1$. To this purpose, we only need verify that $m_2/(k_2 + 1) - m_1/(k_1 + 1) \geq (n_2 - n_1)/2$: indeed as soon as this is accomplished, the desired ordering exists by Theorem 1. Now,

$$\frac{m_2}{k_2 + 1} - \frac{m_1}{k_1 + 1} = \frac{n_2 k_2}{2(k_2 + 1)} - \frac{n_1 k_1}{2(k_1 + 1)} = \frac{(n_2 - n_1)k_1 k_2 + n_2 k_2 - n_1 k_1}{2(k_1 + 1)(k_2 + 1)}.$$

By assumption, $\Delta(G) = k_2 + n_1$, and so $k_2 - k_1 \geq n_2 - n_1$. Now, $(n_2 - n_1)(n_1 + k_2) = (n_2 - n_1)n_1 + (n_2 - n_1)k_2$, and so $(n_2 - n_1)(n_1 + k_2) \leq (k_2 - k_1)n_1 + (n_2 - n_1)k_2 = n_2 k_2 - n_1 k_1$. Then

$$\frac{m_2}{k_2 + 1} - \frac{m_1}{k_1 + 1} \geq \frac{n_2 - n_1}{2} \frac{k_1 k_2 + n_1 + k_2}{(k_1 + 1)(k_2 + 1)}.$$

Since $k_1 k_2 + n_1 + k_2 > (k_1 + 1)(k_2 + 1)$ (because $n_1 > k_1 + 1$), the corollary follows.

Now we are ready to prove our main result.

Theorem 2. *Let $G = G_1 + G_2$ be a join graph with an even number n of vertices such that G_i is a noncomplete k_i-regular graph with n_i vertices, with $n_1 < n_2$. If $\Delta(G) = k_2 + n_1$ then G is Class 1.*

Proof. Write $G = (V, E)$, $G_i = (V_i, E_i)$ $(i = 1, 2)$. By Corollary 1, G contains a $(k_1 + 1)$-regular spanning subgraph $H = (V, F)$, having G_1 as an induced subgraph, and such that $\chi'(H) = k_1 + 1$.

Now, consider the graph $G - H = (V, E - F)$. Note that the set V_1 is an independent set of $G - H$ of size n_1 (because G_1 is an induced subgraph of H); moreover, in the graph $G - H$, every vertex in V_1 is adjacent to precisely $n_2 - 1$ vertices in V_2. Hence, $d_{G-H}(u) = n_2 - 1$ for every vertex u in V_1. Since H is $(k_1 + 1)$-regular it follows that, for every vertex v in V_2, $d_{G-H}(v) = d_G(v) - (k_1 + 1) = k_2 - k_1 + n_1 - 1$, and so $\Delta(G - H) = k_2 - k_1 + n_1 - 1$. Hence

(a) $d_{G-H}(u) = n_2 - 1 \leq \Delta(G - H)$ for every vertex u in V_1,
(b) $d_{G-H}(v) = \Delta(G - H)$ for every vertex v in V_2.

Now, let $G_2' = (V_2, E_2')$ denote the subgraph of $G - H$ induced by V_2. Clearly, the number of edges of G_2' is equal to the number of edges of $G - H$ minus the number of edges in $G - H$ that join vertices in V_1 to vertices in V_2, that is $|E_2'| = |E - F| - n_1(n_2 - 1)$. Now (a) and (b) imply that $G - H$ has precisely $n_1(n_2 - 1)/2 + n_2 \Delta(G - H)/2$ edges, and so

$$|E_2'| = \frac{n_2 \Delta(G - H)}{2} - \frac{n_1(n_2 - 1)}{2}. \tag{13}$$

Since $\Delta(G_2') \leq k_2 < n_2 - 1 \leq \Delta(G - H)$, it follows that we can colour the edges of G_2' with $\Delta(G - H)$ colours. Let $\mathcal{C} = \{c_1, \ldots, c_{\Delta(G-H)}\}$ be an equitable $\Delta(G - H)$-edge-colouring of G_2'.

Set $x = |E_2'|/\Delta(G - H)$; by (13)

$$x = \frac{n_2}{2} - \frac{n_1(n_2 - 1)}{2\Delta(G - H)}.$$

Since $n_2 - 1 \leq \Delta(G - H)$, it follows that $x \geq (n_2 - n_1)/2$, and so $\lfloor x \rfloor \geq (n_2 - n_1)/2$. Then, for every i, $|c_i| \geq (n_2 - n_1)/2$ (because $|c_i| \geq \lfloor x \rfloor$), and so

$$|X(c_i)| = n_2 - 2|c_i| \leq n_1, \quad i = 1, \ldots, \Delta(G - H). \tag{14}$$

For every vertex v_j in V_2 let $t(v_j)$ denote the number of vertices u_i in V_1 that are nonadjacent to v_j in the graph $G - H$; to put it differently, $t(v_j)$ is nothing but the number of vertices u_i that are adjacent to v_j in the graph H. Then, in the graph $G - H$ every vertex v_j is adjacent to precisely $n_1 - t(v_j)$ vertices u_i, and so

$$d_{G_2'}(v_j) = d_{G-H}(v_j) - (n_1 - t(v_j)) = \Delta(G - H) - (n_1 - t(v_j)). \tag{15}$$

Now, we are ready to extend the colouring $\mathcal{C}$ to all the edges of $G - H$ that join vertices in V_1 to vertices in V_2. To this purpose, let Q denote this set of edges. Let B denote the bipartite graph with bipartition C and V_2, and edge set $\{c_i v_j : c_i \in C, v_j \in V_2,$ and v_j is missed by colour $c_i\}$.

Observe that (14) and (15) imply that $d_B(c_i) \leq n_1$ for every $c_i \in C$, and $d_B(v_j) = n_1 - t(v_j) \leq n_1$ for every $v_j \in V_2$. Let $E(B)$ denote the set of edges of B. We have: $|E(B)| = \sum_{v_j \in V_2} d_B(v_j) = \sum_{v_j \in V_2}(n_1 - t(v_j))$. Since $\sum_{v_j \in V_2} t(v_j) = n_1$, it follows that $|E(B)| = n_1(n_2 - 1)$. Note that at least one v_j has degree equal to n_1 in the bipartite graph B: if, for every vertex v_j in V_2, $d_B(v_j) \leq n_1 - 1$ then $\sum_{v_j \in V_2} d_B(v_j) \leq n_2(n_1 - 1)$, and so $|E(B)| \leq n_2(n_1 - 1)$. But then $n_1 \geq n_2$, contradicting the assumption that $n_1 < n_2$. Thus $\Delta(B) = n_1$.

Then, there exists an equitable edge-colouring $\mathcal{D} = \{d_1, \ldots, d_{n_1}\}$ of B. Since $|E(B)| = n_1(n_2 - 1)$, it follows that $|d_i| = n_2 - 1$ for every i, and so each matching d_i misses exactly one of the n_2 vertices of V_2, and such a vertex must have degree smaller than n_1 in B. Now, in the graph $G - H$ each vertex u_i in V_1 is adjacent to every vertex v_j in V_2, but one; let v_{u_i} denote such a vertex. Without loss of generality, we may assume that, for every $i = 1, \ldots, n_1$, the matching d_i misses precisely v_{u_i} (this can always be done because $t(v_{u_i}) \geq 1$ and so $d_B(v_{u_i}) \leq n_1 - 1$).

Finally, consider an arbitrary edge $c_i v_j$ of B (with $c_i \in C$ and $v_j \in V$) and let d_r be its colour. Since v_j is not missed by d_r, it follows that the edge $v_j u_r$ belongs to Q. We claim that we can colour edge $v_j u_r$ with colour c_i. To verify that the colouring so obtained is admissible, assume the contrary: there exist in Q two adjacent edges e and e' that have the same colour c_i. Let $e = v_j u_r$. If $e' = v_h u_r$ then in B both edges $c_i v_j$ and $c_i v_h$ would be coloured d_r, which is impossible; if $e' = v_j u_t$ then in B the edge $c_i v_j$ would be coloured both d_r and d_t, which again is impossible. Thus, $\chi'(G - H) = \Delta(G - H)$. But then, $\chi'(G) \leq \chi'(H) + \chi'(G - H) \leq (k_1 + 1) + (k_2 - k_1 + n_1 - 1) = \Delta(G)$. Thus G is Class 1 and the theorem follows.

An immediate consequence of the above results, is that, under the assumptions of Theorem 2, there exists a polynomial time algorithm for finding a $\Delta(G)$-edge-colouring of a graph G that is the join of two regular graphs. The running time of this algorithm lies on the complexity of finding equitable edge-colourings. Since an equitable edge-colouring of a graph with q vertices can be found in time $O(q^4)$ (see Proposition 1), it follows that the running time of our algorithm is $O(n^4)$.

The edge-colouring algorithm

Input. $G = G_1 + G_2$ such that: G_i noncomplete k_i-regular with n_i vertices, $n_1 < n_2$, $n_1 + n_2$ even, and $\Delta(G) = k_2 + n_1$.

1. Find H satisfying the properties in Corollary 1 and a $(k_1 + 1)$-edge-colouring $\mathcal{F}$ of H.
2. Set G'_2 as the subgraph of $G - H$ spanned by the vertices of G_2.
3. Find an equitable $\Delta(G - H)$-edge-colouring $\mathcal{C}$ of G'_2.
4. Extend $\mathcal{C}$ to a $\Delta(G - H)$-edge-colouring $\mathcal{C}^*$ of $G - H$.

Output. A $\Delta(G)$-edge-colouring of G: $\mathcal{F} \cup \mathcal{C}^*$.

References

1. Berry, R., Modiano, E.: Optimal Transceiver Scheduling in WDM/TDM Networks. IEEE J. on Select. Areas in Comm. 23, 1479–1495 (2005)
2. Chvátal, V.: On certain polytopes associated with graphs. J. Combin. Theory, Ser. B 18, 138–154 (1975)
3. Chetwynd, A.G., Hilton, A.J.W.: Star multigraphs with three vertices of maximum degree. Math. Proc. Cambridge Phil. Soc. 100, 303–317 (1986)
4. Chetwynd, A.G., Hilton, A.J.W.: The edge-chromatic class of graphs with maximum degree at least $|V| - 3$. Annals of Discrete Mathematics 41, 91–110 (1989)
5. Corneil, D.G., Lerchs, H., Burlingham, L.S.: Complement reducible graphs. Discrete Appl. Math. 3, 163–174 (1981)
6. Corneil, D.G., Perl, Y., Stewart, L.K.: A linear recognition algorithm for cographs. SIAM J. of Comput. 14, 926–934 (1985)
7. De Werra, D.: Investigations on an edge-coloring problem. Discrete Math. 1, 167–179 (1972)
8. De Simone, C., Galluccio, A.: Edge-colouring of regular graphs of large degree. Theor. Comp. Sc. 389, 91–99 (2007)
9. De Simone, C., de Mello, C.P.: Edge colouring of join graphs. Theor. Comp. Sc. 355, 364–370 (2006)
10. Holyer, I.: The NP-completeness of edge-colouring. SIAM J. Comput. 14, 718–720 (1981)
11. Hoffman, D.G., Rodger, C.A.: The chromatic index of complete multipartite graphs. J. Graph Theory 16, 159–163 (1992)
12. Möhring, R.H.: Algorithmic aspects of the substitution decomposition in optimization over relations, set systems and Boolean functions. Ann. Oper. Res. 4, 195–225 (1985)
13. McDiarmid, C.J.H.: The solution of a timetabling problem. J. Inst. Math. Appl. 9, 23–34 (1972)
14. Perkovic, L., Reed, B.: Edge coloring regular graphs of high degree. Discrete Math. 165/166, 567–570 (1997)
15. Perkovic, L.: Edge Coloring, Polyhedra and Probability, Ph.D Thesis Carnegie Mellon Univ., US (1998)
16. Vizing, V.G.: On an estimate of the chromatic class of a p-graph. Diskret Analiz 3, 25–30 (1964) (in Russian)

Magic Labelings on Cycles and Wheels

Andrew Baker and Joe Sawada

University of Guelph, Guelph, Ontario, Canada, N1G 2W1
{abaker04,jsawada}@uoguelph.ca

Abstract. We present efficient algorithms to generate all edge-magic and vertex-magic total labelings on cycles, and all vertex-magic total labelings on wheels. Using these algorithms, we extend the enumeration of the total labelings on these classes of graphs.

1 Introduction

Consider a wireless network in which every device must be able to connect to a subset of the other devices in the network using a unique channel to prevent collisions. One way to create such a channel assignment is to give numeric labels to the devices and channels in such a way that the labels of two devices and the communication line between them sum to a consistent value across every pair of devices in the network. In this case, knowing the labels of the two communicating devices gives the identification number of the communication line between them [1].

This solution is an example of an edge-magic total labeling (EMTL). EMTLs are one application of the "magic" concept of magic squares to graphs. Given a simple undirected graph $G = (V, E)$, let λ be a mapping from the numbers 1 through $|V| + |E|$ to the elements (vertices and edges) of G such that each element has a unique label. An *edge-magic total labeling* is a labeling λ in which the weight of each edge is the same. The weight of an edge is obtained by the sum of the label of the edge and the labels of its two endpoints and denoted by $w(e)$. If the weight is the same for every edge, it is termed the *magic constant* of the labeling, and is given by h. For an example of an EMTL with $h = 20$, see Fig 1(a).

A *vertex-magic total labelling* (VMTL) is a labeling λ in which the weight $w(v)$ of each vertex is the same. The weight of a vertex is obtained by adding the sum of the labels of the incident edges to the label of the vertex itself. If the weight is the same for every vertex in the graph, it is called the magic constant and is given by k. For an example of an VMTL with $k = 20$, see Fig 1(b).

A *totally magic labeling* is a labeling λ which is simultaneously both a vertex-magic total labeling and an edge-magic total labeling. The magic constants h and k are not necessarily equal. The class of totally magic graphs (those which admit a totally magic labeling) is much more restricted than the edge-magic or vertex-magic graphs. Figure 2 gives an example of a totally magic labeling on the cycle C_3. The only known connected totally magic graphs are K_1, K_3, and P_3. There are however an infinite number of disconnected totally magic graphs, as any graph consisting of a union of $2n + 1$ ($n \geq 0$, $n \in \mathbb{Z}$) disjoint triangles is a totally magic graph [1]. There are additional types of magic labelings described beyond

B. Yang, D.-Z. Du, and C.A. Wang (Eds.): COCOA 2008, LNCS 5165, pp. 361–373, 2008.
© Springer-Verlag Berlin Heidelberg 2008

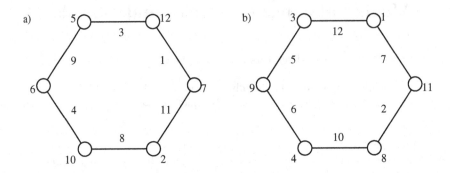

Fig. 1. Two C_6 graphs with corresponding edge-magic and vertex-magic total labelings. a) gives an edge-magic total labeling, and b) gives a vertex-magic total labeling.

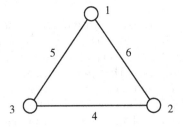

Fig. 2. The graph C_3 with a totally magic labeling. In this case, $h = 9$ (the edge-magic constant), and $k = 12$ (the vertex-magic constant).

EMTLs, VMTLs and totally magic labelings. For a more complete treatment, see Gallian's dynamic survey [2].

Depending on which labels are assigned to vertices and which to edges, it is possible to achieve labelings with different magic constants on the same graph. A lower bound for a VMTL is obtained by applying the largest $|V|$ labels to the vertices, while an upper bound is found by applying the smallest $|V|$ labels to the vertices. Summing the weights of every vertex in a VMTL gives us $\sum_{v \in V} w(v) = |V|k$. Every vertex label contributes to one weight (the weight of that vertex) while every edge label contributes to two weights (the weights of its two end points). Thus $|V|k = \sum_{v \in V} \lambda(v) + 2\sum_{e \in E} \lambda(e)$. By applying either the $|V|$ smallest or largest labels to the vertices, we can obtain the inequality

$$\frac{13n^2 + 11n + 2}{2(n+1)} \le k \le \frac{17n^2 + 15n + 2}{2(n+1)}$$

which gives basic limits on the magic constant of a graph without taking into account the structure of the graph [3]. Once the structure of the graph is taken into account, additional limits may be found. The set of integers which are delimited by these upper and lower bounds is the *feasible range*. The values which are the magic constant for some VMTL of a graph form the graph's *spectrum*. Therefore the spectrum is a subset of the feasible range.

In this paper we focus on finding all non-isomorphic VMTLs for cycles and wheels. Section 2 presents previous results with respect to vertex-magic total labelings on cycles and wheels. Sections 3 and 4 detail the enumeration algorithms and results for cycles and wheels respectively. Open problems for further research are presented in Section 5.

2 Background

Throughout this paper, we focus primarily on two classes of graph, the cycles and the wheels. The cycle C_n is given by the vertex set $v_1, v_2, \ldots, v_n \in V(G)$, and edge set $e_i \in E(G)$ where for $1 \leq i < n$, $e_i = \{v_i, v_{i+1}\}$ and $e_n = \{v_1, v_n\}$. Cycles are regular graphs (graphs in which every vertex has the same degree) as every vertex has degree 2. The wheels W_n consist of a cycle C_n together with an additional dominating vertex. A *dominating vertex* is a vertex which is adjacent to every other vertex in the graph. Figure 3 shows a sample wheel graph (W_6) and illustrates the naming scheme we will use while discussing parts of a wheel. Except for W_3, the wheels are not regular graphs.

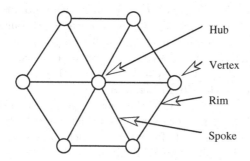

Hub

Vertex

Rim

Spoke

Fig. 3. The wheel graph W_6 demonstrating the naming convention we adopt for the elements of a wheel

2.1 Cycles

Every vertex-magic total labeling on a cycle (and indeed any regular graph) has a mirrored dual labeling. This property allows us, given an original labeling λ on graph G, to obtain the dual labeling λ' given by $\lambda'(v) = |V| + |E| + 1 - \lambda(v)$ for all vertices $v \in V(G)$ and $\lambda'(e) = |V| + |E| + 1 - \lambda(e)$ for all edges $e \in E(G)$. The resulting magic constant k' is given by $k' = 6n + 3 - k$ for cycles [4]. Consequently, the distribution of VMTLs by magic constant is symmetrical over the feasible range, and the presence of a VMTL achieving a magic constant in the upper half of the feasible range may be inferred by the presence of the dual labeling achieving the corresponding magic constant in the lower half of the feasible range.

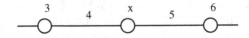

Fig. 4. A partial labeling of a piece of a graph with determined label x

Cycles also have a one-to-one correspondence between their edge- and vertex-magic total labelings. To obtain an EMTL λ_e from a vertex-magic total labeling λ_v, set $\lambda_e(v_i) = \lambda_v(e_i)$ and $\lambda_e(e_i) = \lambda_v(v_{(i+1) \bmod |V|})$ [4]. Figure 1 shows this correspondence graphically. Due to this relationship with EMTLs (which were developed earlier than VMTLs), previous work has been done to enumerate the edge-magic (and therefore also the vertex-magic) total labelings on cycles. The cycles C_3 through C_{10} were completely enumerated by Godbold and Slater [5]. We confirm these calculations, and also count the number of VMTLs/EMTLs on the cycles C_{11} through C_{18}.

Godbold and Slater show that a VMTL exists for every feasible magic constant for C_n when $n > 4$ [5]. Our enumeration breaks down the results for cycles by magic constant.

2.2 Wheels

As the wheel W_n consists of a cycle C_n together with a dominating hub vertex, W_n has $n + 1$ vertices and $2n$ edges. The vertices v_1 through v_n refer to the vertices of the cycle, with the rim edges r_1 through r_n corresponding to the cycle edges e_1 through e_n. The spoke edges are those which connect the hub to a cycle vertex, and are given by $s_i = \{hub, v_i\}$ for $1 \leq i \leq n$. We demonstrate this naming scheme graphically in Figure 3.

A general conjecture on VMTLs is that having vertices in a graph which differ widely with regard to their degrees prevents that graph from having a vertex-magic total labeling. This holds for wheels, which have a high-degree hub, as shown by MacDougall, Miller and Wallis in [3].

MacDougall *et al.* give two different methods of computing a feasible range for wheel graphs, and the true feasible range is given by the most restrictive maximum and minimum values. In addition to the bounds on the feasible range given earlier, the feasible range on wheels can been further bounded from below by $k \geq \frac{(n+1)(n+2)}{2}$ and above by $k \leq 7n + 6$ once you take the structure of the wheel into account. For the wheels W_n with $n > 11$, the minimum magic constant is larger than the maximum magic constant, so no VMTL can exist. MacDougall *et al.* also enumerate the VMTLs on wheels for W_3 (which is also the complete graph K_4), W_4, and W_5 [3]. We extend these results, counting W_6 through W_{10}.

3 Cycle Algorithm

A naïve method to generate all vertex-magic total labelings for a graph is to simply try all $(|V| + |E|)!$ permutations of the mapping of the labels onto the

```
function initializeCycle ()
    for each available label i where i ≤ n + 1 do
        λ(v₁) := i
        avail[i] := false
        for each available label j do
            λ(e₁) := j
            avail[j] := false
            λ(eₙ) := k − λ(v₁) − λ(e₁)
            if λ(e₁) < λ(eₙ) ≤ 2n and avail[λ(eₙ)] then
                avail[λ(eₙ)] := false
                extendCycle (2)
                avail[λ(eₙ)] := true
            avail[j] := true
        avail[i] := true
```

Fig. 5. Pseudocode for the initialization function for cycles. Global variables n and k are set to the desired values before the initializeCycle function is called.

```
function extendCycle (t)
    if t = n then
        λ(vₙ) := k − λ(eₙ) − λ(eₙ₋₁)
        if λ(v₁) < λ(vₙ) ≤ 2n and avail[λ(vₙ)] then
            Print ()
    else
        for each available label i where i > λ(v₁) do
            λ(vₜ) := i
            avail[i] := false
            λ(eₜ) := k − λ(vₜ) − λ(eₜ₋₁)
            if 0 < λ(eₜ) ≤ 2n and avail[λ(eₜ)] then
                avail[λ(eₜ)] := false
                extendCycle (t + 1)
                avail[λ(eₜ)] := true
            avail[i] := true
```

Fig. 6. Pseudocode for the extend function for cycles. Global variables n and k are set to the desired values before the extendCycle function is called.

elements of the graph, and check to see if each result is a VMTL. Not only does this rapidly become infeasible on its own, as the size of the cycle increases it will also allow isomorphic copies of the same labeling to be generated independently. As such, every successfully generated VMTL must be compared to every other previously generated VMTL in order to remove duplicate copies.

Our general approach is to apply vertex and edge labels working iteratively around the cycle. In the cycle VMTL generation algorithm, we remove cases of rotational symmetry by assigning the smallest vertex label to v_1, and then handle the reflective symmetry by making sure that $\lambda(e_1) < \lambda(e_n)$. Since v_1 must receive the smallest vertex label, it cannot be larger than $n + 1$ or there will be insufficient labels to label the remaining vertices.

Table 1. The total number of non-isomorphic VMTLs for cycle graphs C_n $(3 \leq n \leq 18)$

C_n							
n	Unique VMTLs	n	Unique VMTLs	n	Unique VMTLs	n	Unique VMTLs
3	4	7	118	11	36128	15	74931690
4	6	8	282	12	206848	16	613296028
5	6	9	1540	13	1439500	17	5263250382
6	20	10	7092	14	10066876	18	47965088850

Since we are interested in calculating the number of VMTLs for each magic constant, the algorithm we develop takes both n (the size of the cycle) and k (the magic constant) as input.

We say that a label is a *determined label* if it contributes its value to the weight of a vertex for which every other contributing label is known. Assuming we know the magic constant we are trying to reach, there is only one possible value for the determined label. There are three conditions on a determined label $\lambda(x)$ which allow us to terminate the recursion tree at this node and backtrack. These conditions are:

1. $\lambda(x) < 1$,
2. $\lambda(x) > |V| + |E|$, and
3. $\lambda(x)$ has already been used in this labeling.

Figure 4 gives a partial labeling and illustrates a determined label. In this example, if the desired magic constant is 20, then x must be 11. However, if the desired magic constant is 15, then x would have to be 6. Since 6 has already been used in this labeling, it would not be a valid partial labeling for $k = 15$. If in a cycle we have magic constant k, then $\lambda(e_2) = k - \lambda(v_2) - \lambda(e_1)$. Then, once we know $\lambda(v_3)$ and $\lambda(e_2)$, we are able to determine $\lambda(e_3)$.

Our algorithm aims to obtain determined labels as quickly as possible. If a given label being applied in the algorithm is determined and the partial labeling is infeasible, then the entire computation subtree rooted at that partial labeling can immediately be excluded. Even in the worst-case scenario, where none of the determined labels eliminates a partial labeling, the use of a determined label reduces the branching factor at a position in the computation tree from $1 \leq i \leq |V| + |E|$ to 1.

Before the algorithm itself is called, the global variables n and k are set with the size of the cycle and desired magic constant respectively, and the available list is initialized to every label being currently available. The actual algorithm begins with an initialization phase (by a call to **initializeCycle()**) which sets the labels of a vertex and two edges (v_1, e_1, and e_n). The initialize function then calls **extendCycle(2)** which recursively labels the remaining vertices and edges. Execution completes when there is only one vertex (v_n) remaining without a label. A linked list of unused labels is maintained at all times. This way, the more complete the partial labeling becomes, the fewer potential labels must be considered for each non-determined element.

Table 2. The number of unique VMTLs for cycle graphs C_3 through C_{10} broken down by magic constant (k). (Note that the duals have not been included.)

C_3		C_4		C_5		C_6	
k	Unique VMTLs	k	Unique VMTLs	k	Unique VMTLs	k	Unique VMTLs
9	1	12	1	14	1	17	3
10	1	13	2	15	0	18	1
				16	2	19	6

C_7		C_8		C_9		C_{10}	
k	Unique VMTLs	k	Unique VMTLs	k	Unique VMTLs	k	Unique VMTLs
19	9	22	10	24	31	27	125
20	10	23	19	25	43	28	236
21	11	24	57	26	125	29	698
22	29	25	55	27	264	30	1138
				28	307	31	1349

C_{11}		C_{12}		C_{13}		C_{14}	
k	Unique VMTLs	k	Unique VMTLs	k	Unique VMTLs	k	Unique VMTLs
29	308	32	1602	34	3809	37	32077
30	711	33	4111	35	10967	38	91866
31	1781	34	10834	36	33951	39	299525
32	3371	35	19183	37	79234	40	576701
33	4945	36	30877	38	139499	41	977354
34	6948	37	36817	39	202253	42	1427929
				40	250037	43	1627986

C_{15}		C_{16}		C_{17}		C_{18}	
k	Unique VMTLs	k	Unique VMTLs	k	Unique VMTLs	k	Unique VMTLs
39	63995	42	884789	44	1152784	47	26677502
40	284590	43	2706053	45	8660408	48	104169715
41	889063	44	8685625	46	30280605	49	351608789
42	2332807	45	20266824	47	86881643	50	859974262
43	4402572	46	37574150	48	187828262	51	1815449072
44	7339913	47	59829497	49	336981439	52	3082588134
45	10395599	48	83018416	50	511013242	53	4648495519
46	11757306	49	93682660	51	683131331	54	6154283390
				52	785695477	55	6939298042

The initialization function starts the labeling by attempting every possible label for vertex v_1 and edge e_1. We require that $\lambda(v_1)$ be the minimal vertex label in order to remove rotational symmetry. The maximum possible label for v_1 is $n+1$ due to the fact that since v_1 receives the minimum vertex label, we must retain $n-1$ labels greater than $\lambda(v_1)$ for the other vertices. This determines the label for edge e_n. In order to remove reflective symmetry, we require $\lambda(e_n) > \lambda(e_1)$. The initialization function then calls the extend function with parameter 2. The pseudocode for the initialization function can be found in Figure 5.

Table 3. The total number of non-isomorphic VMTLs for wheel graphs W_n ($3 \leq n \leq 8$)

W_n					
n	Unique VMTLs	n	Unique VMTLs	n	Unique VMTLs
3	14	6	859404	9	17804388662
4	2080	7	22063500	10	418858095690
5	31892	8	637402504	11	pending

The extend method takes a single parameter - the position (t) in the cycle which is to be generated. A single loop applies, in turn, every unused label greater than $\lambda(v_1)$ to vertex v_t. Applying a label to v_t determines the label for e_t. The extend method then calls itself with parameter $t + 1$. The recursion terminates when $t = n$. At this point there is only one unlabeled element, v_n, which is obviously determined. If the single remaining label is the required label, then the VMTL is successfully completed and the **Print()** method is called. **Print()** is a generic function which can be used to perform any operation on the completed VMTL. In the case of enumeration, a count of the number of VMTLs is incremented. Figure 6 gives the pseudo-code for the extend function.

In order to obtain results more quickly, the algorithm is parallelized to run on multiple different processors. Each process is given an integer value as a command-line argument which acts as a static value for the first element to be assigned a label. Instead of iterating through all available values, the algorithm simply uses the supplied label. As the runtime increases for larger graphs, the problem is distributed to more processors by supplying two seed values which determine the first two elements to receive labels.

3.1 Results

Table 1 gives the total number of EMTLs/VMTLs on the cycles C_3 through C_{18}, of which C_{11} through C_{18} had not previously been enumerated. Table 2 give the number of unique labelings broken down by magic constant.

4 Wheel Algorithm

Wheel graphs have a clear relationship to the cycles so the algorithm for generating all unique VMTLs on wheel W_n bears a similarity to the algorithm for cycle C_n. However, the extra vertex and additional n edges complicate the process.

As with the cycle algorithm, our wheel VMTL generation algorithm applies vertex and edge labels working iteratively around the edge of the cycle portion of the wheel. We remove rotational symmetry by assigning the smallest spoke label to s_1. As with cycles, reflective symmetry is removed by ensuring that $\lambda(r_1) < \lambda(r_n)$. We use the s_1 instead of the v_1 to remove rotational symmetry for the wheel in order to trim the computation tree of partial labelings which will result in a hub with excessive weight more efficiently. Since s_1 must receive the

Table 4. The number of unique VMTLs for wheel graphs W_3 through W_5 broken down by magic constant (k)

	W_3		W_4		W_5
k	Unique VMTLs	k	Unique VMTLs	k	Unique VMTLs
19	0	26	89	32	239
20	2	27	149	33	1242
21	5	28	522	34	2694
22	0	29	376	35	5180
23	5	30	573	36	7873
24	2	31	211	37	7173
25	0	32	131	38	4124
		33	29	39	2511
				40	776
				41	80

	W_6		W_7		W_8
k	Unique VMTLs	k	Unique VMTLs	k	Unique VMTLs
39	5978	45	24998	52	795294
40	36945	46	204170	53	7352502
41	76335	47	880257	54	28521585
42	158805	48	2198247	55	64090384
43	173887	49	3637665	56	106131735
44	187409	50	4760707	57	132239986
45	116447	51	4425875	58	133415487
46	77827	52	3384967	59	92798616
47	21793	53	1818749	60	53134373
48	3978	54	646233	61	17008206
		55	81632	62	1914336

	W_9		W_{10}		W_{11}
k	Unique VMTLs	k	Unique VMTLs	k	Unique VMTLs
58	0	66	1739667155	78	pending
59	34364364	67	4780216858	79	pending
60	236314351	68	18515045434	80	pending
61	833847423	69	39874554946	81	pending
62	1846542901	70	75518840087	82	162942689359
63	2996328931	71	84888911188	83	8201853531
64	3821193834	72	90187289669		
65	3553033163	73	60230503071		
66	2649033979	74	33425583234		
67	1364327018	75	9122758622		
68	435740211	76	574725426		
69	33662487				

```
function initializeWheel ()
    for each available label i do
        λ(s₁) := i
        hubWeight := λ(s₁)
        avail[i] := false
        for each available label j do
            λ(r₁) := j
            avail[j] := false
            for each available label p where p > λ(r₁) do
                λ(rₙ) := p
                avail[p] := false
                λ(v₁) := k − λ(s₁) − λ(r₁) − λ(rₙ)
                if 0 < λ(v₁) ≤ 3n + 1 and avail[λ(v₁)] then
                    avail[λ(v₁)] := false
                    extendWheel (2)
                    avail[λ(v₁)] := true
                avail[p] := true
            avail[j] := true
        avail[i] := true
```

Fig. 7. Pseudocode for the initialization function for wheels. Global variables n and k are set to the desired values before the initializeWheel function is called.

smallest spoke label, it cannot be larger than $2n + 2$ or there will be insufficient labels to label the remaining spokes.

Determined labels continue to be an asset to remove subtrees of the computation tree. In this case, we require three labels in order to determine a fourth. For example, $\lambda(r_2) = k - \lambda(s_2) - \lambda(v_2) - \lambda(r_1)$.

As in the case of the cycles, before the algorithm itself is called, the global variables n and k are set with the size of the wheel and desired magic constant respectively, and the available list is initialized to every label being currently available. The actual algorithm begins with an initialization phase (by a call to **initializeWheel()**) which labels s_1, e_1, e_n, and v_1 in such a way as to prevent isomorphic labelings from being generated. The initialize function then calls **extendWheel(2)** which recursively labels a spoke, exterior vertex, and rim and calls itself until only s_n, v_n, and the hub remain, which are then labeled by a call to **finalizeWheel()**. Also like the cycle algorithm, a linked list consisting of the unused labels is maintained in order to improve efficiency as the partial labeling becomes more complete.

The initialization function starts the labeling by attempting every possible label for spoke edge s_1 and rim edge r_1. Every possible label greater than $\lambda(r_1)$ is applied to r_n, thus removing reflective symmetry. This determines the label for vertex v_1. The initialization function then calls the extend function with parameter 2. The pseudocode for the initialization function is given in Figure 7.

The extend function takes a single parameter t which gives the position of the wheel currently being expanded and applies every possible label greater than

```
function extendWheel (t)
    if t = n then
        finalizeWheel()
    else
        for each available label i where i > λ(s₁) do
            λ(sₜ) := i
            avail[i] := false
            hubWeight := hubWeight + λ(sₜ)
            potentialHub := the minimum available label
            potentialSpokes := the sum of the n − t smallest available labels > λ(s₁)
            if (hubWeight + potentialHub + potentialSpokes < k) then
                for each available label j do
                    λ(rₜ) := j
                    avail[j] := false
                    λ(vₜ) := k − λ(sₜ) − λ(rₜ) − λ(rₜ₋₁)
                    if 0 < λ(vₜ) ≤ 3n + 1 and avail[λ(vₜ)] then
                        avail[λ(vₜ)] := false
                        extendWheel (t + 1)
                        avail[λ(vₜ)] := true
                    avail[j] := true
            avail[i] := true
```

Fig. 8. Pseudocode for the extend function for wheels. Global variables n and k are set to the desired values before the extendWheel function is called.

```
function finalizeWheel ()
    for each available label i where i > λ(s₁) do
        λ(sₙ) := i
        avail[i] := false
        λ(vₙ) := k − λ(sₙ) − λ(rₙ) − λ(rₙ₋₁)
        if 0 < λ(vₙ) ≤ 3n + 1 and avail[λ(vₙ)] then
            avail[λ(vₙ)] := false
            λ(hub) := k − ∑ⁿᵢ₌₁ λ(sᵢ)
            if 0 < λ(hub) ≤ 3n + 1 and avail[λ(hub)] then
                Print ()
            avail[λ(vₙ)] := true
        avail[i] := true
```

Fig. 9. Pseudocode for the finalize function for wheels. Global variables n and k are set to the desired values before the finalizeWheel function is called.

$\lambda(s_1)$ to s_t. The extend function then applies every possible label to v_t which determines the label for r_t. The finalize function is called when $t = n$.

In order to prune the computation tree more effectively, we keep a close watch on the weight of the hub vertex through the variable $hubWeight$. Due to its high degree, its weight can easily exceed the desired magic constant. Every time a label is applied to a spoke, the partial hub weight variable is updated. Once in each iteration of the extend method, we check to ensure that the minimum

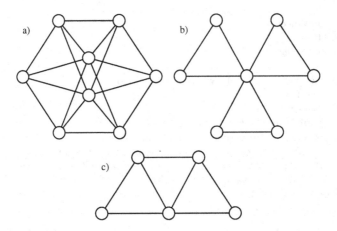

Fig. 10. Examples of graphs in three classes related to the wheels: a) t-fold wheel, b) friendship graph, c) fan

weight the hub can achieve is less than or equal to the desired magic constant. The minimal weight is given by the partial weight plus the smallest unused label (applied to the hub) and the $n - t$ smallest unused labels which are greater than $\lambda(s_1)$. If the minimal hub weight is larger than the desired magic constant, the partial labeling fails and the next set of labels is considered. The pseudocode for the extend function can be found in Figure 8.

The finalize function tries every available label for s_n which is greater than $\lambda(s_1)$. This determines the labels for both v_n and the hub. If these last labels can be successfully applied, then the **Print()** method is called which increments the number of labelings. Figure 9 gives the pseudocode for the finalize function.

4.1   Results

Table 3 gives the total number of VMTLs on the wheels W_3 through W_{10}, of which W_6 through W_{10} had not previously been enumerated. Results on W_{11} are currently pending completion. Table 4 gives the number of unique labelings broken down by magic constant. Of particular note is the fact that W_9 does not have a VMTL for $k = 58$ even though it is in the feasible range as given by MacDougall, Miller, and Wallis in [3]. Goemans gave a counting argument showing why no VMTL on W_9 can have $k = 58$ after we posed the problem of the missing labeling [6].

5   Conclusion and Open Problems

As there are EMTLs/VMTLs for all cycles C_n with $n \geq 3$, the number of unique labelings on larger cycles remain an open problem. It is desirable, however, to determine a formula which gives the number of EMTLs/VMTLs on a cycle of size n without having to actually count them.

In addition to the wheels, MacDougall, Miller and Wallis present other related classes of graphs which have similar size restrictions [3]. Figure 10 gives sample graphs for three of these related classes: fans, t-fold wheels, and friendship graphs.

Acknowledgements

This work was made possible by the facilities of the Shared Hierarchical Academic Research Computing Network (SHARCNET:www.sharcnet.ca).

References

1. Wallis, W.D.: Magic Graphs. Birkhäuser, New York (2001)
2. Gallian, J.A.: A dynamic survey of graph labeling. The Electronic Journal of Combinatorics 15, DS6 (2008)
3. MacDougall, J.A., Miller, M., Wallis, W.D.: Vertex-magic total labelings of wheels and related graphs. Utilitas Mathematica 62, 175–183 (2002)
4. MacDougall, J.A., Miller, M., Slamin, Wallis, W.D.: Vertex-magic total labelings of graphs. Utilitas Mathematica 61, 3–21 (2002)
5. Godbold, R.D., Slater, P.J.: All cycles are edge-magic. Bulletin of the ICA 22, 93–97 (1998)
6. Goemans, M.: Personal Communication (2008)

Minimum Cost Homomorphism Dichotomy for Locally In-Semicomplete Digraphs

A. Gupta[1], M. Karimi[1], E.J. Kim[2], and A. Rafiey[1]

[1] School of Computing Science
Simon Fraser University
Burnaby, B.C., Canada, V5A 1S6
{arvind,mmkarimi,arashr}@cs.sfu.ca
http://cs.sfu.ca
[2] Department of Computer Science
Royal Holloway, University of London
Egham, TW20 0EX, UK
eunjung@cs.rhul.ac.uk
http://www.cs.rhul.ac.uk/

Abstract. For digraphs G and H, a homomorphism of G to H is a mapping $f : V(G) \rightarrow V(H)$ such that $uv \in A(G)$ implies $f(u)f(v) \in A(H)$. In the *minimum cost homomorphism problem* we associate costs $c_i(u)$, $u \in V(G)$, $i \in V(H)$ with the mapping of u to i and the cost of a homomorphism f is defined $\sum_{u \in V(G)} c_{f(u)}(u)$ accordingly. Here the minimum cost homomorphism problem for a fixed digraph H, denoted by MinHOM(H), is to check whether there exists a homomorphism of G to H and to obtain one of minimum cost, if one does exit.

The minimum cost homomorphism problem is now well understood for digraphs with loops. For loopless digraphs only partial results are known. In this paper, we find a full dichotomy classification of MinHom(H), when H is a locally in-semicomplete digraph. This is one of the largest classes of loopless digraphs for which such dichotomy classification has been proved. This paper extends the previous result for locally semicomplete digraphs.

1 Introduction

For digraphs G and H, a mapping $f : V(G) \rightarrow V(H)$ is a *homomorphism of G to H* if $f(u)f(v)$ is an arc of H whenever uv is an arc of G. The problem of deciding whether there is a homomorphism from G to H, called the *homomorphism problem*, is NP-complete in general. However if we fix the target digraph H, the complexity of the homomorphism problem may be polynomial. From this point of view, establishing the computational complexity of the *homomorphism problem* for a fixed digraph H, denoted by HOM(H), has attracted great attention: Given a fixed digraph H, decide whether an input digraph G admits a homomorphism to H.

In the *list homomorphism problem* for a fixed digraph H, denoted by List-HOM(H), the input is a digraph G and the sets $L(u)$, $u \in V(G)$, of vertices of

B. Yang, D.-Z. Du, and C.A. Wang (Eds.): COCOA 2008, LNCS 5165, pp. 374–383, 2008.

H. We wish to decide whether there is a homomorphism f of G to H such that $f(u) \in L(u)$ for each $u \in V(G)$. In the *minimum cost homomorphism problem* we associate costs $c_i(u)$, $u \in V(G)$, $i \in V(H)$ with the mapping of u to i and the cost of a homomorphism f is defined $\sum_{u \in V(G)} c_{f(u)}(u)$ accordingly. Here the minimum cost homomorphism problem for a fixed digraph H, denoted by MinHOM(H), is to check whether there exists a homomorphism of G to H and to obtain one of minimum cost, if one does exit.

The minimum cost homomorphism problem was introduced, in the context of undirected graphs, in [15]. There, it was motivated by a real-world problem in defense logistics; in general, the problem seems to offer a natural and practical way to model many optimization problems.

Our interest is in proving dichotomies: given a class of problems such as HOM(H), we would like to prove that for each digraph H the problem is polynomial-time solvable, or NP-complete. This is, for instance, the case for HOM(H) with undirected graphs H [17]; in that case it is known that HOM(H) is polynomial time solvable when H is bipartite or has a loop, and NP-hard otherwise [17]. This is a dichotomy *classification*, since we specifically classify the complexity of the problems HOM(H), depending on H.

For undirected graphs H, a dichotomy classification for the problem Min-HOM(H) has been provided in [16]. Thus, the minimum cost homomorphism problem for graphs has been handled, and interest shifted to directed graphs. The first studies [12,13,14] focused on irreflexive digraphs (no vertex has a loop), where dichotomies have been obtained for digraphs H such that H is semicomplete or semicomplete multipartite digraph. Later, Gutin et al. could find the dichotomy for oriented cycles, which are fundamental structures in digraphs [8]. The dichotomy is also known for locally-semicomplete and quasi-transitive digraphs [6]. More recently, [10] promoted the study of digraphs with loops allowed; and, in particular, of reflexive digraphs. Hell et al. [5] verified the conjecture of Gutin and Kim [9], and proved the dichotomy for reflexive digraphs, the most general class of digraphs for which the dichotomy is known.

It is still an open problem whether there is a dichotomy classification for the complexity of MinHOM(H) when H is a digraph with possible loops. Gutin, Rafiey, and Yeo [8] conjectured that such a classification exists and, moreover, the following assertion holds:

Conjecture 1. Let H be a digraph with possible loops. Then MinHOM(H) is polynomial time solvable if H has either a Min-Max ordering or a k-Min-Max ordering for some $k \geq 2$. Otherwise, MinHOM(H) is NP-hard.

For the definition of Min-Max ordering and k-Min-Max ordering see Section 2.

In this paper, we verify this conjecture for locally in-semicomplete digraphs (see Section 2 for the definition), which is one of the largest classes of loopless digraphs for which such dichotomy classification has been proved. This paper extends the previous result for locally semicomplete digraphs [6]. Although the class of locally semicomplete digraphs is not very large, the class of locally in-semicomplete digraphs is large enough and contains a wide variety of digraphs from a very sparse digraphs like directed path to very dense like semicomplete digraphs.

Throughout this paper, we always assume that the fixed digraph H is locally in-semicomplete unless stated otherwise. The next section has been devoted to notation and terminology together with some preliminary results. When H is strongly connected, we will show in section 3 that the directed cycles are the only polynomial cases and obviously they have k-MinMax ordering. We will divide the non-strong locally in-semicomplete digraphs to digraphs having a cycle and acyclic digraphs. For digraphs having a cycle, We can find k-MinMax ordering by excluding a few forbidden induced subgraphs for which MinHom(H) are NP-hard. When H is acyclic the situations for which MinHom(H) is NP-hard are too many to classify as a few induced subgraphs. But, roughly speaking, after excluding this NP-hard cases, we will have Min-Max ordering for the remaining digraphs. Due to that, We guess in proving conjecture 1 in general, the most difficult class seems to be the class of acyclic digraphs.

2   Terminology and Preliminaries

A digraph D is *semicomplete* if, for each pair x, y of distinct vertices either x dominates y or y dominates x or both. A digraph D is *locally in-semicomplete* if for every vertex x of D, the in-neighbors of x induce a semicomplete digraph. We assume that all digraphs are loopless and do not have parallel arcs.

An (x, y)-path in a digraph D is a directed path from x to y. A digraph D is *strongly connected* (or, just, *strong*) if, for every pair x, y of distinct vertices in D, there exist an (x, y)-path and a (y, x)-path. A *strong component* of a digraph D is a maximal induced subgraph of D which is strong. A *strong component digraph* of a digraph D, abbreviated by $SCD(D)$, is obtained by contracting each strong component D_i of D into a single vertex v_i and placing an arc from v_i to v_j, $i \neq j$ if and only if there is an arc from D_i to D_j [2]. Observe that $SCD(D)$ is acyclic. We call a strong component an *initial strong component* if its corresponding vertex in $SCD(D)$ is of in-degree zero.

A digraph D is *path-mergeable* if for any choice of vertices x, y of $V(D)$ and any pair of internally disjoint (x, y)-paths P,Q, there exists an (x, y)-path R in D such that $V(R) = V(P) \cup V(Q)$. The following two propositions are due to Bang-Jensen, see [1].

Proposition 1. *Let D be a digraph which is path-mergeable and let $P = xx_1 \ldots x_r y$, $P' = xy_1 \ldots y_s y$, $r, s \geq 0$ be internally disjoint (x,y)-paths in D. The paths P and P' can be merged into one (x,y)-path P^* such that vertices from P(respectively, P') remain in the same order as on that path.*

Proposition 2. *Every locally in-semicomplete digraph is path-mergeable.*

A subgraph T of a digraph D is an *out-branching* if T is a spanning oriented tree of D with only one vertex s of in-degree zero (called *the root*). The following is a basic characterization of digraphs with out-branchings.

Proposition 3. *A connected digraph D contains an out-branching if and only if D has only one initial strong component, or equivalently, $SCD(D)$ has only one vertex of in-degree zero.*

Let D be any digraph. If $xy \in A(D)$, we say x dominates y or y is dominated by x, and denote by $x{\to}y$. For sets $X, Y \subset V(D)$, $X{\to}Y$ means that $x{\to}y$ for each $x \in X, y \in Y$, but no vertex of Y dominates a vertex in X. The *converse* of D is the digraph obtained from D by reversing the directions of all arcs.

A linear ordering $<$ of $V(H)$ is a *Min-Max ordering* if $i < j, s < r$ and $ir, js \in A(H)$ imply that $is \in A(H)$ and $jr \in A(H)$. It is known that if H admits a Min-Max ordering, then the problem MinHOM(H) is polynomial time solvable [12]. However, there are digraphs with polynomial MinHOM(H) which do not have Min-Max ordering [13].

Let H be a digraph and let $k \geq 2$ be an integer. We say that H has a *k-Min-Max ordering* if there is a k-partition of V into subsets $V_1, V_2, \ldots V_k$ and there is an ordering $v_1^i, v_2^i, \ldots, v_{\ell(i)}^i$ of V_i for each i such that

(i) Every arc of H is an arc from V_i to V_{i+1} for some $i \in \{1, 2, \ldots, k\}$,
(ii) $v_1^i, v_2^i, \ldots, v_{\ell(i)}^i v_1^{i+1} v_2^{i+1}, \ldots, v_{\ell(i+1)}^{i+1}$ is a Min-Max ordering of the subgraph of H induced by $V_i \cup V_{i+1}$ for each $i \in \{1, 2, \ldots, k\}$,

where all indices $i + 1$ are taken modulo k.

Due to [14], it is known that if H admits a k-Min-Max ordering, then Min-HOM(H) is polynomial time solvable.

3   Strong Locally In-Semicomplete Digraphs

We start to investigate the complexity of MinHOM(H) by considering the strongly connected case. The next observation is folklore, see [16].

Proposition 4. *Let H' be an induced subgraph of the digraph H. If MinHOM(H') is NP-hard, then MinHOM(H) is NP-hard.* □

Due to Proposition 4, in many cases it suffices to focus on small subgraphs and prove they are NP-hard instead of looking at the whole digraph. In the arguments which will follow, we shall sometimes omit to mention Proposition 4 when it is obvious from the context.

The following two Lemmas follow from [12], [6], and [9].

Lemma 1. *Let H be a semicomplete digraph containing a cycle and let $H \notin \{\overrightarrow{C_2}, \overrightarrow{C_3}\}$. Then MinHom($H$) is NP-hard.*

Lemma 2. *Let H be a digraph obtained from $\overrightarrow{C_k} = 12 \ldots k1, k \geq 2$, and an additional vertex $k + 1$. MinHom(H) is NP-hard if at least one of the following conditions hold.*

(a) $k + 1$ is dominated by at least two vertices of the cycle and no other arc exists.

(b) There are two consecutive vertices $i, i+1$ in $\overrightarrow{C_k}$ such that $i \rightarrow k+1$ and $k+1 \rightarrow i+1$, and no other arc exists.
(c) There are three consecutive vertices $i, i+1, i+2$ such that $i \rightarrow k+1$ and $k+1 \rightarrow \{i+1, i+2\}$, and no other arc exists.

The next theorem is the main result of this section and will use it later for characterizing general locally in-semicomplete digraphs.

Theorem 1. *Let H be a strongly connected locally in-semicomplete digraph. Then MinHom(H) is polynomial time solvable if H is a directed cycle. Otherwise, MinHom(H) is NP-hard.*

Proof: See [7] for the proof.   □

4   Nonstrong Locally In-Semicomplete Digraphs

The next theorem was first proved for locally in-tournament digraphs in [3] and later slightly modified into a more general statement in [4].

Theorem 2. *Let H be a connected non-strong locally in-semicomplete digraph. Then the following holds for H.*

(a) Let A and B be distinct strong components of H. If a vertex $a \in A$ dominates a vertex in B, then $a \rightarrow B$.
(b) H has only one initial strong component, or equivalently $SCD(H)$ has an out-branching.

Corollary 1. *Let H be a connected non-strong locally in-semicomplete digraph and consider the strong components of it. If H has a non-trivial initial strong component other than a directed cycle or a non-trivial non-initial strong component, then MinHOM(H) is NP-hard.*

Proof: By Theorem 1 every strong component of H should be a directed cycle, otherwise, MinHOM(H) is NP-hard. Now, suppose a non-initial strong component B is nontrivial, i.e. $|B| \geq 2$. From Theorem 2 it follows that there exists a vertex a such that $a \rightarrow B$. Choose an induced cycle C from B and let H' be the subgraph induced by $V(C) \cup \{a\}$. Then MinHOM(H') is NP-hard by part (a) of Lemma 2.   □

Theorem 2 and Corollary 1 above tell us that if MinHOM(H) is polynomial time solvable for a non-strong locally in-semicomplete digraph H, the structure of H is globally 'acyclic' once we shrink the initial strong component to a vertex.

In the next two subsections we will show that a locally in-semicomplete digraph H for which MinHOM(H) is polynomial time solvable has a special structure.

4.1   Locally In-Semicomplete Digraphs Having a Cycle

Let $\mathcal{N}$ be the class of connected non-strong locally in-semicomplete digraphs having a directed cycle C as an initial strong component where the other strong components are trivial.

Lemma 3. *Let O_1 be a digraph obtained from a directed cycle $\overrightarrow{C_k} = x_1 x_2 \ldots$ $x_k x_1, k \geq 2$, and the digraph D with vertex set $x_{k+m}, x_{k+m+1}, x_{k+m+2}, m \geq 0$, and arc set $\{x_{k+m}x_{k+m+1}, x_{k+m}x_{k+m+2}, x_{k+m+1}x_{k+m+2}\}$ by joining x_k to x_{k+m} with the directed path $x_k x_{k+1} \ldots x_{k+m}$. Then MinHom($O_1$) is NP-hard.*

Proof: See [7] for the proof. □

Let O_2 be a digraph obtained from a directed cycle of length k by adding an extra vertex dominated by two consecutive vertices of the cycle. MinHom(O_2) is NP-hard by part (a) of Lemma 2.

Theorem 3. *Consider a digraph $H \in \mathcal{N}$. If H does not contain O_1 and O_2 as an induced subgraph then H has a k-Min-Max ordering and thus MinHom(H) is polynomial time solvable. Otherwise, MinHom(H) is NP-hard.*

Proof: See [7] for the proof. □

4.2 Acyclic Locally In-Semicomplete Digraphs

Let $\mathcal{A}$ be the class of connected acyclic locally in-semicomplete digraphs. For any $H \in \mathcal{A}$ we have $SCD(H) = H$. Then, by Theorem 2, H, and any induced subgraph of H has an out-branching, and they have only one vertex of in-degree zero. We make a unique out-branching of H, denoted by $T(H)$, recursively as follows:

Input: $H \in \mathcal{A}$
Output: An out-branching $T(H)$
begin
Find the root of H, denoted by r;
Remove r from H to have different connected components $H_1, H_2, \ldots, H_i$, $i \geq 0$;
Find the roots of $H_1, H_2, \ldots, H_i, i \geq 0$, which are denoted by $r_1, r_2, \ldots, r_i$, $i \geq 0$;
$V(T(H)) = r \cup V(T(H_1)) \cup V(T(H_2)) \ldots \cup V(T(H_i))$;
$E(T(H)) = \{rr_1, rr_2, \ldots, rr_i\} \cup E(T(H_1)) \cup E(T(H_2)) \ldots \cup E(T(H_i))$;
end

We say that a non-trivial H has one stem if it has only one connected component after removing the root r of $T(H)$. Otherwise, H has multi stems. The *level* of a vertex x, denoted by $l(x)$, is the length of the (r, x)-path in $T(H)$. The *parent* of a vertex u, denoted by $P(u)$, is a unique vertex which dominates u in $T(H)$. A *child* of a vertex u is a vertex v which is dominated by u in $T(H)$. A vertex v is an *ancestor* of vertex u, if there is a (v, u)-path from v to u in $T(H)$. For any $u, v \in V(H)$, the *joint* of u and v, denoted by joint(u,v), is the maximal level common ancestor of u and v in $T(H)$ (note that this vertex is unique in $T(H)$). A *s-joint(u,v)* (*s-joint(v,u)* respectively) is a vertex w which is in the directed path between joint(u,v) and u (v) on $T(H)$. In this section, we find the dichotomy for H when it has one stem. Using this result, we will derive the dichotomy for H when it has multi stems.

We need the following lemma proved in [13].

Lemma 4. *Let O_3 be given by $V(O_3) = \{1, 2, 3, 4\}$, $A(O_3) = \{12, 23, 34, 14, 24\}$. Then MinHom($O_3$) is NP-hard.*

Lemma 5. *Let T be an acyclic tournament with vertices $v_1, v_2, \ldots, v_k$, $k \geq 2$. Let O_4 be a digraph containing T and three other vertices u_1, u_2, and u_3 such that $V(T) \rightarrow \{u_1, u_2\}$, $(V(T) - v_1) \rightarrow u_3$, $u_1 \rightarrow u_2$, and there is no other arc in $A(O_4)$. Then MinHom(O_4) is NP-hard.*

Proof: See [7] for the proof.   □

Lemma 6. *Let O_5 be given by $V(O_5) = \{x_1, x_2, x_3, x_4, x_5, x_6\}$, $A(O_5) = A_1 \cup A_2$, where $A_1 = \{x_1x_2, x_2x_3, x_2x_4, x_3x_4, x_2x_5, x_5x_6, x_2x_6\}$ and A_2 is any subset of $\{x_1x_3, x_1x_4, x_1x_5, x_1x_6\}$. Then MinHom($O_5$) is NP-hard.*

Proof: See [7] for the proof.   □

Observation 1. *Let H be in $\mathcal{A}$ and $uv \in A(H)$. Then u is an ancestor of v.*

Proof: Suppose the contrary that u is not an ancestor of v. v is not definitely an ancestor of u on $T(H)$, as otherwise H has a cycle. Thus, neither u nor v is the ancestor of the other one. So, there are two disjoint paths P and Q from joint(u,v) to u and v on $T(H)$, respectively. By definition of $T(H)$, P and Q are the longest paths from joint(u,v) to u and v on H. Since $uv \in A(H)$ and H is path-mergeable, there is a path R in H from joint(u,v) to v such that it includes all vertices of P and Q; hence R is the longest path from joint(u,v) to v in H, contrary to the assumption that Q is the longest path from joint(u,v) to v.   □

We can easily see by Observation 1 that if $l(u) \geq l(v)$ then $uv \notin A(H)$. A vertex v is a minimal dominating ancestor of u, denoted by $MDA(u)$, if it dominates u, and it has a minimal level.

Observation 2. *Let H be in $\mathcal{A}$ and $uv \in A(H)$. If H does not contain O_3 as an induced subgraph, then all ancestors of v on $T(H)$ which are between u and v plus u and v induce and acyclic tournament in H.*

Proof: The proof is trivial since H is acyclic and locally in-semicomplete.   □

As H has one stem then r can not be a joint of any pair u and v in H. So, each joint(u, v) has a parent in H. In the following four Lemmas we assume that H is in $\mathcal{A}$ and it has one stem.

Lemma 7. *MinHom(H) is NP-hard if the following conditions hold for $u, v \in V(H)$:*

- $l(u) - l(x) = l(v) - l(x) = 2$, where x is the joint(u,v);
- $xu \in A(H) \setminus A(T(H))$;
- $l(MDA(P(v))) < l(MDA(P(u)))$

Proof: See [7] for the proof.   □

Lemma 8. *MinHom(H) is NP-hard when there are vertices $u, v \in V(H)$ such that $l(u) = l(v)$, and there is an arc $wu \in A(H) \setminus A(T(H))$, where w is a s-joint(u,v).*

Proof: See [7] for the proof.   □

Lemma 9. *MinHom(H) is NP-hard if the following conditions hold for $u, v \in V(H)$:*

- $l(u) = l(v)$, *and* $P(u) = P(v)$;
- $l(MDA(v)) < l(MDA(u))$;
- u *and* v *lie on path* P *and* Q *of* $T(H)$ *respectively such that there is an arc* $v'v'' \in A(H) \setminus A(T(H))$ *on* Q, *where* $l(v) \le l(v')$, *and a vertex* u' *in* P, *where* $l(v') + 1 = l(u') = l(v'') - 1$.

Proof: See [7] for the proof.   □

Lemma 10. *MinHom(H) is NP-hard if the following conditions hold for $u, v \in V(H)$:*

- $l(u) = l(v)$, *and* $P(u) = P(v)$;
- $P(u)$ *dominates a child of* u;
- u *and* v *lie on path* P *and* Q *of* $T(H)$ *respectively such that there is an arc* $v'v'' \in A(H) \setminus A(T(H))$ *on* Q, *where* $l(v) \le l(v')$, *and a vertex* u' *in* P, *where* $l(v') + 1 = l(u') = l(v'') - 1$.

Proof: See [7] for the proof.   □

Forbidden family. A digraph H belongs to the *forbidden family* F_1 if it is one of the digraphs introduced in the Lemmas $4, \ldots, 6$ and Lemmas $7, \ldots, 10$ for which MinHom(H) is NP-hard.

Theorem 4. *Let H be in $\mathcal{A}$ and it has one stem. IF $H \notin F_1$ then it has a Min-Max ordering and thus MinHom(H) is polynomial time solvable. Otherwise, MinHom(H) is NP-hard.*

Proof: See [7] for the proof.   □

Let $\mathcal{B}$ be a subclass of $\mathcal{A}$ such that for any $H \in \mathcal{B}$, each stem of H has a Min-Max ordering. Note that two stems only share the root r of $T(H)$ and they are different components of H if we remove this root r. In following six Lemmas we assume that $H \in \mathcal{B}$, $\{u, v, w\} \subset V(H)$, joint$(u,v) =$ joint$(u,w) =$ joint$(v,w) = x$, and u, v, w are in different stems of H. It is obvious that $x = r$.

We need the following lemma obtained from Theorem 1.2 in [6].

Lemma 11. *MinHom(H) is NP-hard when $l(u) = l(v) = l(w) = l(x) + 2$, and $\{xu, xv, xw\} \in A(H) \setminus A(T(H))$.*

Lemma 12. *MinHom(H) is NP-hard when $l(y) - l(x) > 2$, where $y \in \{u, v, w\}$, and $\{u'u, v'v, w'w\} \in A(H) \setminus A(T(H))$, where u' is an s-joint(u,v), v' is an s-joint(v,u), and w' is an s-joint(w,u).*

Proof: See [7] for the proof.   □

Lemma 13. *MinHom(H) is NP-hard if the following conditions hold:*

- *$l(y) - l(x) > 2$, where $y \in \{u, v\}$;*
- *$l(w) = max(l(u), l(v))$;*
- *$\{u'u, v'v\} \in A(H) \setminus A(T(H))$, where u' is an s-joint(u,v), v' is an s-joint(v,u).*

Proof: See [7] for the proof.   □

Lemma 14. *MinHom(H) is NP-hard if the following conditions hold:*

- *$l(y) - l(x) > 2$, where $y \in \{u, v\}$;*
- *$l(w) = max(l(u) - 1, l(v))$;*
- *$\{u'u, xu'', v_1v\} \in A(H) \setminus A(T(H))$, where u' is a s-joint(u,v), v' is an s-joint(v,u), and u'' is in the same stem with u', where $l(u'') - l(x) = 2$.*

Proof: See [7] for the proof.   □

Lemma 15. *MinHom(H) is NP-hard if the following conditions hold:*

- *$l(y) - l(x) > 2$, where $y \in \{u, v\}$;*
- *$l(w) = max(l(u) - 1, l(v) - 1)$;*
- *$\{u'u, xu'', v'v, xv''\} \in A(H) \setminus A(T(H))$, where u' is a s-joint(u,v), v' is a s-joint(v,u), and u'', and v'' are in the same stem of u', and v' respectively, where $l(u'') - l(x) = l(v'') - l(x) = 2$.*

Proof: See [7] for the proof.   □

Lemma 16. *MinHom(H) is NP-hard if the following conditions hold:*

- *$l(u) = l(v) = l(x) + 2$;*
- *$\{xu, xv, yw\} \in A(H) \setminus A(T(H))$, where y is an s-joint(w,u).*
- *there are two vertices w_1, and w_2 in the stems, where u, and v belong to respectively such that $l(w_1) = l(w_2) = l(w)$.*

Proof: See [7] for the proof.   □

Forbidden family. A digraph H belongs to the *forbidden family* F_2 if it is one of the digraphs introduced in the Lemmas 11,...,16 for which MinHom(H) is NP-hard.

Theorem 5. *Let H be in $\mathcal{B}$ and it is multi-stem. IF $H \notin F_2$ then it has a Min-Max ordering and thus MinHom(H) is polynomial time solvable. Otherwise, MinHom(H) is NP-hard.*

Proof: See [7] for the proof.   □

References

1. Bang-Jensen, J.: Digraphs with the path-merging property. J. Graph Theory 20, 255–265 (1995)
2. Bang-Jensen, J., Gutin, G.: Digraphs: Theory, Algorithms and Applications. Springer, London (2000)
3. Bang-Jensen, J., Huang, J., Prisner, E.: In-tournament digraphs. J. Combin. Theory Ser. B 59, 267–287 (1993)
4. Bang-Jensen, J., Gutin, G.: Generalizations of tournaments: A survey. J. Graph Theory 28, 171–202 (1998)
5. Gupta, A., Hell, P., Karimi, M., Rafiey, A.: Minimum Cost Homomorphisms to Reflexive Digraphs. In: Proceedings of the 8th Latin American Theoretical Informatics (LATIN 2008) (to appear, 2008)
6. Gupta, A., Gutin, G., Karimi, M., Kim, E.J., Rafiey, A.: Minimum Cost Homomorphisms to Locally semicomplete Digraphs and Quasi-transitive Digraphs (submitted)
7. Gupta, A., Karimi, M., Kim, E.J., Rafiey, A.: Minimum Cost Homomorphisms Dichotomy for Locally In- Semicomplete Digraphs. Journal of Discrete Applied Mathematics (submitted)
8. Gutin, G., Rafiey, A., Yeo, A.: Minimum Cost Homomorphism to Oriented Cycles (submitted)
9. Gutin, G., Kim, E.J.: Complexity of the minimum cost homomorphism problem for semicomplete digraphs with possible loops (submitted)
10. Gutin, G., Kim, E.J.: Introduction to the minimum cost homomorphism problem for directed and undirected graphs. Lecture Notes of the Ramanujan Math. Society (to appear)
11. Gutin, G., Kim, E.J.: On the complexity of the minimum cost homomorphism problem for reflexive multipartite tournaments (submitted)
12. Gutin, G., Rafiey, A., Yeo, A.: Minimum Cost and List Homomorphisms to Semicomplete Digraphs. Discrete Appl. Math. 154, 890–897 (2006)
13. Gutin, G., Rafiey, A., Yeo, A.: Minimum Cost Homomorphisms to Semicomplete Multipartite Digraphs. Discrete Applied Math. (submitted)
14. Gutin, G., Rafiey, A., Yeo, A.: Minimum Cost Homomorphisms to Semicomplete Bipartite Digraphs (submitted)
15. Gutin, G., Rafiey, A., Yeo, A., Tso, M.: Level of repair analysis and minimum cost homomorphisms of graphs. Discrete Appl. Math. 154, 881–889 (2006)
16. Gutin, G., Hell, P., Rafiey, A., Yeo, A.: A dichotomy for minimum cost graph homomorphisms. European J. Combin. (to appear)
17. Hell, P., Nešetřil, J.: On the complexity of H-colouring. J. Combin. Theory B 48, 92–110 (1990)

The Clique Corona Operation and Greedoids

Vadim E. Levit[1,2] and Eugen Mandrescu[2]

[1] Ariel University Center of Samaria, Israel
`levitv@ariel.ac.il`
[2] Holon Institute of Technology, Israel
`eugen_m@hit.ac.il`

Abstract. S is a *local maximum stable set* of G, and we write $S \in \Psi(G)$, if S is a stable set of maximum size in the subgraph induced by $S \cup N(S)$, where $N(S)$ is the neighborhood of S.

It is known that $\Psi(G)$ is a greedoid for every forest G, [10]. Bipartite graphs and triangle-free graphs, whose families of local maximum stable sets form greedoids were characterized in [11] and [12], respectively.

The *clique corona* is the graph $G = H \circ \{H_1, H_2, ..., H_n\}$ obtained by joining each vertex v_k of the graph H with the vertices of some clique H_k, respectively. In this paper we demonstrate that if G is a clique corona, then $\Psi(G)$ forms a greedoid on its vertex set.

1 Introduction

Throughout this paper $G = (V, E)$ is a simple (i.e., a finite, undirected, and without multiple edges) graph with vertex set $V = V(G)$ and edge set $E = E(G)$. The vertices $x, y \in V(G)$ are called *adjacent* if they are the endpoints of some edge in G, and we write $xy \in E(G)$. If $X \subset V$, then $G[X]$ is the subgraph of G induced by X. By $G - W$ we mean the subgraph $G[V - W]$, if $W \subset V(G)$. We also denote by $G - F$ the partial subgraph of G obtained by deleting the edges of F, for $F \subset E(G)$, i.e., $G - F = (V, E - F)$, and we write shortly $G - e$, whenever $F = \{e\}$. K_n, C_n denote, respectively, the *complete graph* on $n \geq 1$ vertices, and the *chordless cycle* on $n \geq 3$ vertices.

The *neighborhood* of a vertex $v \in V$ is the set

$$N(v) = \{w : w \in V \text{ and } vw \in E\},$$

and $N[v] = \{v\} \cup N(v)$. If $|N(v)| = |\{u\}| = 1$, then v is a *pendant vertex* of G. A vertex $v \in V(G)$ is called *simplicial* if $G[N[v]]$ is a complete subgraph of G. By $\text{simp}(G)$ we mean the set of all simplicial vertices of G. Clearly, each pendant vertex is also simplicial, while the converse is not necessarily true. We denote the *neighborhood* of $A \subset V$ by

$$N_G(A) = \{v \in V - A : N(v) \cap A \neq \varnothing\}$$

and its *closed neighborhood* by $N_G[A] = A \cup N(A)$, or shortly, $N(A)$ and $N[A]$, respectively, if no ambiguity.

B. Yang, D.-Z. Du, and C.A. Wang (Eds.): COCOA 2008, LNCS 5165, pp. 384–392, 2008.

A *stable* set in G is a set of pairwise non-adjacent vertices. A stable set of maximum size will be referred to as a *maximum stable set* of G, and by $\alpha(G)$ is denoted the cardinality of a maximum stable set in G. In the sequel, by $\Omega(G)$ we denote the set of all maximum stable sets of the graph G.

A set $A \subseteq V(G)$ is called a *local maximum stable set* of G if A is a maximum stable set in the subgraph induced by $N[A]$, i.e., $A \in \Omega(G[N[A]])$, [10]. Let $\Psi(G)$ stand for the set of all local maximum stable sets of G. Notice that $\Omega(G) \subseteq \Psi(G)$ is true for every graph G.

It is clear that every stable set $S \subseteq \mathrm{simp}(G)$ belongs to $\Psi(G)$ and, of course, there exist local maximum stable sets that do not contain simplicial vertices. For instance, $\{e, g\} \in \Psi(G_2)$, where G_2 is the graph from Figure 1.

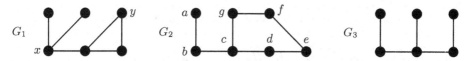

Fig. 1. G_1 is not well-covered, G_2 is well-covered, G_3 is very well-covered

A graph G is *well-covered* if every maximal stable set of G is also a maximum stable set, i.e., it belongs to $\Omega(G)$. If, in addition, G has no isolated vertices and $|V(G)| = 2\alpha(G)$, then G is *very well-covered*, [3]. For example, the graph G_2 depicted in Figure 1 is well-covered, but not very well-covered, while the graph G_3 in Figure 1 is very well-covered. In other words, each stable set of a well-covered graph is contained in a maximum stable set. Since there is no maximum stable set S of G_1 such that $\{a, b\} \subset S$, the graph G_1 from Figure 1 is not well-covered.

Well-covered graphs were defined by Plummer in 1970, [14]. A number of classes of well-covered graphs were completely described (see, for instance, the following references: [3], [4], [5], [7], [16], [17], [19]. A survey on this subject is due to Plummer [15]. In fact, well-covered graphs are exactly those graphs for which the greedy algorithm constructing maximum stable sets vertex by vertex always yields a maximum stable set, no matter how its greediness makes it to chose vertices of a graph.

For general graphs, the problem of finding a maximum stable set, is **NP**-hard. While, in general, it is co-**NP**-complete to determine if a given graph is well-covered (Chvátal and Slater, [2], Sankaranarayana and Stewart, [18]), Tankus and Tarsi showed that claw-free well covered graphs can be recognized in polynomial time, [20], [21].

Let H be a graph with $V(H) = \{v_i : 1 \le i \le n\}$, and $\{H_i : 1 \le i \le n\}$ be a family of graphs, where $H_i = (V_i, E_i) = (\{u_{ij} : 1 \le j \le q_i\}, E_i)$. Joining each $v_i \in V(H)$ to all the vertices of H_i, we obtain a new graph, which we denote by $G = H \circ \{H_1, H_2, ..., H_n\}$.

More precisely, $V(G) = V(H) \cup V_1 \cup ... \cup V_n$ is the vertex set of the graph $G = H \circ \{H_1, H_2, ..., H_n\}$ and its edge set is

$$E(G) = E(H) \cup E_1 \cup ... \cup E_n \cup \{v_1 u_{1j} : 1 \le j \le q_1\} \cup ... \cup \{v_n u_{nj} : 1 \le j \le q_n\}.$$

If $H_i = X, 1 \leq i \leq n$, we write $G = H \circ X$, and in this case, G is called the *corona* of H and X. If all H_i are complete graphs, then $G = H \circ \{H_1, H_2, ..., H_n\}$ is called the *clique corona* of H and $H_1, H_2, ..., H_n$ (see Figure 2 for an example, where $H = K_3 + v_3v_4$).

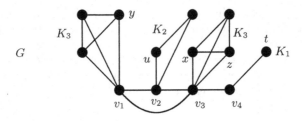

Fig. 2. $G = (K_3 + v_3v_4) \circ \{K_3, K_2, K_3, K_1\}$ is a well-covered graph

Let us notice that $G = H \circ \{H_1, H_2, ..., H_n\}$ has

$$\alpha(G) = \alpha(H_1) + \alpha(H_2) + ... + \alpha(H_n).$$

Theorem 1. *[22] The graph $G = H \circ \{H_1, H_2, ..., H_n\}$ is well-covered if and only if all H_i are complete, i.e., G is a clique corona.*

Moreover, the following result shows that, under certain conditions, every well-covered graph must be of this form.

Theorem 2. *[4] Let G be a connected graph of girth ≥ 6, which is isomorphic to neither C_7 nor K_1. Then G is well-covered if and only if its pendant edges form a perfect matching, i.e., $G = H \circ K_1$ for some graph H.*

In other words, Theorem 2 shows that, apart from K_1 and C_7, connected well-covered graphs of girth ≥ 6 are very well-covered. Consequently, a tree $T \neq K_1$ could be only very well-covered, and consequently, $T = H \circ K_1$ for some tree H (for additional details, see [3], [8], [17]).

Theorem 3. *[13] Every local maximum stable set of a graph is a subset of a maximum stable set.*

The graph G_2 from Figure 1 has the property that every $S \in \Omega(G_2)$ contains some local maximum stable set, but these local maximum stable sets are of different cardinalities: $\{a, d, f\} \in \Omega(G_2)$ and $\{a\}, \{d, f\} \in \Psi(G_2)$, while for $\{b, e, g\} \in \Omega(G_2)$ only $\{e, g\} \in \Psi(G_2)$.

However, there exists a graph G satisfying $\Psi(G) = \Omega(G)$, e.g., $G = C_n$, for $n \geq 4$.

A greedoid is a set system generalizing the notion of matroid.

Definition 1. *[1], [6] A greedoid is a pair $(V, \mathcal{F})$, where $\mathcal{F} \subseteq 2^V$ is a non-empty set system satisfying the following conditions:*

(Accessibility) for every non-empty $X \in \mathcal{F}$ there is $x \in X$ such that $X - \{x\} \in \mathcal{F}$;
(Exchange) for any $X, Y \in \mathcal{F}, |X| = |Y| + 1$, there is $x \in X - Y$ such that $Y \cup \{x\} \in \mathcal{F}$.

Recall that a *matroid* is a set system $(V, \mathcal{F})$ satisfying the "hereditary property", saying that : if $X \in \mathcal{F}$ and $Y \subseteq X$, then $Y \in \mathcal{F}$, and the "exchange property" ([23]). Evidently, any matroid is also a greedoid. It is easy to see that the family of all stable sets of a graph is a matroid if and only if G is a disjoint union of complete graphs, which means that, necessarily, G must be also well-covered, but of very specific form. If G is well-covered, then $\Psi(G)$ is a matroid if and only if each $S \in \Omega(G)$ consists of only simplicial vertices, because $\Omega(G) \subseteq \Psi(G)$ and every $v \in S$, by hereditary property, satisfies $\{v\} \in \Psi(G)$, i.e., $G[N_G[v]]$ must be a complete graph. In other words, G is a disjoint union of complete graphs.

It is worth mentioning that if $(V, \mathcal{F})$ is a greedoid and $X \in \mathcal{F}$, $|X| = k \geq 2$, then according to accessibility property, one can build a chain

$$\{x_1\} \subset \{x_1, x_2\} \subset ... \subset \{x_1, ..., x_{k-1}\} \subset \{x_1, ..., x_{k-1}, x_k\} = X$$

such that $\{x_1, x_2, ..., x_j\} \in \mathcal{F}$, for every $j \in \{1, ..., k-1\}$. Such a chain we call an *accessibility chain* of X.

For example, $\Psi(G_1)$ is a greedoid and $\{a\} \subset \{a, b\} \subset \{a, b, c\}$ is an accessibility chain of $\{a, b, c\} \in \Psi(G_1)$, where G_1 is presented in Figure 3.

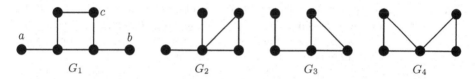

Fig. 3. Graphs whose family of local maximum stable sets form greedoids

In [10] we proved the following result.

Theorem 4. *For every forest T, $\Psi(T)$ is a greedoid on its vertex set.*

The case of bipartite graphs having a unique cycle, whose family of local maximum stable sets forms a greedoid, is studied in [9]. The general case of bipartite graphs was treated in [11], while for triangle-free graphs we refer the reader to [12] for details. Nevertheless, there exist non-bipartite and non-triangle-free graphs whose families of local maximum stable sets form greedoids. The families $\Psi(G_1), \Psi(G_2), \Psi(G_3), \Psi(G_4)$ of the graphs in Figure 3 are greedoids. Let us notice that G_1 is very well-covered and G_3 is well-covered, while G_2, G_4 are not well-covered and also non-triangle-free.

In this paper we prove that if a well-covered graph G is isomorphic to some clique corona, then the family $\Psi(G)$ of local maximum stable sets forms a greedoid on its vertex set.

2   Results

It is easy to see that no maximum stable set of C_5 admits an accessibility chain. The graph G_1 in Figure 4 shows that even if some $S \in \Omega(G_1)$ has an accessibility chain, this is not necessarily true for all maximum stable sets. The case of the graph G_2 from Figure 4 is different: each maximum stable set of G_2 admits an accessibility chain, and the reason is given in Proposition 1. Notice that $G_2 = H \circ \{K_3, K_2, K_1, K_2\}$, where $H = G_2[\{v, x, y, z\}]$.

Fig. 4. $\{a, c, f\}, \{a, g, e\} \in \Omega(G_1)$, but only $\{a, c, f\}$ admits an accessibility chain

Proposition 1. *If $G = H \circ \{K_{q_1}, K_{q_2}, ..., K_{q_n}\}$, then every $S \in \Omega(G)$ has an accessibility chain containing only local maximum stable sets of G.*

Proof. Let us notice that $V(K_{q_1}) \cup V(K_{q_2}) \cup ... \cup V(K_{q_n}) \subseteq simp(G)$ and the equality holds whenever H has no isolated vertex.

If L is a connected component of H, then $G = H \circ \{K_{q_1}, K_{q_2}, ..., K_{q_n}\}$ is the disjoint union of G_1 and G_2, where and G_1 is a clique corona of $(H - L)$ while G_2 is a clique corona of L. Since $\Psi(G) = \{S_1 \cup S_2 : S_1 \in \Psi(G_1), S_2 \in \Psi(G_2)\}$, it is enough to check the existence of accessibility chains for connected components only. Therefore, we may assume that H is connected.

Clearly, $\alpha(G) = n$, where $|V(H)| = |\{v_1, v_2, ..., v_n\}| = n$, and each $S \in \Omega(G)$ satisfies $S \cap simp(G) \neq \varnothing$.

We prove by induction on n that every $S \in \Omega(G)$ has an accessibility chain.

For $n = 1$, the assertion is clearly true.

For $n = 2$, let $S = \{x_1, x_2\} \in \Omega(G)$. Then at least one of x_1, x_2 is simplicial, say x_1. Hence, the chain is $\{x_1\} \subset \{x_1, x_2\} = S$.

Suppose that the assertion is true for every $p < n$.

Let $G = (V, E) = H \circ \{K_{q_1}, K_{q_2}, ..., K_{q_n}\}$ be with $V(H) = \{v_i : 1 \leq i \leq n\}$, and let $S \in \Omega(G)$.

Since $S \cap simp(G) \neq \varnothing$, there exists some $a \in S \cap simp(G)$.

If $N(a) \cap V(H) = \{v_j\}$, then we have

$$G_j = G - (V(K_{q_j}) \cup \{v_j\}) = (H - \{v_j\}) \circ \{K_{q_1}, ..., K_{q_{j-1}}, K_{q_{j+1}}, ..., K_{q_n}\}.$$

Hence, we get that $S_{n-1} = S - \{a\} \in \Omega(G_j)$, and by the induction hypothesis, there is a chain

$$\{x_1\} \subset \{x_1, x_2\} \subset ... \subset \{x_1, x_2, ..., x_{n-2}\} \subset \{x_1, x_2, ..., x_{n-1}\} = S_{n-1}$$

such that $\{x_1, x_2, ..., x_k\} \in \Psi(G_j)$, for every $k \in \{1, ..., n-1\}$. Since $a \in simp(G)$ we have that $N_G[a] = \{v_j\} \cup V(K_{q_j})$. Consequently, we infer that

$$N_G[\{x_1, x_2, ..., x_k\} \cup \{a\}] = N_{G_j}(\{x_1, x_2, ..., x_k\}) \cup V(K_{q_j}) \cup \{v_j\},$$

and therefore $\{x_1, x_2, ..., x_k\} \cup \{a\} \in \Psi(G)$, for every $k \in \{1, ..., n-1\}$. Clearly, $\{a\} \in \Psi(G)$, and consequently, we obtain the chain

$$\{a\} \subset \{a, x_1\} \subset \{a, x_1, x_2\} \subset ... \subset \{a, x_1, x_2, ..., x_{n-2}\} \subset$$
$$\subset \{a, x_1, x_2, ..., x_{n-1}\} = \{a\} \cup S_{n-1} = S,$$

and $\{a, x_1, x_2, ..., x_k\} \in \Psi(G)$ for every $k \in \{1, ..., n-1\}$. In other words, S admits an accessibility chain, and this completes the proof.

Let us remark that Proposition 1 is not valid for C_4, which is well-covered, but no $S \in \Omega(C_4)$ has an accessibility chain.

Lemma 1. *If S is stable in H and $N_H(S) \neq \emptyset$, then $S \notin \Psi(H \circ \{K_{q_1}, ..., K_{q_n}\})$.*

Proof. Let $G = H \circ \{K_{q_1}, K_{q_2}, ..., K_{q_n}\}, V(H) = \{v_1, v_2, ..., v_n\}$ and $v_j \in N_H(S)$. If $N_H(v_j) \cap S = \{v_{j_1}, ..., v_{j_p}\}$, then $A = \{v_j\} \cup \{u_{j_1}, ..., u_{j_p}\} \cup (S - \{v_{j_1}, ..., v_{j_p}\})$ is stable, larger than S, and $A \subseteq N_G[S]$, where

$$u_{j_1} \in V(K_{q_{j_1}}), u_{j_2} \in V(K_{q_{j_2}}), ..., u_{j_p} \in V(K_{q_{j_p}}).$$

Therefore, we infer that $S \notin \Psi(G)$.

Notice that if v_1 is an isolated vertex of H, then $G = H \circ \{K_{q_1}, K_{q_2}, ..., K_{q_n}\}$ is the disjoint union of $(H - v_1) \circ \{K_{q_2}, ..., K_{q_n}\}$ and $K_1 \circ K_{q_1}$, where $K_1 = (\{v_1\}, \emptyset)$, and every $S \in \Omega(G)$ contains exactly one vertex from $\{v_1\} \cup V(K_{q_1})$.

Lemma 2. *If the graph H has no isolated vertices and S is a stable set in the clique corona $G = H \circ \{K_{q_1}, K_{q_2}, ..., K_{q_n}\}$, then the following assertions are equivalent:*

(i) $S \in \Psi(G)$;

(ii) $S = S_1 \cup S_2$, where $\emptyset \neq S_1 \subseteq \mathrm{simp}(G), S_2 \subseteq V(H), N_H(S_2) \subseteq N_G(S_1)$;

(iii) $G[N_G[S]] = H' \circ \{K_{q_{i_1}}, K_{q_{i_2}}, ..., K_{q_{i_p}}\}$, *for some subgraph H' of H, whose* $V(H') = \{v_{i_1}, v_{i_2}, ..., v_{i_p}\}$, *and $S \in \Omega(G[N_G[S]])$.*

Proof. Let us denote:

$$V(H) = \{v_j : 1 \leq j \leq n\}, V(G) = V(H) \cup V(K_{q_1}) \cup ... \cup V(K_{q_n}),$$

where v_j is joined, in G, to all the vertices of $K_{q_j}, 1 \leq j \leq n$.

Notice that $\alpha(G) = n$ and $\mathrm{simp}(G) = V(K_{q_1}) \cup ... \cup V(K_{q_n})$, because H has no isolated vertices.

(i) $\Longrightarrow$ (ii) Assume that $S \in \Psi(G)$.

Let $S_1 = S \cap \mathrm{simp}(G)$ and $S_2 = S \cap V(H)$. According to Lemma 1, the set S_1 is not empty.

If $S_2 = \emptyset$, then the assertion is clearly true.

Suppose now that $S_2 \neq \emptyset$.

If $N_H(S_2) \nsubseteq N_G(S_1)$, then there must be some $v_k \in N_H(S_2)$ such that $V(K_{q_k}) \cap S = \emptyset$, i.e., $V(K_{q_k}) \cap S_1 = \emptyset$. Let $N_H(v_k) \cap S_2 = \{v_{i_1}, v_{i_2}, ..., v_{i_p}\}$. Hence, we infer that

$$(S - N_H(v_k)) \cup \{v_k\} \cup \{u_{i_1}, u_{i_2}, ..., u_{i_p}\},$$

where $u_{i_j} \in V(K_{q_{i_j}})$ for each $1 \leq j \leq p$, is a stable set in $G[N_G[S]]$ larger than S, in contradiction with the choice $S \in \Psi(G)$.

$(ii) \implies (iii)$ Let $S_3 = \{u_k \in V(K_{q_k}) : v_k \in S_2\}$. Then we deduce that

$$G[N_G[S]] = G[S_1 \cup S_3] = H' \circ \{K_{q_{i_1}}, K_{q_{i_2}}, ..., K_{q_{i_p}}\},$$

for some subgraph H' of H, whose $V(H') = \{v_{i_1}, v_{i_2}, ..., v_{i_p}\}$. In addition, we have also that

$$|S| = |S_1| + |S_2| = |S_1| + |S_3| \text{ and } S_1 \cup S_3 \in \Omega(G[S]).$$

Consequently, we get that $S \in \Omega(H' \circ \{K_{q_{i_1}}, K_{q_{i_2}}, ..., K_{q_{i_p}}\})$ as well.

$(iii) \implies (i)$ Since $S \in \Omega(G[N_G[S]])$, it follows, by definition, that $S \in \Psi(G)$.

Theorem 5. *If $G = H \circ \{K_{q_1}, K_{q_2}, ..., K_{q_n}\}$, then $\Psi(G)$ is a greedoid.*

Proof. Let $S_0 \in \Psi(G)$, i.e., S_0 is a maximum stable set, of size say p, in the induced subgraph $H_0 = G[N[S_0]]$.

According to Lemma 2, we have that $G[N[S_0]] = H_{S_0} \circ \{K_{q_{i_1}}, K_{q_{i_2}}, ..., K_{q_{i_p}}\}$ for the subgraph H_{S_0} of H whose $V(H_{S_0}) = \{v_{i_1}, v_{i_2}, ..., v_{i_p}\}$. By Proposition 1, we infer that there exists a chain

$$\{x_1\} \subset \{x_1, x_2\} \subset ... \subset \{x_1, x_2, ..., x_{q-1}\} \subset \{x_1, x_2, ..., x_{q-1}, x_q\} = S_0,$$

such that all $S_k = \{x_1, x_2, ..., x_k\}, 1 \leq k \leq q$, are local maximum stable sets in H_0. The inclusion $S_k \subseteq S_0$ implies the equality $N_{H_0}[S_k] = N_G[S_k]$. Hence, we get that $S_k \in \Psi(G)$, for every $k \in \{1, ..., q\}$. In other words, $\Psi(G)$ satisfies the accessibility property.

We have to show now that $\Psi(G)$ satisfies also the exchange property.

Let us consider $X, Y \in \Psi(G)$ be such that $|Y| = |X| + 1 = m + 1$. According to Lemma 2 *(ii)*, the sets X and Y can be decomposed as follows:

$$X = X_1 \cup X_2 \text{ and } Y = Y_1 \cup Y_2,$$

where X_1, X_2, Y_1, Y_2 satisfy the corresponding conditions, i.e., X_1 and Y_1 are non-empty subsets of $simp(G)$, while X_2, Y_2 are subsets of $V(H)$, such that $N_H(X_2) \subseteq N_G(X_1)$ and $N_H(Y_2) \subseteq N_G(Y_1)$.

Since Y is stable, $X \in \Psi(G)$, and $|X| < |Y|$, it follows that there exists some $y \in Y - X$, such that $y \notin N_G[X]$. In particular, it means that $X \cup \{y\}$ is stable. To check whether $X \cup \{y\} \in \Psi(G)$, we have to analyze the two following cases (see Figure 5).

Case 1. $Y_1 \subseteq X_1$.

Firstly, we deduce that $y \in Y_2$. Lemma 2 *(ii)* implies that $N_H(y) \subseteq N_G(Y_1)$. Since $Y_1 \subseteq X_1$, it follows that $N_G(Y_1) \subseteq N_G(X_1)$. Hence, we obtain that $N_H(y) \subseteq N_G(X_1)$. Therefore, we have that

$$X_1 \subseteq simp(G), \ X_2 \cup \{y\} \subseteq V(H),$$

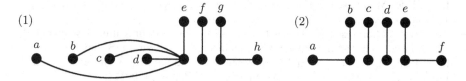

Fig. 5. X, Y are local maximum stable sets illustrating the cases 1 and 2, respectively. (1) $Y = \{a, b, c, d, e\}, X = \{e, f, g, h\}, Y_1 = \{e\} \subset X_1 = \{e, f, g\}$ and (2) $Y = \{a, b, c, d\}, X = \{d, e, f\}, Y_1 = \{b, c, d\} \nsubseteq X_1 = \{d, e\}$

and

$$N_H(X_2 \cup \{y\}) = N_H(X_2) \cup N_H(\{y\}) \subseteq N_G(X_1).$$

Consequently, according to Lemma 2 (ii), we may infer that the stable set $X \cup \{y\}$ is, actually, a local maximum stable set in G.

Case 2. $Y_1 \nsubseteq X_1$.

In this situation, one can choose as y every vertex $z \in Y_1 - X_1$, since clearly, both conditions (namely, $z \in Y - X$ and $X \cup \{z\} \in \Psi(G)$) are satisfied.

Therefore, $\Psi(G)$ satisfies the exchange property as well.

In conclusion, $\Psi(G)$ is a greedoid on the vertex set of G.

Let us notice that $\Psi(C_7)$ is not a greedoid, because every $S \in \Psi(C_7)$ has $|S| \neq 1$.

3   Conclusions

We have shown that $\Psi(G)$ is a greedoid on the vertex set of a well-covered graph G, which is isomorphic to some clique corona. Since there are well-covered graphs different from a clique corona (e.g., C_5 whose $\Psi(C_5)$ is not a greedoid), one can be interested in characterizing well-covered graphs of girth ≤ 5 (see Theorem 2) whose families of local maximum stable sets form greedoids.

References

1. Björner, A., Ziegler, G.M.: Introduction to greedoids. In: White, N. (ed.) Matroid Applications, pp. 284–357. Cambridge University Press, Cambridge (1992)
2. Chvátal, V., Slater, P.J.: A note on well-covered graphs. Quo Vadis, Graph Theory? Annals of Discrete Math. 55, 179–182 (1993)
3. Favaron, O.: Very well-covered graphs. Discrete Mathematics 42, 177–187 (1982)
4. Finbow, A., Hartnell, B., Nowakowski, R.J.: A characterization of well-covered graphs of girth 5 or greater. J. of Combinatorial Theory, Ser. B 57, 44–68 (1993)
5. Hartnell, B., Plummer, M.D.: On 4-connected claw-free well-covered graphs. Discrete Applied Mathematics 64, 57–65 (1996)
6. Korte, B., Lovász, L., Schrader, R.: Greedoids. Springer, Berlin (1991)
7. Levit, V.E., Mandrescu, E.: Well-covered and König-Egerváry graphs. Congressus Numerantium 130, 209–218 (1998)

8. Levit, V.E., Mandrescu, E.: Well-covered trees. Congressus Numerantium 139, 101–112 (1999)

9. Levit, V.E., Mandrescu, E.: Unicycle bipartite graphs with only uniquely restricted maximum matchings. In: Calude, C.S., Dinneen, M.J., Sburlan, S. (eds.) Proceedings of the Third International Conference on Combinatorics, Computability and Logic (DMTCS 2001), pp. 151–158. Springer, Heidelberg (2001)

10. Levit, V.E., Mandrescu, E.: A new greedoid: the family of local maximum stable sets of a forest. Discrete Applied Mathematics 124, 91–101 (2002)

11. Levit, V.E., Mandrescu, E.: Local maximum stable sets in bipartite graphs with uniquely restricted maximum matchings. Discrete Applied Mathematics 132, 163–174 (2003)

12. Levit, V.E., Mandrescu, E.: Triangle-free graphs with uniquely restricted maximum matchings and their corresponding greedoids. Discrete Applied Mathematics 155, 2414–2425 (2007)

13. Nemhauser, G.L., Trotter Jr., L.E.: Vertex packings: structural properties and algorithms. Mathematical Programming 8, 232–248 (1975)

14. Plummer, M.D.: Some covering concepts in graphs. J. of Combinatorial Theory 8, 91–98 (1970)

15. Plummer, M.D.: Well-covered graphs: a survey. Quaestiones Mathematicae 16, 253–287 (1993)

16. Prisner, E., Topp, J., Vestergaard, P.D.: Well-covered simplicial, chordal, and circular arc graphs. J. of Graph Theory 21, 113–119 (1996)

17. Ravindra, G.: Well-covered graphs. J. Combin. Inform. System Sci. 2, 20–21 (1977)

18. Sankaranarayana, R., Stewart, L.K.: Complexity results for well-covered graphs. Networks 22(3), 247–262 (1992)

19. Staples, J.A.: On some sub-classes of well-covered graphs, Ph. D. Thesis, Vanderbilt University (1975)

20. Tankus, D., Tarsi, M.: Well-covered claw-free graphs. J. of Combinatorial Theory Ser. B 66, 293–302 (1996)

21. Tankus, D., Tarsi, M.: The structure of well-covered graphs and the complexity of their recognition problems. J. of Combinatorial Theory Ser. B 69, 230–233 (1997)

22. Topp, J., Volkmann, L.: On the well coveredness of products of graphs. Ars Combinatoria 33, 199–215 (1992)

23. Whitney, H.: On the abstract properties of linear independence. Amer. J. Math. 57, 509–533 (1935)

On the Surface Area of the (n, k)-Star Graph

Zhizhang Shen[1], Ke Qiu[2], and Eddie Cheng[3]

[1] Dept. of Computer Science and Technology, Plymouth State University, USA
zshen@plymouth.edu
[2] Dept. of Computer Science, Brock University, Canada
kqiu@brocku.ca
[3] Dept. of Mathematics and Statistics, Oakland University, USA
echeng@oakland.edu

Abstract. We present an explicit formula of the surface area of the (n, k)-star graphs , i.e., $|\{v|d(e, v) = d\}|$, where e is the identity node of such a graph; by identifying the cyclic structures of all the nodes in the graph, presenting a minimum routing algorithm between any node in the graph and e, and enumerating those nodes v, such that $d(e, v) = d$.

1 Introduction

Given a graph $G(V, E)$ and a node $v \in V$, a question one may ask is *how many nodes are at distance d from v in G* for $d \in [0, D(G)]$, where $D(G)$ is the diameter of G. This quantity is known in the literature as the *surface area of G with radius d* [10]; or the *Whitney numbers of the second kind of the poset associated with G* [12]. One immediate application of a solution to the above problem is in computing various bounds for the problem of k-neighborhood broadcasting [9]. Such a solution can also be used to derive the *transmission of a (node symmetric) graph,* a notion recently suggested in [15] to achieve the generalized Moore bound, an important concept in the extremal graph theory. As a result, this surface area problem has been studied for a variety of graphs, e.g., for the star graph in [12,13,10,16]; for the mesh structures in [2]; for the k-ary n-cubes in [3]; for the rotator graphs in [6]; and for the WK-Recursive and swapped networks in [11].

The star graph was proposed in [1] to be an attractive alternative to the hypercube topology, since it compares favorably with the latter structure in several aspects. A star graph of dimension n, an $n-star$, is a regular graph with degree $n - 1$. It has $n!$ nodes, but both its degree and diameter are $O(n)$, i.e., sub-logarithmic in the number of vertices; while a hypercube with $O(n!)$ vertices has a degree and diameter of $O(\log(n!)) = O(n \log n)$, i.e., logarithmic in the number of vertices. Other attractive properties include their symmetry properties, as well as many desirable fault tolerance characteristics [1]. However, the requirement that the number of nodes in an n-star be $n!$ results in a large gap, in term of the number of nodes, between the n-star and the $(n+1)$-star. To achieve scalability, (n, k)-star graph was proposed, which removes the restriction that the total number of nodes be $n!$ while preserving many ideal properties of the star graph [5,8].

B. Yang, D.-Z. Du, and C.A. Wang (Eds.): COCOA 2008, LNCS 5165, pp. 393–404, 2008.
© Springer-Verlag Berlin Heidelberg 2008

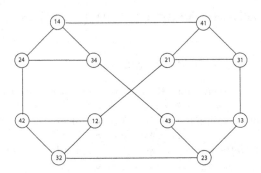

Fig. 1. A (4, 2)-star $S_{4,2}$

Let $\langle n \rangle$ denote $\{1, 2, \ldots, n\}, n \geq 2$, and let Γ_n be the symmetric group of degree n, in a *star graph* $S(\Gamma_n, E)$ *of dimension* n, S_n for short, *for any* $u, v \in \Gamma_n$, $(u, v) \in E$ *iff* v *can be obtained from* u *by applying a transposition* $(1, i)$ *to* $u, i \in [2, n]$. On the other hand, in a (n, k)-*star graph*, $S_{n,k}(V, E)$, $k \in [1, n)$, $S_{n,k}$ for short, $V = \{u_1 \cdots u_k | \forall i \in [1, k], u_i \in \langle n \rangle, u_i \neq u_j$ *if* $i \neq j\}$; *and, for any* $u, v \in V$, $(u, v) \in E$ *iff* v *can be obtained from* u *by either 1) applying* $(1, i)$ *to* u, $i \in [2, k]$ *(i-edge); or 2) for some* $x \in \langle n \rangle - \{u_i | i \in [1, k]\}$, *replacing* u_1 *with* x *in* u *(1-edge)*.

For example, in a $(4, 2)$-star graph as shown in Fig. 1, $(12, 21), (12, 32), (12, 42)$ $\in E$, with the first edge being a 2-edge, and the other two being two 1-edges.

In this paper, we present an explicit formula (a sum of standard and basic operations, including factorials) for the surface area of the (n, k)-star graph. To the best of our knowledge, this is the first result of this nature ever reported in the literature.

The surface area of any node u in a graph G is clearly equal to $|\{v | d_G(u, v) = d\}|$, where $d_G(u, v)$ stands for the distance between u and v in G. Since $S_{n,k}$ is node symmetric [5, Theorem 3], the distance between u and v in a (n, k)-star graph is the same as that between $\phi(v)$ and $e(= 12 \cdots k)$, the identity node of $S_{n,k}$, where ϕ is an automorphism that maps u to e. Therefore, for $n \geq 2, d \in [0, D(S_{n,k})]$, let $B(n, k; d)$ stand for the surface area of $S_{n,k}$ with radius d, we have

$$B(n, k; d) = |\{v | v \in S_{n,k} \text{ and } d(e, v) = d\}| .$$

Since there is exactly one node $v \in S_{n,k}$ such that $d(e, v) = 0$, i.e., e itself, when deriving $B(n, k; d)$, we assume $d \geq 1$, and define $B(n, k; 0) = 1$.

To derive an explicit formula for $B(n, k; d)$, we need to identify all the nodes v in $S_{n,k}$ such that the distance between e and v equals d. We will characterize the *cycle structures* of all the nodes in $S_{n,k}$ in the next section, derive a distance expression between e and any v, via a minimum routing strategy, in Section 3, and then come up with an explicit formula in Section 4. Finally, Section 5 concludes this paper.

2   The Cycle Structure of a Node in the (n, k)-Star Graph

Following [5], for a given $S_{n,k}$, $k \in [1, n)$, we call a symbol $i \in [1, k]$ an *internal symbol,* and a symbol $i \in (k, n]$ an *external symbol.* We also refer to a position in $[1, k]$ an *internal position,* and that in $(k, n]$ an *external position.*

Let e_m be an external symbol occurring in an internal position of $v \in S_{n,k}$, $m \geq 1$, we define E_{e_m}, *the external cycle associated with e_m in v* as

$$E_{e_m} = (e_m; e_0, \cdots, e_{m-1}) \ ,$$

such that 1) for $j \in [1, m]$, the position of e_j in v is e_{j-1}; and 2) the position of e_0 in v is e_m.

For example, in $v = 2968134 \in S_{9,7}$, 8 and 9 are external symbols. It is clear that $v(4) = 8$, and $v(7) = 4$. Although positions 8 and 9 are external, thus undefined in $S_{9,7}$, since we *define* via the above definition that $v(8) = 7$ and $v(9) = 5$, we have $E_8 = (8; 7, 4)$ and $E_9 = (9; 5, 1, 2)$.

We notice that the introduction of the assumed external positions in the above definition leads to an ideal definition for the external cycles, as compared with the original one in [5]:

- Any afore-defined external cycle is now truly a cycle[1].
- In the afore-defined external cycle E_{e_m}, there exists exactly one external symbol, i.e., e_m, and at least one internal symbol(s), i.e., $e_j, j \in [0, m - 1]$; such that, for all $j \in [1, m]$, the position of e_j is $e_{j-1} \in [1, k]$, and the position of e_0 is defined to be e_m, an external position. Thus, $e_0 \notin v$. Therefore, the transposition $(1, e_m), e_m \in (k, n]$, corresponds to a 1-edge, i.e., replacing the symbol in position 1 with e_0.

    On the other hand, for $j \in [0, m - 1]$, if $e_j \neq 1$, then, $(1, e_j)$ corresponds to an i−edge, $i \in [2, k]$.

    Thus, for all the symbols occurring in an external cycle, the two kinds of transpositions as presented in the definition of $S_{n,k}$ are unified in the form of $(1, j), j \in [2, n]$. This fact is crucial for the later discussion of a (minimum) routing algorithm for $S_{n,k}$.

If in an external cycle $E_{e_m} = (e_m; e_0, \cdots, e_{m-1})$, for some $j \in [0, m-1], e_j = 1$, i.e., the position of e_{j+1} is 1, we will then call this cycle the *primary external cycle* and express E_{e_m} in the following equivalent *canonical form:* $(1, e_{j+1}, \ldots, e_m; e_0, \ldots, e_{j-1}.)$ For example, $E_9 = (1, 2, 9; 5)$.

---

[1] For each external symbol, x_m, an *external cycle* is originally defined in [5] as $(x_1, x_2, \ldots, x_m)$ such that the position of x_j in e is held by x_{j+1}, $j \in [1, m]$; and, a *desired symbol, d,* is defined to be an element whose position in e is held by x_1. By this definition, for $v = 2968134$, an external cycle for the external symbol 8 is $(4, 8)$, with its desired symbol being 7.

But, such an "external cycle" is not even a cycle: The desired position of x_m in e, i.e., x_m, is external, thus undefined in $S_{n,k}$, in particular, not held by x_1, a symbol that occurs in an internal position. For example, the (undefined) position 8 in the above v is not held by 4.

The definition of an *internal cycle* for $v \in S_{n,k}$ is given as usual [14, §4.1]:

$$C = (d_1, \cdots, d_l) \ ,$$

such that 1) for $j \in [1, l-1]$, the position of d_{j+1} in v is d_j; and 2) the position of d_1 in v is d_l.

Similar to the previous case, an internal cycle contains at least two symbols; and if for some $j \in [1, l], d_j = 1$, then the position of d_{j+1}, the next element in the cycle, is 1. When this happens, we call C a *primary internal cycle*, and express it as $(1, d_{j+1}, \ldots, d_l, d_1, \ldots, d_{j-1})$.

For example, in $v = 2968134$, there is exactly one internal cycle $(3, 6)$, which is not primary.

Since, for all j, d_j is an internal position, $d_j \in [1, k]$, thus, $(1, d_j), d_j \neq 1$, is also a transposition permitted in $S_{n,k}$.

We now address a process of constructing a cycle structure for any node $v \in S_{n,k}, v \neq e$: If v contains at least one external symbol(s) in an internal position, we arbitrarily select one of them, e_{m_i}, to construct an external cycle $E_{e_{m_i}}$; and repeat this process until we have constructed such cycles for all the external symbols in v. We then proceed to construct internal cycles for those remaining internal symbols in v such that $v(i) \neq i$, i.e., those symbols *not fixed by* v.

If an external cycle is primary, we will put this cycle, in its canonical form, as the first external cycle.

We summarize the above discussion into the following observation, which is quite similar to the factorization result for any permutation in the well-known symmetric group Γ_n.

Lemma 1. *Every node $v \in S_{n,k}, v \neq e$, can be factorized into the following product of disjoint cycles:*

$$v = E_{e_{m_1}} \cdots E_{e_{m_p}} C_1 \cdots C_r, p + r > 0 \ ,$$

where $E_{e_{m_i}}, i \in [1, p]$, are the external cycles and $C_j, j \in [1, r]$, the internal ones. This factorization is unique, except for the order in which the cycles are written.

For any $v \in S_{n,k}$, we refer to its factorization as shown in Lemma 1 as its *cycle structure*, $C(v)$, in the rest of this note. For example, let $v = 2968134 \in S_{9,7}$, its cycle structure, $C(v)$, is $(1, 2, 9; 5)(8; 7, 4)(3, 6)$.

Since a cycle is primary iff it contains 1, there exists at most one primary cycle in the cycle structure of any node v in $S_{n,k}$. A primary cycle contains a transposition $(1, j), j \in [2, n]$, which is the only kind of edge connection permitted in $S_{n,k}$ by definition.

3   A Minimum Routing Strategy

We now discuss a minimum routing strategy that changes any node $v(\neq e) \in S_{n,k}$ to e, based on its cycle structure in terms of a sequence of transpositions. Incidentally, the reverse of such a strategy changes e to v.

Let $C(v)$ be a cycle structure for a node $v \in S_{n,k}$, $g_I (g_E)$ be the number of internal (external) cycles included in $C(v)$, $b_I (b_E)$ be the number of symbols as contained in the respective g_I (g_E) cycles, and let $b = b_I + b_E$ be the total number of symbols that $C(v)$ contains.

- For an internal cycle $C = (d_1, d_2, \cdots, d_l)$, if C is primary, i.e., $d_1 = 1$, then by [7, Lemma 1], a minimum routing strategy for C is

$$(1, d_2) \circ (1, d_3) \circ \cdots \circ (1, d_l) \ ;$$

clearly all permitted in E, with the total number of steps being $l - 1$.
If C is not primary, then a minimum routing strategy for C is

$$(1, d_1) \circ (1, d_2) \circ (1, d_3) \circ \cdots \circ (1, d_l) \circ (1, d_1) \ ;$$

with its total number of steps being $l + 1$.
Since all the internal cycles are disjoint, we have the following result.

Lemma 2. *Let $t_I(v, S_{n,k})$ be a shortest transition sequence, permitted in $S_{n,k}$, that changes the positions of all the symbols as contained in the internal cycles of v to those in e, then*

$$|t_I(v, S_{n,k})| = \begin{cases} b_I + g_I - 2, & \text{if one of the internal cycles is primary} \\ b_I + g_I, & \text{otherwise.} \end{cases}$$

We notice the above distance formula is the same as for the star graph [1].
- For the external cycles, a routing strategy is sketched in [5], which is essentially to consider all the external cycles as a whole. We now fill in the details for such a strategy, discuss its correctness, and prove its optimality.
  Let $E_{e_m^1} = (e_m^1; e_0^1, e_1^1, \ldots, e_{m-1}^1)$ and $E_{e_r^2} = (e_r^2; e_0^2, e_1^2, e_3^2, \ldots, e_{r-1}^2)$ be two external cycles.

  • If one of them is primary, without loss of generality, assume $E_{e_m^1}$ is primary, i.e., for some $j \in [0, m - 1], e_j^1 = 1$. By definition, the position of e_{j+1}^1 in v is 1. We then express $E_{e_m^1}$ in the canonical form: $E_{e_m^1} = (1, e_{j+1}^1, \ldots, e_{m-1}^1, e_m^1; e_0^1, \ldots, e_{j-1}^1)$, and the routing strategy will be the following:

$$\begin{aligned} E_{e_m^1} E_{e_r^2} &= (1, e_{j+1}^1) \circ (1, e_{j+2}^1) \circ \cdots \circ (1, e_{m-1}^1) \circ (1, e_r^2) \circ (1, e_0^2) \circ \cdots \circ \\ &\quad (1, e_{r-1}^2) \circ (1, e_m^1) \circ (1, e_0^1) \cdots \circ (1, e_{j-1}^1) \\ &= (1, e_{j+1}^1, \ldots, e_{m-1}^1, e_r^2; e_0^2, e_1^2, \ldots, e_{r-1}^2, e_m^1; e_0^1, \ldots, e_{j-1}^1) \ . \end{aligned}$$

For example, let $E_{e_3^1} = (e_3^1; e_0^1, 1, e_2^1)$, and $E_{e_2^2} = (e_2^2; e_0^2, e_1^2)$, we have the following routing strategy to change all the internal symbols occurring in these two cycles to their respective positions in e :

$$(1, e_2^1) \circ (1, e_2^2) \circ (1, e_0^2) \circ (1, e_1^2) \circ (1, e_3^1) \circ (1, e_0^1) \ .$$

Table 1 shows the routing process.
At the end, the internal symbols $e_0^1, e_1^1, e_2^1, e_0^2$ and e_1^2 all get to their respectively desired positions in e; while e_3^1 and e_2^2, two external positions in v, now hold e_2^2 and e_3^1, respectively. This certainly is immaterial.

Table 1. An example of the routing strategy

Position	e_2^1	e_3^1	e_0^1	$e_1^1(=1)$	e_1^2	e_2^2	e_0^2
Initial	e_3^1	e_0^1	$e_1^1(=1)$	e_2^1	e_2^2	e_0^2	e_1^2
$(1, e_2^1)$	e_2^1			e_3^1			
$(1, e_2^2)$				e_0^2		e_3^1	
$(1, e_0^2)$				e_1^2			e_0^2
$(1, e_1^2)$				e_2^2	e_1^2		
$(1, e_3^1)$		e_2^2		e_0^1			
$(1, e_0^1)$			e_0^1	e_1^1			
Final	e_2^1	e_2^2	e_0^1	e_1^1	e_1^2	e_3^1	e_0^2

The above strategy can be directly generalized to the case when the cycle structure contains more than two external cycles, by simply combining all the external cycles into one. At the end of the process, all the internal symbols are in the right positions, while the external ones will be cyclically shifted to the right. Further analysis shows that, for each external cycle $E_{e_{m_i}}$ containing m_i+1 external symbols, it takes m_i transpositions, with a total of $b_E - g_E$ transpositions; and for each external cycle other than the primary one, it takes an additional transposition, with the total for this part being $g_E - 1$ transpositions.

Thus, the total number of transpositions it takes to go through all the external cycles, when one of them is a primary cycle, is $b_E - 1$, a minimum by [7, Lemma 1].

- For the other case, when none of the external cycles is primary, we have to somehow convert the involved cycles into a sequence of transpositions permitted in $S_{n,k}$. Let $E_{e_m}^1 = (e_m^1; e_0^1, e_1^1, \ldots, e_{m-1}^1)$ and $E_{e_q}^2 = (e_q^2; e_0^2, e_1^2, \ldots, e_{q-1}^2)$ be two external cycles, it turns out that we can follow the following strategy to route all the symbols to their respective one in e :

$$E_{e_m}^1 E_{e_q}^2 = (1, e_m^1) \circ (1, e_0^1) \circ (1, e_1^1) \circ \cdots \circ (1, e_{m-1}^1) \circ (1, e_q^2)$$
$$\circ (1, e_0^2) \circ (1, e_1^2) \circ \cdots \circ (1, e_{q-1}^2) \circ (1, e_m^1) .$$

It can also be directly generalized, and it is clear that the above strategy takes two extra steps as compared with the previous one, taking a total of $b_E + 1$ routing steps, also a minimum for this case.

We summarize the above discussion into the following result:

Lemma 3. *Let $t_E(v, S_{n,k})$ be a shortest transition sequence, permitted in $S_{n,k}$, that changes the positions of all the symbols as contained in the external cycles in C_v to those in e, then*

$$|t_E(v, S_{n,k})| = \begin{cases} b_E - 1, & \text{if one of the external cycles is primary} \\ b_E + 1, & \text{otherwise.} \end{cases}$$

Based on Lemmas 2 and 3, we can derive the following result.

Theorem 1. *The distance between e and v in $S_{n,k}$ can be expressed as follows:*

1. If v does not contain any external cycle, then

$$d(e,v) = \begin{cases} (a) \ b_I + g_I, & \text{if none of the internal cycles is primary;} \\ (b) \ b_I + g_I - 2, & \text{if one of the internal cycles is primary.} \end{cases} \quad (1)$$

2. Otherwise,

$$d(e,v) = \begin{cases} (a) \ b + g_I + 1 & \text{if none of the cycles is primary,} \\ (b) \ b + g_I - 1, & \text{if one of the cycles is primary.} \end{cases} \quad (2)$$

Table 2 lists all the nodes (including the values in the external positions), their cycle structures, as well as their distance from e, for the graph $S_{4,2}$ as shown in Fig. 1.

Table 2. An example for the distance formula

Node	Cycle form	b_I	b_E	b	g_I	g_E	$d(e,v)$	Justification
21<u>34</u>	(1, 2)	2	0	2	1	0	1	(1).(b)
32<u>14</u>	(1, 3;)	0	2	2	0	1	1	(2).(b)
42<u>31</u>	(1, 4;)	0	2	2	0	1	1	(2).(b)
31<u>24</u>	(1, 3; 2)	0	3	3	0	1	2	(2).(b)
41<u>32</u>	(1, 4; 2)	0	3	3	0	1	2	(2).(b)
23<u>14</u>	(1, 2, 3;)	0	3	3	0	1	2	(2).(b)
24<u>31</u>	(1, 2, 4;)	0	3	3	0	1	2	(2).(b)
13<u>24</u>	(3; 2)	0	2	2	0	1	3	(2).(a)
14<u>32</u>	(4; 2)	0	2	2	0	1	3	(2).(a)
43<u>21</u>	(1, 4;)(3; 2)	0	4	4	0	2	3	(2).(b)
34<u>12</u>	(1, 3;)(4; 2)	0	4	4	0	2	3	(2).(b)

As another example, let $v = 1964237 = (9; 5, 2)(3, 6)$, where $b_E = 3, b_I = 2, g_I = 1$; and let $v(9) = 5$, the routing steps are as follows:

$$196423785 \overset{(1,9)}{\rightarrow} 5796423781 \overset{(1,5)}{\rightarrow} 296453781 \overset{(1,2)}{\rightarrow} 926453781 \overset{(1,9)}{\rightarrow}$$

$$126453789 \overset{(1,3)}{\rightarrow} 621453789 \overset{(1,6)}{\rightarrow} 321456789 \overset{(1,3)}{\rightarrow} 123456789.$$

It takes a total of $7 = (b + g_I + 1 = 5 + 1 + 1)$ steps, consistent with Case a of (2).

4 An Explicit Formula for the Surface Area of $S_{n,k}$

Let $B_1(n,k;d)$ refer to the number of the nodes falling into the case as covered by (1), and $B_2(n,k;d)$ refer to those as covered by (2),

$$B(n,k;d) = B_1(n,k;d) + B_2(n,k;d) \ .$$

Furthermore, let

$$B_2(n, k; d) = B_{21}(n, k; d) + B_{22}(n, k; d) ,$$

where $B_{21}(n, k; d)$ is the number of nodes whose cycle structures do not contain a primary cycle (Case 2(a) in Theorem 1) and $B_{22}(n, k; d)$ is the number of nodes whose cycle structures contain at least one external cycle and zero or more internal cycles with one of them being primary (Case 2(b) in Theorem 1).

When the cycle structure $C(v)$, of a node $v \in S_{n,k}$, contains no external cycles, all the symbols occurring in v are taken from $[1, k]$. The only kind of transpositions of a shortest transition sequence that changes v to e will be in the form of $(1, j), j \in [2, k]$, i.e., the only transposition allowed in S_k, a star graph of dimension k. Therefore, these nodes $v \in S_{n,k}$, where $d(e, v) = d$, are exactly those nodes v in S_k such that $d(e, v) = d$. In other words,

$$B_1(n, k; d) = B_S(k, d) ,$$

where $B_S(k, d)$ refers to the surface area of the star graph S_k with radius d.

$B_1(n, k; d)$ can be calculated as follows:

- If any of the internal cycles is primary, 1 will be included in one of such cycles. We then only select $b_I - 1$ internal symbols out of $k - 1$ such symbols, and use these symbols, together with 1, b_I in total, to construct the g_I internal cycles.
- Otherwise, 1 will not be included in such a cycle. Thus, we will select b_I symbols out of $k - 1$ internal symbols to construct the g_I cycles.

In both cases, we have to form g_I cycles with b_I symbols, each of which contains at least two symbols. The general quantity of $d(n, k)$, *the number of ways of factorizing n symbols into k cycles, each of which contains at least two symbols*, is discussed in [14, §4.4]. Based on Eqs. 4.18 and 4.19 [14]: for $n \geq 2k \geq 1$,

$$d(n, k) = \sum_{j=0}^{n} (-1)^{n+k-j} s(n - j, k - j), \tag{3}$$

where $s(n, k)$ stands for the Stirling numbers of the first kind.

Further analysis shows the following result: for all $k \geq 2, d \in [1, \lfloor \frac{3(k-1)}{2} \rfloor]$,

$$B_1(n, k; d) = B_S(k, d) = \sum_{g_I = \max\{1, d-k+1\}}^{\lfloor \frac{d}{3} \rfloor} \binom{k-1}{d-g_I} d(d - g_I, g_I)$$

$$+ \sum_{g_I = \max\{1, d-k+2\}}^{\lfloor \frac{d+2}{3} \rfloor} \binom{k-1}{d-g_I+1} d(d - g_I + 2, g_I) . \tag{4}$$

Other results of this nature include one as reported in [16], correcting a result derived in [12]; and another recent one obtained via a generating function approach [4].

We now come to the more interesting case that the cycle structure of a node, $v \in S_{n,k}$, does contain at least one external cycle(s). In general, its cycle structure can be the following:

$$C(v) = (e_{m_1}^1; e_0^1, \cdots, e_{m_1-1}^1), \ldots, (e_{m_p}^p; e_0^p, \cdots, e_{m_p-1}^p), (d_1^1, \cdots, d_{l_1}^1),$$
$$\ldots, (d_1^r, \cdots, d_{l_r}^r) \ ,$$

such that $p \geq 1, r \geq 0$, and for all $i \in [1,p], m_i \geq 1$, and for all $i \in [1,r], l_i \geq 2$.

We first construct and enumerate those cycle structures for $v \in S_{n,k}, d(e,v) = d$, that contain b symbols with $g_E (\geq 1)$ external cycle(s), $g_I (\geq 0)$ internal cycles, and one of those cycles is primary. By Case (b) of (2), $d = b + g_I - 1$.

Those cycle structures for this case can be constructed in the following sequence of steps:

1. Select g_E external symbols out of a total of $n - k$ external symbols in $C(n-k, g_E)$ ways. To ensure this binomial is not equal to 0, we require $n - k \geq g_E$, i.e.,

$$1 \leq g_E \leq n - k \ . \tag{5}$$

   Since the order of the external cycles is of no significance, let those g_E external symbols be put down in an arbitrary but fixed manner.
2. Since one of the cycles is primary, 1 has to be chosen as one of the internal symbols, when selecting $b - g_E$ internal symbols out of a total k internal symbols. Hence, we choose $b - g_E - 1$ symbols out of $[2,k]$ in $C(k-1, b-g_E-1)$ ways.

   To ensure that this latter binomial is not equal to 0, we have to require $k - 1 \geq b - g_E - 1$, i.e., $g_E \geq b - k$. Since $b = d - g_I + 1$,

$$g_E \geq d - k + 1 - g_I \ . \tag{6}$$

Since every internal cycle contains at least two symbols, $b_I \geq 2g_I$; and, since there is at least one external cycle, each of which contains at least two symbols, $b_E \geq 2$. Hence, $2g_I \leq b_I = b - b_E \leq b - 2 = d - g_I - 1$. In other words,

$$g_I \leq \frac{d-1}{3} \ . \tag{7}$$

Combine (6) and (7), $g_E \geq d - k + 1 - g_I \geq d - k + 1 - \frac{d-1}{3} = \frac{2d-3k+4}{3}$, i.e., $g_E \geq \lceil \frac{2d-3k+4}{3} \rceil$.

Combining the last lower bound for g_E with (5), we have the following bounds for g_E, the number of external cycles, for this case:

$$\max\left\{1, \left\lceil \frac{2d-3k+4}{3} \right\rceil\right\} \leq g_E \leq n - k \ . \tag{8}$$

We can similarly derive the following bounds for g_I, the number of internal cycles, for this case:

$$\max\{0, d-n+1\} \leq g_I \leq \left\lfloor \frac{d-1}{3} \right\rfloor \ . \tag{9}$$

3. Once those $b - g_E$ internal symbols are chosen, we have to select $b_E - g_E$ out of them, with or without 1, in $C(b - g_E, b_E - g_E)$ ways, to add them into g_E external cycles, each containing one external symbol.

Clearly, both the order of these external cycles and that of the symbols in each cycle matter, and we denote the number of partitioning those $b_E - g_E$ symbols into g_E cycles, such that both the order of these cycles and that of the symbols inside each and every block matter, as $p(b_E - g_E, g_E)$.

A combinatorial argument leads to the following expression for $p(n, k)$, *the number of ways of decomposing n symbols into k blocks, each containing at least one symbol, and both the order of these blocks and those within these blocks are important:*

$$\forall n \geq 1, k \in [1, n), p(n, k) = n! \binom{n-1}{k-1} . \tag{10}$$

We can also get the bounds for b_E, the number of symbols as contained in the g_E external cycles, as follows:

$$2g_E \leq b_E \leq d - 3g_I + 1 . \tag{11}$$

4. We finally use the remaining $b - b_E$ internal symbols, with or without 1, to construct $g_I(\geq 0)$ internal cycles.

There are $d(b - b_E, g_I) = d(d - g_I - b_E + 1, g_I)$ ways to construct those cycles.

Therefore, we have the following expression of the number of nodes whose cycle structures contain at least one external cycle(s) and zero or more internal cycles with one of those cycles being primary.

$$B_{22}(n, k; d)$$
$$= \sum_{g_E, g_I, b_E} \binom{n-k}{g_E} \binom{k-1}{d - g_I - g_E} \binom{d - g_I - g_E + 1}{b_E - g_E}$$
$$p(b_E - g_E, g_E) d(d - b_E - g_I + 1, g_I) ,$$

where the bounds of g_E, g_I and b_E are given in (8), (9), and (11), respectively; and $d(n, k), p(n, k)$ are defined in (3) and (10), respectively.

We can similarly derive the number of the cycle structures containing no primary cycle, namely, $B_{21}(n, k; d)$.

Therefore, let $D(S_{n,k})$ stand for the diameter of $S_{n,k}$, which is given in [5] as follows:

$$D(S_{n,k}) = \begin{cases} 2k - 1, & \text{if } 1 \leq k \leq \lfloor \frac{n}{2} \rfloor, \\ k + \lfloor \frac{n-1}{2} \rfloor, & \text{if } \lfloor \frac{n}{2} \rfloor + 1 \leq k < n ; \end{cases}$$

we have the following result, noticing that $B_1(n, k; d)$ is given in (4):

Theorem 2. *The surface area of $S_{n,k}$, $n \geq 2, k \in [1, n-1]$, is the following:*

$$B(n, k; d) = B_1(n, k; d) + B_{21}(n, k; d) + B_{22}(n, k; d) , \tag{12}$$

where, for $d \in [1, D(S_{n,k})]$,

$$B_{21}(n, k; d)$$
$$= \sum_{g_E, g_I, b_E} \binom{n-k}{g_E} \binom{k-1}{d-g_I-g_E-1} \binom{d-g_I-g_E-1}{b_E-g_E}$$
$$p(b_E - g_E, g_E)d(d - b_E - g_I - 1, g_I) \ ,$$

and the bounds of g_E, g_I and b_E are given as follows:

$$\max\left\{1, \left\lceil \frac{2d - 3k + 3}{3} \right\rceil\right\} \leq g_E \leq n - k, \ \max\{0, d - n\} \leq g_I \leq \left\lfloor \frac{d-3}{3} \right\rfloor, \ and,$$
$$2g_E \leq b_E \leq d - 3g_I - 1 \ ;$$

$$B_{22}(n, k; d)$$
$$= \sum_{g_I, g_E, b_E} \binom{n-k}{g_E} \binom{k-1}{d-g_I-g_E} \binom{d-g_I-g_E+1}{b_E-g_E}$$
$$p(b_E - g_E, g_E)d(d - g_I - b_E + 1, g_I) \ ,$$

where the bounds of g_I, g_E and b_E are given as follows:

$$\max\left\{1, \left\lceil \frac{2d - 3k + 4}{3} \right\rceil\right\} \leq g_E \leq n - k, \ \max\{0, d - n + 1\} \leq g_I \leq \left\lfloor \frac{d-1}{3} \right\rfloor,$$
$$and, \ 2g_E \leq b_E \leq d - 3g_I + 1 \ .$$

Finally, the quantities $d(n, k)$ and $p(n, k)$ in the above expressions are given in (3) and (10), respectively.

For example, we can easily calculate the following with (12): $B(4, 2; 1) = 3$, $B(4, 2; 2) = 4$, and $B(4, 2; 3) = 4$, consistent with the results as shown in Table 2.

We also wrote a simple computer program to calculate the results of (12). For example, the sequence corresponding to $B(8, 3, d), d \in [0, 5]$, is $(1, 7, 22, 81, 145, 80)$, which is not included in the On-line Encyclopedia of Integer Sequences [17].

5   Conclusion

We have studied the minimum distance routing in an (n, k)-star graph by characterizing a node's cycle structure and thus are able to express the minimum routing distance in terms of this cycle structure. Consequently, we use this structure to obtain an explicit formula to compute the surface area centered at the identity node e in the graph for any radius.

This formula will help in establishing various bounds in data communication on the graph, and the techniques applied in deriving such a formula should be also useful elsewhere.

References

1. Akers, S.B., Krishmamurthy, B.: A group theoretic model for symmetric interconnection networks. IEEE Trans. on computers 38(4), 555–566 (1989)
2. Sarbazi-Azad, H.: On some combinatorial properties of meshes. In: Proc. International Symp. on Parallel Architecture, Algorithms and Networks (ISPAN 2004),
   Hong Kong, China, May 2004, pp. 117–122. IEEE Comp. Society, Los Alamitos
   (2004)
3. Sarbazi-Azad, H., Ould-Khaoua, M., Mackenzie, L.M., Akl, S.G.: On the combinatorial properties of k-ary n-cubes. Journal of Interconnection Networks 5(1), 79–91
   (2004)
4. Cheng, E., Qiu, K., Shen, Z.: A short note on the surface area of star graphs.
   Parallel Processing Letters (to appear)
5. Chiang, W., Chen, R.: The (n, k)-star graph: A generalized star graph. Information
   Processing Letters 56, 259–264 (1995)
6. Corbett, P.F.: Rotator graphs: an efficient topology for point-to-point multiproces-
   sor networks. IEEE Trans. on Parallel and Distributed Systems 3(5), 622–626
   (1992)
7. Denés, J.: The representation of permutation as the product of a minimal number
   of transpositions and its connection with the theory of graphs, Magyar Tudományos
   Akadémia. Matematikai Kutatóintézet 4, 63–71 (1959)
8. Duh, D., Lin, T.: Constructing vertex-disjoint paths in (n, k)-star graphs. Information Sciences: an International Journal archive 178(3), 788–801 (2008)
9. Fertin, G., Raspaud, A.: k-Neighbourhood broadcasting. In: 8th International Colloquium on Structural Information and Communication Complexity (SIROCCO
   2001), pp. 133–146 (2001)
10. Imani, N., Sarbazi-Azad1, H., Akl, S.G.: On Some Combinatorial Properties of the
    Star Graph. In: Proc. 2005 International Symp. on Parallel Architecture, Algo-
    rithms and Networks (ISPAN 2005), December 7–9, 2005, pp. 58–65. IEEE Comp.
    Society, Los Alamitos (2005)
11. Imani, N., Sarbazi-Azad1, H., Zomaya, A.Y.: Some properties of WK-recursive
    and swapped networks. In: Stojmenovic, I., Thulasiram, R.K., Yang, L.T., Jia, W.,
    Guo, M., de Mello, R.F. (eds.) ISPA 2007. LNCS, vol. 4742, pp. 856–867. Springer,
    Heidelberg (2007)
12. Portier, F., Vaughan, T.: Whitney numbers of the second kind for the star poset.
    Europ. J. Combinatorics 11, 277–288 (1990)
13. Qiu, K., Akl, S.: On some properties of the star graph. VLSI Design 2(4), 389–396
    (1995)
14. Riordan, J.: An Introduction to Combinatorial Analysis. Wiley, New York (1980)
15. Sampels, M.: Vertex-symmetric generalized Moore graphs. Discrete Applied Math-
    ematics 138, 195–202 (2004)
16. Shen, Z., Qiu, K.: On the Whitney numbers of the second kind for the star poset.
    European Journal of Combinatorics (to appear, 2008)
17. Sloane, N.J.A.: The On-Line Encyclopedia of Integer Sequences,
    http://www.research.att.com/~njas/sequences/

Enumerating Isolated Cliques
in Synthetic and Financial Networks

Falk Hüffner[*], Christian Komusiewicz[**], Hannes Moser[***],
and Rolf Niedermeier

Institut für Informatik, Friedrich-Schiller-Universität Jena,
Ernst-Abbe-Platz 2, D-07743 Jena, Germany
{hueffner,ckomus,moser,niedermr}@minet.uni-jena.de

Abstract. We do computational studies concerning the enumeration of maximal isolated cliques in graphs. Isolation, as recently introduced, measures the degree of connectedness of the cliques to the rest of the graph. Isolation helps both in getting faster algorithms than for the enumeration of maximal general cliques and in filtering out cliques with special semantics. We perform experiments with synthetic graphs (in the $G_{n,m,p}$ model) and financial networks, proposing the enumeration of isolated cliques as a useful instrument in analyzing financial networks.

1 Introduction

We study the generation of maximal cliques of an undirected graph $G = (V, E)$, that is, the enumeration of all vertex subsets $V' \subseteq V$ such that the induced subgraph $G[V']$ is complete and there is no $V'' \supsetneq V'$ such that $G[V'']$ is also complete. Unfortunately, already finding one maximum-cardinality clique is a notoriously hard computational problem, being NP-hard [8] as well as W[1]-hard [7] and hard to approximate [9]. By way of contrast, finding cliques is very important in many practical applications. Recent papers describe applications in computational finance [3, 4] as well as computational biochemistry and genomics [5, 6].

Enumerating all maximal cliques needs exponential time. For instance, a recent paper by Tomita et al. [14] proved a worst-case time complexity of $\Theta(3^{n/3})$ for an n-vertex graph, arguing for its optimality due to the fact that there are example graphs having $3^{n/3}$ maximal cliques. Recently, Ito et al. [10] proposed to restrict the search to certain types of cliques, that is, specifically *isolated cliques*. A clique V' of k vertices is called *c-isolated* in a graph G if there are less than $c \cdot k$ edges leaving the induced subgraph $G[V']$ in G. This concept is interesting for two reasons. First, since one does not search for all maximal cliques any more, faster enumeration algorithms are possible. Second, isolated cliques may be an intrinsically relevant concept, because these cliques can represent structures with

---

[*] Supported by the DFG, Emmy Noether research group PIAF, NI 369/4.
[**] Supported by a PhD fellowship of the Carl-Zeiss-Stiftung.
[***] Supported by the DFG, projects ITKO, NI 369/5 and AREG, NI 369/9.

B. Yang, D.-Z. Du, and C.A. Wang (Eds.): COCOA 2008, LNCS 5165, pp. 405–416, 2008.
© Springer-Verlag Berlin Heidelberg 2008

particularly interesting properties that are detected in this way. Ito et al. [10] stated the linear-time enumerability of isolated cliques (for constant "isolation factor" c) by claiming an algorithm running in $O(4^c \cdot c^5 \cdot m)$ time for an m-edge graph. Unfortunately, their algorithm is flawed. Hence, in our recent theoretical work [12], we presented a nontrivially repaired algorithm with the same running time. Moreover, we introduced two closely related isolation concepts called min-c-isolation and max-c-isolation, respectively.

Here, we present the following results. First, we give a theoretical improvement for the enumeration of Ito et al.'s isolated cliques, now achieving a running time of $O(2.89^c \cdot c^2 \cdot m)$. The main focus of our work, however, is on computational studies, applying the three isolation concepts to random feature graphs (in the $G_{n,m,p}$ model) and financial networks. The random graphs serve as benchmark instances for charting the tractability borderlines of our algorithms. We find for min- and max-isolation that the algorithms are output-sensitive, and hence lead to very fast clique enumeration for lower values of c. For isolation as introduced by Ito et al., however, this is not always the case, and sometimes even for intermediary values of c the enumeration becomes infeasible. Interestingly, we observe that the practical and theoretical bottlenecks of the algorithm differ. The financial networks serve as an example of how isolation can be used to find particularly interesting cliques. In our experiments, we analyze the so-called *clique performance*, which represents the profit/loss of the underlying financial instruments, and observe significant differences between the three isolation concepts.

2   Fundamentals and Algorithms and Implementation Issues

The fundamental strategy and several basic ideas go back to Ito et al. [10]; while their work contains serious flaws as spotted in [12], it initiated the study of isolation in context with the enumeration of maximal cliques. Besides sketching the fundamental algorithmic ideas, we additionally describe a new theoretical result leading to an improved running time.

Fundamentals. Ito et al. [10] introduced the concept of *c-isolation*—which, in the light of the following is called *average-c*-isolation (*avg-c*-isolation for short) in this work—as follows: Let $G = (V, E)$ be an undirected graph and c be a positive integer. A vertex subset $S \subseteq V$ of size k is called *avg-c-isolated* if it has less than $c \cdot k$ outgoing edges, where an outgoing edge is an edge between a vertex in S and a vertex in $V \setminus S$. In follow up-work, we further introduced the concepts of min-c-isolation and max-c-isolation as follows [12]. A vertex set $S \subseteq V$ is *min-c-isolated* if there is at least one vertex in S with less than c neighbors in $V \setminus S$. A vertex set $S \subseteq V$ is *max-c-isolated* if every vertex $v \in S$ has less than c neighbors in $V \setminus S$. Fig. 1 illustrates the three concepts.

For notational simplification we will mostly use the terms min-isolation, avg-isolation, and max-isolation. Note that by definition min-c-isolation is weaker than avg-c-isolation in the sense that every avg-c-isolated clique is also min-c-isolated but not vice versa. The enumeration of maximal min-c-isolated cliques

min-2-isolation   avg-2-isolation   max-2-isolation

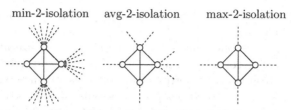

Fig. 1. Isolated 4-vertex cliques for $c = 2$ and the three isolation concepts. The dashed lines denote outgoing edges.

yields cliques that are at least as large as and often larger than avg-c-isolated cliques. By way of contrast, max-c-isolation is stronger than avg-c-isolation. Max-c-isolation is useful when we want to exclude high-degree vertices from the enumerated sets. This can result in the enumeration of smaller cliques than in the other two cases. The theoretical study [12] of these three concepts led to the following theorem. Herein, n denotes the number of graph vertices and m the number of edges.

Theorem 1 ([12]). *Maximal avg-c-isolated cliques can be enumerated in $O(4^c \cdot c^3 \cdot m)$ time, maximal min-c-isolated cliques in $O(2^c \cdot c \cdot m + n \cdot m)$ time, and maximal max-c-isolated cliques in $O(2.44^c \cdot m)$ time.*

Ito et al. [10] only considered avg-c-isolation and claimed a running time of $O(4^c \cdot c^5 \cdot m)$ for the enumeration of maximal avg-c-isolated cliques as their main result. Besides sketching the algorithms behind Theorem 1, we also prove a new result, improving the time bound for avg-c-isolated cliques to $O(2.89^c \cdot c^2 \cdot m)$.

Algorithms. In the following, we focus on describing the algorithm for avg-isolation. The corresponding algorithms for min-isolation and max-isolation have the same basic structure [12]; however, they differ in technical details we cannot go into here. Given a graph $G = (V, E)$ and an isolation factor c, first the vertices are sorted by their degree such that $u < v \Rightarrow \deg(u) \leq \deg(v)$. The *index* of a vertex is its position in this sorted order. Let $N_+[v] := \{u \in N[v] \mid u > v\} \cup \{v\}$ and $N_-(v) := \{u \in N(v) \mid u < v\}$. In an avg-isolated clique, the vertex with the lowest index is called the *pivot* of the clique [10]. Clearly, a pivot has less than c outgoing edges. Since every avg-isolated clique has a pivot, we can enumerate all maximal avg-isolated cliques of a graph by enumerating all maximal avg-isolated cliques with pivot v for each $v \in V$ and then removing those avg-isolated cliques with pivot v that are a subset of an avg-isolated clique with another pivot.

The enumeration of maximal avg-isolated cliques *with pivot v for a $v \in V$* is called the *pivot procedure*. It comprises three successive stages:

Trimming stage. This stage builds in polynomial time a candidate set C that is a superset of all avg-isolated cliques with pivot v. The set C is initialized with $N_+[v]$, and then vertices that obviously cannot be part of an avg-isolated clique with pivot v are removed from C. In particular, we remove vertices from c that have too many neighbors outside of C or too few neighbors in C.

Enumeration stage. This stage enumerates cliques with pivot v. Let C be the candidate set after the trimming stage, and $|N[v] \setminus C| = d$. In total, we can delete at most $c - 1$ vertices from $N[v]$, since otherwise v obtains too many outgoing edges. Therefore, $\tilde{c} := c - 1 - d$ is the number of vertices that we may still remove from C. We can enumerate cliques $C' \subseteq C$ of size *at least* $|C| - \tilde{c}$ by enumerating vertex covers[1] of size *at most* $\tilde{c}$ in the complement graph $\overline{G[C]}$: First, we enumerate all *minimal* vertex covers and thus obtain maximal cliques in the candidate set C. Then, to also capture avg-isolated cliques that are subsets of non-avg-isolated cliques enumerated this way, for each of these cliques, we enumerate all maximal subsets that fulfill the isolation condition. It is possible to show [12, Lemma 1] that given a non-avg-isolated clique C', we may only remove vertices from the set of vertices with the c highest indices in order to obtain a maximal avg-isolated clique that is a subset of C'. This is done in a brute-force way by enumerating subsets of the set of vertices that may be deleted, and then checking for each such subset whether removing this subset yields an avg-isolated clique. This stage has running time $O(2^c \cdot c^6 \cdot m)$ [12].

Screening stage. In the screening stage, all cliques that are either not avg-isolated or that are avg-isolated but not maximal are removed. First, avg-isolation is checked. Next, those cliques that pass the test for isolation are compared pairwise, and we only keep maximal cliques. Finally, we check each clique that is left for pivot v against each clique obtained during calls to pivot(u) with $u \in N_-(v)$, since these are the only cliques that can be superset of a clique obtained for pivot v. The running time of this stage is $O(4^c \cdot c^3 \cdot m)$.

Min-isolation and max-isolation lead to conceptually simpler pivot procedures. The new theoretical contribution provided in this paper when compared to our previous theoretical work [12] is to show an improvement of the screening stage in the case of the avg-isolation concept.

Suppose that an enumerated avg-isolated clique C with pivot v is not maximal. Then there must be a nonempty vertex set S such that $C \cup S$ is an avg-isolated clique. Obviously, $S \subseteq N[v] \backslash C$. Also, S must be a clique and all vertices in S have to be adjacent to all vertices in C. Let $D \subseteq N[v] \backslash C$ such that D contains exactly the vertices that are adjacent to all vertices in C. To test the maximality of C, we first enumerate all maximal cliques $D' \subseteq D$. Then, for each such clique D', the set $C \cup D'$ is a clique. If $C \cup D'$ is also avg-isolated, then C is clearly not maximal and thus removed from the output. If $C \cup D'$ is not avg-isolated, however, then we have to check whether there is an avg-isolated subset of $C \cup D'$ that is also a superset of C. This can be done by removing the vertices of highest degree from D' until either $C \cup D'$ becomes avg-isolated or D' is empty. In the first case, C is not a maximal avg-isolated clique and is thus removed from the output. In the second case, C is a maximal avg-isolated clique in $C \cup D'$. If this can be shown for all maximal cliques $D' \subseteq D$, then C is a maximal avg-isolated clique in G. With this maximality test, we can improve the asymptotic running time bound of the enumeration algorithm (cf. Theorem 1).

---

[1] A vertex cover of a graph is a subset D of vertices such that each graph edge has at least one endpoint in D. See Abu-Khzam et al. [1] for algorithm engineering results in determining minimum-cardinality vertex covers.

Theorem 2. *Maximal avg-c-isolated cliques of an m-edge graph can be enumerated in $O(2.89^c \cdot c^2 \cdot m)$ time.*

Proof. Since the trimming stage and enumeration stage of the algorithm have not changed, their running time amounts to $O(c^4 \cdot m + 2^c \cdot c^2 \cdot c^4 m) = O(2^c \cdot c^6 \cdot m)$ [12]. In the screening stage of the pivot procedure, we have to test each clique for maximality. At most $2^{c-1} \cdot c$ cliques are enumerated during the enumeration stage of the pivot procedure for a pivot v. For any enumerated avg-isolated clique C, we have to enumerate all maximal cliques in a subset of $N[v] \setminus C$. Since $|N[v] \setminus C| \leq c - 1$, this can be done in $O(3^{c/3})$ time [14]. For each pair of an enumerated avg-isolated clique C and a maximal clique D', we decide whether a subset of $C \cup D'$ is avg-isolated by successively removing the vertices with highest degree from D'. Clearly, this can be done in $O(c)$ time. Overall, one execution of the screening stage thus has a worst-case running time of $O(2^c \cdot c) \cdot O(3^{c/3}) \cdot O(c) = O(2.89^c \cdot c^2)$. There are n runs of the screening stage, and together with the running times of the other stages, we achieve a total worst-case running time of $O(2^c \cdot c^6 \cdot m) + O(2.89^c \cdot c^2 \cdot n) = O(2.89^c \cdot c^2 \cdot m)$. □

Implementation Issues. We briefly describe some notable differences between the theoretical algorithms [12] and their actual implementations.[2]

Min-isolation. In the trimming stage, we remove vertices that have lower index than the pivot (this differs from the description in [12]). This does not help in achieving a better worst-case running time, but it speeds up the trimming stage and prevents the algorithm from needlessly entering the enumeration stage for vertices with at least c neighbors of lower index. In many instances this provided a speed-up of factor 3 or more.

Avg-isolation. Since our experiments showed that the enumeration of avg-isolated subsets of non-avg-isolated cliques was a bottleneck, we introduced an additional test: We check whether we can obtain an avg-isolated set by gradually removing the vertices of highest degree. If this is not the case, then no subset of the clique is avg-isolated. Thus, we can avoid unnecessarily enumerating subsets of non-avg-isolated cliques. Furthermore, we perform this test also before entering the enumeration stage, and only enter it when the enumerated cliques have a chance of being c-isolated. Both tests provided a speed-up of approximately two orders of magnitude in our experiments.

Max-isolation. The worst-case running time of $O(2.44^c \cdot c \cdot m)$ can be shown using a maximum clique algorithm in the screening stage (for details see [12]). Running time analysis showed that, unexpectedly, in practice the screening stage was not the bottleneck of the enumeration algorithm. Therefore, in our implementation we instead enumerate all cliques in the set of deleted vertices to check whether an enumerated clique is maximal. This was sufficiently fast, while keeping the implementation simpler.

---

[2] The program is written in Objective Caml and consists of about 1600 lines of code. It is free software and available from http://theinf1.informatik.uni-jena.de/c-isol/.

As maximal clique enumeration algorithm (required for the screening stage of avg-isolation and max-isolation), we used an improved variant of the standard Bron–Kerbosch algorithm by Koch [11]. This algorithm was not a bottleneck, in particular because of its good output-sensitivity (that is, it runs quickly if there are only few maximal cliques). We also use this algorithm as a comparison point for the running times of our clique enumeration algorithms.[3]

3   Experimental Results

Our investigations concentrate on random feature graphs that were created according to the $G_{n,m,p}$ model and on financial networks. All experiments were run on an AMD Athlon 64 3700+ machine with 2.2 GHz, 1 M L2 cache, and 3 GB main memory running under the Debian GNU/Linux 4.0 operating system with the Objective Caml 3.09.2 compiler. Note that for some instances the enumeration of avg-isolated cliques did not terminate because the program exceeded the memory limit of 3 GB or the corresponding run timed out (after half an hour). This causes some missing data points for avg-isolation in the diagrams.

Synthetic Data. We generated random graphs using the $G_{n,m,p}$ model (see Behrisch and Taraz [2] and references therein). The underlying model is that cliques are defined by *features*. More precisely, each of n vertices draws each of m features with probability p, and two vertices are connected by an edge iff they have at least one feature in common (note that here m does not denote the number of edges as elsewhere). Since every nonempty intersection of vertex sets corresponding to some features defines a maximal clique, these graphs contain very many maximal cliques, and are tough inputs for clique enumeration.

Our main finding is that enumerating min- and max-isolated cliques is feasible over a far wider range than enumerating general maximal cliques or avg-isolated cliques, and that the isolation concepts can help keeping the number of enumerated *isolated* cliques in check even in graphs that contain excessively many *maximal* cliques. Furthermore, we observe a difference in output-sensitivity. Whereas min-isolation seems to be output-sensitive in general and max-isolation in most instances, avg-isolation had high running times sometimes even for relatively few enumerated cliques. Starting from a base setting with $c = 40$, $n = 200$, $m = 45$, and $p = 0.1$, we examined the effect of varying parameters. Fig. 2a shows the number of cliques output for varying c averaged over 5 instances. The average number of maximal cliques is 92611. Starting from $c \approx 80$, all maximal cliques are enumerated using min-isolation. For avg- and max-isolation all maximal cliques are found with $c \approx 150$. In Fig. 2b, we see that the running time of the min- and max-isolation concepts closely follows the number of cliques output, that is, the algorithms are output-sensitive. This can not be observed for avg-isolation, since its running time peaks for intermediary values of c. Notably, for all three isolation concepts almost all time is spent in the enumeration stage. Therefore,

---

[3] Note that we could not perform comparisons with the claimed fastest general clique enumeration algorithm by Tomita et al. [14], since the code is unavailable.

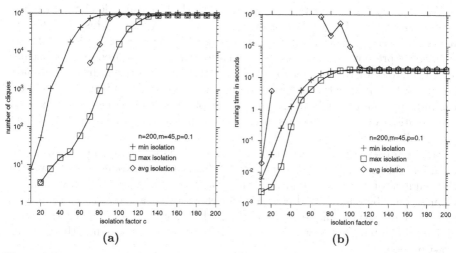

Fig. 2. $G_{n,m,p}$ model with $n = 200$, $m = 45$, and $p = 0.1$. Average running time for Bron–Kerbosch is 5.06 seconds.

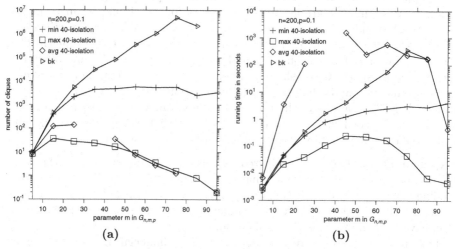

Fig. 3. $G_{n,m,p}$ model with $c = 40$, $n = 200$, and $p = 0.1$. The missing point for avg-isolation is due to the memory limit of the test runs (3 GB).

the increased running time and lack of output-sensitivity for avg-isolation stems from the enumeration of isolated subsets of non-avg-isolated cliques, since this is the only part where the enumeration stages differ. Furthermore, this means that in practice the screening stage, which dominates the overall worst-case running time, is not the bottleneck of the algorithm. Compared to the Bron–Kerbosch algorithm, when enumerating the *whole* set of maximal cliques, all three algorithms are about 4 times slower, but min- and max-isolation are significantly faster when the output is restricted by a small c (see Fig. 2).

We next examine variation of m (Fig. 3). More features lead to an exponential growth of the number of maximal cliques (Fig. 3a). This growth only wears off when the graph becomes very dense ($m = 85$, about 57 % of all possible edges present). In contrast, the number of min-40-isolated cliques reaches a plateau, and for the more stringent criteria, we even notice a drop-off already for $m \geq 30$. While for the Bron–Kerbosch algorithm and min-isolation, we have running times mostly following the number of generated cliques, for max- and avg-isolation, we have a maximum for $m = 35$ and $m = 45$, respectively. Again, almost all time is spent in the enumeration stage.

Similar observations were made for varying values of p and n. For both p and n, increasing the parameter value leads to an exponential growth in the number of maximal cliques of the graph. Again, min- and max-isolated cliques could be enumerated over a wider range of parameter values than avg-isolated and maximal cliques. In particular, the algorithms for enumerating min- and max-isolated cliques were output-sensitive while this was not the case for avg-isolation.

Financial Networks. Many works on financial network analysis are based on market graphs (see, e.g., [13]). We generated market graphs from publicly available stock data[4]. A market graph is constructed as follows. Financial instruments (e.g., stocks or indices) are represented by vertices. For each pair of vertices u, v there is an edge connecting them if the corresponding correlation coefficient C_{uv} based on the price fluctuations of u and v in some prespecified time range exceeds some prespecified threshold θ, where $-1 \leq \theta \leq 1$. Informally speaking, two instruments u and v have a positive correlation coefficient C_{uv} if they show similar daily fluctuations in the prespecified time range, and they have a negative correlation coefficient if their daily fluctuations behave oppositional. Details about the construction of market graphs can be found, e.g., in [3].

Experimental Setup. We considered various market graphs based on the daily fluctuations of several thousand financial instruments during 500 consecutive trading days. Basic properties of such graphs, like degree distribution, edge density, clustering coefficient, maximum clique size, and maximum independent set size, have been analyzed by Boginski et al. [3, 4].

The following diagrams rely on data from 2204 financial instruments beginning at 2003-12-02 over 500 consecutive trading days. However, the experiments were also executed on many other graphs (based on data from other start dates and other threshold values) for which the following observations also hold true (in the qualitative sense). Note that the graphs do not include financial instruments whose values get below one dollar in the considered time period, since such "penny stocks" often show strong daily fluctuations, which are additionally biased by the rounding of the available data. In the experiments with fixed threshold, the threshold is set to $\theta = 0.5$ as proposed by Boginski et al. [4] in order to ensure that only significantly correlated stocks are adjacent. Moreover, our experiments showed that for $\theta = 0.5$ there is a good balance between

---

[4] We used the data from finance.yahoo.com.

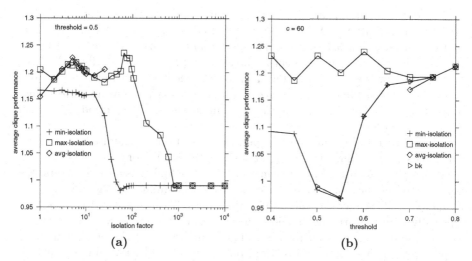

Fig. 4. Average clique performance in a market graph based on 500 consecutive trading days beginning at 2003-12-02. Note that the performance of the NASDAQ in the considered time period is 1.01.

the number of isolated cliques in the graph and the edge density (for very low threshold levels, the graph gets too dense to contain many isolated cliques, and for very high threshold levels, the graph can get too sparse to contain interesting cliques of significant size). For threshold $\theta = 0.5$, the graph contains 2204 vertices and 64376 edges and approximately 70000 maximal cliques.

Basic Results. As for the $G_{n,m,p}$ graphs, we found enumerating min- and max-isolated cliques to be feasible over a wide range of parameters, while the Bron–Kerbosch algorithm and the avg-isolation algorithm are sometimes too slow. For all three isolation concepts and for $c \leq 10$ the running time is around a second. For intermediate isolation factors we observe a peak in the running time of max- and avg-isolation. Surprisingly, we also find that enumerating all maximal cliques using the algorithm for min-∞-isolation is faster than Bron–Kerbosch by one order of magnitude.

The number of enumerated isolated cliques ranges from a few hundred for very low isolation factors up to all maximal cliques (≈ 70000) for high isolation factors, where there are generally much more min-isolated cliques than max- and avg-isolated cliques (up to one order of magnitude). For low isolation factors, max- and avg-isolated cliques have size at most 10, whereas there are already min-1-isolated cliques of size ≈ 50. For high isolation factors, the enumerated cliques have maximum size ≈ 80.

Clique Performance. Boginski et al. [3, 4] suggested the use of clique analysis for classifying stocks, based on the property that cliques represent sets of "similar" financial instruments. However, they do not provide any method to find cliques of good quality. Therefore, we measured the average performance of the enumerated cliques. The *average price* of a financial instrument at some given trading day t is the mean price of the instrument at day t and the 10 trading days before and after t. Average prices are used to balance stronger daily fluctuations of financial instruments. The *performance* in the time interval $[t_1, t_2]$ $(t_1 < t_2)$ of a financial instrument is the average price at day t_2 divided by the average price at day t_1. The performance of a clique is the mean performance of its vertices. The *average performance* of a set of cliques is the mean performance of the cliques. We always measure the performance in the time period the market graph is based on.

We can observe (Fig. 4a) that the performance of the enumerated min-, max-, and avg-isolated cliques is better for lower isolation factors and generally exceeds the performance of all maximal cliques. For higher isolation factors, the min-isolated cliques show a performance which is similar to the average performance of all vertices in the graph. Most notably, max-isolated cliques have especially high performance for intermediate isolation levels; we can observe a peak of the performance for max-isolation around $c = 100$. Avg-isolation seems to perform similarly as max-isolation, but we usually observe running time or memory consumption problems for intermediate isolation levels. For very high isolation factors, all three isolation concepts generate all maximal cliques and therefore obviously yield the same average performance. In general, the described effects depend on the underlying graph and the performance of the overall market and are more or less pronounced. Note that for low isolation factors ($c \leq 20$) we could not observe a significant general difference of the performance of the three isolation concepts. In our example (Fig. 4a), max- and avg-isolation are slightly better for low isolation factors, but there are other graphs (based on other time periods), for which min-isolation performs better. Note that the average performance of all financial instruments in the considered time period is approximately 1.19. Surprisingly, the maximal cliques have an average performance of about 0.99. This is caused by financial instruments with a particularly bad performance that are included in many maximal cliques, but not in *isolated* maximal cliques.

When varying the threshold value, Fig. 4b shows that the performance of max-isolation is relatively independent of the threshold level, whereas min-isolated and all maximal cliques perform better for higher threshold levels. Note that this only holds true for low isolation factors $c \leq 100$, since for higher isolation factors the performance of all three isolation concepts gets closer to the performance of all maximal cliques.

Possible Applications. We believe that especially max-isolated cliques have some interesting properties with respect to the average clique performance: First of all, the average performance of max-isolated cliques is relatively independent from the chosen threshold values. This is beneficial in practice, as finding a good threshold value is usually a relatively difficult task. Moreover, looking more closely at the cliques responsible for the peak of the performance for intermediate isolation levels, we observe that these cliques represent some niche in the market. For instance, in Fig. 4a the peak is caused by American raw material, oil, and energy stocks, and by related industries like transportation, pipeline construction, and refineries. This peak is less pronounced in graphs based on earlier time periods (that is, beginning before 2003-12-02) and becomes even more pronounced for graphs based on later time periods (that is, beginning after 2003-12-02). This indicates that max-isolation can be useful to detect market trends. Finally, isolated cliques performed better than general maximal cliques. Hence, we can employ isolation to filter out financial instruments with bad performance when enumerating cliques. This could provide a new alternative for investors to classify financial instruments (using clique analysis as proposed by Boginski et al. [3]). Here, a more thorough and detailed study is necessary, cooperating with financial experts.

4   Conclusion and Outlook

Our results indicate the relevance of the newly introduced isolation concepts [12] in comparison with the older avg-isolation [10]. For min- and max-isolation, the enumeration algorithms show output-sensitivity. Therefore, for both of these isolation concepts the restricted number of cliques output can make enumeration algorithms for isolated cliques much faster than the standard Bron–Kerbosch algorithm. However, for avg-isolation, further algorithmic improvements have to be made in order to obtain output-sensitivity. In particular, the enumeration of isolated subsets of non-avg-isolated cliques needs to be improved. For certain instances the c-isolation algorithms are faster than Bron–Kerbosch even for $c = \infty$, which results in the same output as the Bron–Kerbosch algorithm has. It would be interesting to see whether we could further optimize our implementations for this goal and for which kind of graphs we see a gain over Bron–Kerbosch. Our findings with financial networks support that isolation provides "interesting" cliques. In particular, max-isolated cliques perform better for intermediary isolation factors. This should be analyzed more thoroughly (with the help of financial experts) to better understand what distinguishes stocks in max-isolated cliques from those in general maximal cliques or min-isolated cliques, and hence what leads to the difference in clique performance. Furthermore, are there any application scenarios in which the relatively weak min-isolation concept is useful? For example, does the pivot element of a min-isolated clique which has the fewest (and thus less than c) neighbors outside of the clique somehow characterize the whole clique?

Acknowledgements. We thank our student assistants Robert Bredereck and Manuel Sorge for their excellent support in gathering experimental data.

References

[1] Abu-Khzam, F.N., Collins, R.L., Fellows, M.R., Langston, M.A., Suters, W.H., Symons, C.T.: Kernelization algorithms for the vertex cover problem: Theory and experiments. In: Proc. 6th ALENEX, pp. 62–69. SIAM, Philadelphia (2004)
[2] Behrisch, M., Taraz, A.: Efficiently covering complex networks with cliques of similar vertices. Theoret. Comput. Sci. 355(1), 37–47 (2006)

[3] Boginski, V., Butenko, S., Pardalos, P.M.: Statistical analysis of financial networks. Comput. Statist. Data Anal. 48(2), 431–443 (2005)

[4] Boginski, V., Butenko, S., Pardalos, P.M.: Mining market data: A network approach. Comput. Oper. Res. 33(11), 3171–3184 (2006)

[5] Butenko, S., Wilhelm, W.E.: Clique-detection models in computational biochemistry and genomics. European J. Oper. Res. 173(1), 1–17 (2006)

[6] Chesler, E.J., Lu, L., Shou, S., Qu, Y., Gu, J., Wang, J., Hsu, H.C., Mountz, J.D., Baldwin, N.E., Langston, M.A., Threadgill, D.W., Manly, K.F., Williams, R.W.: Complex trait analysis of gene expression uncovers polygenic and pleiotropic networks that modulate nervous system function. Nat. Genet. 37(3), 233–242 (2005)

[7] Downey, R.G., Fellows, M.R.: Parameterized Complexity. Springer, Heidelberg (1999)

[8] Garey, M.R., Johnson, D.S.: Computers and Intractability: A Guide to the Theory of NP-Completeness. W.H. Freeman, New York (1979)

[9] Håstad, J.: Clique is hard to approximate within $n^{1-\epsilon}$. Acta Math. 182(1), 105–142 (1999)

[10] Ito, H., Iwama, K., Osumi, T.: Linear-time enumeration of isolated cliques. In: Brodal, G.S., Leonardi, S. (eds.) ESA 2005. LNCS, vol. 3669, pp. 119–130. Springer, Heidelberg (2005)

[11] Koch, I.: Enumerating all connected maximal common subgraphs in two graphs. Theoret. Comput. Sci. 250(1–2), 1–30 (2001)

[12] Komusiewicz, C., Hüffner, F., Moser, H., Niedermeier, R.: Isolation concepts for enumerating dense subgraphs. In: Lin, G. (ed.) COCOON. LNCS, vol. 4598, pp. 140–150. Springer, Heidelberg (2007)

[13] Mantegna, R.N., Stanley, H.E.: Introduction to Econophysics: Correlations and Complexity in Finance. Cambridge University Press, Cambridge (2000)

[14] Tomita, E., Tanaka, A., Takahashi, H.: The worst-case time complexity for generating all maximal cliques and computational experiments. Theoret. Comput. Sci. 363(1), 28–42 (2006)

A Risk-Reward Competitive Analysis for the Recoverable Canadian Traveller Problem*

Bing Su[1,2], Yinfeng Xu[1,3], Peng Xiao[1,3], and Lei Tian[1]

[1] School of Management, Xi'an Jiaotong University,
Xi'an, 710049, P.R. China
[2] School of Economics and Management, Xi'an Technological University,
Xi'an, 710032, P.R. China
[3] The State Key Lab for Manufacturing Systems Engineering,
Xi'an, 710049, P.R. China
{subing,yfxu,xiaopeng}@mail.xjtu.edu.cn, ttianlei@163.com

Abstract. From the online point of view, we study the Recoverable Canadian Traveller Problem (Recoverable-CTP) in a special network, in which the traveller knows in advance the structure of the network and the travel time of each edge. However, some edges may be blocked and the traveller only observes that upon reaching the vertex of the blocked edge, and the blocked edge may be reopened but the traveller doesn't know its recovery time. The goal is to find a least-cost route from the origin node to the destination node, more precisely, to find an adaptive strategy minimizing the ratio of traversed time to the travel time of the optimal offline shortest path (where all blocked edges and their recovery time are known in advance). We present an optimal online strategy - a comparison strategy and prove its competitive ratio. Moreover, with the different forecasts of the recovery time, some online strategies are given under the risk-reward framework, and the rewards and the risks of the different strategies are analysed.

Keywords: Recoverable-CTP, Competitive analysis, Comparison strategy, Risk-reward model.

1 Introduction

The Canadian Traveller Problem (CTP) has been introduced in [1] and is defined as follows: Suppose that a traveller knows in advance the structure of a network and the travel time of each edge. However, some edges may fail and the traveller only observes that upon reaching a vertex of the blocked/failed edge. The problem is to devise a good travel strategy from the origin node to the destination node without any knowledge of future edge blockages. Under this setting, Papadimitriou and Yannakakis proved that devising an online algorithm with a bounded competitive ratio is PSPACE-complete [1].

---

* The authors would like to acknowledge the support of research grant No. 70525004, 60736027, 70121001 from the NSF, No. 20060401003 from the PSF Of China and No. 06JK099 from the Education Department of Shaanxi.

B. Yang, D.-Z. Du, and C.A. Wang (Eds.): COCOA 2008, LNCS 5165, pp. 417–426, 2008.
© Springer-Verlag Berlin Heidelberg 2008

Several variations of the CTP were studied in [2-5]. If there is a given parameter k which bounds the number of blocked edges from above, the resulting problem is called the k-Canadian Traveller Problem (k-CTP) [2]. Bar-Noy and Schieber studied the k-CTP, but they did not consider the problem from a competitive analysis point of view. Instead, they considered the worst-case criterion which aims at a strategy where the maximum cost was minimized [2]. Westphal considered the online version of k-CTP and showed that no deterministic online algorithm can achieve a competitive ratio smaller than $2k + 1$ and gave an easy algorithm which matches this lower bound [3]. The same bound was in fact obtained independently in [4]. Westphal also showed that randomization can not improve the competitive ratio substantially. He showed that by establishing a lower bound of for the competitiveness of randomized online algorithms against an oblivious adversary [3]. Recoverable-CTP is a variation of CTP, in which the blocked edges may be reopened [2]. Under the assumptions that the upper bound on the number of blockages is known in advance and the recovery time are not very long compared with the travel time, Bar-Noy and Schieber presented a polynomial-time travel strategy which guarantees the shortest worst-case travel time. For the Stochastic Recoverable CTP, again when the recovery time are not very long relative to the travel time, they also presented a polynomial-time strategy which minimizes the expected travel time [2]. The online strategies were studied for the Recoverable-CTP in general networks in [5]. Under the assumption that the traveller doesn't know the recovery time upon reaching a vertex of the blocked edge, two adaptive strategies - a waiting strategy and a greedy strategy were presented, and the competitive ratios for the two strategies were given, respectively [5].

In this paper, we focus on the online version of the Recoverable-CTP in a special network, in which the vertex set is $V = \{v_1, v_2, \cdots, v_n\}$ and there are multiple edges between v_i and v_{i+1}, where $i = 1, \cdots, n - 1$. Some edges may be blocked and the traveller only observes that upon reaching the vertex of the blocked edge, and the blocked edge may be reopened but the traveller doesn't know its recovery time. Our goal is to find a least-cost route from the origin node to the destination node by passing through $v_1, v_2, \cdots, v_n$ one by one, more precisely, to find an adaptive strategy minimizing the competitive ratio, which compares the performance of this strategy with that of a hypothetical offline algorithm that knows the entire topology in advance. We present an optimal online strategy - a comparison strategy and prove its competitive ratio for the Recoverable-CTP. Moreover, with the different forecasts of the recovery time, some online strategies are given under the risk-reward framework, and the rewards and the risks of different strategies are analysed.

The organization of this paper is as follows. In Section 2, the problem definition and some assumptions are briefly reviewed. In Section 3, we propose and investigate optimal online strategies for the Recoverable-CTP in a special network. In Section 4, we consider the performance of some strategies under the risk-reward framework. Finally, we conclude the work in Section 5.

2   Problem Statement and Formulation

Let G be an undirected network with $|V| = n$ vertexes, where $V = \{v_1, v_2, \cdots, v_n\}$. Let v_1 be the origin and v_n the destination. Let $e_{i,m^i} = \{e_{i,1}, \cdots, e_{i,j}, \cdots, e_{i,m^i}\}$ denote the set of all edges between v_i and v_{i+1}, where $e_{i,j}$ is the j shortest edge from v_i to v_{i+1}. Let $t_{i,j}$ denote the travel time of $e_{i,j}$. Therefore, $t_{i,1} \leq \cdots \leq t_{i,j} \cdots \leq t_{i,m^j}$. Let $P_1 = \{v_1, e_{1,1}, \cdots, v_i, e_{i,1}, \cdots, v_{n-1}, e_{n-1,1}, v_n\}$ be the shortest path from v_1 to v_n. Denote $\delta = (\delta_1, \delta_2, \cdots, \delta_k)$ as the blockages sequence, and $t_{k,r}$ as the recovery time with respect to δ_k. As shown in Fig. 1, suppose that blockages happen at P_1 and the traveller has to move from v_1 to v_n by passing through $v_2, v_3, \cdots, v_{n-1}$ one by one, then the problem is to design a good travel strategy without any knowledge of future blockages.

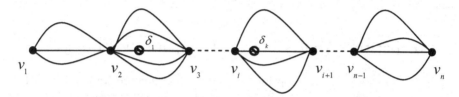

Fig. 1. Graph G

In order to discuss the problem, we make the following assumptions:

(1) The traveller knows the entire network and the travel time of each edge in advance.
(2) Blockages may happen at $P_1 = \{v_1, e_{1,1}, \cdots, v_i, e_{i,1}, \cdots, v_{n-1}, e_{n-1,1}, v_n\}$ and the traveller does not know which one edge will be blocked in advance, and the traveller only observes that upon arriving the vertex of the blocked edge.
(3) The recovery time of a blockage is not known in advance, but the traveller can obtain the recovery time until the blockage is reopened.
(4) Blockages may not happen at $e_{i,j}$ ($j \neq 1$) between v_i and v_{i+1}.

If all of the blockages and their recovery time are known in advance, then the problem becomes an offline problem, and the optimal travel strategy is obtained by following the shortest edge from v_1 to v_n after modifying the travel time of known blocked edges. If the blocked edges are unpredictable, then the problem is obviously an online problem.

Let $C_{OPT}(\delta)$ be the travel time of the optimal offline shortest path from v_1 to v_n, and let $C_A(\delta)$ be the corresponding travel time of the online strategy A for the traveller to go from v_1 to v_n. Strategy A is said to be α-competitive [6-9] if $C_A(\delta) \leq \alpha_A \cdot C_{OPT}(\delta) + b$ holds, where α_A and b are constants not related to δ. Denote $\alpha^\star$ as the optimal competitive ratio for the on-line problem such that $\alpha^\star = \inf_{A \in S}(\alpha_A)$ for any online strategy $A \in S$, where S is the set of all online strategies. If $\alpha_{A^\star} = \alpha^\star$, then $A^\star$ is called the optimal online algorithm.

The above competitive analysis is the most fundamental and significant approach. However, the above competitive analysis is not very flexible, especially in the uncertainty environment. In practice, many travellers hope to manage their risk and willing to take certain risks for more rewards sometimes. Al-Binali [10] first defined the concepts of risk and reward for online financial problems. From the risk-reward point of view, The following definitions are given for our problem. Let I be the range of the recovery time of a blockage and $F \subset I$ be a forecast for the recovery time. If $F \subset I$ is the correct forecast, then denote $\hat{\alpha}_{\hat{A}} = \sup\limits_{t_{k,r} \subset F} \frac{C_{\hat{A}}(\delta)}{C_{OPT}(\delta)}$ as the restricted competitive ratio of $\hat{A}$, and $f_{\hat{A}} = \frac{\alpha^{*}}{\hat{\alpha}_{\hat{A}}}$ the reward of $\hat{A}$. The optimal restricted competitive ratio under the forecast F is $\hat{\alpha}^{*} = \inf\limits_{\hat{A} \in S}(\hat{\alpha}_{\hat{A}})$. If $F \subset I$ is the false forecast, then denote $\alpha_{\hat{A}}$ be the competitive ratio of $\hat{A}$ for any $t_{k,r} \in I$. Define $\tau_{\hat{A}} = \frac{\alpha_{\hat{A}}}{\alpha^{*}}$ as the risk of $\hat{A}$.

3   Competitive Analysis of the Comparison Strategy

In this section, we present an optimal online strategy for the Recoverable-CTP in a special network and analyse its corresponding competitive ratio.

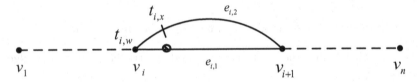

Fig. 2. Analysis of CS

Comparison Strategy: When the traveller reaches v_i and knows that the edge $e_{i,1}$ is blocked, he/she sets a upper bound of waiting time $t_{i,w} = \frac{t_{i,2}-t_{i,1}}{2}$ at v_i. If the recovery time $t_{i,r}$ of $e_{i,1}$ satisfies $t_{i,r} \leq t_{i,w}$, then the traveller follows the shortest edge $e_{i,1}$ after waiting time $t_{i,r}$; otherwise, the traveller follows the second shortest edge $e_{i,2}$ and when he/she knows that the blockage is reopened after traversed time $t_{i,x}$ ($t_{i,x} < t_{i,2}$) in $e_{i,2}$, he/she makes a decision according to the following condition: If $t_{i,1} + t_{i,x} \geq t_{i,2} - t_{i,x}$, he/she continues on the second shortest edge $e_{i,2}$ as intended; If $t_{i,1} + t_{i,x} \leq t_{i,2} - t_{i,x}$, then he/she returns to v_i and follows the shortest edge $e_{i,1}$.

Denote the Comparison Strategy as CS. As shown in Fig.2.

According to the above travelling strategy CS, we have the following lemma and theorems.

Lemma 1. If $t_{i,w} \leq t_{i,r} \leq t_{i,2} - t_{i,1}$ and $t_{i,x} < \frac{t_{i,2}-t_{i,1}}{2}$, he/she returns to v_i and follows the shortest edge $e_{i,1}$ by using CS.

Proof. If $t_{i,r} \geq t_{i,w}$, then the traveller follows the second shortest edge $e_{i,2}$ from v_i to v_{i+1} by using CS. When he/she knows that the blockage is reopened, he/she returns to v_i and follows the shortest edge $e_{i,1}$ under the condition $t_{i,x} + t_{i,1} < t_{i,2} - t_{i,x}$. Since $t_{i,r} = t_{i,x} + \frac{t_{i,2}-t_{i,1}}{2}$, we have $t_{i,r} < t_{i,2} - t_{i,1}$. Hence, if $t_{i,w} \leq t_{i,r} \leq t_{i,2} - t_{i,1}$, then the traveller returns to v_i and follows the shortest edge $e_{i,1}$.

Theorem 1. For the Recoverable-CTP with the blockage sequence $\delta = (\delta_1, \delta_2, \cdots, \delta_k)$, the competitive ratio of CS is $\frac{3-\beta}{2}$, where $\beta = \max \beta_i$, $\beta_i = \frac{t_{i,1}}{t_{i,2}}$.

Proof. Set $t_{i,w} = \frac{t_{i,2}-t_{i,1}}{2}$.

(1) If $t_{i,r} < t_{i,w}$, then the online traveller waits $t_{i,r}$ time at v_i and follows the shortest edge $e_{i,1}$, and we have $C_{CS}(\delta_i) = t_{i,r} + t_{i,1}$. The offline optimal strategy is the same as the online strategy. Hence, we have $C_{OPT}(\delta_i) = t_{i,r} + t_{i,1}$ and $\frac{C_{CS}(\delta_i)}{C_{OPT}(\delta_i)} = 1$.

(2) If $t_{i,w} \leq t_{i,r} < t_{i,2} - t_{i,1}$, then the online traveller waits $t_{i,w}$ time at v_i and follows the second shortest edge $e_{i,2}$. When he/she knows that the blockage is reopened, he/she returns to v_i and follows the shortest edge $e_{i,1}$ by Lemma 1. The offline optimal strategy is that the traveller waits $t_{i,r}$ time at v_i and follows the shortest edge $e_{i,1}$. Hence, we have $C_{CS}(\delta_i) = t_{i,r} + t_{i,1} + t_{i,x}$, $C_{OPT}(\delta_i) = t_{i,r} + t_{i,1}$ and $C_{CS}(\delta_i) = (1 + \frac{t_{i,x}}{t_{i,r}+t_{i,1}})C_{OPT}(\delta_i)$. Since $t_{i,r} < t_{i,2} - t_{i,1}$ and $t_{i,x} < \frac{t_{i,2}-t_{i,1}}{2}$, we have $\frac{C_{CS}(\delta_i)}{C_{OPT}(\delta_i)} \leq \frac{3}{2} - \frac{t_{i,1}}{2t_{i,2}} = \frac{3-\beta_i}{2}$, where $\beta_i = \frac{t_{i,1}}{t_{i,2}}$.

(3) If $t_{i,r} \geq t_{i,2} - t_{i,1}$, then the online traveller waits $t_{i,w}$ time at v_i and follows the second shortest edge $e_{i,2}$. The offline optimal strategy is that the traveller follows the second shortest edge $e_{i,2}$ without any waiting time. Hence, we have $C_{CS}(\delta_i) = t_{i,w} + t_{i,1} = \frac{t_{i,2}+t_{i,1}}{2} \leq \frac{3t_{i,2}-t_{i,1}}{2}$, $C_{OPT}(\delta_i) = t_{i,2}$ and $\frac{C_{CS}(\delta_i)}{C_{OPT}(\delta_i)} \leq \frac{3}{2} - \frac{t_{i,1}}{2t_{i,2}} \leq \frac{3-\beta_i}{2}$, where $\beta_i = \frac{t_{i,1}}{t_{i,2}}$.

From (1), (2) and (3), we have $\frac{C_{CS}(\delta_i)}{C_{OPT}(\delta_i)} = \frac{\sum\limits_{i=1}^{k} C_{CS}(\delta_i)}{\sum\limits_{i=1}^{k} C_{OPT}(\delta_i)} \leq \frac{3-\beta}{2}$, where $\beta = \max \beta_i$ and $\beta_i = \frac{t_{i,1}}{t_{i,2}}$. Therefore, the competitive ratio of CS is $\frac{3-\beta}{2}$.

This concludes the proof of Theorem 1. □

Theorem 2. For the Recoverable-CTP with the blockage sequence $\delta = (\delta_1, \delta_2, \cdots, \delta_k)$, the competitive ratio h of any deterministic online strategy satisfies $h \geq \frac{3-\beta}{2}$.

Proof. For any online deterministic strategy B, let $t_{i,B}$ be the upper bound of waiting time when the traveller reaches v_i of blocked edge $e_{i,1}$. Set $t_{i,w} = \frac{t_{i,2}-t_{i,1}}{2}$, and consider three cases.

(1) $t_{i,B} \in [0, t_{i,w}]$

The worst case is that the online traveller knows the blockage reopened after traversed time $\frac{t_{i,2}-t_{i,1}}{2}$ along $e_{i,2}$, he/she can continue on the second shortest edge $e_{i,2}$ or return to v_i and follow the shortest edge $e_{i,1}$. The travel

time is $C_B(\delta_i) = t_{i,B} + \frac{t_{i,2}-t_{i,1}}{2} + \frac{t_{i,2}-t_{i,1}}{2} + t_{i,1} = t_{i,B} + t_{i,2}$. The offline optimal strategy is that the traveller follows the shortest edge $e_{i,1}$ after waiting time $t_{i,r} = t_{i,B} + \frac{t_{i,2}-t_{i,1}}{2}$, and we have $C_{OPT}(\delta_i) = t_{i,B} + \frac{t_{i,2}-t_{i,1}}{2} + t_{i,1} = t_{i,B} + \frac{t_{i,2}+t_{i,1}}{2}$. Since $t_{i,B} < \frac{t_{i,2}-t_{i,1}}{2}$, we have $\alpha_{i,1} = \frac{C_B(\delta_i)}{C_{OPT}(\delta_i)} = $
$\frac{t_{i,B}+t_{i,2}}{t_{i,B}+\frac{t_{i,2}+t_{i,1}}{2}} > \frac{\frac{t_{i,2}-t_{i,1}}{2}+t_{i,2}}{\frac{t_{i,2}-t_{i,1}}{2}+\frac{t_{i,2}+t_{i,1}}{2}} = \frac{3t_{i,2}-t_{i,1}}{2t_{i,2}} = \frac{3-\beta_i}{2}$, where $\beta_i = \frac{t_{i,1}}{t_{i,2}}$. Hence,

we have $\alpha_{B,1} = \dfrac{\sum\limits_{i=1}^{k} C_B(\delta_i)}{\sum\limits_{i=1}^{k} C_{OPT}(\delta_i)} > \frac{3-\beta}{2}$, where $\beta = \max \beta_i$.

(2) $t_{i,B} = t_{i,w}$

If $t_{i,B} = t_{i,w}$, then by Theorem 1, we have the competitive ratio of B is $\alpha_{B,2} = \frac{3-\beta}{2}$.

(3) $t_{i,B} \in (t_{i,w}, +\infty)$

The worst case is that the online traveller follows the second shortest edge $e_{i,2}$ after waiting time $t_{i,B}$ by using online strategy B. The offline optimal strategy is that the traveller follows the second shortest edge $e_{i,2}$ without any waiting time. Hence, we have $C_B(\delta_i) = t_{i,B} + t_{i,2}$, $C_{OPT}(\delta_i) = t_{i,2}$ and $\alpha_{i,3} = \frac{C_B(\delta_i)}{C_{OPT}(\delta_i)} = 1 + \frac{t_{i,B}}{t_{i,2}}$. Since $t_{i,B} > \frac{t_{i,2}-t_{i,1}}{2}$, we have $\alpha_{i,3} > 1 + \frac{t_{i,2}-t(i,1)}{2t_{i,2}} = $

$\frac{3-\beta_i}{2}$, where $\beta_i = \frac{t_{i,1}}{t_{i,2}}$. Hence, we have $\alpha_{B,3} = \dfrac{\sum\limits_{i=1}^{k} C_B(\delta_i)}{\sum\limits_{i=1}^{k} C_{OPT}(\delta_i)} > \frac{3-\beta}{2}$, where

$\beta = \max \beta_i$.

From the above analysis, we have $\alpha_B = \min\{\alpha_{B,1}, \alpha_{B,2}, \alpha_{B,3}\}$. Then the competitive ratio of any deterministic online strategy is no less than $\frac{3-\beta}{2}$.

This concludes the proof of Theorem 2.   □

From the above analysis, it is known that the comparison strategy is the optimal deterministic online strategy, and $\alpha^* = \frac{3-\beta}{2}$.

4   Competitive Analysis of the Risk-Reward Strategies

When an online traveller is risk-averse, he will use the classical online algorithm A and achieve the optimal competitive ratio. If the online traveller is a risk-seeker, then the risk-reward strategy allows him to benefit from his capability, and allows him to control his risk by using a risk strategy $\hat{A}$. Next, we will give the risk-reward strategy $\hat{A}$ with respect to the Recoverable-CTP and analysis it's competitive.

The online traveller can make three different forecasts of the recovery time upon reaching the blockage: $t_{i,r} < \frac{t_{i,2}-t_{i,1}}{2}$, $\frac{t_{i,2}-t_{i,1}}{2} \leq t_{i,r} < t_{i,2} - t_{i,1}$ and $t_{i,r} > t_{i,2} - t_{i,1}$, we will discuss the three cases as following.

Forecast 1. $t_{i,r} < \frac{t_{i,2}-t_{i,1}}{2}$.

For the forecast 1, we give the following online strategy $\hat{A}_1$.

$\hat{A}_1$: Set the upper bound of waiting time $\hat{t}_{i,w} = \frac{t_{i,2}-t_{i,1}}{2}$. If $t_{i,r} \leq \hat{t}_{i,w}$, then the traveller follows the shortest edge $e_{i,1}$ after waiting time $t_{i,r}$; otherwise, the traveller follows the second shortest edge $e_{i,2}$ and when he/she knows the blockage being reopened after traversed time $t_{i,x}$ $(t_{i,x} < t_{i,2})$ in $e_{i,2}$, he/she makes a decision according to the following condition: If $t_{i,1} + t_{i,x} \geq t_{i,2} - t_{i,x}$, he/she continues on the second shortest edge $e_{i,2}$ as intended; If $t_{i,1} + t_{i,x} < t_{i,2} - t_{i,x}$, then he/she returns to v_i and follows the shortest edge $e_{i,1}$.

Theorem 3. For the Recoverable-CTP with a correct forecast of the recovery time $t_{i,r} < \frac{t_{i,2}-t_{i,1}}{2}$, the restricted optimal competitive ratio of $\hat{A}_1$ is 1.

Proof. Set the upper bound of waiting time $\hat{t}_{i,2} = \frac{t_{i,2}-t_{i,1}}{2}$. If the forecast is correct, then $t_{i,r} < \frac{t_{i,2}-t_{i,1}}{2}$. The online traveller waits $t_{i,r}$ time at v_i and follows the shortest edge $e_{i,1}$, and we have $C_{\hat{A}_1}(\delta_i) = t_{i,r} + t_{i,1}$. The offline optimal strategy is the same as the online strategy, and we have $C_{OPT}(\delta_i) = t_{i,r} + t_{i,1}$. Hence, the restricted competitive ratio for the Recoverable-CTP is

$$\hat{\alpha}_{\hat{A}_1} = \frac{\sum\limits_{i=1}^{k} C_{\hat{A}_1}(\delta_i)}{\sum\limits_{i=1}^{k} C_{OPT}(\delta_i)} = 1.$$

This concludes the proof of Theorem 3.   □

From the above theorem, It is known that the reward of the strategy $\hat{A}_1$ is $f_{\hat{A}_1} = \frac{\alpha^*}{\hat{\alpha}_{\hat{A}_1}} = \frac{3-\beta}{2}$.

If the traveller makes a false forecast regarding the recovery time of the blockage, then $t_{i,r} > \frac{t_{i,2}-t_{i,1}}{2}$, and the competitive ratio of $\hat{A}_1$ is $\alpha_{\hat{A}_1} = \frac{3-\beta}{2}$ by Theorem 1. Hence, $\tau = \frac{\alpha_{\hat{A}_1}}{\alpha^*} = 1$.

Forecast 2. $\frac{t_{i,2}-t_{i,1}}{2} \leq t_{i,r} < t_{i,2} - t_{i,1}$

For the forecast 2, we give the following online strategy $\hat{A}_2$.

$\hat{A}_2$: Set the upper bound of waiting time $\hat{t}_{i,w} = t_{i,2} - t_{i,1}$. If $t_{i,r} \leq \hat{t}_{i,w}$, then the traveller follows the shortest edge $e_{i,1}$ after waiting time $t_{i,r}$; otherwise, the traveller follows the second shortest edge $e_{i,2}$.

Theorem 4. For the Recoverable-CTP with a correct forecast of the recovery time $\frac{t_{i,2}-t_{i,1}}{2} \leq t_{i,r} < t_{i,2} - t_{i,1}$, the restricted optimal competitive ratio of $\hat{A}_2$ is 1.

Proof. Set the upper bound of waiting time $\hat{t}_{i,w} = t_{i,2} - t_{i,1}$. If the forecast is correct, then $\frac{t_{i,2}-t_{i,1}}{2} \leq t_{i,r} < t_{i,2} - t_{i,1}$. The online traveller waits $t_{i,r}$ time at v_i and follows the shortest edge $e_{i,1}$, and we have $C_{\hat{A}_2}(\delta_i) = t_{i,r} + t_{i,1}$. The offline optimal strategy is the same as the online strategy, and we have $C_{OPT}(\delta_i) = t_{i,r} + t_{i,1}$. Hence, the restricted competitive ratio for the Recoverable-

CTP is $\hat{\alpha}_{\hat{A}_2} = \frac{\sum\limits_{i=1}^{k} C_{\hat{A}_2}(\delta_i)}{\sum\limits_{i=1}^{k} C_{OPT}(\delta_i)} = 1.$

This concludes the proof of Theorem 4.   □

From the above proof, we can obtain the reward of the strategy $\hat{A}_2$ is $f_{\hat{A}_2} = \frac{\alpha^*}{\hat{\alpha}_{\hat{A}_2}} = \frac{3-\beta}{2}$.

If the traveller makes a false forecast regarding the recovery time of the blockage, then $t_{i,r} < \frac{t_{i,2}-t_{i,1}}{2}$ or $t_{i,r} \geq t_{i,2} - t_{i,1}$. If $t_{i,r} < \frac{t_{i,2}-t_{i,1}}{2}$, then the online traveller follows the shortest edge $e_{i,1}$ after waiting time $t_{i,r}$ by using online strategy $\hat{A}_2$, the offline optimal strategy is the same as the online strategy. Hence, the competitive ratio is 1, and $\tau_1 = \frac{\alpha_{\hat{A}_2}}{\alpha^*} = \frac{1}{\frac{3-\beta}{2}} = \frac{2}{3-\beta}$. If $t_{i,r} \geq t_{i,2} - t_{i,1}$, then the online traveller follows the second shortest edge $e_{i,2}$ after waiting time $\hat{t}_{i,w}$ by using online strategy $\hat{A}_2$, the offline optimal strategy is that the traveller follows the second shortest edge without any waiting time. The competitive ratio of $\hat{A}_2$

is $\alpha_{\hat{A}_2} = \frac{\sum\limits_{i=1}^{k} C_{\hat{A}_2}(\delta_i)}{\sum\limits_{i=1}^{k} C_{OPT}(\delta_i)} = \frac{\sum\limits_{i=1}^{k}(t_{i,2}-t_{i,1}+t_{i,2})}{\sum\limits_{i=1}^{k} t_{i,2}} \leq 2 - \beta$, and $\tau_2 = \frac{\alpha_{\hat{A}_2}}{\alpha^*} = \frac{2-\beta}{\frac{3-\beta}{2}} = \frac{4-2\beta}{3-\beta}$.

Since $\tau_1 \leq \tau_2$, we have $\tau = \max\{\tau_1, \tau_2\} = \frac{4-2\beta}{3-\beta}$.

Forecast 3. $t_{i,r} > t_{i,2} - t_{i,1}$

For the forecast 3, we give the following online strategy $\hat{A}_3$.

$\hat{A}_3$: Set the upper bound of waiting time $\hat{t}_{i,w} = 0$. The traveller follows the second shortest edge $e_{i,2}$.

Theorem 5. For the Recoverable-CTP with a correct forecast of the recovery time $t_{i,r} > t_{i,2} - t_{i,1}$, the restricted optimal competitive ratio of $\hat{A}_3$ is 1.

Proof. Set the upper bound of waiting time $\hat{t}_{i,w} = 0$. If the forecast is correct, then $t_{i,r} > t_{i,2} - t_{i,1}$. The online traveller follows the second shortest edge $e_{i,2}$, and we have $C_{\hat{A}_3}(\delta_i) = t_{i,2}$. The offline optimal strategy is the same as the online strategy, and we have $C_{OPT}(\delta_i) = t_{i,2}$. Hence, the restricted optimal competitive

ratio for the Recoverable-CTP is $\hat{\alpha}_{\hat{A}_3} = \frac{\sum\limits_{i=1}^{k} C_{\hat{A}_3}(\delta_i)}{\sum\limits_{i=1}^{k} C_{OPT}(\delta_i)} = 1$.

This concludes the proof of Theorem 5. □

From the above proof, we can obtain the reward of the strategy $\hat{A}$ is $f_{\hat{A}_3} = \frac{\alpha^*}{\hat{\alpha}_{\hat{A}_3}} = \frac{3-\beta}{2}$.

If the traveller makes a false forecast regarding the recovery time of the blockage, then $t_{i,r} < t_{i,2}-t_{i,1}$. The traveller follows the second shortest edge $e_{i,2}$ by using strategy $\hat{A}_3$. The worst case is that he/she knows the blockage reopened after traversed time $\frac{t_{i,2}-t_{i,1}}{2}$, he/she can continue on the second shortest edge $e_{i,2}$ or return to v_i and follow the shortest edge $e_{i,1}$. The total travel time is $C_{\hat{A}_3}(\delta_i) = t_{i,2}$. The offline optimal strategy is that the traveller follows the shortest edge $e_{i,1}$ after waiting time $\frac{t_{i,2}-t_{i,1}}{2}$, and we have $C_{OPT}(\delta_i) = \frac{t_{i,2}-t_{i,1}}{2} + t_{i,1} = \frac{t_{i,2}+t_{i,1}}{2}$.

The competitive ratio of $\hat{A}_3$ is $\alpha_{\hat{A}_3} = \dfrac{\sum\limits_{i=1}^{k} C_{\hat{A}_3}(\delta_i)}{\sum\limits_{i=1}^{k} C_{OPT}(\delta_i)} = \dfrac{\sum\limits_{i=1}^{k} t_{i,2}}{\sum\limits_{i=1}^{k} \frac{t_{i,2}+t_{i,1}}{2}} \leq \dfrac{2}{1+\beta}$, and

$\tau = \dfrac{\alpha_{\hat{A}_3}}{\alpha^*} = \dfrac{\frac{2}{1+\beta}}{\frac{3-\beta}{2}} = \dfrac{4}{(3-\beta)(1+\beta)}$.

From the above analysis, we conclude the results as shown in Table 1.

Table 1. Risk-reward strategy and its competitive analysis

Forecast	Strategy $\hat{A}$	$\hat{\alpha}_{\hat{A}} = \hat{\alpha}^*$	$f_{\hat{A}} = \dfrac{\alpha^*}{\hat{\alpha}_{\hat{A}}}$	$\alpha_{\hat{A}}$	$\tau = \dfrac{\alpha_{\hat{A}}}{\alpha^*}$
$t_{i,r} \leq \dfrac{t_{i,2}-t_{i,1}}{2}$	$\hat{t}_{i,w} = \dfrac{t_{i,2}-t_{i,1}}{2}$	1	$\dfrac{3-\beta}{2}$	$\dfrac{3-\beta}{2}$	1
$\dfrac{t_{i,2}-t_{i,1}}{2} < t_{i,r} \leq t_{i,2}-t_{i,1}$	$\hat{t}_{i,w} = t_{i,2}-t_{i,1}$	1	$\dfrac{3-\beta}{2}$	$2-\beta$	$\dfrac{4-2\beta}{3-\beta}$
$t_{i,r} > t_{i,2}-t_{i,1}$	$\hat{t}_{i,w} = 0$	1	$\dfrac{3-\beta}{2}$	$\dfrac{2}{1+\beta}$	$\dfrac{4}{(3-\beta)(1+\beta)}$
$\alpha^* = \dfrac{3-\beta}{2}, \beta = \max\limits_{i} \dfrac{t_{i,1}}{t_{i,2}}$					

5   Conclusions

The Recoverable Canadian Traveller Problem is valuable and important for the traffic congestion problems. Most previous studies are based on classical competitive analysis. The classical competitive analysis is the most fundamental and important framework to study online problems, but it is not very flexible. In this paper, we present an optimal online strategy - a comparison strategy and prove its competitive ratio for Recoverable-CTP in a special network. Moreover, with the different forecasts of the recovery time, some online strategies are given under the risk-reward framework, and the rewards and the risks of different strategies are analysed. For the Recoverable Canadian Traveller Problem, there are some further directions, such as, how to deal with the problems in general networks under the risk-reward framework.

References

1. Papadimitriou, C.H., Yannakakis, M.: Shortest paths without a map. Theoretical Computer Science 84(1), 127–150 (1991)
2. Bar-Noy, A., Schieber, B.: The Canadian traveller problem. In: Proceedings of the second annual ACM-SIAM Symposium on Discrete Algorithms, pp. 261–270 (1991)

3. David, S.B., Borodin, A.: A new measure for the study of the on-line algorithm. Algorithmica 11, 73–91 (1994)
4. Westphal, S.: A note on the k-Canadian traveller problem. Information Processing Letters 106(3), 87–89 (2008)
5. Xu, Y.F., Hu, M.L., Su, B., Zhu, B.H., Zhu, Z.J.: The Canadian Traveller Problem and Its Competitive Analysis. Journal of Combinatorial Optimization, 4 (in press, 2008)
6. Su, B., Xu, Y.F.: Online recoverable Canadian traveller problem. In: Proceedings of the International Conference on Management Science and Engineering, pp. 633–639 (2004)
7. Sleator, D., Tarjan, R.: Amortized efficiency of list update and paging rules. Communications of the ACM 28(2), 202–208 (1985)
8. Borodin, A., El-Yaniv, R.: Online computation and competitive analysis. Cambridge University Press, Cambridge (1998)
9. Fiat, A., Rabani, Y., Ravid, Y.: Competitive k-server algorithms. In: Proceedings of the 22nd IEEE Symposium on Foundation of Computer Science, pp. 454–463 (1990)
10. Fiat, A., Woeginger, G.J.: Online algorithms: The state of art. Springer, Heidelberg (1998)
11. Al-Binali, S.: A risk-reward framework for the competitive analysis of financial games. Algorithmica 25, 99–115 (1999)

Minimizing Total Completion Time in Two-Machine Flow Shops with Exact Delays

Yumei Huo[1], Haibing Li[2], and Hairong Zhao[3]

[1] Department of Computer Science, College of Staten Island, CUNY, 2800 Victory Blvd 1N-215, Staten Island, New York 10314
huo@mail.csi.cuny.edu
[2] Lehman Brothers Inc., New York City, NY 10019, USA
hl27@njit.edu
[3] Department of Mathematics, Computer Science & Statistics, Purdue University Calumet, 2200 169th Street, Hammond, IN 46323
hairong@calumet.purdue.edu

Abstract. We study the problem of minimizing total completion time in the two-machine flow shop with exact delay model. This problem is a generalization of the no-wait flow shop problem which is known to be strongly NP-hard. Our problem has many applications but little results are given in the literature so far. We focus on permutation schedules. We first prove that some simple algorithms can be used to find the optimal schedules for some special cases. Then for the general case, we design some heuristics as well as metaheuristics whose performance are shown to be very well by computational experiments.

1 Introduction

Flow shop is one of the most classic scheduling models that has been studied since 1950's. In the past, most research about flow shop is done under the assumption that the time needed to move a job from one machine to another is negligible. But this may not be true in our real life. For example, in manufacturing there may be a transportation time (delay) from one production facility to another.

In this paper, we study the scheduling problems in the two-machine flow shop with *exact* delays model. This problem arises in chemistry manufacturing where there often may be an exact technological delay between the completion time of some operation and the initial time of the next operation ([1]).

Formally, our scheduling model can be stated as follows. There are two machines, the *upstream* machine M_1 and the *downstream* machine M_2. There are n jobs, denoted by $j = 1, 2, \ldots, n$, for simplicity. Each job j is described by a triple $(p_{1,j}, \bar{l}_j, p_{2,j})$: $p_{1,j}$ is the length of job j's first operation that has to be processed on the upstream machine M_1; $p_{2,j}$ is the length of job j's second operation that has to be processed on the downstream machine M_2; $\bar{l}_j$ is the exact time interval between the completion time of the first operation and the start time of the second operation. For convenience, we also use $p_{1,j}$ and $p_{2,j}$ to represent the first and the second operation of job j, respectively. We assume that both machines

B. Yang, D.-Z. Du, and C.A. Wang (Eds.): COCOA 2008, LNCS 5165, pp. 427–437, 2008.
© Springer-Verlag Berlin Heidelberg 2008

are available from time 0, and that at any time a machine can only process one operation. We also assume that both $p_{1,j}$ and $p_{2,j}$ are positive.

Our interest in this paper is minimizing total completion time which is one of the most common objectives in scheduling. We use $C_{i,j}(S)$ to represent the finish time of operation $p_{i,j}$ in the schedule S. The completion time of job j in S, denoted by $C_j(S)$, is the completion time of its second operation $p_{2,j}$, i.e., $C_{2,j}(S)$. The total completion time of a schedule S, denoted by $\sum C_j(S)$, is simply the sum of the completion times of all jobs. If S is clear from the context, we simply use $C_{i,j}$, C_j, and $\sum C_j$ for short. We use C^* to denote the minimum total completion time.

Related results. One of the special case of our problem is when $\bar{l}_j = 0$ for all j. This case is also known as two-machine no-wait flow shop, denoted by $F_2 \mid$ nowait $\mid \sum C_j$. Much effort has been devoted to the no-wait flow shop scheduling. The no-wait flow shop scheduling problem with the objective of minimizing total completion time was posed by van Deman and Baker [17]. In the same paper, they developed a branch and bound algorithm for the problem. The complexity of the this problem was solved by Röck [13]: he showed that $F2 \mid$ no-wait $\mid \sum C_i$ is NP-hard in the strong sense. Gonzales [7] showed that $F \mid no - wait, p_{i,j} \in \{0,1\} \mid \sum C_j$ is NP-complete in the strong sense. Sriskandarajah and Ladet [14] showed that $F2 \mid no - wait, p_{i,j} \in \{0,1\} \mid \sum C_j$ is polynomial time solvable. Rajendran and Chaudhuri [12] developed a job insertion heuristic for the problem. They compared the performance of their heuristic with that of the other existing heuristics, and showed that their heuristic dominates all the existing heuristics. Chen et al. [4] presented a genetic algorithm for the problem and some computational results. Fink and Voß [5] used metaheuristics to solve the problem with m machines. Some theoretical work has been done by [3], [9], [11], [15], [16].

It is easy to see that for the two-machine no-wait flow shop model every feasible schedule is also a permutation schedule and it contains no forced idle time. But this is not true in general when the exact delay is nonzero. For example, suppose we are given three jobs (1,5,3), (1,9,2), (3,3,3). Let us denote them by job 1, 2, 3 respectively. The optimal schedule for these three jobs is illustrated in Figure 1. We can see that it is not a permutation schedule. Furthermore, there is one unit of forced idle time between $p_{1,1}$ and $p_{1,2}$.

Despite of its important applications, not many results are known when the exact delay is non-zero. For the makespan objective, Yu et.al ([18]) proves that the problem is strongly NP-hard even in the case of unit processing times. Ageev and Baburin [1] and Leung, Li and Zhao ([10]) independently design constant-factor approximation algorithms for the general case. Ageev and Baburin ([1], [2]) also designed better approximation algorithms for some special cases. Forthermore, they proved that the existence of a $(1.5 - \epsilon)$-approximation algorithm implies P=NP. For the objective of total completion time, the only work as we are aware of is from Leung, Li and Zhao ([10]). They have shown that by greedily scheduling the jobs in non-decreasing order of delay, one can obtain the optimal schedule for the problem $F2 \mid \bar{l}_j, p_{1,j} = p_1, p_{2,j} = p_2, p_1 \geq p_2 \mid \sum C_j$, and obtain a 2-approximation schedule for the problem $F2 \mid \bar{l}_j, p_{1,j} = p_1, p_{2,j} = p_2, p_1 < p_2 \mid \sum C_j$.

New contributions. In this paper, we concentrate on permutation schedules of two machines flow shop with exact delay. Following the three-field notation introduced by Graham et al. [8], our problem can be classified as $F2 \mid \text{permu}, \bar{l}_j, \mid \sum C_j$. One can easily show that for permutation schedules, adding forced idle time can only increase the total completion time. Thus, without loss of generality, we will assume that there is no forced idle time in any feasible schedule. It should be noted that even for permutation schedules, our problem is still NP-Hard in the strong sense. Our contributions can be summarized as follows:

- We prove that some very simple algorithms can be used to find the optimal schedules for the following cases:

  - The jobs can be ordered in a way such that $p_{1,j} \leq p_{1,j+1}$, and $p_{1,j+1} + \bar{l}_{j+1} \geq \bar{l}_j + p_{2,j}$ for all $1 \leq j \leq n - 1$.
  - The jobs can be ordered in a way such that $p_{2,j} \leq p_{2,j+1}$, $p_{1,1} + \bar{l}_1 = \min_{1 \leq j \leq n}(p_{1,j} + \bar{l}_j)$ and $p_{1,j+1} + \bar{l}_{j+1} < \bar{l}_j + p_{2,j}$ for all $1 \leq j \leq n - 1$.
  - $\max_j(p_{1,j}) \leq \min_i(p_{2,i})$, and $\bar{l}_j = l$.
  - For all $1 \leq j \leq n$, we have $p_{1,j} = p_1$, $p_{2,j} = p_2$ where p_1 and p_2 are constants.
  - For all $1 \leq j \leq n$, $\bar{l}_j = l$ and $p_{1,i} < p_{1,j} \rightarrow p_{2,i} < p_{2,j}$.

- We design the first metaheuristics, a tabu search algorithm and a simulated annealing algorithm, to solve our problem.
- We show that our metaheuristics perform very well through experiments on randomly generated data.

It should be noted that the purpose of studying the special cases of the problem are two-folded: (1) It is meaningful from the theoretical point of view by giving a boundary on the problems in P and those in NP. (2) While some cases are hypothetical, some other cases are very practical, for example, when all jobs have the same exact delay. When the delay is zero, this is just no wait flow shop.

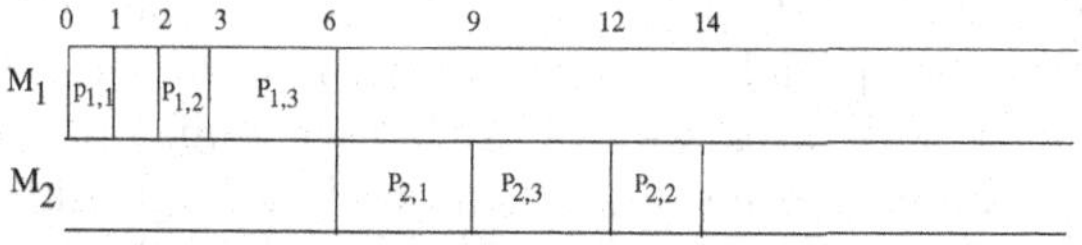

Fig. 1. The optimal schedule for three jobs: (1,5,3), (1,9,2), (3,3,3), denoted by 1, 2 and 3 respectively. It is not a permutation schedule and there is one unit of forced idle time between $p_{1,1}$ and $p_{1,2}$.

The paper is organized as follows. In Section 2 we state some preliminary results. In Section 3, we show that optimal schedules for some special cases can be found in polynomial time. In Section 4 we design some heuristics as well as metaheuristics for the general case. In Section 5 we give a short summary.

2   Preliminary Results

The NP-hardness of $F2 \mid$ no-wait $\mid \sum C_j$ immediately implies the following.

Lemma 1. *The problems $F2 \mid \bar{l}_j = l \mid \sum C_j$ and $F2 \mid permu, \bar{l}_j \mid \sum C_j$ are both strongly NP-hard.*

We now consider the lower bounds of the minimum total completion time of any schedule. Let P_1, P_2 be the total processing time of the first and second operation of all jobs, respectively, i.e. $P_1 = \sum_{j=1}^{n} p_{1,j}$, $P_2 = \sum_{j=1}^{n} p_{2,j}$. Suppose that $p_{1,j_1} \leq p_{1,j_2} \leq \ldots \leq p_{1,j_n}$ and $p_{2,i_1} \leq p_{2,i_2} \leq \ldots \leq p_{2,i_n}$. Let $P_1^* = \sum_{i=1}^{n}(n-i+1)p_{1,j_i}$, $P_2^* = \sum_{k=1}^{n}(n-k+1)p_{2,i_k}$, and $L = \sum_{j=1}^{n} \bar{l}_j$. Then we have the following lower bounds for the optimal total completion time C^*.

Lemma 2. *For the problem $F2 \mid \bar{l}_j \mid \sum C_j$, we have*

$$C^* \geq P_1^* + L + P_2 \tag{1}$$
$$C^* \geq n \cdot \min_{1 \leq j \leq n} (p_{1,j} + \bar{l}_j) + P_2^* \tag{2}$$

The following are some properties of any feasible schedule S. The correctness follows from the assumption that there is no forced idle time in a feasible schedule.

Property 3. Let S be a feasible schedule for $F2 \mid permu, \bar{l}_j \mid \sum C_j$. Suppose that S schedules the jobs in the order of 1, 2, $\ldots$, n. Then for any job j and $j+1$, $1 \leq j \leq n-1$, we have

 - either $p_{1,j+1} + \bar{l}_{j+1} \geq p_{2,j} + \bar{l}_j$, so there is no idle time between $p_{1,j}$ and $p_{1,j+1}$;
 - or $p_{1,j+1} + \bar{l}_{j+1} \leq p_{2,j} + \bar{l}_j$, so there is no idle time between $p_{2,j}$ and $p_{2,j+1}$.

3   Simple Heuristics and Solvable Special Cases

Since the general problem $F2 \mid permu, \bar{l}_j \mid \sum C_j$ is strongly NP-hard, we would like to investigate some special cases whose optimal schedules can be found in polynomial time. For most cases, we will use list scheduling to schedule the jobs. That is, start with an empty schedule, then one by one greedily insert the jobs to the partial schedules in an order given by a priority rule (list) subject to the exact delay constraint and the permutation schedule constraint. The priority rules we will use are the following:

 - SPT: order the jobs in non-decreasing order of $p_{1,j} + \bar{l}_j + p_{2,j}$.
 - SPT(p_1): order the jobs in non-decreasing order of $p_{1,j}$.
 - SPT(p_2): order the jobs in non-decreasing order of $p_{2,j}$.

By Property 3 and the lower bounds in Lemma 2, we can prove the following two theorems. The proof is omitted due to space limit.

Theorem 4. *Suppose the jobs are numbered such that $p_{1,j} \leq p_{1,j+1}$. If $p_{1,j+1} + \bar{l}_{j+1} \geq \bar{l}_j + p_{2,j}$ for all $1 \leq j \leq n-1$, then $SPT(p_1)$ rule schedules the jobs optimally.*

Theorem 5. *Suppose the jobs are numbered such that $p_{2,j} \leq p_{2,j+1}$. If $p_{1,1} + \bar{l}_1 = \min_{1 \leq j \leq n}(p_{1,j} + \bar{l}_j)$ and $p_{1,j+1} + \bar{l}_{j+1} < \bar{l}_j + p_{2,j}$ for all $1 \leq j \leq n-1$, then $SPT(p_2)$ rule schedules the jobs optimally.*

Our next special case assumes that all jobs have the same delay, i.e. $\bar{l}_j = l$, where $l \geq 0$ is a constant.

Theorem 6. *If $\max_j(p_{1,j}) \leq \min_i(p_{2,i})$, and $\bar{l}_j = l$, the optimal schedule can ge obtained in polynomial time.*

Proof. It is easy to see that every schedule must be a permutation schedule in this case. Suppose the first job j is fixed, to minimize the total completion time, the remaining jobs should be scheduled using $SPT(p_2)$. To find the optimal schedule, we do the following: for each job j, we generate a schedule that schedules j first, and schedule the remaining jobs in $SPT(p_2)$ order; the optimal schedule will be the schedule with minimum total completion time among the n schedules produced.

Theorem 7. *The optimal schedule for $F2 \mid permu, \bar{l}_j, p_{1,j} = p_1, p_{2,j} = p_2 \mid \sum C_j$ can be obtained by SPT rule.*

Proof. Note that since $p_{1,j} = p_1$, and $p_{2,j} = p_2$, SPT rule is actually the same as ordering the jobs in non-decreasing order of delays. Depending on the values of p_1 and p_2, we consider two cases: $p_1 \geq p_2$ and $p_1 < p_2$.

Case1: $p_1 \geq p_2$. In this case, using similar arguments as in Theorem 4, one can easily show that if we schedule the jobs in non-decreasing order of the delays, we obtain an optimal schedule.

Case2: $p_1 < p_2$. The proof is quite involved, we sketch the main idea here. Let $\hat{S}$ be a schedule that schedules the jobs in non-decreasing order of delays. Let S be an optimal schedule that is different from $\hat{S}$. Suppose that S schedules the jobs in the order of $1, 2, \ldots, n$. We compare the jobs in S with jobs in $\hat{S}$ one by one and find the first job i in S that is different from the corresponding job j in $\hat{S}$. Now we modify S by rescheduling the jobs $i, i+1, \ldots, j-1, j$ as follows: move all jobs between i and $j-1$ inclusively forward by an amount of p_2, then schedule job j before i so that there is no forced idle time between job j and i. All other jobs are kept unchanged. One can show that the obtained schedule is feasible and the total completion time of the new schedule S' is not greater than that of S.

By repeatedly doing the modification as described, one can get a schedule that schedules the jobs in the same order of $\hat{S}$, i.e. in the non-decreasing order of the delays. This completes the proof.

Using similar approach as we prove the case 2 in Theorem 7, we can prove the following theorem.

Theorem 8. *SPT rule solves* $F2 \mid \bar{l}_j = l, p_{1,i} < p_{1,j} \rightarrow p_{2,i} < p_{2,j} \mid \sum C_j$
optimally.

4   Metaheuristics

Although certain special cases can be solved by simple priority rules, their performance for the general case could be very bad. In this section, we design two metaheuristics, namely tabu search algorithm and simulated annealing algorithm, to solve our problem. Both metaheuristics have been applied to solve a variety of combinatorial optimization problems. However, this is the first time that they are applied to two machine flow shop with exact delay problem.

4.1   Tabu Search

Tabu search is classified as a local search technique and it enhances the local search performance by using memory. The basic idea of tabu search is to explore the solution space using a local search procedure by iteratively moving from a solution S to a new best one $\widehat{S}$ in its *neighborhood* $\mathcal{N}(S)$, until certain *stopping criterion* is satisfied. To avoid being trapped in a local optima, $\widehat{S}$ is allowed be worse than S. To avoid cycling, explored solutions are marked as *tabu* in memory and excluded from being the candidate for $\widehat{S}$. For more information about tabu search, the reader is referred to Glover [6]. In the following, we describe our design of the three important components in the tabu search algorithm, neighborhood, stopping criterion and the tabu structure.

Neighborhood Generation. In the literature, two of the popular neighborhood generation methods are exchange method and insertion method. In the exchange method, the neighbors are generated by selecting a job in the schedule and exchanging it with another job in the schedule. In the insertion method, the neighbors are generated by selecting a job in the schedule and inserting it into different position in the schedule. In this paper we use a variation of insertion method, which generates a larger neighborhood set and gives better error ratio. Given a schedule sequence $S = <j_1, j_2, \ldots, j_n>$ and a parameter K, we generate the neighborhood of S with respect to K, denoted by $\mathcal{N}_K(S)$, as follows: 1) Set $\mathcal{N}_{K(S)} = \emptyset$. 2) For each position $i = 1, 2, \ldots, n$ in S and each length $k = 1, 2, \ldots, K$, move the subsequence $<j_i, j_{i+1}, \ldots, j_{i+k-1}>$ to each position in the set $\{1, 2, \ldots, i-1, i+k, i+k+1, \ldots, n\}$, to obtain a new sequence S'. Add S' into $\mathcal{N}_K(S)$. 3) Return $\mathcal{N}_K(S)$.

Stopping Criterion. We combine the number of consecutive non-improving steps and the total number of restarts. Whenever the number of consecutive non-improving steps reaches I_{max}, we restart the search from the randomly generated schedule. If the total number of restarts during the course of search reaches R_{max}, we stop the search process and return the current best solution. Here R_{max} is introduced to avoid the local optima.

Tabu structure. Ideally we would like to store all the explored schedules into memory in the tabu list in order to prevent the algorithm from cycling. However, storing all these schedules would require too much memory. One popular solution in the literature is that only recently marked tabus are kept in the memory and the old tabus will be removed from tabu list as searching moves on. Here we propose a new method to resolve this problem, we map a permutation schedule $S = < j_1, j_2, \ldots, j_n >$ to a tabu structure which only contains six attributes: $M(S) = < j_1, j_{\lceil n/4 \rceil}, j_{\lceil n/2 \rceil}, j_{\lceil 3n/4 \rceil}, j_n, \sum C_j(S) >$. Such a compact structure can be encoded to take only several bytes in physical memory. With such a mapping scheme, we keep the tabu structure of an explored solution in memory for the life time of tabu search. If schedule S_1 and schedule S_2 have the same tabu structure, i.e. $M(S_1) = M(S_2)$, we simply treat them as one. It should be noted that $M(S_1) = M(S_2)$ means the six attributes of S_1 and those of S_2 are exactly the same. This tabu structure strictly avoids cycling with possible penalty that some solutions could not be explored. But our tests show that the possibility that $S_1 \neq S_2$ and $M(S_1) = M(S_2)$ is quite small (only several times before the algorithm stops), so we can reasonably ignore this side-effect.

Now we are ready to describe our algorithm. The notations used are: $\mathcal{L}$ – tabu list; S^* – the best schedule found so far; R_{max} – the maximum number of multi-restarts; I_{max} – the maximum number of consecutively non-improving steps for restart; c_1 – the counter for number of non-improving steps; c_2 – the counter for number of multi-restarts.

Algorithm *TabuSearch*(S)

(1) Set tabu list $\mathcal{L} = \emptyset$; Let $S^* = S$. Configure the value of K, $R_{\max}$, and $I_{\max}$; Set $c_2 = 0$;
(2) Repeat the following until $c_2 > R_{\max}$
   $c_1 = 0$
   found $=$ true
   While found and $c_1 < I_{\max}$
      Search in $\mathcal{N}_K(S)$ a best neighbor $\widehat{S}$ such that $M(\widehat{S})$ is not in tabu list $\mathcal{L}$.
      If $\widehat{S}$ is found
         If $\sum C_j(\widehat{S}) < \sum C_j(S^*)$
            $S^* = \widehat{S}$, $c_1 = 0$, $\mathcal{L} = \mathcal{L} \cup \{M(S)\}$, $S = \widehat{S}$
         else
            $c_1 = c_1 + 1$;
      else found $=$ false
   generate a random schedule S that is not in the tabu list, $c_2 = c_2 + 1$
(3) return S^*.

4.2   Simulated Annealing

Simulated annealing is another metaheuristic that has been widely used to solve combinatorial optimization problems. It was inspired by annealing in metallurgy. By simulating this physical process, the simulated annealing algorithm moves in each step from the current solution to a random neighbor solution, which is chosen with a probability that depends on the difference between the corresponding function values and on a global temperature parameter T, which is gradually decreased during the search process. The dependency is such that the current solution changes almost randomly when T is large, but almost always moves to a better solution as T goes to zero. Accepting worse solutions prevents the algorithm from being trapped at local optima. A standard simulated annealing algorithm works as follows: (1) Randomly choose an initial solution S and set the initial temperature T_0; (2) Randomly generate a solution S' from the neighborhood of the current solution S; (3) If $\sum C_j(S') < \sum C_j(S)$, then accept S', otherwise accept S' with probability $\rho = e^{\frac{\sum C_j(S) - \sum C_j(S')}{T}}$, let $S = S'$ if S' is accepted; (4) update the temperature T; if it is less than the stop temperature, then stop, else go to(2).

In order to apply the simulated annealing to our problem, the main components, including neighborhood structure, temperature, and termination criterion need to be determined. We use the same neighborhood structure as in the tabu search algorithm. The temperature includes initial temperature T_0 , the cooling ratio δ for annealing and the stop temperature ϵ. We set the initial temperature T_0 at which there is 95% chance to accept a random neighbor of the initial solution. The cooling ratio δ, which is less than and close to 1, is a parameter tuned by our experiments. We set the stop temperature ϵ to be 0.000000001. Furthermore, at each temperature, instead of one move, we allow at most n^2 moves. Therefore we can search a larger solution space. Following is the description of our algorithm, where S^* represents the best schedule found so far.

Algorithm $SimulatedAnnealing(S)$

(1) Configure T_0, δ, and ϵ; let $S^* = S$ and $T = T_0$.
(2) Repeat n^2 times
       randomly choose a neighbor solution $\widehat{S}$ of S
       if $\sum C_j(\widehat{S}) < \sum C_j(S)$
       then accept $\widehat{S}$
       otherwise accept $\widehat{S}$ with probability $\rho = e^{\frac{\sum C_j(S) - \sum C_j(\widehat{S})}{T}}$
       let $S = \widehat{S}$ if an $\widehat{S}$ is accepted and let $S^* = S$ if $\sum C_j(S) < \sum C_j(S^*)$.
(3) Cool down the temperature to δT and goto 2) until $T < \epsilon$.
(4) Return S^*.

5   Computation Results

Since there is no benchmark instances for our problem yet, we generated our own instances for $n = 20, 40, 60, 80, 100$ and 120. For each n, 40 instances of n jobs are generated; and for each job $j = 1, 2, \ldots, n$, $p_{1,j}, \bar{l}_j, p_{2,j}$ are randomly chosen in the range $[1, 100]$.

Besides the heuristics (SPT, $SPT(p_1)$, $SPT(p_2)$) and the metaheuristics (tabu search and simulated annealing) we mentioned above, we also tested the following heuristics:

- $SPT(p_1 + \bar{l}_j)$: Order the jobs in non-decreasing order of $p_{1,j} + \bar{l}_j$.
- $SPT(p_2 + \bar{l}_j)$: Order the jobs in non-decreasing order of $p_{2,j} + \bar{l}_j$.
- ECT: Greedily insert the jobs one by one, so the completion time of the inserted job has the smallest completion time.

The algorithms are coded in C++ and the running environment is a Linux cluster–Typhon, which consists of 32 Dell Power Edge 2650 machines and each machine has two Intel(R) Xeon(TM) CPU of 2.80GHz and 2GB RAM.

The error ratio is estimated by using the lower bound described in Lemma 2. Let $r_{err}(H)$ denote the error ratio of algorithm H against the lower bound and let S_H be the schedule returned by H, then

$$ r_{err}(H) = \frac{\sum C_j(S_H)}{\max\{P_1^* + L + P_2, n \cdot \min_{1 \leq j \leq n}(p_{1,j} + \bar{l}_j) + P_2^*\}} - 1, $$

the notations P_1^*, L, and P_2, P_2^* are referred to Lemma 2. Clearly, this error ratio can be computed in linear time and gives an upper bound for the actual error ratio calculated against the optimal solution.

Set the parameters for Tabu Search. We investigate the effect of changing K, which determines the size of neighborhood, I_{max}, the number of consecutive non-improvement steps and R_{max}, the total number of restart. We divide the instances into two groups, the small instances where n is 20, 40, and 60; and the large instances where n is 80, 100 and 120.

For small instances, we use the instances for $n = 40$ to find the values of K, I_{max} and R_{max}. For large instances, we use the instances for $n = 100$. Based on the results, for small instances, we set $K = 3$, $R_{max} = 6$ and $I_{max} = 600$. For large instances, we set $K = 10$, $R_{max} = 12$ and $I_{max} = 1000$.

Set the parameters for Simulated Annealing. We used the same parameters K as found through experiments in tabu search. Then we investigate the effect of changing the cooling ratio. Again, we use the instances for $n = 40$ to find the best cooling ratio for small instances and we use $n = 100$ to find the best cooling ratio for large instances. It turns out in both cases we choose 0.98 by considering both the time and error ratio.

Computational results. To save space, we omit the result tables, interested reader can contact the authors for detailed results. From the experiment results, we have the following findings: 1) The average error ratios show that the metaheuristics produce solutions that are very close to the lower bounds hence closer to the optimal solutions. 2) On average, simulated annealing performs better in terms of both the average error ratio and the running time. However, simulated annealing does not dominate tabu search. 3) Both metaheuristics perform better than the six simple heuristics. 4) Among the six simple heuristics, ECT performs the best, SPT is the second best.

6   Conclusion

In this paper, we considered the two-machine flow shop problem with exact delays to minimize total completion time. The problem is strongly NP-hard in general. We studied some polynomially solvable special cases and developed some simple heuristics as well as two metaheuristics for the general case. Computational results showed that the ECT performs the best among the six simple heuristics. It is not surprise that the two metaheuristics, a tabu search algorithm and a simulated annealing algorithm, perform better than all of the simple heuristics. In our configuration of parameters, simulated annealing performs better than tabu search.

References

1. Ageev, A.A., Kononov, A.V.: Approximation Algorithms for Scheduling Problems with Exact Delays. In: Erlebach, T., Kaklamanis, C. (eds.) WAOA 2006. LNCS, vol. 4368, pp. 1–14. Springer, Heidelberg (2007)
2. Ageev, A.A., Kononov, A.V.: Approximation Algorithms for the Single and Two- Machine Scheduling Problems with Exact Delays. Operations Research Letters 35(4), 533–540 (2007)
3. Adiri, I., Pohoryles, D.: Flowshop/no-idle or no-wait scheduling to minimize the sum of completion times. Naval Research Logistics Quarterly 29, 495–504 (1982)
4. Chen, C.L., Neppalli, R.V., Aljaber, N.: Genetic algorithms applied to the continuous flow shop problem. Computers and Industrial Engineering 30, 919–929 (1996)
5. Fink, A., Voß, S.: Solving the continuous flow-shop scheduling problem by mataheuristics. European Journal of Operational Research 151, 400–414 (2003)
6. Glover, F.: Tabu Search - Part II. ORSA J. on Comp. 2, 4–32 (1990)
7. Gonzales, T.: Unit Execution Time Shop Problems. Mathematics of Operations Research 7, 57–66 (1982)
8. Grham, R.L., Lenstra, J.K., Rinnooy Kan, A.H.G.: Optimization and approximation in deterministic sequence and scheduling: A survey. Annals of Discrete Math. 5, 287–326 (1979)
9. Gupta, J.N.D.: Optimal flowshop schedules with no intermediate storage space. Naval Research Logistics Quarterly 23, 235–243 (1976)
10. Leung, J.Y.-T., Li, H., Zhao, H.: Scheduling Two-Machine Flow Shops with Exact Delay. International Journal of Foundations of Computer Science 18(2), 341–360 (2007)
11. Papadimitriou, C.H., Kanellakis, P.C.: Flowshop scheduling with limited temporary storage. Journal of the ACM 27, 533–549 (1980)
12. Rajendran, C., Chaudhuri, D.: Heursitic algorithms for coninuos flow-shop problem. Naval Research Logistics 37, 695–705 (1990)
13. Röck, H.: Some new results in flow shop scheduling. Mathematical Methods of Operations Research (ZOR) 28(1), 1–16 (1984)
14. Sriskandarajah, C., Ladet, P.: Some No-wait Shops Scheduling Problems: Complexity Aspects. European Journal of Operational Research 24(3), 424–438 (1986)
15. Szwarc, W.: A note on the fflow-shop problem without interruptions in job processing. Naval Research Logistics Quarterly 28, 665–669 (1981)

16. Van der Veen, J.A.A., van Dal, R.: Solvable cases of the no-wait flow-shop scheduling problem. Journal of the Oper. Res. Society 42, 971–980 (1991)
17. Van Deman, J.M., Baker, K.R.: Minimizing mean flowtime in the flow shop with no intermediate queues. AIIE Transactions 6, 28–34 (1974)
18. Yu, W., Hoogeveen, H., Lenstra, J.K.: Minimizing makespan in a two-machine flow shop with delays and unit- time operations is NP-hard. Journal of Scheduling 7, 333–348 (2004)

Efficient Method for Periodic Task Scheduling with Storage Requirement Minimization

Karine Deschinkel and Sid-Ahmed-Ali Touati

University of Versailles Saint-Quentin-en-Yvelines, France

Abstract. In this paper, we study the general problem of one-dimensional periodic task scheduling under storage requirement, irrespective of machine constraints. We have already presented in [9] a theoretical framework that allows an optimal optimization of periodic storage requirement in a periodic schedule. This problem is used to optimize processor register usage in embedded systems. Our storage optimization problem being NP-complete [8], solving an exact integer linear programming formulation is too expensive in practice. In this article, we propose an efficient two-steps heuristic using model's properties that allows fast resolution times while providing nearly optimal results. This method includes the resolution of a integer linear program with a totally unimodular constraints matrix in first step, then the resolution of a linear assignment problem. Our solution has been implemented and included inside a compiler for embedded processors.

1 Introduction

This article addresses the problem of storage optimization in cyclic data dependence graphs (DDG), which is for instance applied in the practical problem of periodic register allocation for innermost loops on modern Instruction Level Parallelism (ILP) processors [10]. The massive introduction of ILP processors since the last two decades makes us re-think new ways of optimizing register/storage requirement in assembly codes before starting the instruction scheduling process under resource constraints. In such processors, instructions are executed in parallel thanks to the existence of multiple small computation units (adders, multipliers, load-store units, etc.). The exploitation of this new fine grain parallelism (at the assembly code level) asks to completely revisit the old classical problem of register allocation initially designed for sequential processors. Nowadays, register allocation has not only to minimize the storage requirement, but has also to take care of parallelism and total schedule time. In this research article, we do not assume any resource constraints (except storage requirement); Our aim is to analyze the trade-off between memory (register pressure) and parallelism in a periodic task scheduling problem. Note that this problem is abstract enough to be considered in other scheduling disciplines that worry about conjoint storage and time optimization in repetitive tasks (manufacturing, transport, networking, etc.).

Existing techniques in this field usually apply a periodic instruction scheduling under resource constraints that is sensitive to register/storage requirement. Therefore a great amount of work tries to schedule the instructions of a loop (under resource and time constraints) such that the resulting code does not use more than R values simultaneously alive. Usually they look for a schedule that minimizes the storage requirement

B. Yang, D.-Z. Du, and C.A. Wang (Eds.): COCOA 2008, LNCS 5165, pp. 438–447, 2008.

under a fixed scheduling period while considering resource constraints [3,4,6,1]. In this paper, we satisfy register/storage constraints early before instruction scheduling under resource constraints: we directly handle and modify the DDG in order to fix the storage requirement of any further subsequent periodic scheduling pass while taking care of not altering parallelism exploitation if possible. This idea uses the concept of reuse vector used for multi-dimensional scheduling [11, 12].

This article is on continuation on our previous work on register allocation [10]. In that paper, we showed that register allocation implies a loop unrolling. However, the general problem of storage optimization does not require such loop unrolling. So the current paper is an abstraction of our previous results on register optimization. Furthermore, it extends it with a new heuristic and experimental results.

Our article is organized as follows. Section 2 recalls our task model and notations already presented in [9]. Section 3 recalls the exact problem of optimal periodic scheduling under storage constraints with integer linear programming: our detailed results on the optimal resolution of this problem have been presented in [9]. Since the exact model is not practical (too expensive in terms of resolution time), our current article provides a new look by writing an efficient approximate method in Section 4, that we call *SIR-ALINA*. Before concluding, Section 5 presents the results of our experimental evaluation of SIRALINA, providing practical evidence of its efficiency.

2   Tasks Model

Our task model is similar to [2]. We consider a set of l generic tasks (instructions inside a program loop) $T_0, \ldots, T_{l-1}$. Each task T_i should be executed n times, where n is the number of loop iterations. n is an unknown, unbounded, but finite integer. This means that each task T_i has n instances. The kth occurrence of task T_i is noted $T\langle i, k \rangle$, which corresponds to task i executed at the kth iteration of the loop, with $0 \leq k < n$.

The tasks (instructions) may be executed in parallel. Each task may produce a result that is read/consumed by other tasks. The considered loop contains some data dependences represented with a graph G such that:

- V is the set of the generic tasks of the loop body, $V = \{T_0, \ldots, T_{l-1}\}$.
- E is the set of edges representing precedence constraints (flow dependences or other serialization constraints). Any edge $e = (T_i, T_j) \in E$ has a latency $\delta(e) \in \mathbb{N}$ in terms of processor clock cycles and a distance $\lambda(e) \in \mathbb{N}$ in terms of number of loop iterations. The distance $\lambda(e)$ means that the edge $e = (T_i, T_j)$ is a dependence between the task $T\langle i, k \rangle$ and $T\langle j, k + \lambda(e) \rangle$ for any $k = 0, \ldots, n - 1 - \lambda(e)$.

We make a difference between tasks and precedence constraints depending whether they refer to data to be stored into registers or not

1. V_R is the set of tasks producing data to be stored into registers.
2. E_R is the set of flow dependence edges through registers. An edge $e = (T_i, T_j) \in E_R$ means that the task $T\langle i, k \rangle$ produces a result stored into a register and read/consumed by $T\langle j, k + \lambda(e) \rangle$. The set of consumers (readers) of a generic task T_i is then the set:

$$Cons(T_i) = \{T_j \in V \mid e = (T_i, T_j) \in E_R\}$$

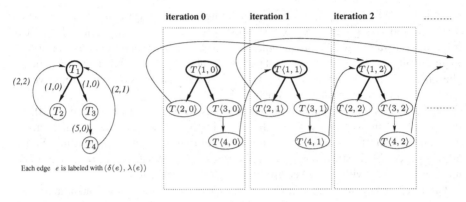

(a) Example of a DDG with Generic Taks   (b) Loop iterations and instruction/task instances

Fig. 1. Example of Data Dependence Graphs with Recurrent Tasks

Figure 1 is an example of a data dependence graph (DDG) where bold circles represent V_R the set of generic tasks producing data to be stored into registers. Bold edges represent flow dependences (each sink of such edge reads/consumes the data produced by the source and stored in a register). Tasks that are not in bold circles are instructions that do not write into registers (write the data into memory or simply do not produce any data). Non-bold edges are other data or precedence constraints different from flow dependences. Every edge e in the DDG is labeled by the pair $(\delta(e), \lambda(e))$.

In our generic processor model, we assume that the reading and writing from/into registers may be delayed from the starting time of task execution. Let assume $\sigma(T\langle i, k\rangle)$ $\in \mathbb{N}$ as the starting execution time of task $T\langle i, k\rangle$. We thus define two delay functions δ_r and δ_w in which

$\delta_w : V_R \to \mathbb{N}$
$\qquad T_i \mapsto \delta_w(T_i) | \ 0 \leq \delta_w(T_i)$
$\qquad$ the writing time of data produced by $T\langle i, k\rangle$ is $\sigma(T\langle i, k\rangle) + \delta_w(T_i)$
$\delta_r : V \to \mathbb{N}$
$\qquad T_i \mapsto \delta_r(T_i) | \ 0 \leq \delta_r(T_i)$
$\qquad$ the reading time of the data consumed by $T\langle i, k\rangle$ is $\sigma(T\langle i, k\rangle) + \delta_r(T_i)$

These two delays functions depend on the target processor and model almost all regular hardware architectures (VLIW, EPIC/IA64 and superscalar processors).

3   Exact Problem Formulation

This section recalls the exact integer linear model for solving the problem of Periodic Scheduling with Storage Minimisation (PSSM). It is built for a fixed desired period $p \in \mathbb{N}$. For more details on this problem, please refer to [9].

3.1 Basic Variables

- A schedule variable $\sigma_i \geq 0$ for each task $T_i \in V$, including σ_{K_i} for each killing node K_i. We assume a finite upper bound L for such schedule variables (L sufficiently large, $L = \sum_{e \in E} \delta(e)$); The schedule variables are integer. As our scheduling is periodic, we only consider the integer execution date of the first task occurrence $\sigma_i = T\langle i, 0 \rangle$ and the execution date of any other occurrence $T\langle i, k \rangle$ becomes equal to $\sigma(T\langle i, k \rangle) = \sigma_i + k * p$.
- A binary variables $\theta_{i,j}$ for each $(T_i, T_j) \in V_R^2$. It is set to 1 iff (T_i, T_j) is a reuse edge;
- A reuse distance $\mu_{i,j}$ for all $(T_i, T_j) \in V_R^2$; The reuse distance are nonnegative integer variables.

3.2 Linear Constraints

- **Data dependences**

  The schedule must at least satisfy the precedence constraints defined by the DDG.

$$\forall e = (T_i, T_j) \in E : \sigma_j - \sigma_i \geq \delta(e) - p \times \lambda(e) \tag{1}$$

- **Flow dependences**

  Each flow dependence $e = (T_i, T_j) \in E_R$ means that the task occurrence $T\langle j, k + \lambda(e) \rangle$ reads the data produced by $T\langle i, k \rangle$ at time $\sigma_j + \delta_r(T_j) + (\lambda(e) + k) \times p$. Then, we should schedule the killing node K_i of the task T_i after all T_i's consumers.

$$\forall T_i \in V_R, \ \forall T_j \in Cons(T_i)|e = (T_i, T_j) \in E_R : \ \sigma_{K_i} \geq \sigma_j + \delta_r(T_j) + p \times \lambda(e) \tag{2}$$

- **Storage dependences**

  There is a storage dependence between K_i and T_j if (T_i, T_j) is a reuse edge.

$$\forall (T_i, T_j) \in V_R^2 : \theta_{i,j} = 1 \implies \sigma_{K_i} - \delta_w(T_j) \leq \sigma_j + p \times \mu_{i,j}$$

This involvement can result in the following inequality :

$$\forall (T_i, T_j) \in V_R^2 : \sigma_j - \sigma_{K_i} + p \times \mu_{i,j} + M_1(1 - \theta_{ij}) \geq -\delta_w(T_j) \tag{3}$$

where M_1 is an arbitrarily large constant.
If there is no register reuse between two tasks T_i and T_j, then $\theta_{i,j} = 0$ and the storage dependence distance $\mu_{i,j}$ must be set to 0.

$$\forall (T_i, T_j) \in V_R^2 : \mu_{i,j} \leq M_2 \theta_{i,j} \tag{4}$$

where M_2 is an arbitrarily large constant.

- **Reuse Relations**

  The reuse relation must be a bijection from V_R to V_R. A register can be reused by one task and a task can reuse one released register:

$$\forall T_i \in V_R : \sum_{T_j \in V_R} \theta_{i,j} = 1 \tag{5}$$

$$\forall T_j \in V_R : \sum_{T_i \in V_R} \theta_{i,j} = 1 \tag{6}$$

3.3 Objective Function

As proved in [10], the storage requirement is equal to $\sum \mu_{i,j}$. In our periodic scheduling problem, we want to minimize the storage requirement: Minimize $z = \sum_{(T_i,T_j) \in V_R^2} \mu_{i,j}$

Using the above integer linear program to solve an NP-problem problem as PSSM is not efficient in practice. With a classical Branch and Bound method, we are only able to solve small instances (DDG sizes), in practice arround 12 nodes. For this reason, we suggest to make use of the problem structure to propose an efficient heuristic as follows.

4 SIRALINA: A Two Steps Approximate Resolution Method

Our resolution strategy is based on the analysis of the model constraints. As the problem involves scheduling constraints and assignment constraints, and the reuse distances are the link between these two sets of constraints, we attempt to decompose the problem into two subproblems :

- A scheduling problem : to find a scheduling for which the potential reuse distances are as small as possible.
- An assignment problem : to select which pairs of tasks will share the same register.

4.1 Preliminaries

If edge $e = (T_i, T_j) \in V_R^2$ is a reuse edge, its reuse distance should satisfy the inequality given by 3, where $\theta_{ij} = 1$. This inequality gives a lower bound for each reuse distance. If $(T_i, T_j) \in V_R^2$ is a reuse edge (E_r denotes the set of reuse edge) then :

$$\forall (T_i, T_j) \in E_r : \mu_{i,j} \geq \frac{1}{p}(\sigma_{K_i} - \delta_w(T_j) - \sigma_j) \tag{7}$$

If $(T_i, T_j) \in V_R^2$ is not a reuse edge then $\mu_{ij} = 0$ according to the inequality 4.

$$\forall (T_i, T_j) \notin E_r : \mu_{i,j} = 0$$

The aggregation of constraint 7 for each reuse edge provides a lower bound of the objective function value. $z = \sum_{(T_i,T_j) \in V_R^2} \mu_{i,j} \geq \frac{1}{p}(\sum_{(T_i,T_j) \in E_r} \sigma_{K_i} - \delta_w(T_j) - \sigma_j)$ As the reuse relation is a bijection from V_R to V_R, the left sum of the inequality can be separated into two parts.

$$\sum_{(T_i,T_j) \in E_r} \sigma_{K_i} - \delta_w(T_j) - \sigma_j = \sum_{i \in V_R} \sigma_{K_i} - \sum_{j \in V_R} (\delta_w(T_j) + \sigma_j)$$

$$= \sum_{i \in V_R} \sigma_{K_i} - \sum_{j \in V_R} \sigma_j - \sum_{j \in V_R} \delta_w(T_j)$$

We deduce from this inequality a lower bound for the number of required registers. In this context, it may be useful to find an appropriate scheduling for which this value is minimal. As $\sum_{j \in V_R} \delta_w(T_j)$ is a constant for the problem, we could ignore it in the following optimization problem.

We consider *the scheduling problem (P)*:

$$\begin{cases} \min \sum_{i \in V_R} \sigma_{K_i} - \sum_{j \in V_R} \sigma_j \\ \text{subject to :} \\ \sigma_j - \sigma_i \geq \delta(e) - p \times \lambda(e), \quad \forall e = (T_i, T_j) \in E \\ \sigma_{K_i} - \sigma_j \geq \delta_r(T_j) + p \times \lambda(e), \forall T_i \in V_R, \forall T_j \in Cons(T_i) | e = (T_i, T_j) \in E_R \end{cases} \tag{8}$$

As the constraints matrix of the integer linear program of System 8 is totally unimodular, *i.e.*, the determinant of each square sub-matrix is equal to 0 or to ± 1, we can use polynomial algorithms to solve this problem [7]. This would allow us to consider huge DDG. The resolution of problem (P) by a simplex method will provide optimal values σ_i^* for each task $T_i \in V_R$ and the optimal values $\sigma_{K_i}^*$ for each killing node K_i.

Once the scheduling variables have been fixed, the minimal value of each potential reuse distance would be equal to $\overline{\mu_{ij}} = \lceil \frac{\sigma_{K_i}^* - \delta_w(T_j) - \sigma_j^*}{p} \rceil$ according to 7 . Knowing the reuse distance values $\overline{\mu_{ij}}$ if T_j reuses the register freed by T_i, the storage allocation which consists of choosing which task reuses which released register can be modeled as a linear assignment problem.

We consider *the linear assignment problem (A)*:

$$\begin{cases} \min \quad \sum_{(T_i, T_j) \in V_R^2} \overline{\mu_{i,j}} \theta_{ij} \\ \text{Subject to} \\ \sum_{T_j \in V_R} \theta_{i,j} = 1, \qquad \forall T_i \in V_R \\ \sum_{T_i \in V_R} \theta_{i,j} = 1, \qquad \forall T_j \in V_R \\ \theta_{ij} \in \{0, 1\} \end{cases} \tag{9}$$

where $\overline{\mu_{i,j}}$ is a fixed value for each edge $e = (T_i, T_j) \in V_R^2$.

4.2 Heuristic

We suggest to solve the problem with the following heuristic :

- Solve the problem (P) to deduce the optimal values σ_i^* for each task $T_i \in V_R$ and the optimal values $\sigma_{K_i}^*$ for each killing node K_i,
- Compute the cost $\overline{\mu_{ij}} = \max \left(\lceil \frac{\sigma_{K_i}^* - \delta_w(T_j) - \sigma_j^*}{p} \rceil \right)$ for each edge $e = (T_i, T_j) \in V_R^2$,
- Solve the linear assignment problem (A) with the Hungarian algorithm [5] which solves assignment problems in polynomial time ($O(n^3)$) to deduce the optimal values $\theta_{i,j}^*$,
- If $\theta_{i,j}^* = 1$ for the edge $e = (T_i, T_j) \in V_R^2$, then (T_i, T_j) is a reuse edge and the reuse distance is equal to $\overline{\mu_{ij}}$.

5 Experiments

We now present the results obtained on several DDG extracted from many well known benchmarks (Spec95, whetstone, livermore, lin-ddot, DSP filters, etc.). The data dependence graphs of all these loops are present in [8]. The small test instances have 2 nodes

Table 1. SIRALINA and optimal Results

Benchmark	$|V|$	$|E|$	S_{opt}	$S_{siralina}$	T_{opt}	$T_{siralina}$
lin-ddot	4	4	7	7	0.007	0.066
liv-loop1	9	11	5	5	0.364	0.067
liv-loop5	5	5	3	3	0.005	0.066
liv-loop23	20	26	10	12	605.548	0.069
spec-dod-loop1	13	15	5	6	198.472	0.067
spec-dod-loop2	10	10	3	3	0.084	0.067
spec-dod-loop3	11	11	3	4	0.257	0.067
spec-dod-loop7	4	4	35	35	0.004	0.066
spec-fp-loop1	5	6	2	2	0.006	0.067
spec-spice-loop1	2	2	3	3	0.004	0.067
spec-spice-loop2	9	10	15	15	2.757	0.067
spec-spice-loop3	4	5	2	2	0.005	0.067
spec-spice-loop4	12	51	8	8	0.088	0.068
spec-spice-loop5	2	2	1	1	0.003	0.067
spec-spice-loop6	6	7	14	14	0.016	0.067
spec-spice-loop7	5	5	40	40	0.005	0.067
spec-spice-loop8	4	4	7	7	0.005	0.067
spec-spice-loop9	11	17	7	7	26.242	0.067
spec-spice-loop10	4	4	2	2	0.005	0.069
spec-tom-loop1	15	18	5	7	604.278	0.068
test-christine	18	17	230	230	600.847	0.068
Elliptic	36	59	NA	10	NA	0.074
whet-cycle4-1	4	4	1	1	0.005	0.066
whet-cycle4-2	4	4	2	2	0.006	0.067
whet-cycle4-4	4	4	4	4	0.01	0.067
whet-cycle4-8	4	4	8	8	0.013	0.069
whet-loop1	16	28	5	6	0.2	0.068
whet-loop2	7	10	5	5	0.006	0.067
whet-loop3	5	16	4	4	0.006	0.067

and 2 edges, the large instances have multiples hundreds of nodes and edges. We use the ILOG CPLEX 10.2 to solve the integer linear program. The experiment was run on PC under linux, equipped with a Pentium IV 2.13 Ghz processor, and 2 Giga bytes of memory. In practice, the optimal method [9] can solve small instances, around 10 nodes. As far as we know about our problem, it is still NP-complete even for DDG chains and trees [8]: we do not have simple DDG instances larger than 10 nodes. We are able to check the efficiency of our heuristic on small DDG instances by comparing its results against the optimal ones. The theoretical computation of the approximation ratio constitutes an additional problem which is not studied here.

Table 1 presents the results of SIRALINA against optimal method using common benchmarks. This table presents the results for the minimal period of each benchmark. Note that every benchmark has its own minimal period, defined as the critical circuit of the DDG, which is inherent to the data dependences [2]. The first column represents the name of the benchmark. The second and third column represent the instance size

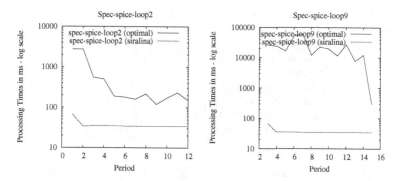

Fig. 2. Processing Times of some Benchmarks vs. Period

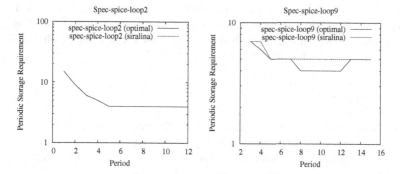

Fig. 3. Storage Requirement of some Benchmarks vs. Period

(numbers of DDG nodes and edges). Columns number 4 and 5 give the storage requirement (objective function values) computed by the optimal and SIRALINA methods (some instances could not be solved). The two lasts columns give the resolution times in seconds. As can be seen in this table, SIRALINA is fast and nearly optimal. Sometimes SIRALINA is slightly longer than the optimal method for two reasons: 1) the timer is too precise (milliseconds) and the interactions with operating system disturbs our timing measurements, and 2) SIRALINA performs in two steps while the optimal method performs in one step (resolving a unique integer linear program). In our context, we consider that a time difference which is less than 0.1 seconds is negligible. Another interesting remark is that the processing time of SIRALINA is relatively constant compared to the processing time of the optimal method. Another improvement of SIRALINA compared to the optimal method is that SIRALINA performs in relatively a constant resolution time (irrespective of the considered period p). Figure 2 illustrates some examples, where we can see that the optimal method performs in a high variable processing time in function of the period p while SIRALINA is more stable. Figure 3 shows that SIRALINA is still nearly optimal with various period values. This remark has been checked for all other benchmarks and periods: in almost all benchmarks, SIRALINA computes nearly optimal results for all periods in a satisfactory (fast) processing time.

Table 2. SIRALINA and other heuristics

Benchmark	S_{f3}	S_{f_5}	$S_{siralina}$	$T_{siralina}$
lin-ddot-10	40	27	7	0.072
liv-loop1-10	80	60	5	0.094
spec-dod-loop3-10	100	75	4	0.107
spec-spice-loop1-10	20	19	3	0.069
spec-spice-loop3-10	30	29	2	0.071
spec-spice-loop6-10	60	48	14	0.081

For large instances we compare our results against the ones obtained by two heuristics (f3 and f5) proposed in [10]. Results are reported in table 2, where S_{f3} and S_{f_5} design the storage requirement (objective function values) computed by heuristics f3 and f5 and the last column gives the resolution time in seconds with SIRALINA. Benchmarks presented in table 2 correspond to instances where loop bodies are duplicated ten times (number of DDG nodes mutliplied by 10) and for which results with heuristic f3 and f5 have been presented in [10].

6 Conclusion

This article presents an efficient heuristic for the periodic task scheduling problem under storage constraints. Our model is based on the theoretical approach of reuse graphs studied in [9]. Storage allocation is expressed in terms of reuse edges and reuse distances to model the fact that two tasks use the same storage location.

Since computing an optimal periodic storage allocation is intractable in large data dependence graphs (larger than 12 nodes for instance), we have identified a two steps resolution method. We call this simplified method as SIRALINA. A first optimal step provides scheduling variables and allows to compute the potential reuse distances if the corresponding reuse edge is added. Then a second step solves a linear assignment problem using the Hungarian method in order to select the appropriate reuse edges.

Our practical experiments on many DDGs show that SIRALINA provides satisfactory solutions with fast resolution times. Consequently, this method is included inside a compiler for embedded systems (in collaboration with STmicroelectronics).

Finally our future work will concentrate on the particular structure of the model constraints to consider the application of lagrangean relaxation to produce a bound stronger than the bound obtained by continuous relaxation and/or to find another heuristic. Furthermore it will be interesting to investigate how some hardware specificities could be take into account. For instance, the use of a rotating register file (implemented inside some processors) implies the presence of a Hamiltonian reuse cycle. The handling of these kind of specificities is an additional challenge.

Acknowledgement

This work has been partially supported by the ANR MOPUCE project (ANR number 05-JCJC-0039).

References

1. de Dinechin, B.D.: Parametric Computation of Margins and of Minimum Cumulative Register Lifetime Dates. In: Sehr, D., Banerjee, U., Gelernter, D., Nicolau, A., Padua, D.A. (eds.) LCPC 1996. LNCS, vol. 1239, pp. 231–245. Springer, Heidelberg (1997)
2. Hanen, C., Munier, A.: A Study of the Cyclic Scheduling Problem on Parallel Processors. Discrete Applied Mathematics 57(2-3), 167–192 (1995)
3. Eichenberger, A.E., Davidson, E.S., Abraham, S.G.: Minimizing Register Requirements of a Modulo Schedule via Optimum Stage Scheduling. International Journal of Parallel Programming 24(2), 103–132 (1996)
4. Fimmel, D., Muller, J.: Optimal Software Pipelining Under Resource Constraints. International Journal of Foundations of Computer Science (IJFCS) 12(6), 697–718 (2001)
5. Kuhn, H.W.: The Hungarian Method for the assignment problem. Naval Research Logistics Quarterly 2, 83–97 (1955)
6. Janssen, J.: Compilers Strategies for Transport Triggered Architectures. PhD thesis, Delft University, Netherlands (2001)
7. Schrijver, A.: Theory of Linear and Integer Programming. John Wiley and Sons, New York (1987)
8. Touati, S.-A.-A.: Register Pressure in Instruction Level Parallelisme. PhD thesis, Université de Versailles, France (June 2002),
   ftp.inria.fr/INRIA/Projects/a3/touati/thesis
9. Touati, S.-A.-A.: Periodic Task Scheduling under Storage Constraints. In: Proceedings of the Multidisciplinary International Scheduling Conference: Theory and Applications (MISTA 2007) (August 2007)
10. Touati, S.-A.-A., Eisenbeis, C.: Early Periodic Register Allocation on ILP Processors. Parallel Processing Letters 14(2) (June 2004)
11. Strout, M.M., Carter, L., Ferrante, J., Simon, B.: Schedule-Independent Storage Mapping for Loops. ACM SIG-PLAN Notices 33(11), 24–33 (1998)
12. Thies, W., Vivien, F., Sheldon, J., Amarasinghe, S.: A Unified Framework for Schedule and Storage Optimization. ACM SIGPLAN Notices 36(5), 232–242 (2001)

Stochastic Online Scheduling Revisited

Andreas S. Schulz

Sloan School of Management, Massachusetts Institute of Technology,
E53-361, 77 Massachusetts Avenue, Cambridge, MA 02139, USA

Abstract. We consider the problem of minimizing the total weighted completion time on identical parallel machines when jobs have stochastic processing times and may arrive over time. We give randomized as well as deterministic online and off-line algorithms that have the best known performance guarantees in either setting, deterministic and off-line or randomized and online. Our analysis is based on a novel linear programming relaxation for stochastic scheduling problems, which can be solved online.

1 Introduction

We study approximation algorithms for stochastic and online versions of the following deterministic, off-line scheduling problem. There is a set of n jobs to be processed on m identical parallel machines. Each job j has a nonnegative weight w_j, processing time p_j, and release date r_j. After its release, a job has to be processed on some machine, and each machine can handle at most one job at a time. The objective is to assign jobs to machines and to determine a feasible sequence on each machine so as to minimize the total weighted completion time, $\sum_{j=1}^{n} w_j C_j$. Here, C_j denotes the completion time of job j in the schedule. The deterministic problem is well understood: It is known to be strongly NP-hard (Lenstra, Rinnooy Kan, and Brucker 1977), and it has a polynomial-time approximation scheme (Afrati et al. 1999); a simpler 2-approximation algorithm, which is of particular relevance to the work described here, was earlier given by Schulz and Skutella (20002b).

In stochastic scheduling (Möhring, Radermacher, and Weiss 1984), job processing times are modeled as random variables, each specified by some probability distribution (with expected value μ_j and standard deviation σ_j). The actual processing time of a job does not become known before it is completed. Research has traditionally focused on nonanticipative policies that aim at minimizing the objective function in expectation. Moreover, it is typically assumed that job processing times are stochastically independent. These views are adopted here as well. A scheduling policy is nonanticipative if its decisions about which jobs to schedule at any given time t depend only on the jobs that are already completed by that time and on the conditional distributions of the remaining processing times of jobs that are still active at time t.

For the single-machine problem without nontrivial release dates ($m = 1, r_j = 0$ for all jobs j), Rothkopf (1966) showed that the WSEPT rule is optimal, which schedules the jobs in order of nonincreasing ratios of weight to expected

B. Yang, D.-Z. Du, and C.A. Wang (Eds.): COCOA 2008, LNCS 5165, pp. 448–457, 2008.

processing time. For unit weights and exponentially distributed processing times, the Shortest Expected Processing Time rule remains optimal on identical parallel machines (Weiss and Pinedo 1980). In fact, Weber, Varaiya, and Walrand (1986) showed that it suffices when the processing time distributions are stochastically comparable in pairs. For arbitrary weights, WSEPT is optimal for exponentially distributed processing times if the WSEPT order of jobs coincides with sequencing the jobs in the order of nonincreasing weights (Kämpke 1987). Under minor technical assumptions, Weiss (1990) showed that the WSEPT rule is asymptotically optimal.

The stochastic scheduling problem considered here, when jobs may have individual release dates, was first addressed by Möhring, Schulz, and Uetz (1999). For processing time distributions whose coefficients of variation σ_j/μ_j are bounded from above by $\sqrt{\Delta}$, they gave a static priority policy whose expected objective function value is within a factor of $\max\{4, 3 + \Delta\}$ of that of an optimal policy.[1] In addition, they showed that the WSEPT rule has a performance guarantee of $1 + (\Delta + 1)/2$ for the problem with identical release dates. This marked the first time that the use of approximation algorithms was proposed in the realm of stochastic scheduling. The analysis as well as the algorithm for the general case is based on a linear programming relaxation, which provides a lower bound on the expected value of an optimal policy.

A different way of dealing with incomplete information is that of online algorithms and competitive analyses. In our context, jobs arrive over time and are completely unknown prior to their arrival. However, a job's processing time and weight are fully revealed at the time of its arrival. The performance of an online algorithm is usually compared to that of an optimal off-line algorithm, which has full hindsight. This quotient is known as the competitive ratio. For randomized online algorithms, we compare the expected objective function value of the solution generated by the algorithm to the value of an off-line optimum. This corresponds to the so-called oblivious adversary model. We refer the reader to Borodin and El-Yaniv (1998) for a general introduction to online algorithms, and to Sgall (1998) for a survey of online scheduling models and results. In the context of the identical parallel machine scheduling problem considered here, online algorithms were designed and analyzed by Hall et al. (1997), Chakrabarti et al. (1996), Schulz and Skutella (20002b), Megow and Schulz (2004), and Correa and Wagner (2008). The currently best deterministic online algorithm has a competitive ratio of 2.618 (Correa and Wagner 2008), while the best known randomized algorithm is 2-competitive (Schulz and Skutella 20002b).

Chou et al. (2006) proposed to study stochastic online scheduling, where jobs arrive over time, as in online scheduling, but when a job arrives only its weight and processing time distribution become known. The expected total weighted completion time of the schedule computed by an online policy is then compared to that of an optimal stochastic policy, which has access to all job release dates,

---

[1] The performance guarantee of this algorithm is actually slightly better than this; to make for an improved reading, we generally suppress terms of order $1/m$ from this extended abstract.

weights, and processing time distributions at time 0. In other words, the adversary controls the arrival of jobs, their weights, and their processing time distributions, but he cannot influence the actual realization of processing times. While Chou et al. (2006) considered single-machine and flow-shop problems, Megow, Uetz, and Vredeveld (2006) looked at the identical parallel machine model considered here. They introduced δ-NBUE distributions[2] and gave a deterministic online algorithm with performance guarantee $3/2 + \delta + \sqrt{4\delta^2 + 1}/2$. Their analysis uses the linear programming relaxation introduced by Möhring, Schulz, and Uetz (1999), when their algorithm does not.

In this paper, we present deterministic and randomized approximation algorithms that have the best known performance guarantees for stochastic (off-line and online) scheduling on identical parallel machines with the sum of weighted completion times objective. The key is a new, stronger linear programming relaxation for this problem. Moreover, this linear program can be solved by a simple online rule, which constructs a preemptive single-machine schedule. We use this schedule to define an online policy for the original, stochastic problem. This approach has previously been used successfully for various deterministic online problems, including nonpreemptive scheduling on a single machine (Goemans et al. 2002), preemptive single-machine scheduling (Schulz and Skutella 20002a), identical parallel machine scheduling (Schulz and Skutella 20002b), and uniform parallel machine scheduling (Chou, Queyranne, and Simchi-Levi 2006).

We present one randomized and one deterministic algorithm; both work online and run in polynomial time.[3] Their respective performance ratios are $2 + \Delta$ and $\max\{2.618, 2.309 + 1.309\Delta\}$, respectively. The randomized algorithm can be derandomized, which results in a deterministic $(2 + \Delta)$-approximation algorithm for the stochastic (off-line) scheduling problem. Table 1 compares the new results from this paper to earlier results.

The algorithms proposed here are derived from earlier algorithms for deterministic scheduling problems, as were previous algorithms for stochastic scheduling. In our case, we manipulate a randomized online algorithm of Schulz and Skutella (20002b) as well as a deterministic online algorithm by Correa and Wagner (2008). Previously, Möhring, Schulz, and Uetz (1999) built on deterministic algorithms by Hall et al. (1997); Skutella and Uetz (2005) used techniques of Chekuri et al. (2001); and Megow, Uetz, and Vredeveld (2006) drew on ideas

---

[2] For $\delta = 1$, one recaptures the well-known NBUE distributions, "New Better than Used in Expectation," which include, among others, exponential, Erlang, uniform, and Weibull distributions. In the context of stochastic scheduling, an NBUE distribution would imply that the expected remaining processing time of a job in process is never more than the expected processing time of that job before it was started. NBUE distributions satisfy $\Delta \leq 1$ (Hall and Wellner 1981). In general, $\Delta \leq 2\delta - 1$ (Megow, Uetz, and Vredeveld 2006).

[3] A general definition of the input size of a stochastic scheduling problem would need to deal with the way in which arbitrary probability distributions are specified. The running times of the algorithms proposed here depend only on the input size of the corresponding deterministic problems where job processing times are replaced by expected values.

Table 1. Overview of the development of performance guarantees/competitive ratios for stochastic scheduling with the total weighted completion time objective. To allow for a comparison, we assume that the processing time of each job follows an NBUE distribution.

Model	Performance Guarantee		Reference
	deterministic	randomized	
off-line, all $r_j = 0$	2	–	Möhring et al. (1999)
off-line, general r_j	4	–	Möhring et al. (1999)
	3.618	–	Megow et al. (2006)
	3	–	this paper
online	3.618		Megow et al. (2006)
	3.618	3	this paper

from Megow and Schulz (2004). In each case, the challenge is to refine the algorithm and its analysis such that they still work, even though job processing times are random. In contrast to the previous approximation and online algorithms for stochastic scheduling problems, which all relied on the lower bounds introduced by Möhring, Schulz, and Uetz (1999), we use a linear programming relaxation that is new in the context of stochastic scheduling.

2   A Linear Programming Relaxation

Möhring, Schulz, and Uetz (1999) showed that the vector of expected completion times of any nonanticipative policy satisfies the following inequalities:

$$\sum_{j\in S}\mu_j C_j \geq \frac{1}{2m}\Big(\sum_{j\in S}\mu_j\Big)^2 - \frac{\Delta-1}{2}\sum_{j\in S}\mu_j^2 \quad \text{for all } S\subseteq N.$$

As mentioned before, Δ is an upper bound on the squared coefficients of variation; i.e., $\sigma_j^2/\mu_j^2 \leq \Delta$ for all $j \in N$, where $N := \{1, 2, \ldots, n\}$ denotes the set of all jobs. One can strengthen these inequalities by observing that none of the jobs in a subset S can be processed before time $r_{\min}(S) := \min\{r_j : j \in S\}$.

Lemma 1. *Let Π be a nonanticipative policy for the stochastic identical parallel machine scheduling problem. Then, the corresponding vector $E[C^\Pi]$ of expected completion times satisfies the following inequalities:*

$$\sum_{j\in S}\mu_j\Big(C_j + \frac{\Delta-1}{2}\mu_j\Big) \geq r_{\min}(S)\sum_{j\in S}\mu_j + \frac{1}{2m}\Big(\sum_{j\in S}\mu_j\Big)^2 \quad \text{for all } S\subseteq N. \quad (1)$$

A similar observation was made earlier in the context of deterministic scheduling; see Queyranne and Schulz (1995). Its relevance in our situation is a consequence of the fact that the associated linear programming relaxation, when we minimize $\sum_{j \in N} w_j C_j$ over (1), is equivalent to that of a deterministic single-machine problem. In fact, setting $M_j := C_j + \frac{\Delta - 1}{2} \mu_j$, leads to the following, equivalent linear program:

$$\min \quad \sum_{j \in N} w_j M_j \tag{2a}$$

$$\text{s.t.} \quad \sum_{j \in S} \frac{\mu_j}{m} M_j \geq \frac{\sum_{j \in S} \mu_j}{m} \left(r_{\min}(S) + \frac{\sum_{j \in S} \mu_j}{2m} \right) \quad \text{for all } S \subseteq N. \tag{2b}$$

Note that in (2) we have dropped the term $\frac{1-\Delta}{2} \sum_{j \in N} w_j \mu_j$ from the objective function, as it is constant anyway.

The linear program (2) can be interpreted as a relaxation of a single-machine scheduling problem where jobs have (deterministic) processing times μ_j / m, and the formulation makes use of mean busy time variables M_j. The mean busy time M_j of job j is the average point in time at which the (single) machine is busy with processing job j. In other words, if $I_j(t)$ is 1 if the machine is processing job j at time t, and 0 otherwise, then

$$M_j = \frac{1}{p_j} \int_{r_j}^{\infty} I_j(t) \, t \, dt \ .$$

Here and henceforth, we use p_j to denote the processing time of job j on the "fast" single machine; i.e., $p_j = \mu_j / m$.

Theorem 2 (Goemans et al. 2002). *The mean busy time vector of the preemptive single-machine schedule that is constructed by the following online algorithm is an optimal solution to the linear programming relaxation (2):*

> *At any point in time, schedule from the available (and as of yet not completed) jobs one with the highest ratio of weight to processing time.*

As was done before (Goemans et al. 2002; Schulz and Skutella 20002a; Chou et al. 2006; Correa and Wagner 2008), we refer to this preemptive schedule as the "LP schedule." It is worth pointing out that Theorem 2 effectively implies that one can solve the linear programming relaxation of minimizing $\sum_{j \in N} w_j C_j$ over (1) online. So, not only it provides a lower bound on the expected value of an optimal off-line policy, but also it can be used to design an online algorithm for the stochastic scheduling problem itself.

3 A Randomized Algorithm

In the spirit of all previous approximation algorithms for nonpreemptive stochastic scheduling problems, which are based on existing algorithms for deterministic scheduling problems, we will now extend an algorithm of Schulz and Skutella (20002b) to stochastic online scheduling. To describe the algorithm, we need the notion of α-points. For $0 < \alpha \leq 1$, the α-point $t_j(\alpha)$ of a job j is the first moment in time when an α-fraction of j has been completed in the LP schedule. α-points were introduced by Sousa (1989), and have since been used in the design of a variety of approximation and online algorithms for scheduling problems.

The algorithm that we analyze here works as follows. It maintains, side-by-side with the actual schedule on m machines, the preemptive LP schedule on the (virtual) single machine. For this, we create a priority list L of jobs, sorted by nonincreasing ratios of weight to expected processing time. Initially, L is empty. Whenever a new job j arrives, we draw some value $\alpha_j \in (0, 1]$ uniformly at random (independent from the drawings for other jobs). Moreover, job j is inserted into the correct position in L. (Ties are broken arbitrarily.) We obtain the LP schedule by scheduling, at any point in time, the first job in L on the virtual machine. As soon as job j has reached its α_j-point in the LP schedule; i.e., when it has been processed for $\alpha_j p_j$ units of time on the virtual machine, it is randomly assigned to one of the m machines (independent of the assignments of other jobs). It then enters another priority list on that machine, which is arranged by nondecreasing α-points. (As before, ties are broken arbitrarily.) On each real machine, jobs are then scheduled nonpreemptively in that order. Note that, by construction, no job can start before its α-point. Finally, whenever a job j is completed on the virtual machine (i.e., it has been processed for p_j periods of time), it is removed from L. And when it is completed on its real machine, it is deleted from the priority list of that machine. This concludes the description of the algorithm, which we call RSOS (for Randomized Stochastic Online Scheduling).

Theorem 3. *Let I be an instance of the stochastic scheduling problem to minimize the total weighted completion time on identical parallel machines in which jobs arrive over time, and let Δ be an upper bound on the squared coefficients of variation of the jobs' processing times. Moreover, let $OPT(I)$ be the objective function value of an optimal off-line nonanticipative scheduling policy for I. Finally, let $RSOS(I)$ be the value of the schedule produced by the randomized online policy RSOS. Then, $E[RSOS(I)] \leq (\Delta + 2)E[OPT(I)]$.*

Proof. Let us consider an arbitrary, but fixed job j. Initially, let us also fix the index i of the machine to which j has been assigned, as well as a value of α_j. Note that j is ready to start at time $t_j(\alpha_j)$ on machine i; in particular, $r_j \leq t_j(\alpha_j)$. If j is not started at time $t_j(\alpha_j)$, then it is delayed by jobs with a smaller α-point that have been assigned to the same machine i. We denote by $E_{i,\alpha_j}[C_j]$ the conditional expected completion time of job j, where the expectation is taken both over the random choices of the algorithm, except for i and α_j, which are still fixed, and the processing times. We then have

$$E_{i,\alpha_j}[C_j] \leq t_j(\alpha_j) + \mu_j + \sum_{k \neq j} \mu_k \, P(k \text{ on } i \text{ before } j)$$

$$\leq t_j(\alpha_j) + \mu_j + \sum_{k \neq j} \mu_k \frac{1}{m} \frac{1}{p_k} \int_0^{t_j(\alpha_j)} I_k(t)\, dt$$

$$\leq t_j(\alpha_j) + \mu_j + t_j(\alpha_j) = 2\, t_j(\alpha_j) + \mu_j \; .$$

In the first inequality, $P(k$ on i before $j)$ is the probability that job $k \neq j$ is assigned to the same machine as j and will be started before j. The probability that k is assigned to machine i is $1/m$. The integral in the second inequality captures the fraction of job k that is processed in the LP schedule before $t_j(\alpha_j)$, which, by the choice of α_k, is precisely the probability of $t_k(\alpha_k)$ being smaller than $t_j(\alpha_j)$. The remaining two inequalities are straightforward. We finally get rid of the conditional expectation by noting that the average α_j-point is equal to the mean busy time M_j in the LP schedule (Goemans et al. 2002). Therefore,

$$E[C_j] \leq 2 \int_0^1 t_j(\alpha_j)\, d\alpha_j + \mu_j = 2\, M_j + \mu_j \ .$$

The result now follows from our earlier observation that $\sum_{j \in N} w_j M_j - \frac{\Delta-1}{2} \sum_{j \in N} w_j \mu_j$ is a lower bound on the expected value of an optimal policy (Lemma 1), and so is $\sum_{j \in N} w_j \mu_j$. Hence,

$$
\begin{aligned}
E\Big[\sum_{j \in N} w_j C_j\Big] &\leq 2 \sum_{j \in N} w_j M_j + \sum_{j \in N} w_j \mu_j \\
&= 2 \sum_{j \in N} w_j M_j - (\Delta - 1) \sum_{j \in N} w_j \mu_j + \Delta \sum_{j \in N} w_j \mu_j \\
&\leq (2 + \Delta)\mathrm{OPT}(I) \ . \qquad \qquad \square
\end{aligned}
$$

The crucial observation, which makes this proof work, is that the set of jobs that is scheduled on machine i before job j does not depend on the actual realization of processing times. The order of jobs is determined by the LP schedule, which depends only on the expected processing times.

Corollary 4. *There exists a deterministic $(2 + \Delta)$-approximation algorithm for the stochastic (off-line) problem of minimizing the weighted sum of completion times on identical parallel machines subject to release dates.*

We omit the proof from this extended abstract, but note that this algorithm can be obtained from RSOS by the method of conditional probabilities (Spencer 1987). Of course, this implies that the derived algorithm does not work in an online context. This will be fixed, to some extent, in the next section.

4   A Deterministic Algorithm

A simple, though somewhat less effective way of derandomizing the RSOS policy, yet one that does not destroy its online nature, is to choose α_j deterministically beforehand. The rest of the algorithm, to which we will refer as *DSOS*, remains unchanged, except that jobs are not randomly assigned to machines. Instead, we employ a list scheduling strategy: whenever a machine becomes available, we start a job with the smallest α-point of all not-yet-processed jobs whose α-points have already passed. Let ϕ denote the golden ratio, and let us choose $\alpha_j = \phi - 1$ for all $j \in N$.

Theorem 5. *Let I be an instance of the stochastic scheduling problem to minimize the total weighted completion time on identical parallel machines in which jobs arrive over time, and let Δ be an upper bound on the squared coefficients of variation of the jobs' processing times. Moreover, let $OPT(I)$ be the objective function value of an optimal off-line nonanticipative scheduling policy for I. Finally, let $DSOS(I)$ be the value of the schedule produced by the deterministic online policy DSOS. Then, $E[DSOS(I)] \leq \max\{\phi + 1, \frac{\phi+1}{2}\Delta + \frac{\phi+3}{2}\}E[OPT(I)]$.*

Proof. The proof is saved for the full version of this paper; it is based on careful modifications of the proof of Correa and Wagner (2008, Theorem 3.2), which itself is based on that of Goemans et al. (2002, Theorem 3.3). Apart from Lemma 1, the key insight is that the start of any job j is always delayed by the same set of jobs, regardless of the actual instantiation of processing times.   □

5   Concluding Remarks

In this paper, we have taken the design and analysis of approximation and online algorithms for nonpreemptive stochastic scheduling problems a step further. The main ingredient is a new linear programming relaxation for stochastic scheduling problems on identical parallel machines that is provably stronger than the one that was used in the design of all previously proposed approximate policies.

While the algorithms studied here do have deterministic counterparts that were analyzed before, it is important to recognize that the extension of algorithms designed for deterministic scheduling problems to stochastic problems is not automatic. In fact, many approaches that work well for deterministic scheduling problems cannot be modified to handle random processing times.

In the course of this work, we have obtained the first randomized policy for stochastic online scheduling as well as the best performance guarantee for stochastic (off-line) scheduling on identical parallel machines with release dates. Looking beyond the realm of stochastic scheduling, this paper provides additional proof of the versatility of the LP schedule, which had previously been used to derive a series of best known performance guarantees, competitive ratios, and asymptotic optimality results for a variety of scheduling problems; see Goemans et al. (2002), Schulz and Skutella (20002a, 20002b), Chou et al. (2006), and Correa and Wagner (2008), among others.

It is also worth mentioning that both RSOS and DSOS need information on expected processing times only, even though they are compared to optimal policies that have full access to the entire distribution. Moreover, as was the case for the previous linear programming relaxation by Möhring, Schulz, and Uetz (1999), the new linear programming relaxation remains valid for preemptive schedules. In particular, the nonpreemptive schedules generated by the algorithms considered here and in previous papers are approximate solutions for preemptive stochastic scheduling as well. While their worst-case performance guarantees are not as good as the one in Megow and Vredeveld (2006), the policies are simple, can be implemented in polynomial time, and require little information about the distribution of processing times.

Acknowledgments

The author is grateful to the organizers of Dagstuhl Seminar 05031, where this work was originally conceived, and to José Correa for sending him a preliminary version of Correa and Wager (2008), which inspired him to revisit stochastic scheduling problems. This research was supported by NSF awards #0426686 and #0700044, and by ONR grant N00014-08-1-0029.

References

Afrati, F., Bampis, E., Chekuri, C., Karger, D., Kenyon, C., Khanna, S., Milis, I., Queyranne, M., Skutella, M., Stein, C., Sviridenko, M.: Approximation schemes for minimizing average weighted completion time with release dates. In: Proceedings of the 40th Annual IEEE Symposium on Foundations of Computer Science, pp. 32–43 (1999)

Borodin, A., El-Yaniv, R.: Online Computation and Competitive Analysis. Cambridge University Press, Cambridge (1998)

Chakrabarti, S., Phillips, C., Schulz, A., Shmoys, D., Stein, C., Wein, J.: Improved scheduling algorithms for minsum criteria. In: auf der Heide, F., Monien, B. (eds.) ICALP 1996. LNCS, vol. 1099, pp. 646–657. Springer, Heidelberg (1996)

Chekuri, C., Motwani, R., Natarajan, B., Stein, C.: Approximation techniques for average completion time scheduling. SIAM Journal on Computing 31, 146–166 (2001)

Chou, M., Liu, H., Queyranne, M., Simchi-Levi, D.: On the asymptotic optimality of a simple on-line algorithm for the stochastic single-machine weighted completion time problem and its extensions. Operations Research 54, 464–474 (2006)

Chou, M., Queyranne, M., Simchi-Levi, D.: The asymptotic performance ratio of an on-line algorithm for uniform parallel machine scheduling with release dates. Mathematical Programming 106, 137–157 (2006)

Correa, J., Wagner, M.: LP-based online scheduling: From single to parallel machines. Mathematical Programming (in press, 2008)

Goemans, M., Queyranne, M., Schulz, A., Skutella, M., Wang, Y.: Single machine scheduling with release dates. SIAM Journal on Discrete Mathematics 15, 165–192 (2002)

Hall, L., Schulz, A., Shmoys, D., Wein, J.: Scheduling to minimize average completion time: Off-line and on-line approximation algorithms. Mathematics of Operations Research 22, 513–544 (1997)

Hall, W., Wellner, J.: Mean residual life. In: Csörgö, M., Dawson, D., Rao, J., Saleh, A.E. (eds.) Proceedings of the International Symposium on Statistics and Related Topics, pp. 169–184 (1981)

Kämpke, T.: On the optimality of static priority policies in stochastic scheduling on parallel machines. Journal of Applied Probability 24, 430–448 (1987)

Lenstra, J., Rinnooy Kan, A., Brucker, P.: Complexity of machine scheduling problems. Annals of Discrete Mathematics 1, 343–362 (1977)

Megow, N., Schulz, A.: On-line scheduling to minimize average completion time revisited. Operations Research Letters 32, 485–490 (2004)

Megow, N., Uetz, M., Vredeveld, T.: Models and algorithms for stochastic online scheduling. Mathematics of Operations Research 31, 513–525 (2006)

Megow, N., Vredeveld, T.: Approximation results for preemptive stochastic online scheduling. In: Azar, Y., Erlebach, T. (eds.) ESA 2006. LNCS, vol. 4168, pp. 516–527. Springer, Heidelberg (2006)

Möhring, R., Radermacher, F., Weiss, G.: Stochastic scheduling problems I: General strategies. Zeitschrift für Operations Research 28, 193–260 (1984)

Möhring, R., Schulz, A., Uetz, M.: Approximation in stochastic scheduling: The power of LP-based priority policies. Journal of the ACM 46, 924–942 (1999)

Queyranne, M., Schulz, A.: Scheduling unit jobs with compatible release dates on parallel machines with nonstationary speeds. In: Balas, E., Clausen, J. (eds.) IPCO 1995. LNCS, vol. 920, pp. 307–320. Springer, Heidelberg (1995)

Rothkopf, M.: Scheduling with random service times. Management Science 12, 703–713 (1966)

Schulz, A., Skutella, M.: The power of α-points in preemptive single machine scheduling. Journal of Scheduling 5, 121–133 (2002a)

Schulz, A., Skutella, M.: Scheduling unrelated machines by randomized rounding. SIAM Journal on Discrete Mathematics 15, 450–469 (2002b)

Sgall, J.: On-line scheduling. In: Fiat, A., Woeginger, G. (eds.) Online Algorithms: The State of the Art. LNCS, vol. 1442, ch. 9, pp. 196–231. Springer, Heidelberg (1998)

Skutella, M., Uetz, M.: Stochastic machine scheduling with precedence constraints. SIAM Journal on Computing 34, 788–802 (2005)

Sousa, J.: Time Indexed Formulations of Non-Preemptive Single-Machine Scheduling Problems. Ph.D. thesis, Université Catholique de Louvain, Belgium (1989)

Spencer, J.: Ten Lectures on the Probabilistic Method. CBMS-NSF Regional Conference Series in Applied Mathematics, vol. 52. SIAM, Philadelphia (1987)

Weber, R., Varaiya, P., Walrand, J.: Scheduling jobs with stochastically ordered processing times on parallel machines to minimize expected flowtime. Journal of Applied Probability 23, 841–847 (1986)

Weiss, G.: Approximation results in parallel machines stochastic scheduling. Annals of Operations Research 26, 195–242 (1990)

Weiss, G., Pinedo, M.: Scheduling tasks with exponential service times on nonidentical processors to minimize various cost functions. Journal of Applied Probability 17, 187–202 (1980)

Delay Management Problem: Complexity Results and Robust Algorithms*

Serafino Cicerone[1], Gianlorenzo D'Angelo[1], Gabriele Di Stefano[1], Daniele Frigioni[1], and Alfredo Navarra[2]

[1] Department of Electrical and Information Engineering,
University of L'Aquila, Poggio di Roio, 67040 L'Aquila Italy
{cicerone,gdangelo,gabriele,frigioni}@ing.univaq.it
[2] Department of Mathematics and Informatics, University of Perugia,
Via Vanvitelli 1, 06123 Perugia, Italy
navarra@dipmat.unipg.it

Abstract. In this paper, we study the problem of planning a *timetable* for passenger trains considering that possible delays might occur due to unpredictable (but bounded) circumstances. Once arrival and departure events are scheduled, if the timetable cannot be respected since an external event has determined a delay to a train, the so called *delay management* problem occurs. Delays might be managed in several ways and the usual objective function considered for such purpose is the minimization of the overall waiting time caused to passengers.

We analyze the interaction between timetable planning and delay management in terms of the *recoverable robustness* model, where a timetable is said to be *robust* if it is able to absorb small delays by possibly applying given recovery capabilities. The quality of a robust timetable is measured by the *price of robustness* that is the ratio between the cost of the robust timetable and that of a non-robust optimal timetable.

We consider the problem of designing robust timetables subject to bounded delays. We show that finding an optimal solution for this problem is NP-hard. Hence, we propose robust algorithms and evaluate their prices of robustness. Moreover, we show that such algorithms are optimal with respect to particular assumptions.

1 Introduction

Many real world applications are characterized by a *strategic planning* phase and an *operational planning* phase. The main difference among the two planning phases resides in the time in which they are applied. The strategic planning phase aims to plan how to optimize the use of the available resources according to some objective function before the system starts operating. The operational planning phase aims to have immediate reaction to disturbing events that can

---

* This work was partially supported by the Future and Emerging Technologies Unit of EC (IST priority - 6th FP), under contract no. FP6-021235-2 (project ARRIVAL).

B. Yang, D.-Z. Du, and C.A. Wang (Eds.): COCOA 2008, LNCS 5165, pp. 458–468, 2008.

occur when the system is running. In general, the objectives of strategic and operational planning might be in conflict with each other. As disturbing events are unavoidable in large and complex systems, it is fundamental to understand the interaction between the objectives of the two phases.

A concrete example of real world systems, where this interaction is important, is the *timetable planning* in railways systems. It arises in the strategic planning phase of railways systems, and it requires to compute a timetable for passenger trains that determines minimal passenger waiting times. However, many disturbing events might occur during the operational phase, and they might completely change the scheduled activities. The main effect of such disturbing events is the arising of delays. These might be caused by malfunctioning infrastructure/devices, special events, or extreme weather conditions. The conflicting objectives of strategic against operational planning are evident in timetable optimization. In fact, a train schedule that lets trains sit in stations for some time will not suffer from small delays of arriving trains, because delayed passengers can still catch potential connecting trains. On the other hand, big delays can cause passengers to loose trains and hence imply extra traveling time.

The problem of deciding when to guarantee connections from a delayed train to a connecting train is known in the literature as *delay management problem* [1,2,3] and it has a twofold impact. On the one hand, the passengers arriving late still catch their connection and do not have to wait for the next train. On the other hand, passengers in the connecting train now face a delay and may miss subsequent connections. The latter implies that the delay can propagate through the railway network. The trade-off between these two effects leads to the natural objective of minimizing the overall delay faced by the total passenger population. Although its natural formalization, the problem turns out to be very complicated to be optimally solved. In fact it has been shown to be NP-hard in the general case, while it is polynomial in some particular cases (see [2,3,4,5,6,7]).

In order to cope with the delay management problem we can follow two possible approaches: 1) to apply a recovery strategy to the timetable defined in the strategic planning phase and try to rapidly obtain a new feasible timetable which considers the occurred delays; 2) to design the timetable in the strategic planning phase in order to be "prepared" to react against possible disruptions.

The second approach is known in the literature as *robust optimization*. In the last years, several attempts have been done in order to formalize the notion of robustness for optimization problems (see, e.g., [8,9,10,11]). In such models, the basic idea of robustness is given by a problem and some kind of disturbing events. That is, the solution provided for a given instance of the problem must hold even though a disturbing event occurs. This approach is not always suitable in practical scenarios as it does not allow to use *recovery strategies*. In fact, in real cases, we are allowed to modify the planned solution by using (possibly) limited resources during the operational phase. Considering robustness and recoverability in a unified way has lead to the *recoverable robustness* model. This model has been recently introduced in [12] and it has been extended and applied to shunting problems in [13].

In this paper, we apply the recoverable robustness model in the context of timetable planning and delay management problems. In detail, we take a particular timetabling problem TT and turn it into a recoverable robustness problem that we call *Robust Delay Management* problem ($\mathcal{RDM}$). We show that finding a solution for $\mathcal{RDM}$ which minimizes the objective function of TT is NP-hard. Hence, we propose robust algorithms and we show that such algorithms are optimal with respect to some restrictions.

2   The Recoverable Robustness Model

In this section, we report the model of recoverable robustness given in [12] and modified in [13]. Such a model describes how an optimization problem P can be turned into a *robustness problem* $\mathcal{P}$. Hence, concepts like *robust solution*, *robust algorithm* for $\mathcal{P}$ and *price of robustness* are defined. In the remainder, an optimization problem P is characterized by the following parameters.

- I, the set of instances of P;
- F, the function that associates to any instance $i \in I$ the set of all feasible solutions for i;
- $f \colon S \to \mathbb{R}$, the objective function of P, where $S = \bigcup_{i \in I} F(i)$ is the set of all feasible solutions for P.

Without loss of generality from now on we consider minimization problems. Additional concepts to introduce robustness requirements for a minimization problem P are needed:

- $M : I \to 2^I$ – a *modification* function for instances of P. This function models the following case. Let $i \in I$ be the considered input to the problem P, and let $s \in S$ be the planned solution for i. A *disruption* is meant as a modification to the input i, and such a modification can be seen as a new input $j \in I$. Typically, the modification j depends on the current input i, and this fact is modeled by the constraint $j \in M(i)$. Hence, given $i \in I$, $M(i)$ represents the set of instances of P that can be obtained by applying all possible modifications to i. Of course, when a disruption $j \in M(i)$ occurs, a new solution $s' \in F(j)$ has to be recomputed for P.
- $\mathbb{A}$ – a class of *recovery algorithms* for P. Algorithms in $\mathbb{A}$ represent the capability of recovering against disruptions. Since in a real-world problem the capability of recovering is limited, the class $\mathbb{A}$ can be defined in terms of some kind of *restrictions*, such as feasibility or algorithmic restrictions. An element $A_{rec} \in \mathbb{A}$ works as follows: given $(i, s) \in I \times S$, an instance/solution pair for P, and $j \in M(i)$, a modification of the current instance i, then $A_{rec}(i, s, j) = s'$, where $s' \in F(j)$ represents the recovered solution for P.

Definition 1. *A recoverable robustness problem* $\mathcal{P}$ *is defined by the triple* $(P, M, \mathbb{A})$. *All the recoverable robustness problems form the class* RRP.

Definition 2. *Let* $\mathcal{P} = (P, M, \mathbb{A}) \in$ RRP. *Given an instance* $i \in I$ *for* P, *an element* $s \in F(i)$ *is a* feasible solution *for* i *with respect to* $\mathcal{P}$ *if and only if the following relationship holds:*

$$\exists A_{rec} \in \mathbb{A} : \forall j \in M(i), \; A_{rec}(i, s, j) \in F(j)$$

In other words, $s \in F(i)$ is feasible for i with respect to $\mathcal{P}$ if it can be *recovered* by applying some algorithm $A_{rec} \in \mathbb{A}$ for each possible disruption $j \in M(i)$. We use the notation $F_{\mathcal{P}}(i)$ to represent all the feasible solutions for i with respect to $\mathcal{P}$. Formally $F_{\mathcal{P}}(i)$ is defined as:

$$F_{\mathcal{P}}(i) = \{s \in F(i) : s \text{ is a feasible solution for } i \text{ with respect to } \mathcal{P}\}.$$

Notice that, $F_{\mathcal{P}}(i)$ can be also considered as the set of *robust solutions* for i with respect to the original problem P.

Definition 3. *Let* $\mathcal{P} = (P, M, \mathbb{A}) \in$ RRP. *A* robust algorithm *for* $\mathcal{P}$ *is any algorithm* A_{rob} *such that, for each* $i \in I$, $A_{rob}(i)$ *is a robust solution for* i *with respect to* P.

It is worth to mention that, if $\mathbb{A}$ is the class of algorithms that do not change the solution s, that is, if each algorithm $A_{rec} \in \mathbb{A}$ fulfills the following condition

$$\forall (i, s) \in I \times S, \forall j \in M(i), \; A_{rec}(i, s, j) = s,$$

then the robustness problem $\mathcal{P} = (P, M, \mathbb{A})$ represents the so called *strict robustness problem*. In this case, a robust algorithm A_{rob} for $\mathcal{P}$ must provide a solution s for i such that, for each possible modification $j \in M(i)$, $s \in F(j)$. This means that, since A_{rec} has no capability of recovering against possible disruptions, then A_{rob} has to find solutions that "absorb" *any* possible disruption.

The following definition introduces the concepts of *price of robustness* of both a robust algorithm and a recoverable robustness problem.

Definition 4. *Let* $\mathcal{P} \in$ RRP. *The* price of robustness *of a robust algorithm* A_{rob} *for* $\mathcal{P}$ *is given by*

$$P_{rob}(\mathcal{P}, A_{rob}) = \max_{i \in I} \left\{ \frac{f(A_{rob}(i))}{\min\{f(x) : x \in F(i)\}} \right\}.$$

Definition 5. *Let* $\mathcal{P} \in$ RRP. *The* price of robustness *of* $\mathcal{P}$ *is given by*

$$P_{rob}(\mathcal{P}) = \min\{P_{rob}(\mathcal{P}, A_{rob}) : A_{rob} \text{ is a robust algorithm for } \mathcal{P}\}.$$

Definition 6. *Let* $\mathcal{P} \in$ RRP *and let* A_{rob} *be a robust algorithm for* $\mathcal{P}$. *Then,*

- A_{rob} *is* exact *if* $P_{rob}(\mathcal{P}, A_{rob}) = 1$;
- A_{rob} *is* $\mathcal{P}$-optimal *if* $P_{rob}(\mathcal{P}, A_{rob}) = P_{rob}(\mathcal{P})$.

3   Delay Management Problem

In this section we first consider a particular timetable problem and then we turn it into a recoverable robustness problem, the *Robust Delay Management* problem ($\mathcal{RDM}$), according to the model of Section 2.

3.1   The Timetabling Problem

We use an *event activity network* defined in [3] and reported in the following.

An arrival of a vehicle g at a station v is denoted as the arrival event (g, v, arr), while the departure event (g, v, dep) describes the departure of some vehicle g at some station v. The sets of arrival and departure events are denoted by $\mathcal{E}_{arr}$ and $\mathcal{E}_{dep}$, respectively. At a station v a train g might wait some time before departing. The *waiting* activity is represented by an arc from (g, v, arr) to (g, v, dep), while a *driving* activity, i.e., the activity performed by a departing train g from a station v to a station u is represented by an arc from (g, v, dep) to (g, u, arr). Another activity, called *changing* activity, can be performed by passengers that need/want to move from a train g to a train h, and it is represented by an arc from (g, v, arr) to (h, v, dep). The sets of waiting, driving and changing activities are denoted by $\mathcal{A}_{wait}$, $\mathcal{A}_{drive}$ and $\mathcal{A}_{change}$, respectively. More formally, the event activity network is a graph $\mathcal{N} = (\mathcal{E}, \mathcal{A})$ where:

- $\mathcal{E} = \mathcal{E}_{arr} \cup \mathcal{E}_{dep}$ is a set of nodes, called *events*;
- $\mathcal{A} = \mathcal{A}_{wait} \cup \mathcal{A}_{drive} \cup \mathcal{A}_{change}$ is a set of directed arcs, called *activities*, where:

$$\mathcal{A}_{wait} = \{((g, v, arr), (g, v, dep)) \in \mathcal{E}_{arr} \times \mathcal{E}_{dep}\}$$
$$\mathcal{A}_{drive} = \{((g, v, dep), (g, u, arr)) \in \mathcal{E}_{dep} \times \mathcal{E}_{arr} :$$
$$\text{vehicle } g \text{ goes directly from station } v \text{ to } u\}$$
$$\mathcal{A}_{change} = \{((g, v, arr), (h, v, dep)) \in \mathcal{E}_{arr} \times \mathcal{E}_{dep} : \text{ a changing}$$
$$\text{possibility from vehicle } g \text{ into } h \text{ at station } v \text{ is required}\}.$$

The driving and waiting activities are performed by vehicles, while the changing activities are performed by passengers. Notice that a precedence relation $\preceq$ between events is canonically given, where $u \preceq v$ indicates that there exists a direct path from u to v. A minimal element with respect to $\preceq$ always exists, but it may be not unique. Moreover, if $u \in \mathcal{E}$, then $\mathcal{E}(u) = \{v \in \mathcal{E} : u \preceq v\}$ represents the set of all events that can be reached from u.

A solution for a timetabling problem requires to assign a time to each event in such a way that all the constraints provided by the set of activities are respected. Given a function $L : \mathcal{A} \to \mathbb{N}$ that assigns to each activity its minimal duration time, a timetable $\Pi \in \mathbb{R}_+^{|\mathcal{E}|}$ for $\mathcal{N}$ is given by assigning a time $\Pi(u)$ to each event $u \in \mathcal{E}$ such that $\Pi(v) - \Pi(u) \geq L(a)$, for all $a = (u, v) \in \mathcal{A}$.

Given a function $w : \mathcal{A} \to \mathbb{N}$ that assigns to each activity a number of passengers, we are interested to a particular timetabling problem TT that requires to compute Π by also minimizing the *total travel time of all passengers*. Formally, TT can be defined as follows:

$$(TT) \qquad \min f = \sum_{a=(u,v)\in\mathcal{A}} w(a)\,(\Pi(v) - \Pi(u))$$

$$\text{subject to: } \Pi(v) - \Pi(u) \geq L(a) \quad \text{for all} \quad a = (u, v) \in \mathcal{A} \qquad (1)$$
$$\Pi(u) \in \mathbb{R}_+ \quad \text{for all} \quad u \in \mathcal{E} \qquad (2)$$

More precisely, an instance i of TT is specified by a triple $(\mathcal{N}, L, w)$, where:

- $\mathcal{N} = (\mathcal{E}, \mathcal{A})$ is the event activity network,
- $L : \mathcal{A} \to \mathbb{N}$ associates to each activity the minimal duration time,
- $w : \mathcal{A} \to \mathbb{N}$ associates to each activity the number of passengers.

The set of feasible solutions for i is

$$F(i) = \{\Pi : \Pi(u) \in \mathbb{R}_+, \ \forall u \in \mathcal{E} \text{ and } \Pi(v) - \Pi(u) \geq L(a), \ \forall a = (u, v) \in \mathcal{A}\}.$$

A solution Π for TT may produce a positive slack time $s(a)$ for each $a \in \mathcal{A}$. In particular, since the planned duration of an activity $a = (u, v)$ is given by $\Pi(v) - \Pi(u)$, then $s(a) = \Pi(v) - \Pi(u) - L(a)$.

In general, a feasible solution for TT cannot cope with possible delays occurring to the activities. A delay, in fact might affect many activities and the possible slack times spread around the event activity network are not enough in order to absorb it. Recovery (on-line) strategies might be necessary, and according to f, a new possible solution might also consider the possibility to skip some constrains provided by changing activities. In practice, the skip of a changing activity from a train g to a train h means that all passengers involved in such activity have to wait for the next train of "type" h, i.e., it is assumed that there exists a period of time after which all the events are repeated (see [3] for details). In this paper we are interested in solutions that respect all the constraints, hence we do not consider the possibility to skip some changing activities.

3.2   The Robust Delay Management ($\mathcal{RDM}$) Problem

We now transform the timetabling problem TT defined in the previous section into a robust recoverable problem $\mathcal{RDM} = (TT, M, \mathbb{A})$, where M represents a modification function and $\mathbb{A}$ is a class of recovery algorithms (see Section 2).

Given an instance $i = (\mathcal{N}, L, w)$ for TT, and a constant $\alpha \in \mathbb{N}$, we limit the modifications on i by admitting a single delay of at most α time. We model it as an increase on the minimal duration time of the delayed activity. Formally, $M(i)$ is defined as follows:

$$M(i) = \{(\mathcal{N}, L', w) : \exists \, \bar{a} \in \mathcal{A} : L(\bar{a}) \leq L'(\bar{a}) \leq L(\bar{a}) + \alpha, \ L'(a) = L(a) \ \forall a \neq \bar{a}\}.$$

We define the class $\mathbb{A}$ by introducing the concept of *events affected by one delay* as follows. Assume that $\Pi \in F(i)$, and that $j = (\mathcal{N}, L', w) \in M(i)$. Notice that, the increase of the initial lower bound $L(\bar{a})$ can be meant as the modeling of a *delay* on the activity $\bar{a} = (u, v)$. In this case, if $L'(\bar{a}) > \Pi(v) - \Pi(u)$, then the solution Π is not feasible for j, and hence a new solution $\Pi' \in F(j)$ must be computed. Note that, function M admits at most one delay occurring at some activity (u, v), and hence v is the first node affected by such delay.

Let $\bar{a} = (u, v)$ be the delayed activity, we define $d(v) = L'(\bar{a}) - (\Pi(v) - \Pi(u))$. If $d(v) > 0$, then the event v has been affected by the delay on $\bar{a}$. As side effect, other nodes in $\mathcal{E}(v)$ may be affected by the delay. The set of nodes affected by a delay

$d(v)$ is denoted as $\mathrm{Aff}(v)$, and it can be computed by algorithm AFFECTEDEVENTS shown in Figure 1. The algorithm starts from v and visits each node $u_k \in \mathcal{E}(v)$ in order to check whether the delay $d(v)$ is propagated. This is realized by means of function $d' : \mathcal{E} \to \mathbb{R}_+$ defined as $d'(u_k) = \max_{a=(u,u_k)\in\mathcal{A}}\{d'(u) - s(a)\}$ if $u_k \in \mathcal{E}(v)$, $d'(u_k) = d(v)$ if $u_k = v$, $d'(u_k) = 0$ otherwise. If $d'(u_k) > 0$ then u_k is included in $\mathrm{Aff}(v)$.

Algorithm AFFECTEDEVENTS

Input: $(\mathcal{N}, L, w)$, Π, v and the delay $d(v) > 0$
Output: $\mathrm{Aff}(v)$, set of events affected by the delay $d(v)$

1. $\mathrm{Aff}(v) := \{v\}$
2. sort $\mathcal{E}(v) = \{u_1, u_2, \ldots, u_{|\mathcal{E}(v)|}\}$ according to $\preceq$
3. **for** $k = 1$ to $|\mathcal{E}(v)|$ **do**
4. $d'(u_k) := \max_{a=(u,u_k)\in\mathcal{A}}\{d'(u) - s(a)\}$
5.   **if** $d'(u_k) > 0$ **then** $\mathrm{Aff}(v) := \mathrm{Aff}(v) \cup \{u_k\}$
6. **return** $\mathrm{Aff}(v)$

Fig. 1. Computing the events in $\mathcal{E}(v)$ affected by a delay

As limitation of the recovery algorithms in $\mathbb{A}$, we require that an algorithm in $\mathbb{A}$ can change the time of at most Δ events. Formally, each algorithm in $\mathbb{A}$ is able to compute a solution $\Pi' \in F(j)$ if $|\mathrm{Aff}(v)| \leq \Delta$, where $\Delta \in \mathbb{N}$ is a constant. This implies that a robust solution for $\mathcal{RDM}$ must guarantee that, if a delay of at most α time occurs, then it affects at most Δ events.

4   Complexity Analysis

In this section we discuss the computational complexity of $\mathcal{RDM}$. In particular, we first show in Section 4.1 the NP-completeness of the *Bounded Delay Management* problem (denoted as *BDM*) which is a decision problem derived by $\mathcal{RDM}$. Then, in Section 4.2 we discuss the NP-hardness of finding an optimal solution for $\mathcal{RDM}$ by showing the differences with *BDM*.

4.1   Complexity of *BDM* Problem

In order to study the complexity of $\mathcal{RDM}$, we formulate it as a more general decision problem. Instead of an event activity network, a generic DAG is considered. Functions L and w and the number α are allowed to assume positive real values. Moreover, for the sake of simplicity, we consider $\sum_{a\in A} w(a)s(a)$ as the objective function, which differs from f only by a constant term.

In order to formulate *BDM*, we need to extend the concept of events affected by a delay, given in Section 3 for event activity networks, to general DAGs.

Definition 7. *Given a DAG $G = (V, A)$, a function $s : A \to \mathbb{R}_+$, and a number $\alpha \in \mathbb{R}_+$, a vertex x is α-affected by $a = (u, v) \in A$ (a α-affects x) if there exists a path $p = (u \equiv v_0, v \equiv v_1, \ldots, v_k \equiv x)$ in G, such that $\sum_{i=1}^{k} s((v_{i-1}, v_i)) < \alpha$.*

Remark 1: If x is α-affected by a according to a path p, then all the vertices belonging to p but the first are α-affected by a.

BOUNDED DELAY MANAGEMENT PROBLEM (BDM)

GIVEN: A DAG $G = (V, A)$, a function $L : A \rightarrow \mathbb{R}_+$, a function $w : A \rightarrow \mathbb{R}_+$, and three numbers $\alpha \in \mathbb{R}_+, \Delta \in \mathbb{N}, K \in \mathbb{R}_+$.

PROBLEM: Is there a function $\Pi : V \rightarrow \mathbb{R}_+$ such that each edge in A α-affects at most Δ nodes, according to the function $s : A \rightarrow \mathbb{R}_+$ defined as $s(a = (i, j)) = \Pi(j) - \Pi(i) - L(a)$, and such that $\sum_{a \in A} w(a)s(a) \leq K$?

The next theorem, shows that BDM is NP-complete, the proof is given in [14].

Theorem 1. *BDM is NP-complete for $\Delta \geq 3$.*

4.2   Complexity of $\mathcal{RDM}$

In this section we discuss the complexity of $\mathcal{RDM}$. Note that $\mathcal{RDM}$ only requires to find a feasible solution. Nevertheless, it is worth to find a solution that minimizes the total travel time of all passengers, that is the objective function of TT. We call $\mathcal{RDM}_{opt}$ the problem of finding such a solution.

We will show that $\mathcal{RDM}_{opt}$ is NP-hard by showing that its corresponding decision problem is NP-complete. We do not give a formal proof, being the proof very similar to that provided for Theorem 1. There are only three differences between instances of $\mathcal{RDM}_{opt}$ and BDM: the functions L and w and the number α, which in $\mathcal{RDM}_{opt}$ assume only natural values; the underlying graph, which is an event activity network in $\mathcal{RDM}_{opt}$, whereas it is a general DAG in BDM; and the objective functions which differ only by a constant value. The way how the proof of Theorem 1 can be used to prove the NP-hardness of $\mathcal{RDM}_{opt}$ is given in [14]. However, due to the restrictions on the topology of a event activity network with respect to a general DAG, the proof holds only when $\Delta \geq 5$. It follows that, we can state the following theorem and corollary.

Theorem 2. *$\mathcal{RDM}_{opt}$ is NP-hard for $\Delta \geq 5$.*

Corollary 1. *Computing $P_{rob}(\mathcal{RDM})$ is NP-hard.*

5   Particular Cases

As $\mathcal{RDM}_{opt}$ is NP-hard, there do not exist polynomial $\mathcal{RDM}$-optimal algorithms unless $P = NP$. Then, in this section we provide solutions for particular instances of $\mathcal{RDM}$. All the formal proofs of the claimed results can be found in [14]. Given an event activity network $\mathcal{N} = (\mathcal{E}, \mathcal{A})$ with a single source v_0 and minimal duration function $L : \mathcal{A} \to \mathbb{N}$, the *Critical Path Method* (see e.g. [15]) solves TT when w is constant. *CPM* works as follows. Given $i = (\mathcal{N}, L, w)$,

$$CPM(i) = \begin{cases} \Pi(v) = 0 & \text{if } v = v_0 \\ \Pi(v) = \max\left\{\Pi(u) + L(a) \ : \ a = (u,v) \in \mathcal{A}\right\} & \text{otherwise} \end{cases}$$

Let w_{min} and w_{max} (L_{min} and L_{max}, resp.) be the minimum and maximum values assigned by the function w (L, resp.), with respect to all the possible instances of $\mathcal{RDM}$. Given $i = (\mathcal{N}, L, w)$, we denote $\gamma = (1 + \frac{\alpha}{L_{min}})$, $i_\gamma = (\mathcal{N}, \gamma L, w)$ and $i_\alpha = (\mathcal{N}, L + \alpha, w)$, where α is in the definition of the modification function M of $\mathcal{RDM}$. We use the critical path method to find robust solutions for $\mathcal{RDM}$. In particular, we use the algorithm CPM_γ defined as $CPM_\gamma(i) = CPM(i_\gamma)$. CPM_γ is clearly a robust algorithm for $\mathcal{RDM}$ since for each $a = (u, v) \in \mathcal{A}$, $\Pi(v) - \Pi(u) \geq \left(1 + \frac{\alpha}{L_{min}}\right) L(a) = L(a) + \alpha \frac{L(a)}{L_{min}} \geq L(a) + \alpha$, hence a delay of at most α does not affect any node. CPM_γ outputs a timetable with a constant factor slack time to each activity, such slack time is known as *proportional buffering* [16]. The next theorem provides the price of robustness of CPM_γ.

In the following, $\mathbb{A}_\Delta$ denotes the class of recovery algorithms limited to compute a timetable which changes the time of at most Δ events.

Theorem 3. *For any $\Delta \geq 0$, the price of robustness of CPM_γ for $\mathcal{RDM} = (TT, M, \mathbb{A}_\Delta)$ is bounded by:* $P_{rob}(\mathcal{RDM}, CPM_\gamma) \leq \gamma \frac{w_{max}}{w_{min}}$.

The following theorem shows the price of robustness of $\mathcal{RDM}$ when $\Delta = 0$ and the function L is constant.

Theorem 4. *Let $\mathcal{RDM} = (TT, M, \mathbb{A}_0)$ and $L(a) = \ell$ for each $a \in \mathcal{A}$, then $P_{rob}(\mathcal{RDM}) \geq \gamma \frac{w_{min}}{w_{max}}$.*

Corollary 2. *Let $\mathcal{RDM} = (TT, M, \mathbb{A}_0)$, if w and L are constant, then $P_{rob}(\mathcal{RDM}, CPM_\gamma) = \gamma$ and CPM_γ is $\mathcal{RDM}$-optimal.*

$CPM_\alpha(i) = CPM(i_\alpha)$ constitutes another algorithm (better then CPM_γ) for solving $\mathcal{RDM}$.

Theorem 5. *For each instance i of $\mathcal{RDM}$, $f(CPM_\alpha(i)) \leq f(CPM_\gamma(i))$.*

6   Conclusions

In this paper, we have studied the interaction between the timetabling problem, and the delay management problem. This interaction has been analyzed in the recoverable robustness model and the quality of a robust timetable has been measured in terms of price of robustness.

In particular, we have considered the problem of designing robust timetables such that a delay can affect only a limited number of subsequent events. We have shown that finding an optimal solution for this problem is NP-hard. Hence, we have proposed robust algorithms and evaluated their prices of robustness.

Moreover, we have shown that such algorithms are optimal with respect to some restrictions.

The field of robust optimization is still in its beginning. This work can be considered as a first step in the study of robust timetables. Several directions for future works deserve investigation as the analysis of different recovery strategies and the application of other modification functions to the expected input. To this aim, it would be interesting to follow practical experiences from real world scenarios.

References

1. De Giovanni, L., Heilporn, G., Labbé, M.: Optimization models for the delay management problem in public transportation. European Journal of Operational Research 189(3), 762–774 (2007)
2. Schöbel, A.: A model for the delay management problem based on mixed integer programming. Electronic Notes in Theoretical Computer Science 50(1), 1–10 (2004)
3. Schöbel, A.: Integer Programming Approaches for Solving the Delay Management Problem. In: Geraets, F., Kroon, L.G., Schoebel, A., Wagner, D., Zaroliagis, C.D. (eds.) Railway Optimization 2004. LNCS, vol. 4359, pp. 145–170. Springer, Heidelberg (2007)
4. Gatto, M., Glaus, B., Jacob, R., Peeters, L., Widmayer, P.: Railway delay management: Exploring its algorithmic complexity. In: Hagerup, T., Katajainen, J. (eds.) SWAT 2004. LNCS, vol. 3111, pp. 199–211. Springer, Heidelberg (2004)
5. Gatto, M., Jacob, R., Peeters, L., Schöbel, A.: The Computational Complexity of Delay Management. In: Kratsch, D. (ed.) WG 2005. LNCS, vol. 3787, pp. 227–238. Springer, Heidelberg (2005)
6. Gatto, M., Jacob, R., Peeters, L., Widmayer, P.: Online Delay Management on a Single Train Line. In: Geraets, F., Kroon, L.G., Schoebel, A., Wagner, D., Zaroliagis, C.D. (eds.) Railway Optimization 2004. LNCS, vol. 4359, pp. 306–320. Springer, Heidelberg (2007)
7. Ginkel, A., Schöbel, A.: The bicriteria delay management problem. Transportation Science 41(4), 527–538 (2007)
8. Bayer, H.G., Sendhoff, B.: Robust Optimization - A Comprehensive Survey. Computer Methods in Applied Mechanics and Engineering 196(33-34), 3190–3218 (2007)
9. Ben-Tal, A., El Ghaoui, L., Nemirovski, A.: Mathematical Programming: Special Issue on Robust Optimization, vol. 107. Springer, Berlin (2006)
10. Bertsimas, D., Sim, M.: The price of robustness. Operations Research 52(1), 35–53 (2004)
11. Fischetti, M., Monaci, M.: Robust optimization through branch-and-price. In: Proceedings of the 37th Annual Conference of the Italian Operations Research Society (AIRO) (2006)
12. Liebchen, C., Lüebbecke, M., Möhring, R.H., Stiller, S.: Recoverable robustness. Technical Report ARRIVAL-TR-0066, ARRIVAL Project (2007)
13. Cicerone, S., D'Angelo, G., Di Stefano, G., Frigioni, D., Navarra, A.: Robust Algorithms and Price of Robustness in Shunting Problems. In: Proc. of the 7th Workshop on Algorithmic Approaches for Transportation Modeling, Optimization, and Systems (ATMOS), pp. 175–190 (2007)

14. Cicerone, S., D'Angelo, G., Di Stefano, G., Frigioni, D., Navarra, A.: On the interaction between robust timetable planning and delay management. Technical Report ARRIVAL-TR-0116, ARRIVAL project (2007)
15. Levy, F., Thompson, G., Wies, J.: The ABCs of the Critical Path Method. Graduate School of Business Administration. Harvard University (1963)
16. Liebchen, C., Stiller, S.: Delay resistant timetabling. Technical Report ARRIVAL-TR-0056, ARRIVAL Project (2006) Presented at CASPT (2006)

Clustered SplitsNetworks

Lichen Bao and Sergey Bereg

Department of Computer Science
Erik Jonsson School of Engineering & Computer Science
The University of Texas at Dallas
Richardson, TX 75080, USA
{lxb042000,besp}@utdallas.edu

Abstract. We address the problem of constructing phylogenetic networks using two criteria: the number of cycles and the fit value of the network. Traditionally the fit value is the main objective for evaluating phylogenetic networks. However, a small number of cycles in a network is desired and pointed out in several publications.

We propose a new phylogenetic network called *CS-network* and a method for constructing it. The method is based on the well-known split-stree method. A CS-network contains a face which is k-cycle, $k \geq 3$ (not as splitstree). We discuss difficulties of using non-parallelogram faces in splitstree networks. Our method involves clustering and optimization of weights of the network edges.

The algorithm for constructing the underlying graph (except the optimization step) has a polynomial time. Experimental results show a good performance of our algorithm.

1   Introduction

A phylogenetic tree is a commonly used tool for showing the evolutionary relationships among various biological species. However phylogenetic trees have important limitations in representing recombination, recurrent and back mutation, horizontal gene transfer and cross-species hybridization. Phylogenetic networks are often considered for this task.

A phylogenetic network is a generalization of a phylogenetic tree, allowing structural properties that are not tree-like [14]. Phylogenetic networks can represent the relationships between the gene sequences better since they are able to show the recombination, hybridization, which are the most important phenomena in understanding genomic role [14]. There are several methods for phylogenetic network construction [2,4,5,11,13]. A popular program SplitsTree [7,8] incorporates many methods of phylogenetic network construction and is based on the results of metric decomposition theory. The package SplitsTree could be found at http://www.splitstree.org

We assume that evolutionary distances between species are computed and sorted in a distance matrix $D = (d(i,j))$. To evaluate a phylogenetic network N, we use a *fit value* (based on the least square fit value)

B. Yang, D.-Z. Du, and C.A. Wang (Eds.): COCOA 2008, LNCS 5165, pp. 469–478, 2008.

$$fit(D, N) = \left(1 - \frac{\sum_{i<j}(d_N(i,j) - d(i,j))^2}{\sum_{i<j} d(i,j)^2}\right) \cdot 100,$$

where $d_N(i,j)$ is the distance between taxa i and j in the network N.

The fit value is an important objective in constructing phylogenetic networks. Another objective is to minimize the number of cycles in the network. The problem of minimizing the number of recombinations in a rooted directed phylogenetic network has been studied both theoretically and practically [9,5,12,6]. Galled networks studied in [9] have minimum reticulate nodes and, thus, have fewer cycles. We study the problem of minimizing the number of cycles in unrooted phylogenetic networks.

Phylogenetic networks constructed using SplitsTree have a property that all bounded faces are 4-cycles and can be drawn as parallelograms. In this paper we propose a new network called *CS-network* that uses splits as the splitstree and clustering with k clusters. It is constructed starting from a k-cycle (a polygon with k vertices), with the vertices connecting to the taxa clusters. We develop a method of constructing CS-networks and optimizing the edge weights.

2   Construction of CS-Network

Our method has an integer parameter k which is the number of clusters. We find k clusters and introduce a face which is a k-cycle. We borrow the computation of a circular order of taxa $X = \{x_1, x_2, \ldots, x_n\}$ and a set $\mathcal{S} = \{S_1, S_2, \ldots, S_m\}$ of circular splits from NeighborNet [2]. The order of splits has a property that the sizes of corresponding split sets containing x_1 are non-decreasing. We assume that all trivial splits (where one of the sets is a singleton) are present in the set of splits. A high level description of our method has 3 steps.

1. Find a circular order of taxa and m splits. Find k clusters $C_1, \ldots, C_k$ of taxa consistent with the circular order, see Fig. 1.

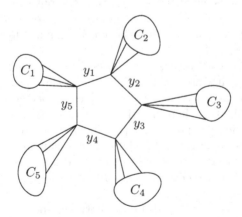

Fig. 1. Starting figure of the network

2. Insert m splits into the network.

3. Optimize the edge weights of the network.

In the rest of the paper we assume that the circular order is computed and we use indices of taxa in the circular order.

3   Clustering

Our clustering method is similar to the neighbor joining algorithm or Neighbor-Net [2]. In the main step the algorithm finds two clusters and joins them into a new cluster. Two matrices are used here: the distance matrix and Q-matrix. For current clusters $C_1, C_2, \ldots, C_r$, the distance between two clusters C_i and C_j is defined as

$$d(C_i, C_j) = \frac{\sum_{x \in C_i} \sum_{y \in C_j} d(x, y)}{|C_i||C_j|} \tag{1}$$

and Q-distance is calculated as

$$Q(i, j) = (r - 2)d(i, j) - \sum_{t=1}^{r} d(i, t) - \sum_{t=1}^{r} d(j, t) \tag{2}$$

where r is the number of clusters.

The clustering in NeighborNet [2] takes $O(n^3)$ time. Our clustering is based on the circular order of taxa. We modify the neighbor joining algorithm and show that it can can be implemented using $O(n^2)$ time only.

1. Make n clusters $C_1, C_2, \ldots, C_n$, one per taxon, using the clockwise order.
2. For all pairs of clusters, compute $d(C_i, C_j)$ using Equation (1). For n pairs of clusters in the circular order, compute $Q(C_i, C_{i+1})$ using Equation (2). Find two clusters C_i and C_{i+1} with minimum $Q(C_i, C_{i+1})$ and combine them a new cluster.[1]
3. Repeat step 2 until the number of clusters is k.

Note that, using the NeighborNet algorithm for maintaining all intercluster distances and Q-distances, our algorithm can be implemented in $O(n^3)$ time. The running time can be reduced to $O(n^2)$ by computing only n Q-distances in the circular order.

4   Graph Construction

In this Section we develop a method for constructing an underlying graph of the clustered splits network.

---

[1] We use indices of clusters in the circular order. For example, $C_{n+1} = C_1$.

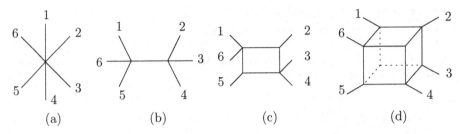

Fig. 2. Splitstree network construction. (a) Start from a star. (b) Add the split $\{1,5,6\}\{2,3,4\}$. (c) Add the split $\{1,2,6\}\{3,4,5\}$. (d) Add the split $\{1,2,3\}\{4,5,6\}$. Note: dotted lines represent an alternative way to add it.

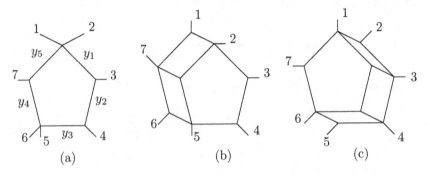

Fig. 3. Adding the split $\{1,6,7\}, \{2,3,4,5\}$ to the network shown in (a). b) The network in the case $y_4+y_5 \leq y_1+y_2+y_3$. (c) The network in the case $y_4+y_5 > y_1+y_2+y_3$.

4.1 Difficulty

The topology of a splits network can be found using only splits and the circular order of taxa [4]. The splits can be added incrementally starting with a star, see Fig. 2 for an example. Weights of the splits can be decided later (the only condition is that the weights should be non-negative). It turns out that this is not true if we start with a network different from a star.

Suppose that we start with a network that contains one k-cycle, say $k = 5$. Let $y_1,\ldots,y_5$ be the weights of its edges in clockwise order. We want to add a split (X_1,X_2) of weight w and increase the network distances between all pairs $x_1 \in X_1$ and $x_2 \in X_2$ by w. The new network actually depends on the weights y_i, see Fig. 3 for an example.

We impose k conditions for the weights of the k-cycle to be able to construct a network. We assume for simplicity that k is odd[2]. The ith condition is

$$y_i + y_{i+1} + \cdots + y_{i+\lfloor k/2 \rfloor} \leq y_{i+\lceil k/2 \rceil} + \cdots + y_{i+k-1} \tag{3}$$

where $i=1,\ldots,k$ and indices are in the circular order (i.e. for example $y_{k+1}=y_1$).

---

[2] Even k will be considered in the final version.

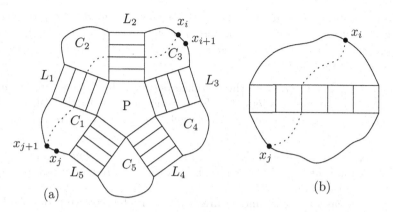

Fig. 4. (a) $span(X') = \{C_3, C_4, C_5, C_1\}$ and $span(X'') = \{C_1, C_2, C_3\}$. Since $|span(X')| > |span(X'')|$, $x_p = x_i$ and $x_q = x_{j+1}$. The shortest path $x_p x_q$ crosses ladders L_1 and L_2. (b) Splits always separate the corresponding two taxa on two sides of the ladder.

4.2 Construction of CS-Network Using Splits

For a circular subset of taxa $X' = \{x_i, x_{i+1}, \ldots, x_j\}$, we define $span(X')$ as the chain of clusters $C_{i'}, C_{i'+1}, \ldots, C_{j'}$ such that $C_s \cap X' \neq \emptyset$ for each $i' \leq s \leq j'$.

With the conditions above, we can start the building of the phylogenetic network. The following is the algorithm:

1. Construct the k-cycle with the clusters as in Figure 1.
2. In each cluster, add the trivial splits by connecting taxa directly to the k-cycle vertices.
3. For each split X', X'' from $\mathcal{S}$ (using the order of splits in $\mathcal{S}$), change the graph as follows. Let $X' = \{x_{i+1}, \ldots, x_j\}$. Then $X'' = \{x_{j+1}, \ldots, x_i\}$. Compute $span(X')$ and $span(X'')$.
   (a) *Find two taxa* x_p, x_q. If the size of $span(X')$ is smaller than the size of $span(X'')$ then $p = i + 1$ and $q = j$; otherwise $p = i$ and $q = j + 1$, see Fig. 4 (a).
   (b) Find the shortest path (using minimum number of edges) δ from x_p to x_q in the current (unweighted) graph.
   (c) Insert a ladder by doubling edges of the path δ except the first and last edge as in Fig. 2.

We show how the phyletic distances can be computed in CS-network. Let $\pi(i, j)$ be the shortest path (with minimum number of edges[3]) in the k-cycle between clusters containing x_i and x_j.

Lemma 1 (Phyletic Distance). *Let N be a CS-network with assigned non-negative weights satisfying the boundary condition (3). Let x_i and x_j be two*

---
[3] If k is even and $(i - j) \equiv k/2 \mod k$ then there are two shortest paths and we select either one.

taxa from clusters $C_{i'}$ and $C_{j'}$. Let $s(i,j)$ be the sum of the weights of splits such that x_i and x_j are in different sets. Let $w(i,j)$ be the sum of y-weights of the path $\pi(i',j')$. The network distance between two taxa x_i and x_j is equal to $s(i,j)+w(i,j)$ (note that $w(i,j)=0$ if $i'=j'$).

Proof (Sketch). Let $d_N(i,j)$ be the distance between the taxa x_i and x_j in N. There are two types of ladders: S-ladders corresponding to the splits and L-ladders $L_1, L_2, \ldots, L_k$ attached to the k-cycle, see Fig. 4 (a). We consider two cases.

1. The taxa x_i and x_j are in the same cluster, i.e. $i'=j'$. The argument is the same as for splits graphs [4] since the taxa in $C_{i'}$ are connected as a splits graph, see Fig. 4 (a).
2. In general case where x_i and x_j are in different clusters, we prove the lower bound $d_N(i,j) \geq s(i,j)+w(i,j)$ and the upper bound $d_N(i,j) \leq s(i,j)+w(i,j)$. Any path between x_i and x_j in N crosses every ladder corresponding to a split separating x_i and x_j, see Fig. 4 (b). Also this path intersects L-ladders whose total weight is $w(i,j)$. Therefore $d_N(i,j) \geq s(i,j)+w(i,j)$.
   We can prove by induction that the following properties are maintained during the construction of N: (i) any shortest path (using minimum number of edges) intersects at most $\lfloor k/2 \rfloor$ L-ladders, and (ii) any two ladders cross at most one time. □

5   Optimization of the Weights

5.1   Linear Programming Formulation

Lemma 1 allows us to optimize the weights. There are $m+k$ unknowns: m weights of the splits and k weights of the k-cycle edges. We denote them by unknown variables $z_i, 1 \leq i \leq m$ is a weight of the split S_i and $z_{m+i} = y_i, 1 \leq i \leq k$. We represent them by a $(m+k)$-dimensional vector $\mathbf{z} = (z_1, z_2, \ldots, z_{m+k})'$.

To find phyletic distances we introduce a 0/1 matrix A of size $n(n-1)/2 \times (m+k)$. The rows of A are indexed by pairs of taxa. The columns of A are indexed by m splits and k edges of the k-cycle. An entry $A_{(ij)l}$ is given by

$$A_{(ij)l} = \begin{cases} 1 & \text{If } l \leq m \text{ and } x_i, x_j \text{ are in different sets of the split } S_l. \\ 1 & \text{If } l > m \text{ and the path } \pi(i',j') \text{ contains vertices corresponding} \\ & \text{to clusters } C_{l-m} \text{ and } C_{l-m+1} \text{ where } x_i \in C_{i'} \text{ and } x_j \in C_{j'}. \\ 0 & \text{otherwise} \end{cases}$$

The phyletic distances can be computed as $A\mathbf{z}$.

We represent input distances between taxa by a $\binom{n}{2}$-dimensional vector

$$\mathbf{d} = (d(x_1, x_2), d(x_1, x_3), \ldots, d(x_{n-1}, x_n))'. \tag{4}$$

Ideally, the optimal weights satisfy the linear system $\mathbf{d} = A\mathbf{z}$. We can use OLS to estimates z. However, general OLS could result in negative values, which are not suitable to use for us. So in the following section, we use the gradient method for the optimization.

5.2   Least Squares Optimization

The objective of the least squares method is to minimize $f(\mathbf{z}) = (A\mathbf{z} - \mathbf{d})^2$. The vector $A\mathbf{z} - \mathbf{d}$ is a $\binom{n}{2}$-dimensional vector. An ith element of $A\mathbf{z} - \mathbf{d}$ is equal to $A_i\mathbf{z} - d_i$ where A_i is the ith row of A. Then

$$f(\mathbf{z}) = \sum_{i=1}^{\binom{n}{2}}(A_i\mathbf{z} - d_i)^2$$

where A_i is the ith row of A. Suppose that we start with an initial vector $\mathbf{z}$. We want to optimize z_r for some r assuming that all $z_i, i \neq r$ are fixed. Then

$$f(\mathbf{z}) = c_2 z_r^2 + c_1 z_r + c_0 \tag{5}$$

where the constants c_1, c_2 are equal to

$$c_2 = \sum_{i=1}^{\binom{n}{2}} a_{i,r}^2 \tag{6}$$

$$c_1 = 2 \sum_{i=1}^{\binom{n}{2}} a_{i,r} \cdot \left(\sum_{j=1, j \neq r}^{m+k} a_{i,j} z_j - d_i \right). \tag{7}$$

Let $t = -c_1/2c_2$. The optimal value is $z_r = t$ if there would be no boundary conditions on z_r. If $r \leq m$ then the boundary condition is $z_r \geq 0$ and $z_r = \max(0, t)$.

Let $y_1, y_2, \ldots, y_k$ be the weights of the k-cycle edges, i.e. $y_i = z_{m+i}$. We have $z_r = y_i, i = r - m$. Each of the k boundary conditions (3) simplifies to either an upper or lower bound of y_i. Using the smallest upper bound and the largest lower bound it reduces to $c' \leq y_i \leq c''$.

If $c'' < 0$ or then we assign $z_r = 0$. Suppose that $c'' > 0$. If $t > c''$ then we set $z_r = c''$. If $t < \max(0, c')$ then we set $z_r = \max(0, c')$. In the remaining case $\max(0, c') \leq t \leq c''$ and we set $z_r = t$.

We repeat the optimization for variables $z_1, \ldots, z_{m+k}$ in a loop while $f(\mathbf{z})$ decreases.

5.3   Experiments and Results

According to the metric decomposition theory [3,1], the graph $K_{2,3}$ is the only split prime for five taxa, which means it is not totally decomposable. The graph $K_{3,3}$ is also one split prime for six taxa [10]. We select these metrics to compare NeighborNet and our method.

The result of splitstree with NeighborNet is shown in Fig. 5. The network has 3 cycles and fit value 95.45. Using split filter one can reduce the number of cycles to one at the price of the fit value which is 87.3.

The results for $K_{3,3}$ are shown in Fig. 6. The network shown in Fig. 6 (a) is produced by splitstree and has 6 cycles and fit value 93.94. The split filter does not help much here and produces a network with 6 cycles again, see Fig. 6 (b).

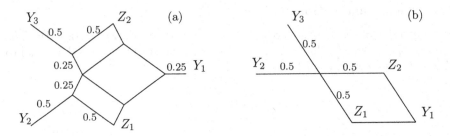

Fig. 5. Splitstree network for the graph $K_{2,3}$. (a) 3 cycles, fit value 95.45, and (b) 1 cycle, fit value 87.3.

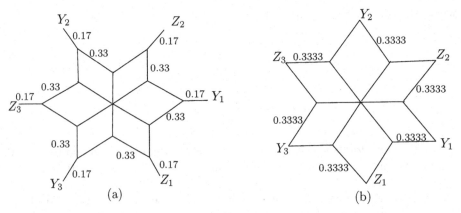

Fig. 6. Splitstree network for the graph $K_{3,3}$. (a) 6 cycles, fit value 93.94, and (b) 6 cycles, fit value 88.89.

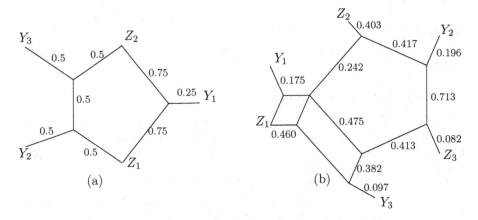

Fig. 7. (a) CS-network for $K_{2,3}$: 1 cycle, fit value 95.45. (b) CS-network for $K_{3,3}$: 3 cycles, fit value 93.31.

We use a pentagon $k = 5$ to construct CS-networks for $K_{2,3}$ and $K_{3,3}$, see Fig. 7. The network for $K_{2,3}$ has only one cycle and the same fit value as the splitstree network in Fig. 5 (a). For $K_{3,3}$ the CS-network has only 3 cycles and the fit value is still good, Fig. 7 (b).

Finally, we construct CS-network for $K_{3,3}$ starting from a triangle. The network is shown in Fig. 8. It contains fewer cycles (only 4) and has the same fit value as the splitstree network in Fig. 6 (a).

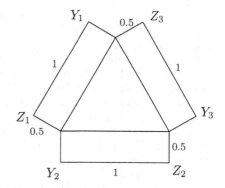

Fig. 8. CS-network for $K_{3,3}$ constructed with $k = 3$ contains 4 cycles and has fit value 80.95

6   Conclusion

We studied the problem of constructing a phylogenetic network using two objectives: the fit value and the number of cycles. Based on the methods in splitstree, we proposed a new method of building phylogenetic networks called CS-networks starting with a k-cycle and by inserting splits and optimizing the weights of the splits and edges of the k-cycle.

We run our method on distance matrices considered as "difficult" metrics for split decomposition methods. Preliminary results show a good performance in terms of both the number of cycles and the fit value.

References

1. Bandelt, H., Dress, A.: A canonical decomposition theory for metrics on a finite set. Advances in Mathematics 92, 47–105 (1992)
2. Bryant, D., Moulton, V.: Neighbornet: An agglomerative method for the construction of planar phylogenetic networks. In: Guigó, R., Gusfield, D. (eds.) WABI 2002. LNCS, vol. 2452, pp. 375–391. Springer, Heidelberg (2002)
3. Dress, A.: Trees, tight extensions of metric spaces, and the cohomological dimension of certain groups: A note on combinatorial properties of metric spaces. Advances in Mathematics 53, 321–402 (1984)
4. Dress, A., Huson, D.: Constructing splits graphs. IEEE/ACM Transactions in Computational Biology and Bioinformatics 1, 109–115 (2004)
5. Gusfield, D., Eddhu, S., Langley, C.: The fine structure of galls in phylogenetic networks. INFORMS Journal on Computing 16, 459–469 (2004)
6. Gusfield, D., Hickerson, D., Eddhu, S.: An efficiently computed lower bound on the number of recombinations in phylogenetic networks: Theory and empirical study. Discrete Applied Mathematics 155, 806–830 (2007)
7. Huson, D.: Splitstree: a program for analyzing and visualizing evolutionary data. Bioinformatics 14, 68–73 (1998)

8. Huson, D., Bryant, D.: Application of phylogenetic networks in evolutionary studies. Molecular Biology and Evolution 23, 254–267 (2006)
9. Huson, D., Klöpper, T.: Beyond galled trees - decomposition and computation of galled networks. In: Speed, T., Huang, H. (eds.) RECOMB 2007. LNCS (LNBI), vol. 4453, pp. 211–225. Springer, Heidelberg (2007)
10. Koolen, J., Moulton, V., Tönges, U.: A classification of the six-point prime metrics. Europ. J. Combinatorics 21, 815–829 (2000)
11. Moret, B.M.E., Nakhleh, L., Warnow, T., Linder, C.R., Tholse, A., Padolina, A., Sun, J., Timme, R.E.: Phylogenetic networks: Modeling, reconstructibility, and accuracy. IEEE/ACM Trans. Comput. Biology Bioinformatics 155(1), 13–23 (2004)
12. Myers, S., Griffiths, R.: Bounds on the minimum number of recombination events in a sample history. Genetics 163, 375–394 (2003)
13. Nakhleh, L., Warnow, T., Linder, C.R., John, K.St.: Reconstructing reticulate evolution in species: Theory and practice. Journal of Computational Biology 12(6), 796–811 (2005)
14. Song, Y.S., Wu, Y., Gusfield, D.: Efficient computation of close lower and upper bounds on the minimum number of recombinations in biological sequence evolution. Advances in Mathematics 21, 413–422 (2005)

Author Index

Lecture Notes in Computer Science

Sublibrary 1: Theoretical Computer Science and General Issues

For information about Vols. 1– 4878
please contact your bookseller or Springer